Mobile Communications

Jochen H. Schiller

Mobile Communications

Second Edition

An imprint of Pearson Education

London • Boston • Indianapolis • New York • Mexico City
Toronto • Sydney • Tokyo • Singapore • Hong Kong • Cape Town
New Delhi • Madrid • Paris • Amsterdam • Munich • Milan

PEARSON EDUCATION LIMITED
Edinburgh Gate
Harlow CM20 2JE
Tel:+44 (0)1279 623623
Fax:+44 (0)1279 431059

Website: www.pearsoned.co.uk

First Published in Great Britain 2000
Second edition 2003

ISBN 0 321 12381 6

British Library Cataloguing-in-Publication Data
A catalogue record for this book is available from the British Library.

Library of Congress Cataloging in Publication Data
A catalog record for this book is available from the Library of Congress.

10 9 8 7 6 5 4 3 2 1
08 07 06 05 04

Text design by barker/hilsdon @ compuserve.com
Typeset by Pantek Arts Ltd., Maidstone, Kent
Printed and bound in Great Britain by Biddles Ltd, www.biddles.co.uk

The publishers' policy is to use paper manufactured from sustainable forests.

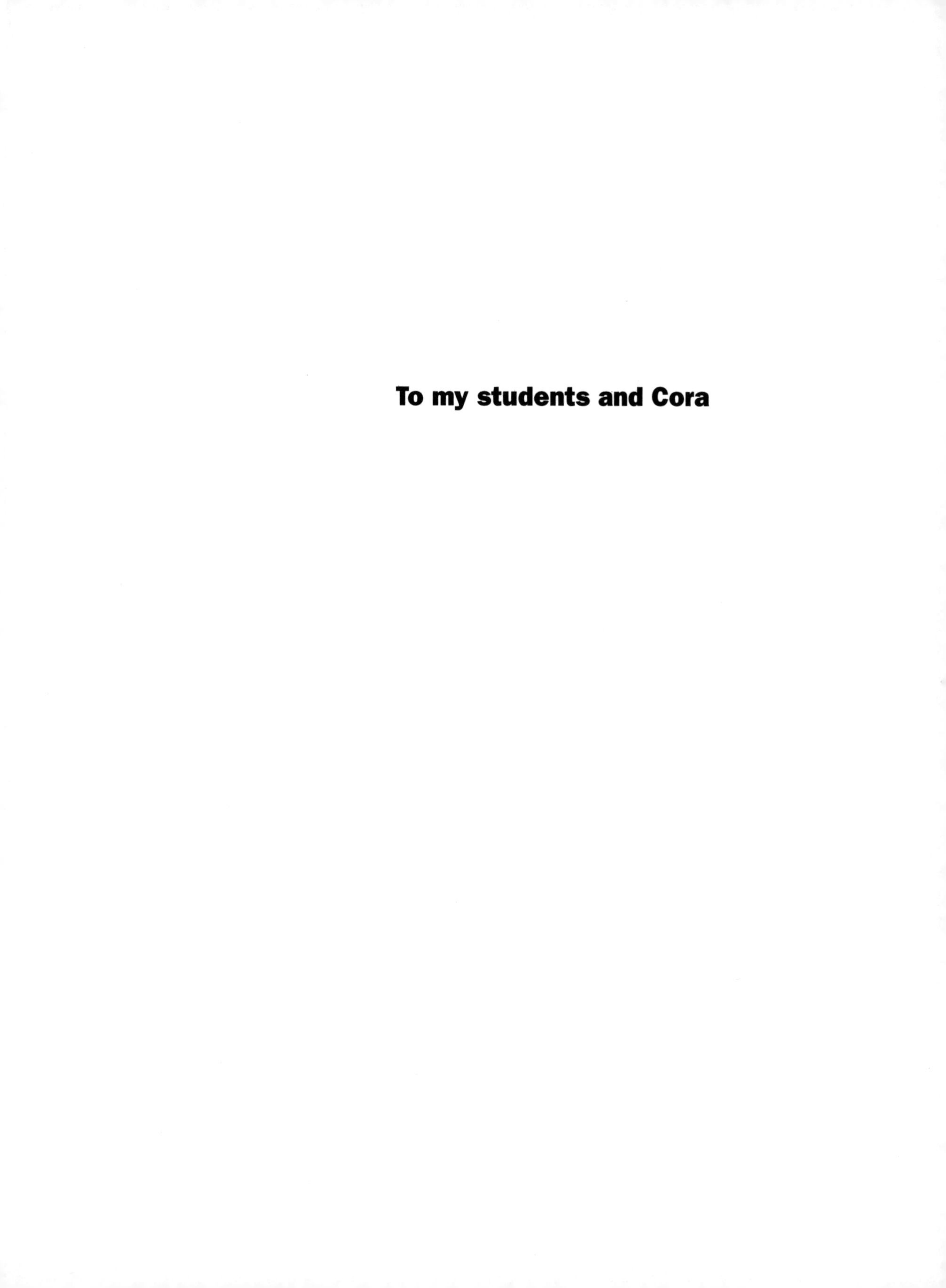

To my students and Cora

Contents

About the author

Jochen Schiller is head of the working group Computer Systems & Telematics at the Institute of Computer Science, FU Berlin, Germany. He received his MS and PhD degrees in Computer Science from the University of Karlsruhe, Germany, in 1993 and 1996, respectively. As a postdoc he joined Uppsala University, Sweden, and worked in several industry co-operations and European projects. Since April 2001 he has been a full professor at FU Berlin. The focus of his research is on mobile and wireless communications, communication architectures and operating systems for embedded devices, and quality of service aspects in communication systems. He is a member of IEEE and GI and acts as consultant for several companies in the networking and communication business.

Preface

Welcome to the second edition of Mobile Communications – and welcome to the confusing, complex, but very interesting world of wireless and mobile technologies! In the last few years, we have all experienced the hype and frustration related to mobile technology. Once praised as the Internet on the mobile phone, the frustration with third generation mobile phone systems came at the same time the dotcoms crashed. The reader should remember that all technologies need their time to develop.

Nevertheless, we are experiencing huge growth rates in mobile communication systems (mainly in Asia), increasing mobility awareness in society, and the worldwide deregulation of former monopolized markets. While traditional communication paradigms deal with fixed networks, mobility raises a new set of questions, techniques, and solutions. For many countries, mobile communication is the only solution due to the lack of an appropriate fixed communication infrastructure. Today, more people use mobile phones (over one billion!) than traditional fixed phones. The trends mentioned above create an ever-increasing demand for well-educated communication engineers who understand the developments and possibilities of mobile communication. What we see today is only the beginning. There are many new and exciting systems currently being developed in research labs. The future will see more and more mobile devices, the merging of classical voice and data transmission technologies, and the extension of today's Internet applications (e.g., the world wide web) onto mobile and wireless devices. New applications and new mobile networks will bring ubiquitous multimedia computing to the mass market; radios, personal digital assistants (PDAs), laptops and mobile phones will converge and many different functions will be available on one device – operating on top of Internet technologies.

This book is an introduction to the field of mobile communications and focuses on digital data transfer. The book is intended for use by students of EE or CS in computer networking or communication classes, engineers working with fixed networks who want to see the future trends in networking, as well as managers who need a comprehensible overview in mobile communication. The reader requires a basic understanding of communication and a rough knowledge of the Internet or networking in general. While resources are available which focus on a particular technology, this book tries to cover many aspects of mobile communications from a computer science point of view. Furthermore, the book points out common properties of different technical solutions and shows the integration of services and applications well-known from fixed networks into networks supporting mobility of end systems and wireless access. If the reader is interested in more detailed information regarding a certain topic, he or she will find many pointers to research publications or related websites.

Teachers will find this book useful for a course that follows a general data communication or computer networking class. The book can also replace parts of more general courses if it is used together with other books covering fixed networks or aspects of high-speed networks. It should be straightforward to teach a mobile networking class using this book together with the course material provided online via the following link:

http://www.jochenschiller.de/

The material comprises all of the figures, over 500 slides in English and German as PDF and PowerPoint™ files, a list of all acronyms, and many links to related sites. Additionally, the questions included in the book can provide a good self-test for students. Solutions to all the questions in the book can be found at the publisher's password-protected site:

http://www.booksites.net/schiller

This book addresses people who want to know how mobile phone systems work, what technology will be next in wireless local area networks, and how mobility will influence applications, security, or IP networks. Engineers working in fixed networks can see paths of migration towards mixed fixed/mobile networks.

The book follows a 'tall and thin' approach. It covers a whole course in mobile communication, from signals, access protocols, up to application requirements and security, and does not stress single topics to the neglect of others. It focuses on digital mobile communication systems, as the future belongs to digital systems such as CDMA, GSM, DECT, W-CDMA, cdma2000, UMTS, DAB. New and important topics in the higher layers of communication, like the wireless application protocol (WAP), i-mode, and wireless TCP are included.

Chapter 1 introduces the field of mobile and wireless communication, presents a short history and challenges for research, and concludes with a market vision, which shows the potential of mobile technology. Chapter 2 follows the classical layers of communication systems and explains the basics of wireless technology from a computer science point of view. Topics in this chapter are signal propagation, multiplexing, and modulation. Profound electrical engineering knowledge is not required; however, it is necessary to comprehend the basic principles of wireless transmission to understand the design decisions of higher layer communication protocols and applications. Chapter 3 presents several media access schemes and motivates why the standard schemes from fixed networks fail if used in a wireless environment.

Chapters 4–7 present different wireless communication systems and may be read in any order. All the systems involve wireless access to a network and they can transfer arbitrary data between communication partners. Chapter 4 comprises the global system for mobile communications (GSM) as today's most successful public mobile phone system, cordless phone technology, trunked radios, and the future development with the universal mobile telecommunications system (UMTS). Satellite systems are covered in chapter 5, while chapter 6 discusses digital broadcast systems such as digital audio broadcasting (DAB) which can be one component of a larger communication system providing end-users with mass data. Wireless LANs as replacement for cabling inside buildings are presented in chapter 7. Examples are IEEE 802.11, HiperLAN2, and Bluetooth. A special feature of HiperLAN2 is the provisioning of quality of service (QOS), i.e., the system can give guarantees for certain parameters, such as bandwidth or error rates.

Chapter 8 mainly presents mobile IP, the extension of the Internet protocol (IP) into the mobile domain. Ad-hoc networks with their requirements for specific routing protocols are also covered. The subsequent layer, the transport layer, is covered in chapter 9. This chapter discusses several approaches of adapting the current transmission control protocol (TCP), which is well known from the Internet, to the special requirements of mobile communication systems. Chapter 10 presents the wireless application protocol (WAP) standard that enables wireless and mobile devices to use parts of the world wide web (www) from today's fixed Internet. Additionally, this chapter shows the migration to WAP 2.0, which includes components from i-mode and the Internet. The book closes with an outlook to fourth generation systems in chapter 11.

The book is based on a course I have taught several times at the University of Karlsruhe and the Free University of Berlin. The course typically consists of 14 lectures of 90 minutes each (typically, not every topic of the book is covered during the lecture in the same detail). Over 100 universities, colleges, and other institutions around the world have already used the material compiled for this book. Teachers may include the online material in their courses or even base whole courses on the material.

What is new in the second edition?

Over three years have passed since the publication of the first edition. During this time, many new ideas showed up, several ideas were dropped, and many systems have been improved. The main changes, besides updates of all references and links, are the following:

- Integration of higher data rates for GSM (HSCSD, GPRS).
- Complete new section about third generation systems with in-depth discussion of UMTS/W-CDMA.
- Addition of the new WLAN standards for higher data rates: 802.11a, .11b, .11g and HiperLAN2.
- Extension of the Bluetooth section: IEEE 802.15, profiles, applications.

- More on ad-hoc networking and wireless profiled TCP.
- Migration of WAP 1.x and i-mode towards WAP 2.0.

You are encouraged to send any comments regarding the book or course material to schiller@computer.org. Finally, I hope you enjoy reading this completely revised book and forgive me for simplifications I have used to avoid blurring the big picture of mobile communications. Many such details may change as research and standards evolve over time. As this book covers many aspects of mobile communications it cannot dig into each and every detail with the same scientific depth. Finally, the famous quote from Goethe is valid for this book, too:

Properly speaking, such work is never finished; one must declare it so when, according to time and circumstances, one has done one's best.

Acknowledgements

First of all, I want to thank the numerous students from my Mobile Communications courses who pushed me towards writing this book by requesting more information regarding wireless and mobile communication. The students' questions and answers were a great help to shaping the contents. I also want to thank the numerous readers around the world for their comments and reviews. You all helped me a lot fixing unclear explanations.

For the first edition, many of my former colleagues at the University of Karlsruhe gave generously of their intellect and time, reading, commenting, and discussing the chapters of the book. I am especially grateful to Marc Bechler, Stefan Dresler, Jochen Seitz, and Günter Schäfer. I want to thank Hartmut Ritter for his support by taking over some of my daily routine work. I also had help from Verena Rose and Elmar Dorner during the early stages of the course material. I also want to thank the former head of the Institute of Telematics, Prof. Gerhard Krüger, for giving me the freedom and support to set up the Mobile Communications course in an inspiring environment.

For many insightful comments to the book, I thank Per Gunningberg from Uppsala University, Sweden. I am profoundly grateful to Angelika Rieder and Kerstin Risse for their help in polishing the first edition. Without their help, it would not be as easy to read.

For the second edition, which I wrote at the Free University of Berlin, I am particularly grateful to Thiemo Voigt from SICS, Sweden, who helped me a lot integrating the new ideas of mobile and wireless technology. Furthermore, I want to thank all the anonymous reviewers, colleagues, and students from many different places around the world for their valuable contributions to this edition.

As for the first edition, I have to thank the Addison-Wesley team in the UK for all their support during the making of the book, special thanks to Bridget Allen, Tessa Fincham and Michael Strang who firmly pushed me towards the final manuscript for this edition.

1 Introduction

What will **computers** look like in ten years? No one can make a wholly accurate prediction, but as a general feature, most computers will certainly be **portable**. How will users **access** networks with the help of computers or other communication devices? An ever-increasing number without any wires, i.e., **wireless**. How will people spend much of their time at work, during vacation? Many **people** will be **mobile** – already one of the key characteristics of today's society. Think, for example, of an aircraft with 800 seats. Modern aircraft already offer limited network access to passengers, and aircraft of the next generation will offer easy Internet access. In this scenario, a mobile network moving at high speed above ground with a wireless link will be the only means of transporting data to and from passengers. Think of cars with Internet access and billions of embedded processors that have to communicate with, for instance, cameras, mobile phones, CD-players, headsets, keyboards, intelligent traffic signs and sensors. This plethora of devices and applications show the great importance of mobile communications today.

Before presenting more applications, the terms 'mobile' and 'wireless' as used throughout this book should be defined. There are two different kinds of mobility: user mobility and device portability. **User mobility** refers to a user who has access to the same or similar telecommunication services at different places, i.e., the user can be mobile, and the services will follow him or her. Examples for mechanisms supporting user mobility are simple call-forwarding solutions known from the telephone or computer desktops supporting roaming (i.e., the desktop looks the same no matter which computer a user uses to log into the network).

With **device portability**,[1] the communication device moves (with or without a user). Many mechanisms in the network and inside the device have to make sure that communication is still possible while the device is moving. A typical example for systems supporting device portability is the mobile phone system, where the system itself hands the device from one radio transmitter (also called a base station) to the next if the signal becomes too weak. Most of the scenarios described in this book contain both user mobility and device portability at the same time.

1 Apart from the term 'portable', several other terms are used when speaking about devices (e.g., 'mobile' in the case of 'mobile phone'). This book mainly distinguishes between wireless access to a network and mobility of a user with a device as key characteristics.

With regard to devices, the term **wireless** is used. This only describes the way of accessing a network or other communication partners, i.e., without a wire. The wire is replaced by the transmission of electromagnetic waves through 'the air' (although wireless transmission does not need any medium).

A communication device can thus exhibit one of the following characteristics:

- **Fixed and wired**: This configuration describes the typical desktop computer in an office. Neither weight nor power consumption of the devices allow for mobile usage. The devices use fixed networks for performance reasons.
- **Mobile and wired**: Many of today's laptops fall into this category; users carry the laptop from one hotel to the next, reconnecting to the company's network via the telephone network and a modem.
- **Fixed and wireless**: This mode is used for installing networks, e.g., in historical buildings to avoid damage by installing wires, or at trade shows to ensure fast network setup. Another example is bridging the last mile to a customer by a new operator that has no wired infrastructure and does not want to lease lines from a competitor.
- **Mobile and wireless**: This is the most interesting case. No cable restricts the user, who can roam between different wireless networks. Most technologies discussed in this book deal with this type of device and the networks supporting them. Today's most successful example for this category is GSM with more than 800 million users.

The following section highlights some application scenarios predestined for the use of mobile and wireless devices. An overview of some typical devices is also given. The reader should keep in mind, however, that the scenarios and devices discussed only represent a selected spectrum, which will change in the future. As the market for mobile and wireless devices is growing rapidly, more devices will show up, and new application scenarios will be created. A short history of wireless communication will provide the background, briefly summing up the development over the last 200 years. Section 1.3 shows wireless and mobile communication from a marketing perspective. While there are already over a billion users of wireless devices today and the wireless business has experienced some problems in the last few years, the market potential is still tremendous.

Section 1.4 shows some open research topics resulting from the fundamental differences between wired and wireless communication. Section 1.5 presents the basic reference model for communication systems used throughout this book. This chapter concludes with an overview of the book, explaining the 'tall and thin' approach chosen. Tall and thin means that this book covers a variety of different aspects of mobile and wireless communication to provide a complete picture. Due to this broad perspective, however, it does not go into all the details of each technology and systems presented.

1.1 Applications

Although many applications can benefit from wireless networks and mobile communications, particular application environments seem to be predestined for their use. The following sections will enumerate some of them – it is left to you to imagine more.

1.1.1 Vehicles

Today's cars already comprise some, but tomorrow's cars will comprise many wireless communication systems and mobility aware applications. Music, news, road conditions, weather reports, and other broadcast information are received via digital audio broadcasting (DAB) with 1.5 Mbit/s. For personal communication, a universal mobile telecommunications system (UMTS) phone might be available offering voice and data connectivity with 384 kbit/s. For remote areas, satellite communication can be used, while the current position of the car is determined via the global positioning system (GPS). Cars driving in the same area build a local ad-hoc network for the fast exchange of information in emergency situations or to help each other keep a safe distance. In case of an accident, not only will the airbag be triggered, but the police and ambulance service will be informed via an emergency call to a service provider. Cars with this technology are already available. In the future, cars will also inform other cars about accidents via the ad-hoc network to help them slow down in time, even before a driver can recognize an accident. Buses, trucks, and trains are already transmitting maintenance and logistic information to their home base, which helps to improve organization (fleet management), and saves time and money.

Figure 1.1 shows a typical scenario for mobile communications with many wireless devices. Networks with a fixed infrastructure like cellular phones (GSM, UMTS) will be interconnected with trunked radio systems (TETRA) and wireless LANs (WLAN). Satellite communication links can also be used. The networks between cars and inside each car will more likely work in an ad-hoc fashion. Wireless pico networks inside a car can comprise personal digital assistants (PDA), laptops, or mobile phones, e.g., connected with each other using the Bluetooth technology.

This first scenario shows, in addition to the technical content, something typical in the communication business – many acronyms. This book contains and defines many of these. If you get lost with an acronym, please check the appendix, which contains the complete list, or check the terms and definitions database interactive (TEDDI) of ETSI (2002).

Think of similar scenarios for air traffic or railroad traffic. Different problems can occur here due to speed. While aircraft typically travel at up to 900 km/h and current trains up to 350 km/h, many technologies cannot operate if the relative speed of a mobile device exceeds, e.g., 250 km/h for GSM or 100 km/h for AMPS. Only some technologies, like DAB work up to 900 km/h (unidirectional only).

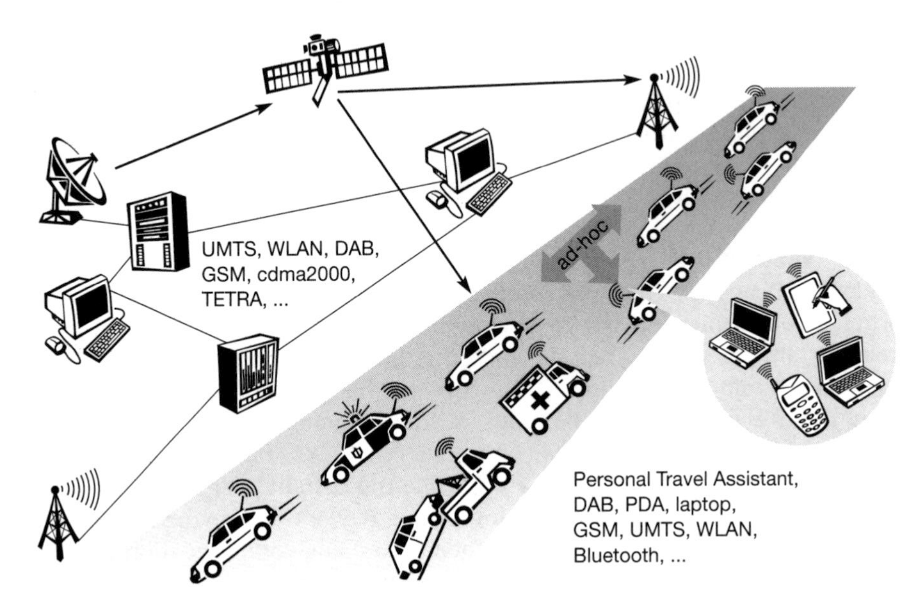

Figure 1.1
A typical application of mobile communications: road traffic

1.1.2 Emergencies

Just imagine the possibilities of an ambulance with a high-quality wireless connection to a hospital. Vital information about injured persons can be sent to the hospital from the scene of the accident. All the necessary steps for this particular type of accident can be prepared and specialists can be consulted for an early diagnosis. Wireless networks are the only means of communication in the case of natural disasters such as hurricanes or earthquakes. In the worst cases, only decentralized, wireless ad-hoc networks survive. The breakdown of all cabling not only implies the failure of the standard wired telephone system, but also the crash of all mobile phone systems requiring base stations!

1.1.3 Business

A travelling salesman today needs instant access to the company's database: to ensure that files on his or her laptop reflect the current situation, to enable the company to keep track of all activities of their travelling employees, to keep databases consistent etc. With wireless access, the laptop can be turned into a true mobile office, but efficient and powerful synchronization mechanisms are needed to ensure data consistency. Figure 1.2 illustrates what may happen when employees try to communicate off base. At home, the laptop connects via a WLAN or LAN and DSL to the Internet. Leaving home requires a handover to another technology, e.g., to an enhanced version of GSM, as soon as the WLAN coverage ends. Due to interference and other factors discussed in chapter 2, data rates drop while cruising at higher speed. Gas stations may offer WLAN hot spots as well as gas. Trains already offer support for wireless connectivity. Several more handovers to different technologies might be necessary before reaching the office. No matter

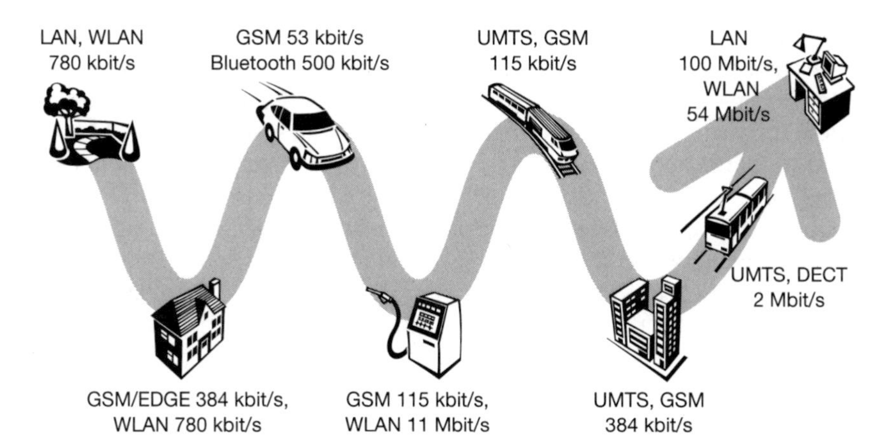

Figure 1.2
Mobile and wireless services – always best connected

when and where, mobile communications should always offer as good connectivity as possible to the internet, the company's intranet, or the telephone network.

1.1.4 Replacement of wired networks

In some cases, wireless networks can also be used to replace wired networks, e.g., remote sensors, for tradeshows, or in historic buildings. Due to economic reasons, it is often impossible to wire remote sensors for weather forecasts, earthquake detection, or to provide environmental information. Wireless connections, e.g., via satellite, can help in this situation. Tradeshows need a highly dynamic infrastructure, but cabling takes a long time and frequently proves to be too inflexible. Many computer fairs use WLANs as a replacement for cabling. Other cases for wireless networks are computers, sensors, or information displays in historical buildings, where excess cabling may destroy valuable walls or floors. Wireless access points in a corner of the room can represent a solution.

1.1.5 Infotainment and more

Internet everywhere? Not without wireless networks! Imagine a travel guide for a city. Static information might be loaded via CD-ROM, DVD, or even at home via the Internet. But wireless networks can provide up-to-date information at any appropriate location. The travel guide might tell you something about the history of a building (knowing via GPS, contact to a local base station, or triangulation where you are) downloading information about a concert in the building at the same evening via a local wireless network. You may choose a seat, pay via electronic cash, and send this information to a service provider (Cheverst, 2000). Another growing field of wireless network applications lies in entertainment and games to enable, e.g., ad-hoc gaming networks as soon as people meet to play together.

1.1.6 Location dependent services

Many research efforts in mobile computing and wireless networks try to hide the fact that the network access has been changed (e.g., from mobile phone to WLAN or between different access points) or that a wireless link is more error prone than a wired one. Many chapters in this book give examples: Mobile IP tries to hide the fact of changing access points by redirecting packets but keeping the same IP address (see section 8.1), and many protocols try to improve link quality using encoding mechanisms or retransmission so that applications made for fixed networks still work.

In many cases, however, it is important for an application to 'know' something about the location or the user might need location information for further activities. Several services that might depend on the actual location can be distinguished:

- **Follow-on services**: The function of forwarding calls to the current user location is well known from the good old telephone system. Wherever you are, just transmit your temporary phone number to your phone and it redirects incoming calls.[2] Using mobile computers, a follow-on service could offer, for instance, the same desktop environment wherever you are in the world. All e-mail would automatically be forwarded and all changes to your desktop and documents would be stored at a central location at your company. If someone wanted to reach you using a multimedia conferencing system, this call would be forwarded to your current location.
- **Location aware services**: Imagine you wanted to print a document sitting in the lobby of a hotel using your laptop. If you drop the document over the printer icon, where would you expect the document to be printed? Certainly not by the printer in your office! However, without additional information about the capabilities of your environment, this might be the only thing you can do. For instance, there could be a service in the hotel announcing that a standard laser printer is available in the lobby or a color printer in a hotel meeting room etc. Your computer might then transmit your personal profile to your hotel which then charges you with the printing costs.
- **Privacy**: The two service classes listed above immediately raise the question of privacy. You might not want video calls following you to dinner, but maybe you would want important e-mails to be forwarded. There might be locations and/or times when you want to exclude certain services from reaching you and you do not want to be disturbed. You want to utilize location dependent services, but you might not want the environment to know exactly who you are. Imagine a hotel monitoring all guests and selling these profiles to companies for advertisements.
- **Information services**: While walking around in a city you could always use

2 Actually, this is already done with the phone network – your phone just handles some signalling.

your wireless travel guide to 'pull' information from a service, e.g., 'Where is the nearest Mexican restaurant?' However, a service could also actively 'push' information on your travel guide, e.g., the Mexican restaurant just around the corner has a special taco offer.

- **Support services**: Many small additional mechanisms can be integrated to support a mobile device. Intermediate results of calculations, state information, or cache contents could 'follow' the mobile node through the fixed network. As soon as the mobile node reconnects, all information is available again. This helps to reduce access delay and traffic within the fixed network. Caching of data on the mobile device (standard for all desktop systems) is often not possible due to limited memory capacity. The alternative would be a central location for user information and a user accessing this information through the (possibly large and congested) network all the time as it is often done today.

1.1.7 Mobile and wireless devices

Even though many mobile and wireless devices are available, there will be many more in the future. There is no precise classification of such devices, by size, shape, weight, or computing power. Currently, laptops are considered the upper end of the mobile device range.[3] The following list gives some examples of mobile and wireless devices graded by increasing performance (CPU, memory, display, input devices etc.). However, there is no sharp line between the categories and companies tend to invent more and more new categories.

- **Sensor**: A very simple wireless device is represented by a sensor transmitting state information. One example could be a switch sensing the office door. If the door is closed, the switch transmits this to the mobile phone inside the office which will not accept incoming calls. Without user interaction, the semantics of a closed door is applied to phone calls.
- **Embedded controllers**: Many appliances already contain a simple or sometimes more complex controller. Keyboards, mice, headsets, washing machines, coffee machines, hair dryers and TV sets are just some examples. Why not have the hair dryer as a simple mobile and wireless device (from a communication point of view) that is able to communicate with the mobile phone? Then the dryer would switch off as soon as the phone starts ringing – that would be a nice application!
- **Pager**: As a very simple receiver, a pager can only display short text messages, has a tiny display, and cannot send any messages. Pagers can even be integrated into watches. The tremendous success of mobile phones, has made the pager virtually redundant in many countries. Short messages have replaced paging. The situation is somewhat different for emergency services

3 Putting a mainframe on a truck does not really make it a mobile device.

where it may be necessary to page a larger number of users reliably within short time.

- **Mobile phones**: The traditional mobile phone only had a simple black and white text display and could send/receive voice or short messages. Today, mobile phones migrate more and more toward PDAs. Mobile phones with full color graphic display, touch screen, and Internet browser are easily available.
- **Personal digital assistant**: PDAs typically accompany a user and offer simple versions of office software (calendar, note-pad, mail). The typical input device is a pen, with built-in character recognition translating handwriting into characters. Web browsers and many other software packages are available for these devices.
- **Pocket computer**: The next steps toward full computers are pocket computers offering tiny keyboards, color displays, and simple versions of programs found on desktop computers (text processing, spreadsheets etc.).
- **Notebook/laptop**: Finally, laptops offer more or less the same performance as standard desktop computers; they use the same software – the only technical difference being size, weight, and the ability to run on a battery. If operated mainly via a sensitive display (touch sensitive or electromagnetic), the devices are also known as notepads or tablet PCs.

The mobile and wireless devices of the future will be more powerful, less heavy, and comprise new interfaces to the user and to new networks. However, one big problem, which has not yet been solved, is the energy supply. The more features that are built into a device, the more power it needs. The higher the performance of the device, the faster it drains the batteries (assuming the same technology). Furthermore, wireless data transmission consumes a lot of energy.

Although the area of mobile computing and mobile communication is developing rapidly, the devices typically used today still exhibit some major drawbacks compared to desktop systems in addition to the energy problem. Interfaces have to be small enough to make the device portable, so smaller keyboards are used. This makes typing difficult due to their limited key size. Small displays are often useless for graphical display. Higher resolution does not help, as the limiting factor is the resolution capacity of the human eye. These devices have to use new ways of interacting with a user, such as, e.g., touch sensitive displays and voice recognition.

Mobile communication is greatly influenced by the merging of telecommunication and computer networks. We cannot say for certain what the telephone of the future will look like, but it will most probably be a computer. Even today, telephones and mobile phones are far from the simple 'voice transmission devices' they were in the past.[4] Developments like 'voice over IP' and the general trend toward packet-oriented networks enforce the metamorphosis of telephones (although voice services still guarantee good revenue). While no one

4 Chapter 4 will present more features of modern mobile phone systems, including the growing demand for bandwidth to use typical Internet applications via the mobile 'phone'.

can predict the future of communication devices precisely, it is quite clear that there will still be many fixed systems, complemented by a myriad of small wireless computing devices all over the world. More people already use mobile phones than fixed phones!

1.2 A short history of wireless communication

For a better understanding of today's wireless systems and developments, a short history of wireless communication is presented in the following section. This cannot cover all inventions but highlights those that have contributed fundamentally to today's systems.

The use of light for wireless communications reaches back to ancient times. In former times, the light was either 'modulated' using mirrors to create a certain light on/light off pattern ('amplitude modulation') or, for example, flags were used to signal code words ('amplitude and frequency modulation', see chapter 2). The use of smoke signals for communication is mentioned by Polybius, Greece, as early as 150 BC. It is also reported from the early (or western) Han dynasty in ancient China (206 BC–24 AD) that light was used for signaling messages along a line of signal towers towards the capitol Chang'an (Xi'an). Using light and flags for wireless communication remained important for the navy until radio transmission was introduced, and even today a sailor has to know some codes represented by flags if all other means of wireless communication fail. It was not until the end of the 18th century, when **Claude Chappe** invented the optical telegraph (**1794**), that long-distance wireless communication was possible with technical means. Optical telegraph lines were built almost until the end of the following century.

Wired communication started with the first commercial telegraph line between Washington and Baltimore in 1843, and **Alexander Graham Bell's** invention and marketing of the telephone in 1876 (others tried marketing before but did not succeed, e.g., **Philip Reis**, 1834–1874, discovered the telephone principle in **1861**). In Berlin, a public telephone service was available in **1881**, the first regular public voice and video service (multimedia!) was already available in 1936 between Berlin and Leipzig.

All optical transmission systems suffer from the high frequency of the carrier light. As every little obstacle shadows the signal, rain and fog make communication almost impossible. At that time it was not possible to focus light as efficiently as can be done today by means of a laser, wireless communication did not really take off until the discovery of electromagnetic waves and the development of the equipment to modulate them. It all started with **Michael Faraday** (and about the same time **Joseph Henry**) demonstrating electromagnetic induction in 1831 and **James C. Maxwell** (1831–79) laying the theoretical foundations for electromagnetic fields with his famous equations (**1864**). Finally, **Heinrich Hertz** (1857–94) was the first to demonstrate the wave character of electrical transmission through space (**1886**), thus proving Maxwell's equations. Today the unit Hz reminds us of this discovery. **Nikola Tesla** (1856–1943) soon increased the distance of electromagnetic transmission.

The name, which is most closely connected with the success of wireless communication, is certainly that of **Guglielmo Marconi** (1874–1937). He gave the first demonstration of wireless telegraphy in 1895 using long wave transmission with very high transmission power (> 200 kW). The first transatlantic transmission followed in 1901. Only six years later, in **1907**, the first **commercial transatlantic connections** were set up. Huge base stations using up to thirty 100 m high antennas were needed on both sides of the Atlantic Ocean. Around that time, the first **World Administration Radio Conference (WARC)** took place, coordinating the worldwide use of radio frequencies. The first **radio broadcast** took place in **1906** when **Reginald A. Fessenden** (1866–1932) transmitted voice and music for Christmas. In 1915, the first wireless voice transmission was set up between New York and San Francisco. The first **commercial radio** station started in **1920** (KDKA from Pittsburgh). Sender and receiver still needed huge antennas and high transmission power.

This changed fundamentally with the discovery of **short waves**, again by Marconi, in **1920** (In connection with wireless communication, short waves have the advantage of being reflected at the ionosphere.) It was now possible to send short radio waves around the world bouncing at the ionosphere – this technique is still used today. The invention of the electronic **vacuum tube** in **1906** by **Lee DeForest** (1873–1961) and **Robert von Lieben** (1878–1913) helped to reduce the size of sender and receiver. Vacuum tubes are still used, e.g., for the amplification of the output signal of a sender in today's radio stations. One of the first **'mobile' transmitters** was on board a Zeppelin in **1911**. As early as **1926**, the first **telephone in a train** was available on the Berlin-Hamburg line. Wires parallel to the railroad track worked as antenna. The first car radio was **commercially** available in **1927** ('Philco Transitone'); but George Frost an 18-year-old from Chicago had integrated a radio into a Ford Model T as early as 1922.

Nineteen twenty-eight was the year of many field trials for **television broadcasting. John L. Baird** (1888–1946) transmitted TV across the Atlantic and demonstrated **color TV**, the station **WGY** (Schenectady, NY) started **regular TV broadcasts** and the first **TV news**. The first **teleteaching** started in **1932** from the **CBS** station W2XAB. Up until then, all wireless communication used amplitude modulation (see section 2.6), which offered relatively poor quality due to interference. One big step forward in this respect was the invention of **frequency modulation** in **1933** by **Edwin H. Armstrong** (1890–1954). Both fundamental modulation schemes are still used for today's radio broadcasting with frequency modulation resulting in a much better quality. By the early 1930s, many radio stations were already broadcasting all over the world.

After the Second World War, many national and international projects in the area of wireless communications were triggered off. The first network in Germany was the analog A-Netz from 1958, using a carrier frequency of 160 MHz. Connection setup was only possible from the mobile station, no handover, i.e., changing of the base station, was possible. Back in 1971, this system had coverage of 80 per cent and 11,000 customers. It was not until 1972 that the B-Netz followed in Germany, using the same 160 MHz. This network could initiate the connection setup from a station in the fixed telephone network, but, the current location of the mobile receiver had to be known. This system was also available in Austria, The Netherlands, and Luxembourg. In 1979, the B-Netz had 13,000 customers in West Germany and needed a heavy sender and receiver, typically built into cars.

At the same time, the northern European countries of Denmark, Finland, Norway, and Sweden (the cradle of modern mobile communications) agreed upon the **nordic mobile telephone (NMT)** system. The analogue NMT uses a 450 MHz carrier and is still the only available system for mobile communication in some very remote places (NMT at 900 MHz followed in 1986). Several other national standards evolved and by the early 1980s Europe had more than a handful of different, completely incompatible analog mobile phone standards. In accordance with the general idea of a European Union, the European countries decided to develop a pan-European mobile phone standard in **1982**. The new system aimed to:

- use a new spectrum at 900 MHz;
- allow roaming[5] throughout Europe;
- be fully digital; and
- offer voice and data service.

The **'Groupe Spéciale Mobile' (GSM)** was founded for this new development.

In **1983** the US system **advanced mobile phone system (AMPS)** started (EIA, 1989). AMPS is an analog mobile phone system working at 850 MHz. Telephones at home went wireless with the standard **CT1 (cordless telephone)** in **1984**, (following its predecessor the **CT0** from **1980**). As digital systems were not yet available, more analog standards followed, such as the German C-Netz at 450 MHz with analog voice transmission. Hand-over between 'cells' was now possible, the signalling system was digital in accordance with the trends in fixed networks (SS7), and automatic localization of a mobile user within the whole network was supported. This analog network was switched off in 2000. Apart from voice transmission the services offered fax, data transmission via modem, X.25, and electronic mail. **CT2**, the successor of CT1, was embodied into British Standards published in **1987** (DTI, 1987) and later adopted by ETSI for Europe (ETS, 1994). CT2 uses the spectrum at 864 MHz and offers a data channel at a rate of 32 kbit/s.

The early 1990s marked the beginning of **fully digital systems**. In **1991**, ETSI adopted the standard **digital European cordless telephone (DECT)** for digital cordless telephony (ETSI, 1998). DECT works at a spectrum of 1880–1900 MHz with a range of 100–500 m. One hundred and twenty duplex channels can carry

---

5 Roaming here means a seamless handover of a telephone call from one network provider to another while crossing national boundaries.

up to 1.2 Mbit/s for data transmission. Several new features, such as voice encryption and authentication, are built-in. The system supports several 10,000 users/km^2 and is used in more than 110 countries around the world (over 150 million shipped units). Today, DECT has been renamed **digital enhanced cordless telecommunications** for marketing reasons and to reflect the capabilities of DECT to transport multimedia data streams. Finally, after many years of discussions and field trials, **GSM** was standardized in a document of more than 5,000 pages in **1991**. This first version of GSM, now called **global system for mobile communication**, works at 900 MHz and uses 124 full-duplex channels. GSM offers full international roaming, automatic location services, authentication, encryption on the wireless link, efficient interoperation with ISDN systems, and a relatively high audio quality. Furthermore, a short message service with up to 160 alphanumeric characters, fax group 3, and data services at 9.6 kbit/s have been integrated. Depending on national regulations, one or several providers can use the channels, different accounting and charging schemes can be applied etc. However, all GSM systems remain compatible. Up to now, over 400 providers in more than 190 countries have adopted the GSM standard (over 70 per cent of the world's wireless market).

It was soon discovered that the analog AMPS in the US and the digital GSM at 900 MHz in Europe are not sufficient for the high user densities in cities. While in the US, no new spectrum was allocated for a new system, in Europe a new frequency band at 1800 MHz was chosen. The effect was as follows. In the US, different companies developed different new, more bandwidth-efficient technologies to operate side-by-side with AMPS in the same frequency band. This resulted in three incompatible systems, the analog narrowband AMPS (IS-88, (TIA, 1993a)), and the two digital systems **TDMA** (IS-136, (TIA, 1996)) and **CDMA** (IS-95, (TIA, 1993b)). The Europeans agreed to use GSM in the 1800 MHz spectrum. These GSM–1800 networks (also known as **DCS 1800**, digital cellular system) started with a better voice quality due to newer speech codecs. These networks consist of more and smaller cells (see chapters 2 and 4). GSM is also available in the US as GSM–1900 (also called **PCS 1900**) using spectrum at 1900 MHz like the newer versions of the TDMA and CDMA systems.

Europe believes in standards, while the US believes in market forces – GSM is one of the few examples where the approach via standardization worked. So, while Europe has one common standard, and roaming is possible even to Australia or Singapore, the US still struggles with many incompatible systems. However, the picture is different when it comes to more data communication-oriented systems like local area networks. Many proprietary wireless local area network systems already existed when ETSI standardized the high performance radio local area network (HIPERLAN) in **1996**. This was a family of standards and recommendations. HIPERLAN type 1 should operate at 5.2 GHz and should offer data rates of up to 23.5 Mbit/s. Further types had been specified with type 4 going up to 155 Mbit/s at 17 GHz. However, although coming later than HIPERLAN in **1997**, the IEEE standard **802.11** was soon the winner for local area networks. It works at the license-free Industrial, Science, Medical (ISM) band at 2.4 GHz and infra red offering 2 Mbit/s in the beginning (up to 10 Mbit/s with proprietary solutions already at that time). Although HIPERLAN has better performance figures, no products were available while many companies soon offered 802.11 compliant equipment.

Nineteen ninety-eight marked the beginning of mobile communication using satellites with the **Iridium** system (Iridium, 2002). Up to this time, satellites basically worked as a broadcast distribution medium or could only be used with big and heavy equipment – Iridium marked the beginning of small and truly portable mobile satellite telephones including data service. Iridium consists of 66 satellites in low earth orbit and uses the 1.6 GHz band for communication with the mobile phone. In 1998 the Europeans agreed on the **universal mobile telecommunications system** (UMTS) as the European proposal for the International Telecommunication Union (ITU) **IMT-2000 (international mobile telecommunications)**. In the first phase, UMTS combines GSM network technology with more bandwidth-efficient CDMA solutions.

The IMT-2000 recommendations define a common, worldwide framework for future mobile communication at 2 GHz (ITU, 2002). This includes, e.g., a framework for services, the network architecture including satellite communication, strategies for developing countries, requirements of the radio interface, spectrum considerations, security and management frameworks, and different transmission technologies.

Nineteen ninety nine saw several more powerful WLAN standards. IEEE published 802.11b offering 11 Mbit/s at 2.4 GHz. The same spectrum is used by **Bluetooth**, a short-range technology to set-up wireless personal area networks with gross data rates less than 1 Mbit/s. The ITU dropped the plan of a single, worldwide standard for third generation mobile phone systems and decided on the IMT-2000 family concept that includes several technologies (UMTS, cdma2000, DECT etc. see chapter 4). The **wireless application protocol** (WAP) started at the same time as **i-mode** in Japan. While WAP did not succeed in the beginning, i-mode soon became a tremendous success (see chapter 10).

The year **2000**, came with higher data rates and packet-oriented transmission for GSM (HSCSD, GPRS – see chapter 4). It should not be forgotten that the late nineties was the time when a lot of hype about the communications business started. Thus it was relatively easy for marketing people to portray third generation technology as high-performance Internet on mobile phones. In Europe, UMTS was announced as capable of handling live, interactive video streaming for all users at 2 Mbit/s. All technically-oriented people knew that this promise could not be fulfilled by the system, but the **auctions and beauty contests** for licensing 3G spectrum started. In Europe alone more than €100 billion had been paid before the disillusionment set in. Companies that had never run a network before paid billions for licenses. Many of these companies are now bankrupt and the remaining companies suffer from the debts.

Most of the hype is over, but the **third generation** of mobile communication

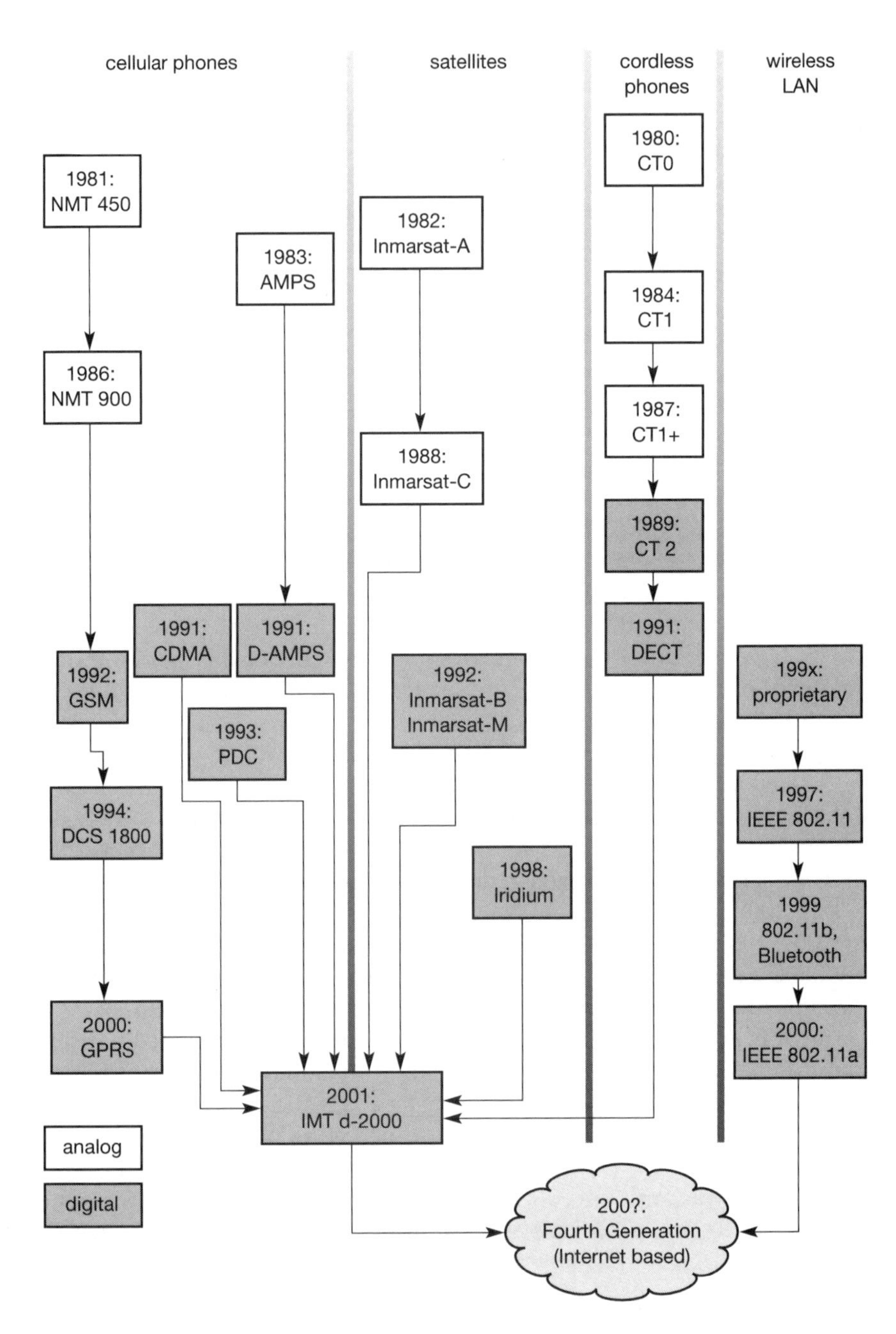

Figure 1.3 Overview of some wireless communication systems

started in **2001** in Japan with the FOMA service, in Europe with several field trials, and in, e.g., Korea with cdma2000 (see Figure 4.2 for the evolution of 3G systems). IEEE released a new WLAN standard, **802.11a**, operating at 5 GHz and offering gross data rates of 54 Mbit/s. This standard uses the same physical layer as **HiperLAN2** does (standardized in 2000), the only remaining member of the HIPERLAN family. In **2002** new WLAN developments followed. Examples are **802.11g** offering up to 54 Mbit/s at 2.4 GHz and many new Bluetooth applications (headsets, remote controls, wireless keyboards, hot syncing etc.). The network providers continued to deploy the infrastructure for 3G networks as many licensing conditions foresee a minimum coverage at a certain date. While digital TV via satellite has existed for several years, digital terrestrial TV (DVB-T, see chapter 6) started as regular service in Berlin in November 2002. This system allows for high-quality TV on the move and requires only an antenna of a few centimeters.

Figure 1.3 gives an overview of some of the networks described above, and shows the development of cellular phone systems and cordless phones together with satellites and LANs. While many of the classical mobile phone systems converged to IMT-2000 systems (with cdma2000 and W-CDMA/UMTS being the predominant systems), the wireless LAN area developed more or less independently. No one knows exactly what the next generation of mobile and wireless system will look like, but, there are strong indicators that it will be widely Internet based – the system will use Internet protocols and Internet applications. While the current third generation systems still heavily rely on classical telephone technology in the network infrastructure, future systems will offer users the choice of many different networks based on the internet (see chapter 11). However, no one knows exactly when and how this common platform will be available. Companies have to make their money with 3G systems first.

The dates shown in the figure typically indicate the start of service (i.e., the systems have been designed, invented, and tested earlier). The systems behind the acronyms will be explained in the following chapters (cellular and cordless phones in chapter 4, satellites in chapter 5, WLANs in chapter 7).[6]

1.3 A market for mobile communications

Although the growth in wireless and mobile communication systems has slowed down, these technologies have still a huge market potential. More and more people use mobile phones, wireless technology is built into many cars, wireless data services are available in many regions, and wireless local area networks are used in many places.

Figure 1.4 shows the increasing number of subscribers to mobile phone services worldwide (GSM World, 2002). This figure shows the tremendous growth rates up to 2000. That growth continues today, mainly due to China that has the largest number of users.

Figure 1.5 shows the cellular subscribers per region (GSM World, 2002).

6 Note that analog systems are not described.

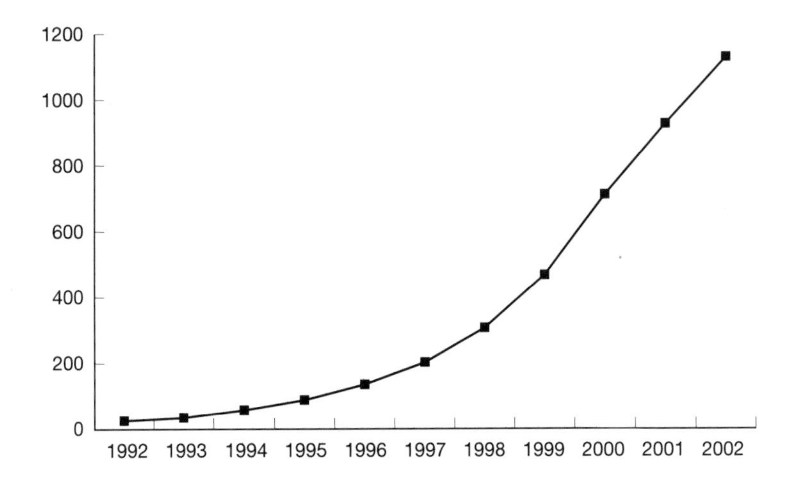

Figure 1.4
Mobile phone service subscribers worldwide (in millions)

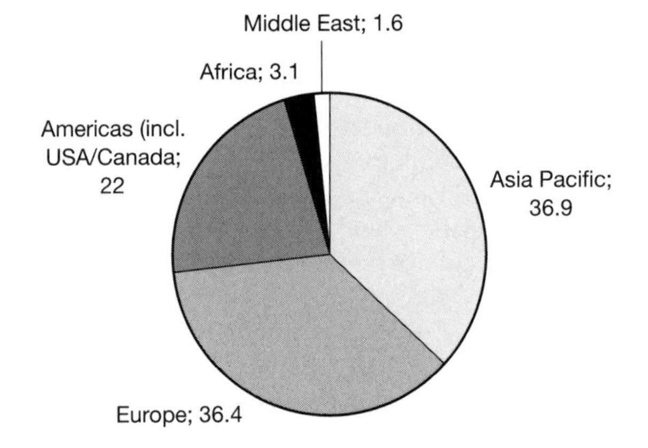

Figure 1.5
Cellular subscribers per region (June 2002)

While the shares of Europe and China are almost equal, the market in Europe is saturated with second-generation GSM systems (mobile penetration is about 70 per cent). Countries such as Germany and France exhibited growth rates of 40 per cent or more in 1998. Europe's share will decrease compared to China, the Americas, and Africa.

1.4 Some open research topics

Although this book explains many systems supporting mobility and explores many solutions for wireless access, a lot remains to be done in the field. We are only at the beginning of wireless and mobile networking. The differences between wired, fixed networks and wireless networks open up various topics. The reader may find even more in, e.g., the book of the **wireless world research forum** (WWRF, 2002):

- **Interference**: Radio transmission cannot be protected against interference using shielding as this is done in coaxial cable or shielded twisted pair. For example, electrical engines and lightning cause severe interference and result in higher loss rates for transmitted data or higher bit error rates respectively.
- **Regulations and spectrum**: Frequencies have to be coordinated, and unfortunately, only a very limited amount of frequencies are available (due to technical and political reasons). One research topic involves determining how to use available frequencies more efficiently, e.g., by new modulation schemes (see chapter 2) or demand-driven multiplexing (see chapter 3). Further improvements are new air interfaces, power aware ad-hoc networks, smart antennas, and software defined radios (SDR). The latter allow for software definable air interfaces but require high computing power.
- **Low bandwidth**: Although they are continuously increasing, transmission rates are still very low for wireless devices compared to desktop systems. Local wireless systems reach some Mbit/s while wide area systems only offer some 10 kbit/s. One task would involve adapting applications used with high-bandwidth connections to this new environment so that the user can continue using the same application when moving from the desktop outside the building. Researchers look for more efficient communication protocols with low overhead.
- **High delays, large delay variation**: A serious problem for communication protocols used in today's Internet (TCP/IP) is the big variation in link characteristics. In wireless systems, delays of several seconds can occur, and links can be very asymmetrical (i.e., the links offer different service quality depending on the direction to and from the wireless device). Applications must be tolerant and use robust protocols.
- **Lower security, simpler to attack**: Not only can portable devices be stolen more easily, but the radio interface is also prone to the dangers of eavesdropping. Wireless access must always include encryption, authentication, and other security mechanisms that must be efficient and simple to use.
- **Shared medium**: Radio access is always realized via a shared medium. As it is impossible to have a separate wire between a sender and each receiver, different competitors have to 'fight' for the medium. Although different medium access schemes have been developed, many questions are still unanswered, for example how to provide quality of service efficiently with different combinations of access, coding, and multiplexing schemes (Fitzek, 2002).
- **Ad-hoc networking**: Wireless and mobile computing allows for spontaneous networking with prior set-up of an infrastructure. However, this raises many new questions for research: routing on the networking and application layer, service discovery, network scalability, reliability, and stability etc.

A general research topic for wireless communication (and a source for endless discussion) is its effect on the human body or organisms in general. It is unclear if, and to what extent, electromagnetic waves transmitted from wireless devices can influence organs. Microwave ovens and WLANs both operate at the same frequency of 2.4 GHz. However, the radiation of a WLAN is very low (e.g.,

100 mW) compared to a microwave oven (e.g., 800 W inside the oven). Additionally, as chapter 2 shows in more detail, propagations characteristics, absorption, directed antennas etc. play an important role. Users, engineers, researchers and politicians need more studies to understand the effect of long-term low-power radiation (Lin, 1997), BEMS (2002), COST (2000), NIEHS (2002). The World Health Organization (WHO) has started a worldwide project on electromagnetic fields (WHO, 2002).

1.5 A simplified reference model

This book follows the **basic reference model** used to structure communication systems (Tanenbaum, 2003). Any readers who are unfamiliar with the basics of communication networks should look up the relevant sections in the recommended literature (Halsall, 1996), (Keshav, 1997), (Tanenbaum, 2003), (Kurose, 2003). Figure 1.6 shows a personal digital assistant (PDA) which provides an example for a wireless and portable device. This PDA communicates with a base station in the middle of the picture. The base station consists of a radio transceiver (sender and receiver) and an interworking unit connecting the wireless link with the fixed link. The communication partner of the PDA, a conventional computer, is shown on the right-hand side.

Underneath each network element (such as PDA, interworking unit, computer), the figure shows the **protocol stack** implemented in the system according to the reference model. **End-systems**, such as the PDA and computer in the example, need a full protocol stack comprising the application layer, transport layer, network layer, data link layer, and physical layer. Applications

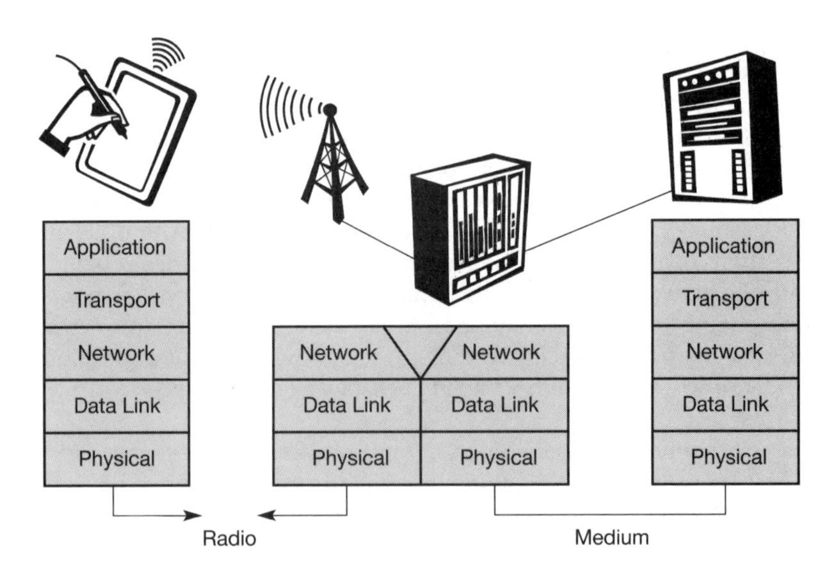

Figure 1.6 Simple network and reference model used in this book

on the end-systems communicate with each other using the lower layer services. **Intermediate systems**, such as the interworking unit, do not necessarily need all of the layers. Figure 1.6 only shows the network, data link, and physical layers. As (according to the basic reference model) only entities at the same level communicate with each other (i.e., transport with transport, network with network) the end-system applications do not notice the intermediate system directly in this scenario. The following paragraphs explain the functions of each layer in more detail in a wireless and mobile environment.

- **Physical layer**: This is the lowest layer in a communication system and is responsible for the conversion of a stream of bits into signals that can be transmitted on the sender side. The physical layer of the receiver then transforms the signals back into a bit stream. For wireless communication, the physical layer is responsible for frequency selection, generation of the carrier frequency, signal detection (although heavy interference may disturb the signal), modulation of data onto a carrier frequency and (depending on the transmission scheme) encryption. These features of the physical layer are mainly discussed in chapter 2, but will also be mentioned for each system separately in the appropriate chapters.
- **Data link layer**: The main tasks of this layer include accessing the medium, multiplexing of different data streams, correction of transmission errors, and synchronization (i.e., detection of a data frame). Chapter 3 discusses different medium access schemes. A small section about the specific data link layer used in the presented systems is combined in each respective chapter. Altogether, the data link layer is responsible for a reliable point-to-point connection between two devices or a point-to-multipoint connection between one sender and several receivers.
- **Network layer**: This third layer is responsible for routing packets through a network or establishing a connection between two entities over many other intermediate systems. Important topics are addressing, routing, device location, and handover between different networks. Chapter 8 presents several solutions for the network layer protocol of the internet (the Internet Protocol IP). The other chapters also contain sections about the network layer, as routing is necessary in most cases.
- **Transport layer**: This layer is used in the reference model to establish an end-to-end connection. Topics like quality of service, flow and congestion control are relevant, especially if the transport protocols known from the Internet, TCP and UDP, are to be used over a wireless link.
- **Application layer**: Finally, the applications (complemented by additional layers that can support applications) are situated on top of all transmission-oriented layers. Topics of interest in this context are service location, support for multimedia applications, adaptive applications that can handle the large variations in transmission characteristics, and wireless access to the world wide web using a portable device. Very demanding applications are video (high data rate) and interactive gaming (low jitter, low latency).

1.6 Overview

The whole book is structured in a bottom-up approach as shown in Figure 1.7. Chapter 2 presents some basics about wireless transmission technology. The topics covered include: frequencies used for communication, signal characteristics, antennas, signal propagation, and several fundamental multiplexing and modulation schemes. This chapter does not require profound knowledge of electrical engineering, nor does it explore all details about the underlying physics of wireless communication systems. Its aim is rather to help the reader understand the many design decisions in the higher layers of mobile communication systems.

Chapter 3 presents a broad range of media access technologies. It explains why media access technologies from fixed networks often cannot be applied to wireless networks, and shows the special problems for wireless terminals accessing 'space' as the common medium. The chapter shows access methods for different purposes, such as wireless mobile phones with a central base station that can control the access, or completely decentralized ad-hoc networks without any dedicated station. This chapter shows how the multiplexing schemes described in chapter 2 can now be used for accessing the medium. Special focus is on code division multiple access (CDMA), which is one of the important access methods for many new systems. Further topics are variants of Aloha and reservation schemes known from satellite networks.

Figure 1.7 Overview of the book's structure

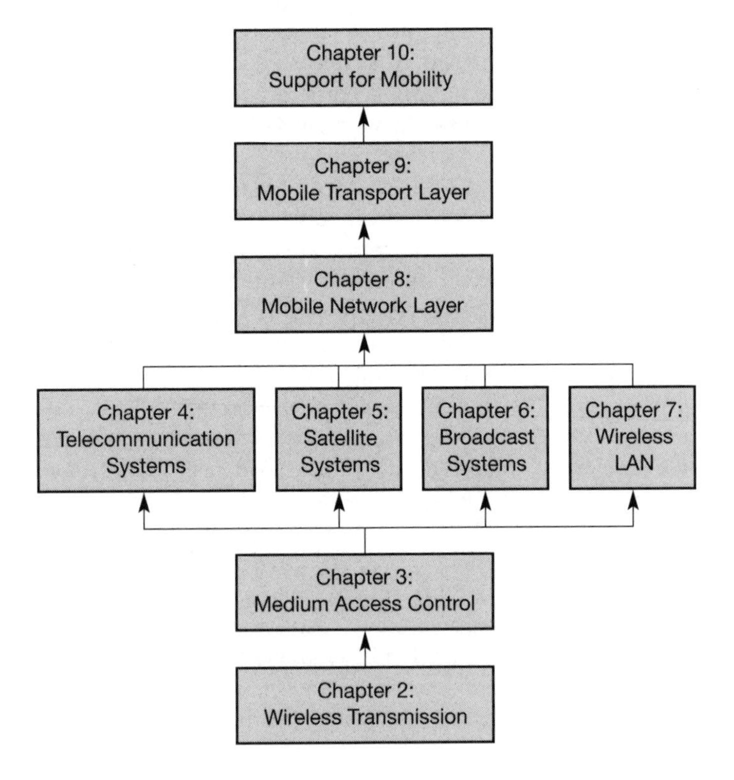

After chapter 3, the reader can select any of the chapters from 4 to 7. These present selected wireless transmission systems using the basic technologies shown in chapters 2 and 3 and offer a data transport service to higher layers. Chapter 4 covers wireless communication systems originating from the telecommunication industry and standardization bodies. On the one hand GSM, which is currently the most successful digital mobile phone system worldwide, and on the other digital enhanced cordless telecommunications (DECT), an example for a local wireless telecommunication system. The chapter presents standards, frequencies, services, access methods, and the architecture of the systems. It shows the migration from voice orientation toward packet transmission. UMTS as one candidate for future mobile systems concludes the chapter.

Chapter 5 gives a basic introduction to satellite communication. The myriad technical details are not of interest, but the potential of satellite systems to exchange data worldwide is.

Chapter 6 covers the relatively young field of wireless communication: digital broadcast systems. Broadcast systems allow only unidirectional data distribution but can offer new services that come along with digital radio (digital audio broadcasting, DAB) and digital television digital video broadcasting (DVB). The ability to distribute arbitrary multi-media data using standard formats known from the Internet is an extremely important design factor for these systems.

Chapter 7 presents wireless LANs, i.e., the extension of today's fixed local area networks into the wireless domain. The examples presented are the standards IEEE 802.11, HiperLAN2, and Bluetooth. After a general introduction, the design goals of wireless LANs, physical layer, medium access control layer, and services of all three LANs are presented. In this chapter, a comparison of the different approaches taken for IEEE 802.11 and HiperLAN2 is of particular interest. While the first system offers only best effort traffic up to 54 Mbit/s, the latter tries to give quality of service (QoS) guarantees at similar data rates. However, while many IEEE 802.11 compliant products are available, no HiperLAN2 product has been released. Bluetooth is different because it was primarily designed for short range ad-hoc networking using very cheap hardware and offers only some hundred kbit/s (as replacement for infrared).

Chapter 8 presents solutions for a mobile network layer. This layer can be used on top of different transmission technologies as presented in chapters 4 through 7. While mobile IP is the main topic of the chapter, it discusses also mechanisms such as the dynamic host configuration protocol (DHCP) and routing in ad-hoc networks. Because IP is clearly dominating data communication networks (it is the basis for the internet), it is only natural to extend this protocol to the mobile domain. The chapter discusses the problems associated with IP and mobility and shows some solutions. From chapter 8 onwards the term 'mobility' is used more often, because it does not matter for higher layer protocols if the mobility is supported by wireless transmission technologies or if the user has to plug-in a laptop wherever she or he currently is. The problems of interrupted connections and changing access points to the network remain almost identical.

Chapter 9 focuses on changes needed for TCP so it can be used in a mobile environment. To work with well-known applications from the internet, e.g., file transfer or remote login, a user might need a reliable end-to-end communication service as provided by the transmission control protocol (TCP). It is shown that today's TCP fails if it is not adapted or enhanced. Several solutions are discussed and compared. Each solution exhibits specific strengths and weaknesses, so up to now there is no standard for a 'Mobile TCP'. However, several communication systems already use TCP with a 'wireless profile'.

All mobile networks are worthless without applications using them. Chapter 10 shows problems with current applications, as they are known from fixed networks, and presents some new developments. Among other topics, it deals with file systems. Here consistency in a distributed system brings about major problems. How is consistency to be maintained in the case of disconnection? Is it better to deny access to data if consistency cannot be granted? The chapter presents systems dealing with these problems.

The big topic in today's Internet is the world wide web (www), the logical structure made by multi-media documents and hyperlinks. In connection with the www, the problems can be summarized as follows: the www of today assumes connections with higher bandwidth, desktop computers with a lot of memory, a powerful CPU, high-resolution graphics, and dozens of plug-ins installed. Mobile devices have scarce energy resources, therefore less powerful CPUs and less memory. These devices have to be portable, and consequently, input devices are very limited and displays are of a low resolution, not to mention the hi-fi capabilities. This obvious mismatch between two technologies, both with huge growth rates, has resulted in a variety of solutions for bringing them together. One approach in this respect, the wireless application protocol (WAP), which is supported by many companies, is explained in more detail. WAP was not a commercial success in the beginning. However, it is expected that the new version, WAP 2.0, which integrates many Internet technologies and i-mode components, will be more successful.

Chapter 11, gives a short outlook to next generation mobile and wireless systems. While no one exactly knows what the future will look like, some indicators may be given.

Many important security aspects will be explained together with the technology in all chapters. Security mechanisms are important in all layers of a communication system. Different users, even different nations, have different ideas about security. However, it is quite clear that a communication system transmitting personal information through the air must offer special security features to be accepted. Companies do not want competitors to listen to their communications and people often do not like the idea that their neighbor might hear their private conversation, which is possible with older analog cordless phone systems. Wireless systems are especially vulnerable in this respect due to air interface. Wire-tapping is not needed to listen in a data stream. Special encryption methods must be applied to guarantee privacy. Further security mechanisms are authentication, confidentiality, anonymity, and replay protection.

The reader can find review exercises and references at the end of each chapter. A complete list of acronyms used throughout the chapters and the index conclude the book.

1.7 Review exercise

1. Discover the current numbers of subscribers for the different systems. As mobile communications boom, no printed number is valid for too long!
2. Check out the strategies of different network operators while migrating towards third generation systems. Why is a single common system not in sight?

1.8 References

BEMS (2002) The Bioelectromagnetics Society, http://www.bioelectromagnetics.org/.

Cheverst, K., Davies, N., Mitchell, K., Friday, A. (2000) *'Experiences of Developing and Deploying a Context-Aware Tourist Guide: The GUIDE Project,'* proc. MOBICOM'2000, ACM Press, Boston, USA.

COST (2000) *Biomedical Effects of Electromagnetic Fields*, European Cooperation in the field of Scientific and Technical Research, COST 244bis, http://www.cordis.lu/.

DTI (1987) *Performance Specification: Radio equipment for use at fixed and portable stations in the cordless telephone service*, Department of Trade and Industry, MPT 1334.

EIA (1989) *Mobile land station compatibility specification*, Electronic Industries Association, ANSI/EIA/TIA Standard 553.

ETSI (1994) *Common air interface specification to be used for the interworking between cordless telephone apparatus in the frequency band 864.1 MHz to 868.1 MHz, including public access services*, European Telecommunications Standards Institute, I-ETS 300 131 (1994–11).

ETSI (1998) *Digital Enhanced Cordless Telecommunications (DECT), Common Interface (CI)*, European Telecommunications Standards Institute, EN 300 175, V1.4.1 (1998–02).

ETSI (2002) Terms and Definitions Database Interactive (TEDDI), European Telecommunications Standards Institute, http://webapp.etsi.org/Teddi/.

Fitzek, F., Köpsel, A., Wolisz, A., Krishnam, M., Reisslein, M. (2002) 'Providing Application-Level QoS in 3G/4G Wireless Systems: A Comprehensive Framework on Multirate CDMA,' *IEEE Wireless Communications*, 9(2).

GSM World (2002) GSM Association, http://www.gsmworld.com/.

Halsall, F. (1996) *Data communications, computer networks and open systems*. Addison-Wesley.

Iridium (2002) Iridium Satellite LLC, Leesburg, VA, USA, http://www.iridium.com/.

ITU (2002) *International Mobile Telecommunications*, International Telecommunication Union, set of recommendations, http://www.itu.int/imt/.

Keshav, S. (1997) *An engineering approach to computer networking*. Addison-Wesley.

Kurose, J., Ross, K. (2003) *Computer Networking*. Addison-Wesley.

Lin, J.C. (1997) 'Biological aspects of mobile communication fields,' (series of articles), *Wireless Networks*, J.C. Baltzer, 3(6).

NIEHS (2002) The National Institute of Environmental Health Sciences, http://www.niehs.nih.gov/emfrapid/.

Tanenbaum, A. (2003) *Computer Networks*. Prentice-Hall.

TIA (1993a) *Mobile station land station compatibility specification for dual-mode narrowband analogue cellular technology*, Telecommunications Industries Association, Interim Standard 88.

TIA (1993b) *Mobile station base station compatibility standard for dual-mode wideband spread spectrum cellular systems*, Telecommunications Industries Association, Interim Standard 95.

TIA (1996) *800 MHz TDMA cellular radio interface mobile station base station compatibility*, Telecommunications Industries Association, Interim Standard 136A.

WHO (2002) *The International EMF Project*, World Health Organization, http://www.who.int/peh-emf/en/.

WWRF (2002) Wireless World Research Forum, http://www.wireless-world-research.org/, http://www.ww-rf.org

2 Wireless transmission

This book focuses on higher layer aspects of mobile communications, the computer science element rather than on the radio and transmission aspects, the electrical engineering part. This chapter introduces only those fundamental aspects of wireless transmission which are necessary to understand the problems of higher layers and the complexity needed to handle transmission impairments. Wherever appropriate, the reader is referred to literature giving a deeper insight into the topic. To avoid too many details blurring the overall picture, this chapter sometimes simplifies the real-world characteristics of wireless transmission. Readers who are more interested in the details of wireless transmission, calculation of propagation characteristics etc. are referred to Pahlavan (2002) or Stallings (2002).

While transmission over different wires typically does not cause interference,[1] this is an important topic in wireless transmission. The frequencies used for transmission are all regulated. The first section gives a general overview of these frequencies. The following sections recall some basic facts about signals, antennas, and signal propagation. The varying propagation characteristics create particular complications for radio transmission, frequently causing transmission errors. Multiplexing is a major design topic in this context, because the medium is always shared. Multiplexing schemes have to ensure low interference between different senders.

Modulation is needed to transmit digital data via certain frequencies. A separate section of this chapter presents standard modulation schemes that will reoccur together with the wireless communication systems presented in chapters 4 to 7. The next section discusses spread spectrum, a special transmission technique that is more robust against errors. A short introduction to cellular systems concludes this chapter.

1 However, if the transmitted frequencies are too high for a certain wire crosstalk takes place. This is a common problem, e.g., for DSL or Powerline installations, especially if many wires are bundled.

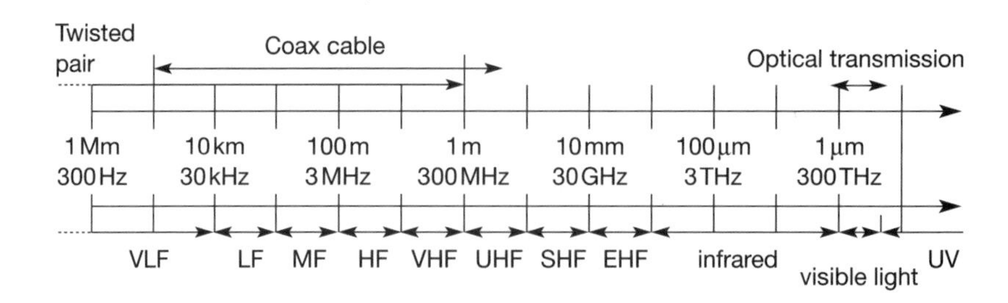

Figure 2.1 Frequency spectrum

2.1 Frequencies for radio transmission

Radio transmission can take place using many different frequency bands. Each frequency band exhibits certain advantages and disadvantages. Figure 2.1 gives a rough overview of the frequency spectrum that can be used for data transmission. The figure shows frequencies starting at 300 Hz and going up to over 300 THz.

Directly coupled to the frequency is the wavelength λ via the equation:

$$\lambda = c/f,$$

where $c \cong 3 \cdot 10^8$ m/s (the speed of light in vacuum) and f the frequency. For traditional wired networks, frequencies of up to several hundred kHz are used for distances up to some km with twisted pair copper wires, while frequencies of several hundred MHz are used with coaxial cable (new coding schemes work with several hundred MHz even with twisted pair copper wires over distances of some 100 m). Fiber optics are used for frequency ranges of several hundred THz, but here one typically refers to the wavelength which is, e.g., 1500 nm, 1350 nm etc. (infra red).

Radio transmission starts at several kHz, the **very low frequency (VLF)** range. These are very long waves. Waves in the **low frequency (LF)** range are used by submarines, because they can penetrate water and can follow the earth's surface. Some radio stations still use these frequencies, e.g., between 148.5 kHz and 283.5 kHz in Germany. The **medium frequency (MF)** and **high frequency (HF)** ranges are typical for transmission of hundreds of radio stations either as amplitude modulation **(AM)** between 520 kHz and 1605.5 kHz, as short wave **(SW)** between 5.9 MHz and 26.1 MHz, or as frequency modulation **(FM)** between 87.5 MHz and 108 MHz. The frequencies limiting these ranges are typically fixed by national regulation and, vary from country to country. Short waves are typically used for (amateur) radio transmission around the world, enabled by reflection at the ionosphere. Transmit power is up to 500 kW – which is quite high compared to the 1 W of a mobile phone.

As we move to higher frequencies, the TV stations follow. Conventional analog TV is transmitted in ranges of 174–230 MHz and 470–790 MHz using the very high frequency **(VHF)** and ultra high frequency **(UHF)** bands. In this range, digital audio broadcasting (DAB) takes place as well (223–230 MHz and 1452–1472 MHz) and digital TV is planned or currently being installed (470–862 MHz), reusing some of the old frequencies for analog TV. UHF is also used for mobile phones with analog technology (450–465 MHz), the digital GSM (890–960 MHz, 1710–1880 MHz), digital cordless telephones following the DECT standard (1880–1900 MHz), 3G cellular systems following the UMTS standard (1900–1980 MHz, 2020–2025 MHz, 2110–2190 MHz) and many more. VHF and especially UHF allow for small antennas and relatively reliable connections for mobile telephony.

Super high frequencies (SHF) are typically used for directed microwave links (approx. 2–40 GHz) and fixed satellite services in the C-band (4 and 6 GHz), Ku-band (11 and 14 GHz), or Ka-band (19 and 29 GHz). Some systems are planned in the **extremely high frequency (EHF)** range which comes close to infra red. All radio frequencies are regulated to avoid interference, e.g., the German regulation covers 9 kHz–275 GHz.

The next step into higher frequencies involves optical transmission, which is not only used for fiber optical links but also for wireless communications. **Infra red (IR)** transmission is used for directed links, e.g., to connect different buildings via laser links. The most widespread IR technology, infra red data association (IrDA), uses wavelengths of approximately 850–900 nm to connect laptops, PDAs etc. Finally, visible light has been used for wireless transmission for thousands of years. While light is not very reliable due to interference, but it is nevertheless useful due to built-in human receivers.

2.1.1 Regulations

As the examples in the previous section have shown, radio frequencies are scarce resources. Many national (economic) interests make it hard to find common, worldwide regulations. The International Telecommunications Union (ITU) located in Geneva is responsible for worldwide coordination of telecommunication activities (wired and wireless). ITU is a sub-organization of the UN. The ITU Radiocommunication sector (ITU-R) handles standardization in the wireless sector, so it also handles frequency planning (formerly known as Consultative Committee for International Radiocommunication, CCIR).

To have at least some success in worldwide coordination and to reflect national interests, the ITU-R has split the world into three regions: **Region 1** covers Europe, the Middle East, countries of the former Soviet Union, and Africa. **Region 2** includes Greenland, North and South America, and **region 3** comprises the Far East, Australia, and New Zealand. Within these regions, national agencies are responsible for further regulations, e.g., the Federal Communications Commission (FCC) in the US. Several nations have a common agency such as European Conference for Posts and Telecommunications (CEPT) in Europe. While CEPT is still responsible for the general planning, many tasks have been transferred to other agencies (confusing anybody following the regulation process). For example, the European Telecommunications Standards Institute (ETSI) is responsible for standardization and consists of national standardization bodies, public providers, manufacturers, user groups, and research institutes.

To achieve at least some harmonization, the ITU-R holds, the World Radio Conference (WRC), to periodically discuss and decide frequency allocations for all three regions. This is obviously a difficult task as many regions or countries may have already installed a huge base of a certain technology and will be reluctant to change frequencies just for the sake of harmonization. Harmonization is, however, needed as soon as satellite communication is used. Satellites, especially the new generation of low earth-orbiting satellites (see chapter 5) do not 'respect' national regulations, but should operate worldwide. While it is difficult to prevent other nations from setting up a satellite system it is much simpler to ban the necessary devices or the infrastructure needed for operation. Satellite systems should operate on frequencies available worldwide to support global usage with a single device.

Table 2.1 gives some examples for frequencies used for (analog and digital) mobile phones, cordless telephones, wireless LANs, and other radio frequency (RF) systems for countries in the three regions representing the major economic power. Older systems like Nordic Mobile Telephone (NMT) are not available all over Europe, and sometimes they have been standardized with different national frequencies. The newer (digital) systems are compatible throughout Europe (standardized by ETSI).

Table 2.1 Example systems and their frequency allocations (all values in MHz)

	Europe	**US**	**Japan**
Mobile phones	**NMT** 453–457 463–467	**AMPS, TDMA, CDMA** 824–849 869–894	**PDC** 810–826 940–956 1429–1465 1477–1513
	GSM 890–915 935–960 1710–1785 1805–1880	**GSM, TDMA, CDMA** 1850–1910 1930–1990	
	UMTS (FDD)/ W-CDMA 1920–1980 2110–2190		**FOMA/ W-CDMA** 1920–1980 2110–2170

	UMTS (TDD) 1900–1920 2020–2025		
Cordless telephones	**CT1+** 885–887 930–932	**PACS** 1850–1910 1930–1990	**PHS** 1895–1918
	CT2 864–868	**PACS–UB** 1910–1930	**JCT** 254–380
	DECT 1880–1900		
Wireless LANs	**IEEE 802.11** 2400–2483	**IEEE 802.11** 902–928 2400–2483	**IEEE 802.11** 2400–2497
	HiperLAN2, IEEE 802.11a 5150–5350 5470–5725	**HiperLAN2, IEEE 802.11a** 5150–5350 5725–5825	**HiperLAN2, IEEE 802.11a** 5150–5250
Others	**RF-Control** 27, 128, 418, 433, 868	**RF-Control** 315, 915	**RF-Control** 426, 868
	Satellite (e.g., Iridium, Globalstar) 1610–1626, 2483–2500		

While older analog **mobile phone** systems like NMT or its derivatives at 450 MHz are still available, Europe is heavily dominated by the fully digital GSM (see chapter 4.1) at 900 MHz and 1800 MHz (also known as DCS1800, Digital Cellular System). In contrast to Europe, the US FCC allowed several cellular technologies in the same frequency bands around 850 MHz. Starting from the analog advanced mobile phone system (AMPS), this led to the co-existence of several solutions, such as dual mode mobile phones supporting digital time division multiple access (TDMA) service and analog AMPS according to the standard IS-54. All digital TDMA phones according to IS-136 (also known as NA-TDMA, North American TDMA) and digital code division multiple access (CDMA) phones according to IS-95 have been developed. The US did not adopt a common mobile phone system, but waited for market forces to decide. This led to many islands of different systems and, consequently, as in Europe, full coverage, is not available in the US. The long discussions about the pros and cons of TDMA and CDMA also promoted the worldwide success of GSM. GSM is available in over 190 countries and used by more than 800 million people (GSM World, 2002). A user can roam with the same mobile phone from Zimbabwe, via Uzbekistan, Sweden, Singapore, USA, Tunisia, Russia, Canada, Italy, Greece, Germany, China, and Belgium to Austria.

Another system, the personal digital cellular (**PDC**), formerly known as Japanese digital cellular (JDC) was established in Japan. Quite often mobile phones covering many standards have been announced, however, industry is still waiting for a cheap solution. Chapter 11 will discuss this topic again in the context of software defined radios (SDR). New frequency bands, e.g., for the universal mobile telecommunications system (**UMTS**) or the freedom of mobile multi-media access (**FOMA**) are located at 1920–1980 MHz and 2110–2170/2190 MHz (see chapter 4).

Many different **cordless telephone** standards exist around the world. However, this is not as problematic as the diversity of mobile phone standards. Some older analog systems such as cordless telephone (**CT1+**) are still in use, but digital technology has been introduced for cordless telephones as well. Examples include **CT2**, the first digital cordless telephone introduced in the UK, digital enhanced cordless telecommunications (**DECT**) as a European standard (see section 4.2), personal access communications system (PACS) and PACS-Unlicensed Band (**PACS-UB**) in the US, as well as personal handyphone system (PHS) as replacement for the analog Japanese cordless telephone (**JCT**) in Japan. Mobile phones covering, e.g., DECT and GSM are available but they have not been a commercial success.

Finally, the area of **WLAN** standards is of special interest for wireless, mobile computer communication on a campus or in buildings. Here the computer industry developed products within the license-free **ISM** band, of which the most attractive is located at 2.4 GHz and is available for license-free operation almost everywhere around the world (with national differences limiting frequencies, transmit power etc.). The most widespread standard in this area is **IEEE 802.11b**, which is discussed in chapter 7 (together with other members of the 802.11 family). The wireless LAN standards **HiperLAN2** and **IEEE 802.11a** operate in the 5 GHz range, but depending on the region on different frequencies with different restrictions.

Many more frequencies have been assigned for trunk radio (e.g., trans-European trunked radio (TETRA), 380–400 MHz, 410–430 MHz, 450–470 MHz – depending on national regulations), paging services, terrestrial flight telephone system (TFTS), 1670–1675 MHz and 1800–1805 MHz, satellite services (Iridium: 1610–1626 MHz, Globalstar: 1610–1626 MHz and 2483–2500 MHz, see chapter 5) etc. Higher frequencies are of special interest for high bit-rate transmission, although these frequencies face severe shadowing by many obstacles. License-free bands at 17.2, 24 and even 61 GHz are under consideration for commercial use. Additionally, a lot of license-free wireless communication takes place at lower frequencies. Garage openers, car locks, wireless headsets, radio frequency identifications (RFID) etc. operate on, e.g., 433 or 868 MHz.

2.2 Signals

Signals are the physical representation of data. Users of a communication system can only exchange data through the transmission of signals. Layer 1 of the ISO/OSI basic reference model is responsible for the conversion of data, i.e., bits, into signals and vice versa (Halsall, 1996), (Stallings, 1997 and 2002).

Signals are functions of time and location. Signal parameters represent the data values. The most interesting types of signals for radio transmission are **periodic signals**, especially **sine waves** as carriers. (The process of mapping of data onto a carrier is explained in section 2.6.) The general function of a sine wave is:

$$g(t) = A_t \sin(2\,\pi f_t\, t + \varphi_t)$$

Signal parameters are the **amplitude** A, the **frequency** f, and the **phase shift** φ. The amplitude as a factor of the function g may also change over time, thus A_t, (see section 2.6.1). The frequency f expresses the periodicity of the signal with the period $T = 1/f$. (In equations, ω is frequently used instead of $2\pi f$.) The frequency f may also change over time, thus f_t, (see section 2.6.2). Finally, the phase shift determines the shift of the signal relative to the same signal without a shift. An example for shifting a function is shown in Figure 2.2. This shows a sine function without a phase shift and the same function, i.e., same amplitude and frequency, with a phase shift φ. Section 2.6.3 shows how shifting the phase can be used to represent data.

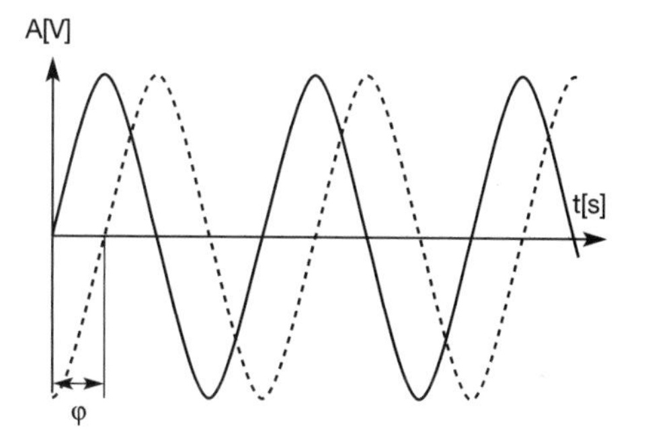

Figure 2.2 Time domain representation of a signal

Sine waves are of special interest, as it is possible to construct every periodic signal g by using only sine and cosine functions according to a fundamental equation of **Fourier**:

$$g(t) = \frac{1}{2}c + \sum_{n=1}^{\infty} a_n \sin(2\pi nft) + \sum_{n=1}^{\infty} b_n \cos(2\pi nft)$$

In this equation the parameter c determines the **Direct Current (DC)** component of the signal, the coefficients a_n and b_n are the amplitudes of the nth sine and cosine function. The equation shows that an infinite number of sine and cosine functions is needed to construct arbitrary periodic functions. However, the frequencies of these functions (the so-called **harmonics**) increase with a growing parameter n and are a multiple of the **fundamental frequency** f. The bandwidth of any medium, air, cable, transmitter etc. is limited and, there is an upper limit for the frequencies. In reality therefore, it is enough to consider a limited number of sine and cosine functions to construct periodic functions – all real transmitting systems exhibit these bandwidth limits and can never transmit arbitrary periodic functions. It is sufficient for us to know that we can think of transmitted signals as composed of one or many sine functions. The following illustrations always represent the example of one sine function, i.e., the case of a single frequency.

A typical way to represent signals is the time domain (see Figure 2.2). Here the amplitude A of a signal is shown versus time (time is mostly measured in seconds s, amplitudes can be measured in, e.g., volt V). This is also the typical representation known from an oscilloscope. A phase shift can also be shown in this representation.

Representations in the time domain are problematic if a signal consists of many different frequencies (as the Fourier equation indicates). In this case, a better representation of a signal is the **frequency domain** (see Figure 2.3). Here the amplitude of a certain frequency part of the signal is shown versus the frequency. Figure 2.3 only shows one peak and the signal consists only of a single frequency part (i.e., it is a single sine function). Arbitrary periodic functions would have many peaks, known as the frequency spectrum of a signal. A tool to display frequencies is a spectrum analyzer. Fourier transformations are a mathematical tool for translating from the time domain into the frequency domain and vice versa (using the inverse Fourier transformation).

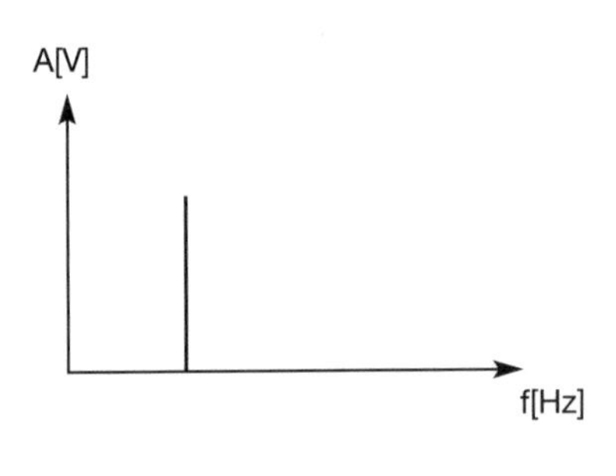

Figure 2.3 Frequency domain representation of a signal

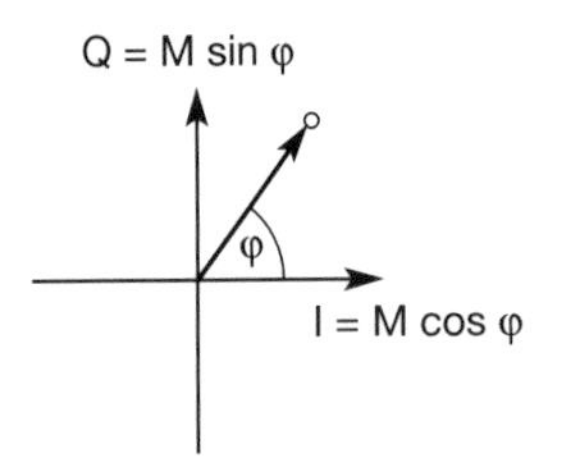

Figure 2.4 Phase domain representation of a signal

A third way to represent signals is the **phase domain** shown in Figure 2.4. This representation, also called phase state or signal constellation diagram, shows the amplitude M of a signal and its phase φ in polar coordinates. (The length of the vector represents the amplitude, the angle the phase shift.) The x-axis represents a phase of 0 and is also called **In-Phase (I)**. A phase shift of 90° or $\pi/2$ would be a point on the y-axis, called **Quadrature (Q)**.

2.3 Antennas

As the name wireless already indicates, this communication mode involves 'getting rid' of wires and transmitting signals through space without guidance. We do not need any 'medium' (such as an ether) for the transport of electromagnetic waves. Somehow, we have to couple the energy from the transmitter to the outside world and, in reverse, from the outside world to the receiver. This is exactly what **antennas** do. Antennas couple electromagnetic energy to and from space to and from a wire or coaxial cable (or any other appropriate conductor).

A theoretical reference antenna is the **isotropic radiator**, a point in space radiating equal power in all directions, i.e., all points with equal power are located on a sphere with the antenna as its center. The **radiation pattern** is symmetric in all directions (see Figure 2.5, a two dimensional cross-section of the real three-dimensional pattern).

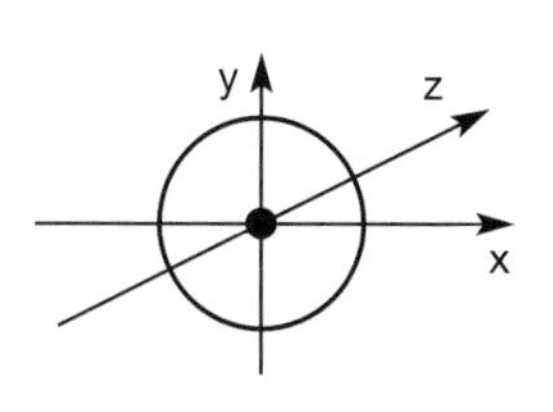

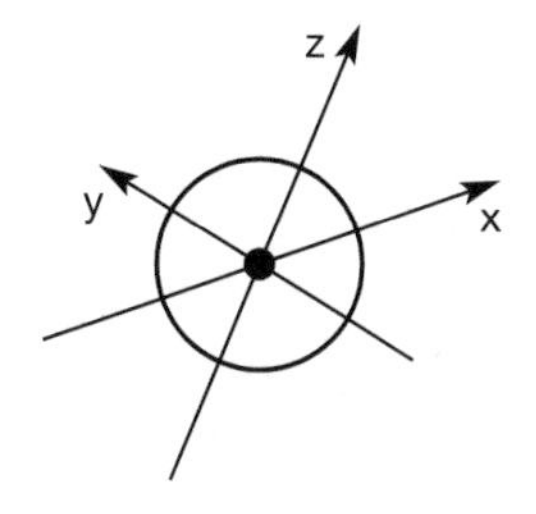

Figure 2.5
Radiation pattern of an isotropic radiator

However, such an antenna does not exist in reality. Real antennas all exhibit **directive effects**, i.e., the intensity of radiation is not the same in all directions from the antenna. The simplest real antenna is a thin, center-fed **dipole**, also called Hertzian dipole, as shown in Figure 2.6 (right-hand side). The dipole consists of two collinear conductors of equal length, separated by a small feeding gap. The length of the dipole is not arbitrary, but, for example, half the wavelength λ of the signal to transmit results in a very efficient radiation of the energy. If mounted on the roof of a car, the length of $\lambda/4$ is efficient. This is also known as Marconi antenna.

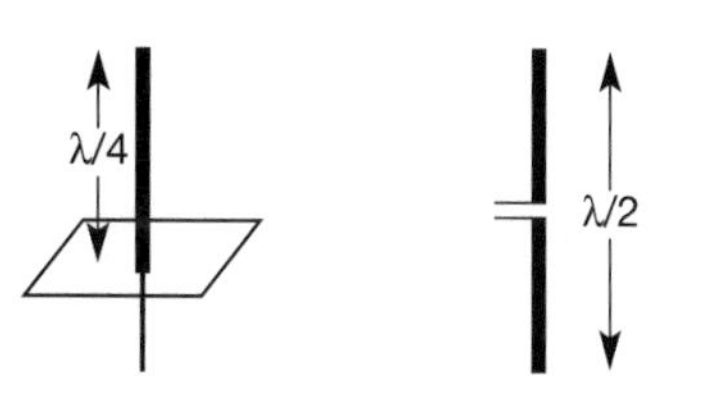

Figure 2.6
Simple antennas

A $\lambda/2$ dipole has a uniform or **omni-directional** radiation pattern in one plane and a figure eight pattern in the other two planes as shown in Figure 2.7. This type of antenna can only overcome environmental challenges by boosting the power level of the signal. Challenges could be mountains, valleys, buildings etc.

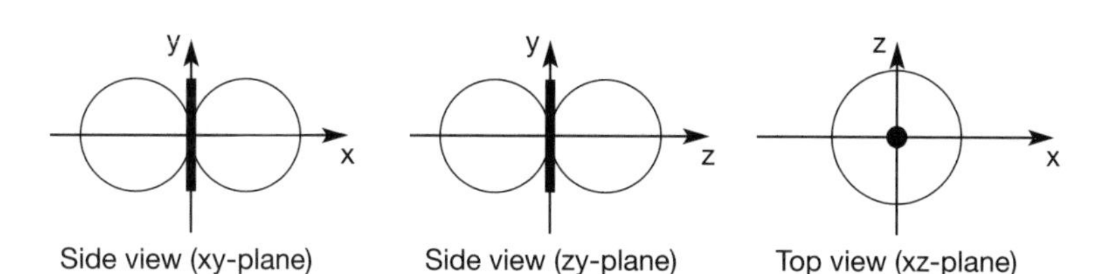

Figure 2.7
Radiation pattern of a simple dipole

If an antenna is positioned, e.g., in a valley or between buildings, an omnidirectional radiation pattern is not very useful. In this case, **directional antennas** with certain fixed preferential transmission and reception directions can be used. Figure 2.8 shows the radiation pattern of a directional antenna with the main lobe in the direction of the x-axis. A special example of directional antennas is constituted by satellite dishes.

Figure 2.8
Radiation pattern of a directed antenna

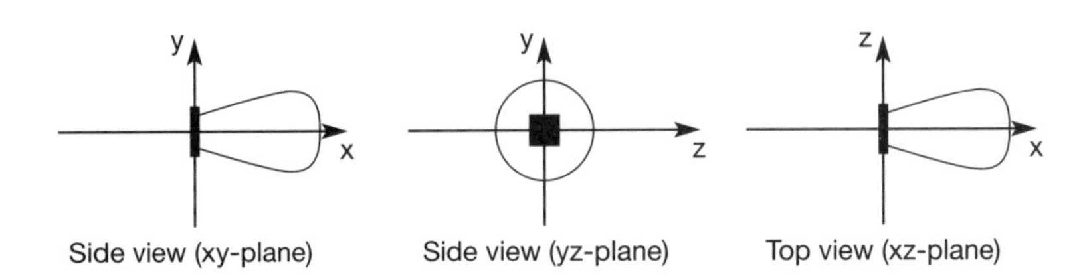

Directed antennas are typically applied in cellular systems as presented in section 2.8. Several directed antennas can be combined on a single pole to construct a **sectorized antenna**. A cell can be sectorized into, for example, three or six sectors, thus enabling frequency reuse as explained in section 2.8. Figure 2.9 shows the radiation patterns of these sectorized antennas.

Figure 2.9
Radiation patterns of sectorized antennas

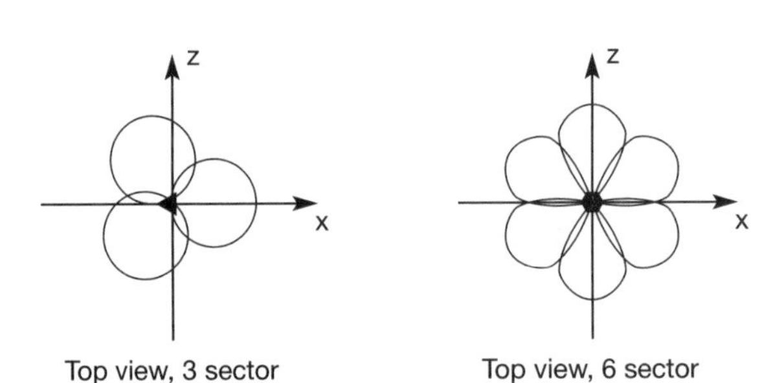

Two or more antennas can also be combined to improve reception by counteracting the negative effects of multi-path propagation (see section 2.4.3). These antennas, also called **multi-element antenna arrays**, allow different diversity schemes. One such scheme is **switched diversity** or **selection diversity**, where the receiver always uses the antenna element with the largest output. **Diversity combining** constitutes a combination of the power of all signals to produce gain. The phase is first corrected (cophasing) to avoid cancellation. As shown in Figure 2.10, different schemes are possible. On the left, two $\lambda/4$ antennas are combined with a distance of $\lambda/2$ between them on top of a ground plane. On the right, three standard $\lambda/2$ dipoles are combined with a distance of $\lambda/2$ between them. Spacing could also be in multiples of $\lambda/2$.

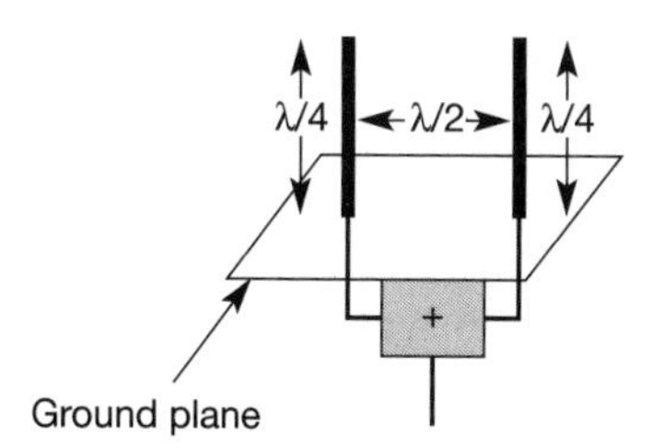

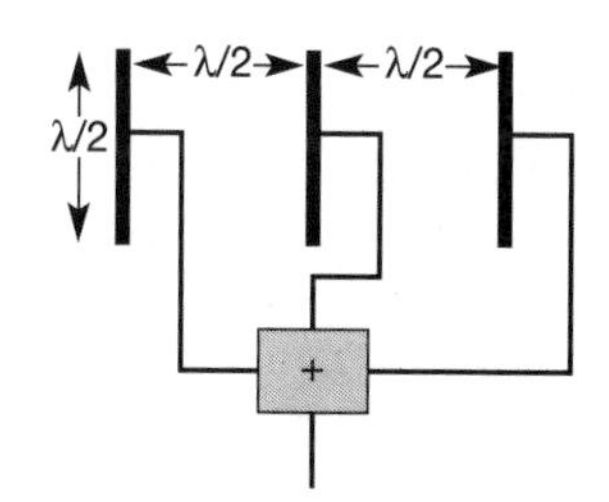

Figure 2.10
Diversity antenna systems

A more advanced solution is provided by **smart antennas** which combine multiple antenna elements (also called antenna array) with signal processing to optimize the radiation/reception pattern in response to the signal environment. These antennas can adapt to changes in reception power, transmission conditions and many signal propagation effects as discussed in the following section. Antenna arrays can also be used for beam forming. This would be an extreme case of a directed antenna which can follow a single user thus using space division multiplexing (see section 2.5.1). It would not just be base stations that could follow users with an individual beam. Wireless devices, too, could direct their electromagnetic radiation, e.g., away from the human body towards a base station. This would help in reducing the absorbed radiation. Today's handset antennas are omni-directional as the integration of smart antennas into mobiles is difficult and has not yet been realized.

2.4 Signal propagation

Like wired networks, wireless communication networks also have senders and receivers of signals. However, in connection with signal propagation, these two networks exhibit considerable differences. In wireless networks, the signal has no wire to determine the direction of propagation, whereas signals in wired networks only travel along the wire (which can be twisted pair copper wires, a coax cable, but also a fiber etc.). As long as the wire is not interrupted or damaged, it typically exhibits the same characteristics at each point. One can precisely determine the behavior of a signal travelling along this wire, e.g., received power depending on the length. For wireless transmission, this predictable behavior is only valid in a vacuum, i.e., without matter between the sender and the receiver. The situation would be as follows (Figure 2.11):

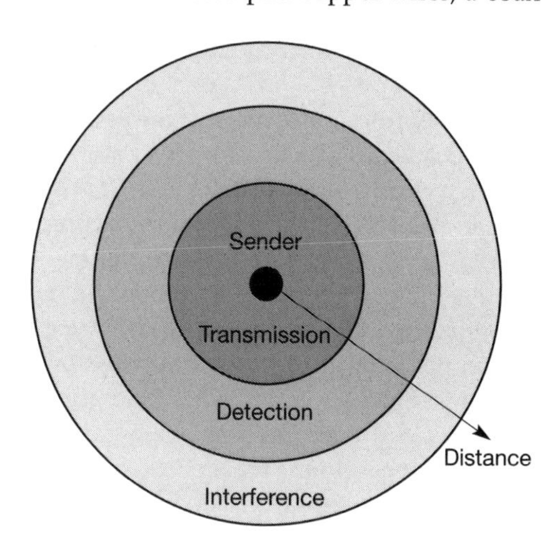

Figure 2.11
Ranges for transmission, detection, and interference of signals

- **Transmission range**: Within a certain radius of the sender transmission is possible, i.e., a receiver receives the signals with an error rate low enough to be able to communicate and can also act as sender.
- **Detection range**: Within a second radius, detection of the transmission is possible, i.e., the transmitted power is large enough to differ from background noise. However, the error rate is too high to establish communication.
- **Interference range**: Within a third even larger radius, the sender may interfere with other transmission by adding to the background noise. A receiver will not be able to detect the signals, but the signals may disturb other signals.

This simple and ideal scheme led to the notion of **cells** around a transmitter (as briefly discussed in section 2.8). However, real life does not happen in a vacuum, radio transmission has to contend with our atmosphere, mountains, buildings, moving senders and receivers etc. In reality, the three circles referred to above will be bizarrely-shaped polygons with their shape being time and frequency dependent. The following paragraphs discuss some problems arising in this context, thereby showing the differences between wireless and wired transmission.

2.4.1 Path loss of radio signals

In free space radio signals propagate as light does (independently of their frequency), i.e., they follow a straight line (besides gravitational effects). If such a straight line exists between a sender and a receiver it is called **line-of-sight (LOS)**. Even if no matter exists between the sender and the receiver (i.e., if there is a vacuum), the signal still experiences the **free space loss**. The received power P_r is proportional to $1/d^2$ with d being the distance between sender and receiver (**inverse square law**). The reason for this phenomenon is quite simple. Think of the sender being a point in space. The sender now emits a signal with certain energy. This signal travels away from the sender at the speed of light as a wave with a spherical shape. If there is no obstacle, the sphere continuously grows with the sending energy equally distributed over the sphere's surface. This surface area s grows with the increasing distance d from the center according to the equation $s = 4\pi d^2$.

Even without any matter between sender and receiver, additional parameters are important. The received power also depends on the wavelength and the gain of receiver and transmitter antennas. As soon as there is any matter between sender and receiver, the situation becomes more complex. Most radio transmission takes place through the atmosphere – signals travel through air, rain, snow, fog, dust particles, smog etc. While the **path loss** or **attenuation** does not cause too much trouble for short distances, e.g., for LANs (see chapter 7), the atmosphere heavily influences transmission over long distances, e.g., satellite transmission (see chapter 5). Even mobile phone systems are influenced by weather conditions such as heavy rain. Rain can absorb much of the radiated energy of the antenna (this effect is used in a microwave oven to cook), so communication links may break down as soon as the rain sets in.

Depending on the frequency, radio waves can also penetrate objects. Generally the lower the frequency, the better the penetration. Long waves can be transmitted through the oceans to a submarine while high frequencies can be blocked by a tree. The higher the frequency, the more the behavior of the radio waves resemble that of light – a phenomenon which is clear if one considers the spectrum shown in Figure 2.1.

Radio waves can exhibit three fundamental propagation behaviors depending on their frequency:

- **Ground wave** (<2 MHz): Waves with low frequencies follow the earth's surface and can propagate long distances. These waves are used for, e.g., submarine communication or AM radio.
- **Sky wave** (2–30 MHz): Many international broadcasts and amateur radio use these short waves that are reflected[2] at the ionosphere. This way the waves can bounce back and forth between the ionosphere and the earth's surface, travelling around the world.
- **Line-of-sight** (>30 MHz): Mobile phone systems, satellite systems, cordless telephones etc. use even higher frequencies. The emitted waves follow a (more or less) straight line of sight. This enables direct communication with satellites (no reflection at the ionosphere) or microwave links on the ground. However, an additional consideration for ground-based communication is that the waves are bent by the atmosphere due to refraction (see next section).

Almost all communication systems presented in this book work with frequencies above 100 MHz so, we are almost exclusively concerned with LOS communication. But why do mobile phones work even without an LOS?

2.4.2 Additional signal propagation effects

As discussed in the previous section, signal propagation in free space almost follows a straight line, like light. But in real life, we rarely have a line-of-sight between the sender and receiver of radio signals. Mobile phones are typically used in big cities with skyscrapers, on mountains, inside buildings, while driving through an alley etc. Hare several effects occur in addition to the attenuation caused by the distance between sender and receiver, which are again very much frequency dependent.

An extreme form of attenuation is **blocking** or **shadowing** of radio signals due to large obstacles (see Figure 2.12, left side). The higher the frequency of a signal, the more it behaves like light. Even small obstacles like a simple wall, a truck on the street, or trees in an alley may block the signal. Another effect is the **reflection** of signals as shown in the middle of Figure 2.12. If an object is large compared to the wavelength of the signal, e.g., huge buildings, mountains,

---

2 Compared to, e.g., the surface of a building, the ionosphere is not really a hard reflecting surface. In the case of sky waves the 'reflection' is caused by refraction.

or the surface of the earth, the signal is reflected. The reflected signal is not as strong as the original, as objects can absorb some of the signal's power. Reflection helps transmitting signals as soon as no LOS exists. This is the standard case for radio transmission in cities or mountain areas. Signals transmitted from a sender may bounce off the walls of buildings several times before they reach the receiver. The more often the signal is reflected, the weaker it becomes. Finally, the right side of Figure 2.12 shows the effect of **refraction**. This effect occurs because the velocity of the electromagnetic waves depends on the density of the medium through which it travels. Only in vacuum does it equal *c*. As the figure shows, waves that travel into a denser medium are bent towards the medium. This is the reason for LOS radio waves being bent towards the earth: the density of the atmosphere is higher closer to the ground.

Figure 2.12
Blocking (shadowing), reflection and refraction of waves

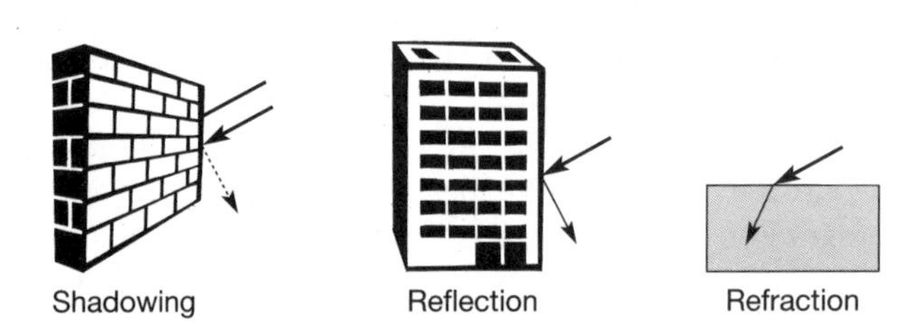

While shadowing and reflection are caused by objects much larger than the wavelength of the signals (and demonstrate the typical 'particle' behavior of radio signals), the following two effects exhibit the 'wave' character of radio signals. If the size of an obstacle is in the order of the wavelength or less, then waves can be **scattered** (see Figure 2.13, left side). An incoming signal is scattered into several weaker outgoing signals. In school experiments, this is typically demonstrated with laser light and a very small opening or obstacle, but here we have to take into consideration that the typical wavelength of radio transmission for, e.g., GSM or AMPS is in the order of some 10 cm. Thus, many objects in the environment can cause these scattering effects. Another effect is **diffraction** of waves. As shown on the right side of Figure 2.13, this effect is very similar to scattering. Radio waves will be deflected at an edge and propagate in different directions. The result of scattering and diffraction are patterns with varying signal strengths depending on the location of the receiver.

Effects like attenuation, scattering, diffraction, and refraction all happen simultaneously and are frequency and time dependent. It is very difficult to predict the precise strength of signals at a certain point in space. How do mobile phone operators plan the coverage of their antennas, the location of the antennas, the direction of the beams etc.? Two or three dimensional maps are used with a resolution down to several meters. With the help of, e.g., ray tracing or radiosity techniques similar to rendering 3D graphics, the signal quality can roughly be calculated in advance. Additionally, operators perform a lot of measurements during and after installation of antennas to fill gaps in the coverage.

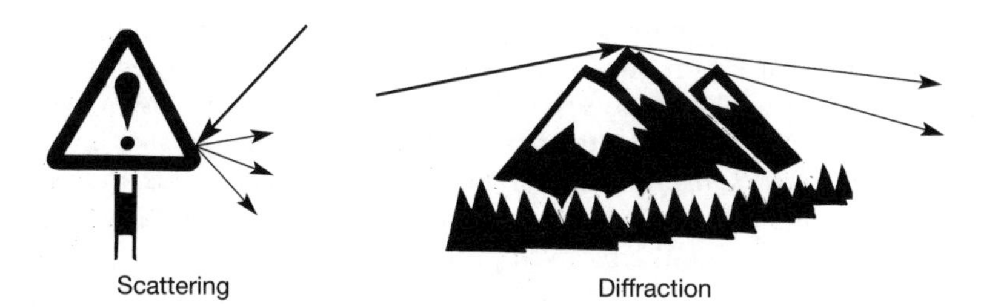

Figure 2.13
Scattering and diffraction of waves

2.4.3 Multi-path propagation

Together with the direct transmission from a sender to a receiver, the propagation effects mentioned in the previous section lead to one of the most severe radio channel impairments, called **multi-path propagation**. Figure 2.14 shows a sender on the left and one possible receiver on the right. Radio waves emitted by the sender can either travel along a straight line, or they may be reflected at a large building, or scattered at smaller obstacles. This simplified figure only shows three possible paths for the signal. In reality, many more paths are possible. Due to the finite speed of light, signals travelling along different paths with different lengths arrive at the receiver at different times. This effect (caused by multi-path propagation) is called **delay spread**: the original signal is spread due to different delays of parts of the signal. This delay spread is a typical effect of radio transmission, because no wire guides the waves along a single path as in the case of wired networks (however, a similar effect, dispersion, is known for high bit-rate optical transmission over multi-mode fiber, see Halsall, 1996, or Stallings, 1997). Notice that this effect has nothing to do with possible movements of the sender or receiver. Typical values for delay spread are approximately 3 μs in cities, up to 12 μs can be observed. GSM, for example, can tolerate up to 16 μs of delay spread, i.e., almost a 5 km path difference.

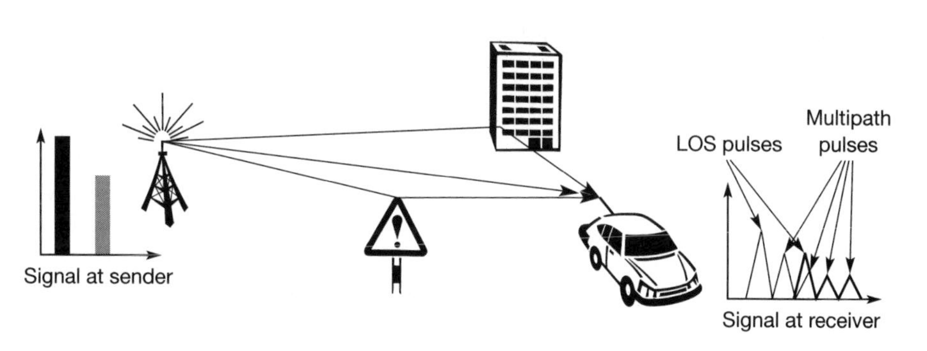

Figure 2.14
Multi-path propagation and intersymbol interference

What are the **effects** of this delay spread on the signals representing the data? The first effect is that a short impulse will be smeared out into a broader impulse, or rather into several weaker impulses. In Figure 2.14 only three possible paths are shown and, thus, the impulse at the sender will result in three smaller impulses at the receiver. For a real situation with hundreds of different paths, this implies that a single impulse will result in many weaker impulses at the receiver. Each path has a different attenuation and, the received pulses have different power. Some of the received pulses will be too weak even to be detected (i.e., they will appear as noise).

Now consider the second impulse shown in Figure 2.14. On the sender side, both impulses are separated. At the receiver, both impulses interfere, i.e., they overlap in time. Now consider that each impulse should represent a symbol, and that one or several symbols could represent a bit. The energy intended for one symbol now spills over to the adjacent symbol, an effect which is called **intersymbol interference (ISI)**. The higher the symbol rate to be transmitted, the worse the effects of ISI will be, as the original symbols are moved closer and closer to each other. ISI limits the bandwidth of a radio channel with multi-path propagation (which is the standard case). Due to this interference, the signals of different symbols can cancel each other out leading to misinterpretations at the receiver and causing transmission errors.

In this case, knowing the channel characteristics can be a great help. If the receiver knows the delays of the different paths (or at least the main paths the signal takes), it can compensate for the distortion caused by the channel. The sender may first transmit a **training sequence** known by the receiver. The receiver then compares the received signal to the original training sequence and programs an **equalizer** that compensates for the distortion (Wesel, 1998), (Pahlavan, 2002), (Stallings, 2002).

While ISI and delay spread already occur in the case of fixed radio transmitters and receivers, the situation is even worse if receivers, or senders, or both, move. Then the channel characteristics change over time, and the paths a signal can travel along vary. This effect is well known (and audible) with analog radios while driving. The power of the received signal changes considerably over time. These quick changes in the received power are also called **short-term fading**. Depending on the different paths the signals take, these signals may have a different phase and cancel each other as shown in Figure 2.15. The receiver now has to try to constantly adapt to the varying channel characteristics, e.g., by changing the parameters of the equalizer. However, if these changes are too fast, such as driving on a highway through a city, the receiver cannot adapt fast enough and the error rate of transmission increases dramatically.

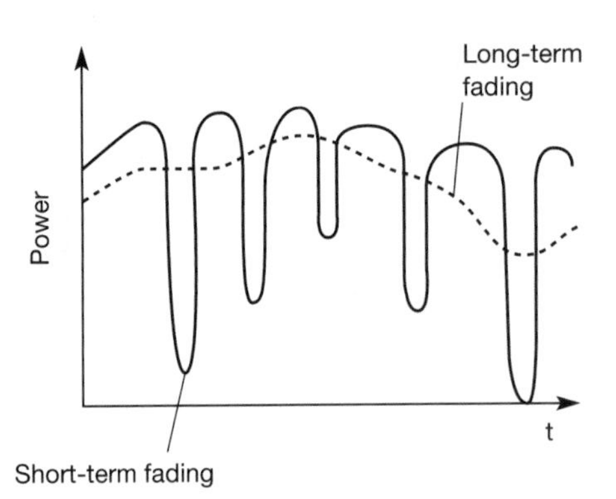

Figure 2.15 Short-term and long-term fading

An additional effect shown in Figure 2.15 is the **long-term fading** of the received signal. This long-term fading, shown here as the average power over time, is caused by, for example, varying distance to the sender or more remote obstacles. Typically, senders can compensate for long-term fading by increasing/decreasing sending power so that the received signal always stays within certain limits.

There are many more effects influencing radio transmission which will not be discussed in detail – for example, the **Doppler shift** caused by a moving sender or receiver. While this effect is audible for acoustic waves already at low speed, it is also a topic for radio transmission from or to fast moving transceivers. One example of such a transceiver could be a satellite (see chapter 5) – there Doppler shift causes random frequency shifts. The interested reader is referred to Anderson (1995), (Pahlavan, 2002), and (Stallings, 2002) for more information about the characteristics of wireless communication channels. For the present it will suffice to know that multi-path propagation limits the maximum bandwidth due to ISI and that moving transceivers cause additional problems due to varying channel characteristics.

2.5 Multiplexing

Multiplexing is not only a fundamental mechanism in communication systems but also in everyday life. Multiplexing describes how several users can share a medium with minimum or no interference. One example, is highways with several lanes. Many users (car drivers) use the same medium (the highways) with hopefully no interference (i.e., accidents). This is possible due to the provision of several lanes (space division multiplexing) separating the traffic. In addition, different cars use the same medium (i.e., the same lane) at different points in time (time division multiplexing).

While this simple example illustrates our everyday use of multiplexing, the following examples will deal with the use of multiplexing in wireless communications. Mechanisms controlling the use of multiplexing and the assignment of a medium to users (the traffic regulations), are discussed in chapter 3 under the aspect of medium access control.

2.5.1 Space division multiplexing

For wireless communication, multiplexing can be carried out in four dimensions: **space**, **time**, **frequency**, and **code**. In this field, the task of multiplexing is to assign space, time, frequency, and code to each communication channel with a minimum of interference and a maximum of medium utilization. The term communication channel here only refers to an association of sender(s) and receiver(s) who want to exchange data. Characteristics of communication channels (e.g., bandwidth, error rate) will be discussed together with certain technologies in chapters 4 to 7.

Figure 2.16 shows six channels k_i and introduces a three dimensional coordinate system. This system shows the dimensions of code c, time t and frequency f. For this first type of multiplexing, **space division multiplexing (SDM)**, the (three dimensional) space s_i is also shown. Here space is represented via circles indicating the interference range as introduced in Figure 2.11. How is the separation of the different channels achieved? The channels k_1 to k_3 can be mapped onto the three 'spaces' s_1 to s_3 which clearly separate the channels and prevent the interference ranges from overlapping. The space between the interference ranges is sometimes called **guard space**. Such a guard space is needed in all four multiplexing schemes presented.

Figure 2.16
Space division multiplexing (SDM)

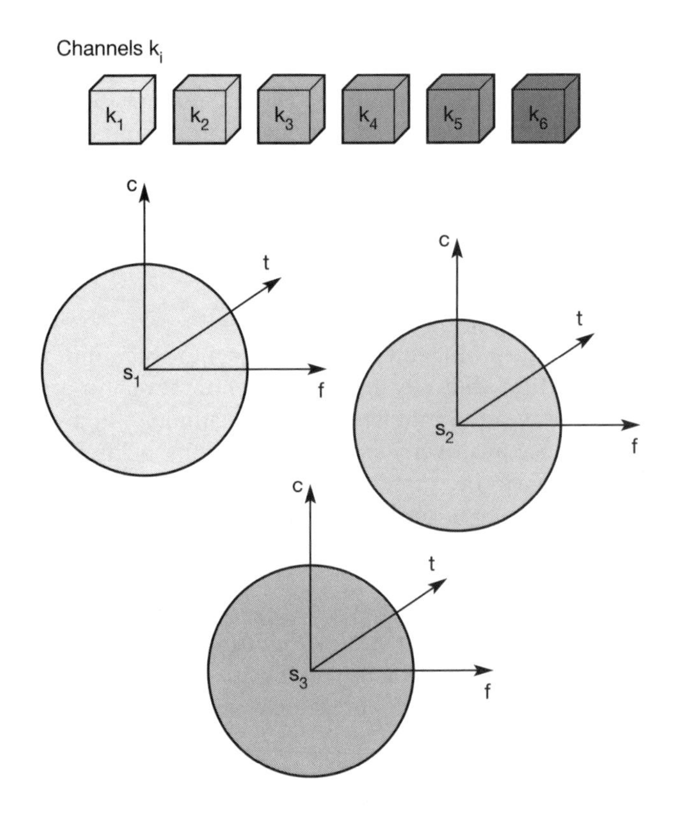

For the remaining channels (k_4 to k_6) three additional spaces would be needed. In our highway example this would imply that each driver had his or her own lane. Although this procedure clearly represents a waste of space, this is exactly the principle used by the old analog telephone system: each subscriber is given a separate pair of copper wires to the local exchange. In wireless transmission, SDM implies a separate sender for each communication channel with a wide enough distance between senders. This multiplexing scheme is used, for example, at FM radio stations where the transmission range is limited to a certain region –

many radio stations around the world can use the same frequency without interference. Using SDM, obvious problems arise if two or more channels were established within the same space, for example, if several radio stations want to broadcast in the same city. Then, one of the following multiplexing schemes must be used (frequency, time, or code division multiplexing).

2.5.2 Frequency division multiplexing

Frequency division multiplexing (FDM) describes schemes to subdivide the frequency dimension into several non-overlapping frequency bands as shown in Figure 2.17. Each channel k_i is now allotted its own frequency band as indicated. Senders using a certain frequency band can use this band continuously. Again, **guard spaces** are needed to avoid frequency band overlapping (also called **adjacent channel interference**). This scheme is used for radio stations within the same region, where each radio station has its own frequency. This very simple multiplexing scheme does not need complex coordination between sender and receiver: the receiver only has to tune in to the specific sender.

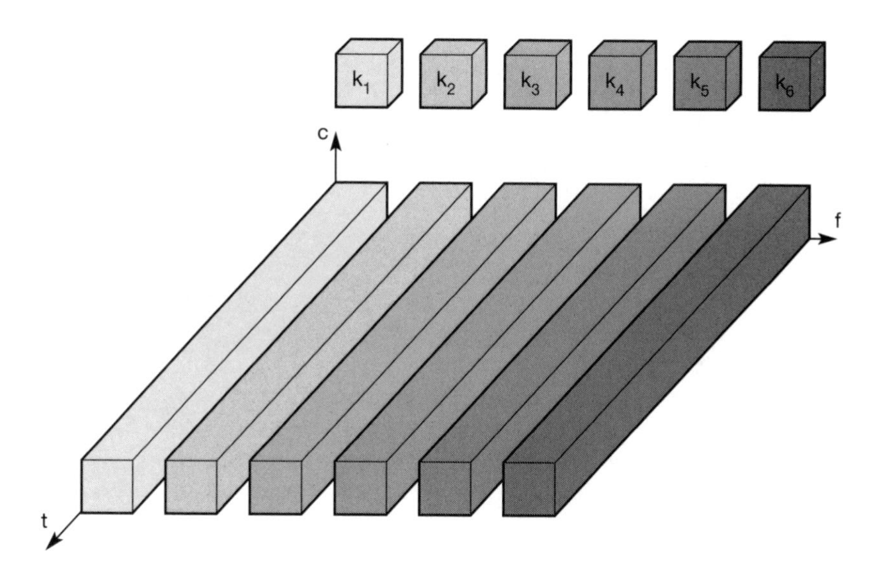

Figure 2.17
Frequency division multiplexing (FDM)

However, this scheme also has disadvantages. While radio stations broadcast 24 hours a day, mobile communication typically takes place for only a few minutes at a time. Assigning a separate frequency for each possible communication scenario would be a tremendous waste of (scarce) frequency resources. Additionally, the fixed assignment of a frequency to a sender makes the scheme very inflexible and limits the number of senders.

2.5.3 Time division multiplexing

A more flexible multiplexing scheme for typical mobile communications is **time division multiplexing (TDM)**. Here a channel k_i is given the whole bandwidth for a certain amount of time, i.e., all senders use the same frequency but at different points in time (see Figure 2.18). Again, **guard spaces**, which now represent time gaps, have to separate the different periods when the senders use the medium. In our highway example, this would refer to the gap between two cars. If two transmissions overlap in time, this is called co-channel interference. (In the highway example, interference between two cars results in an accident.) To avoid this type of interference, precise synchronization between different senders is necessary. This is clearly a disadvantage, as all senders need precise clocks or, alternatively, a way has to be found to distribute a synchronization signal to all senders. For a receiver tuning in to a sender this does not just involve adjusting the frequency, but involves listening at exactly the right point in time. However, this scheme is quite flexible as one can assign more sending time to senders with a heavy load and less to those with a light load.

Figure 2.18
Time division multiplexing (TDM)

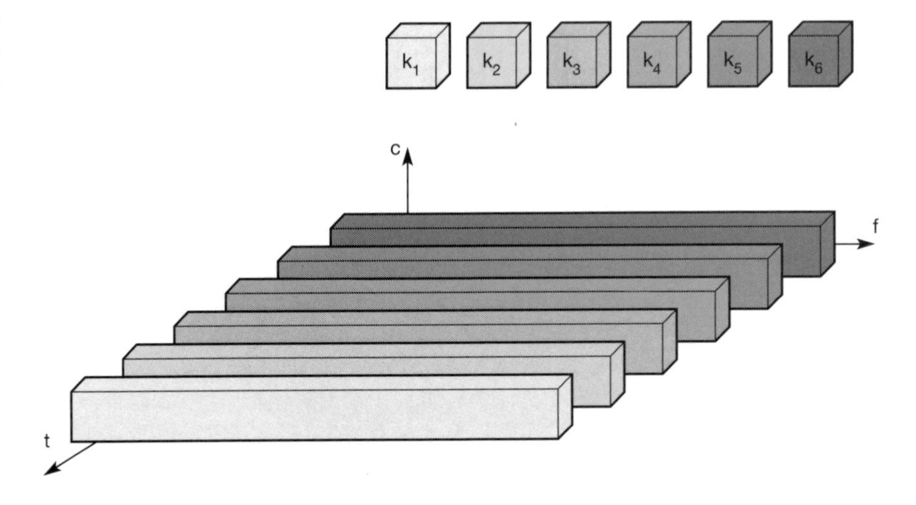

Frequency and time division multiplexing can be combined, i.e., a channel k_i can use a certain frequency band for a certain amount of time as shown in Figure 2.19. Now guard spaces are needed both in the time and in the frequency dimension. This scheme is more robust against frequency selective interference, i.e., interference in a certain small frequency band. A channel may use this band only for a short period of time. Additionally, this scheme provides some (weak) protection against tapping, as in this case the sequence of frequencies a sender uses has to be known to listen in to a channel. The mobile phone standard GSM uses this combination of frequency and time division multiplexing for transmission between a mobile phone and a so-called base station (see section 4.1).

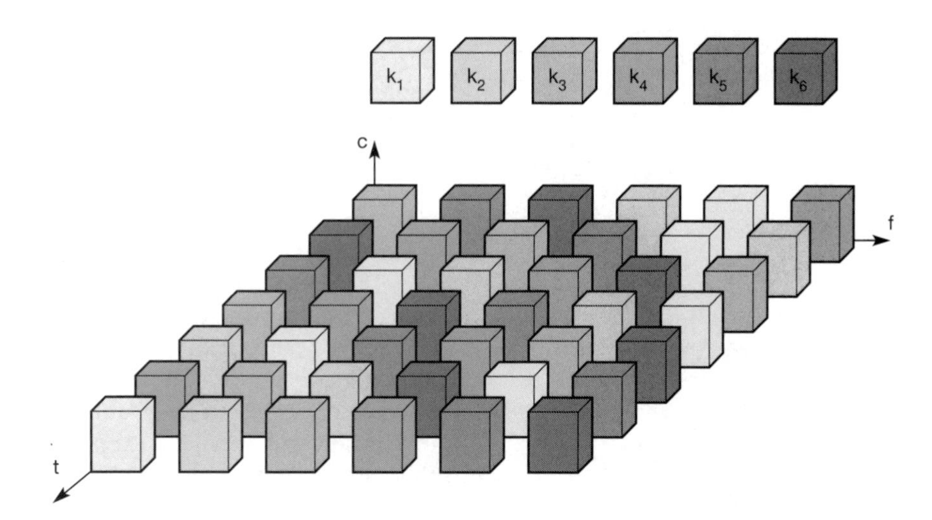

Figure 2.19
Frequency and time division multiplexing combined

A disadvantage of this scheme is again the necessary coordination between different senders. One has to control the sequence of frequencies and the time of changing to another frequency. Two senders will interfere as soon as they select the same frequency at the same time. However, if the frequency change (also called frequency hopping) is fast enough, the periods of interference may be so small that, depending on the coding of data into signals, a receiver can still recover the original data. (This technique is discussed in section 2.7.2.)

2.5.4 Code division multiplexing

While SDM and FDM are well known from the early days of radio transmission and TDM is used in connection with many applications, **code division multiplexing (CDM)** is a relatively new scheme in commercial communication systems. First used in military applications due to its inherent security features (together with spread spectrum techniques, see section 2.7), it now features in many civil wireless transmission scenarios thanks to the availability of cheap processing power (explained in more detail in section 3.5). Figure 2.20 shows how all channels k_i use the same frequency at the same time for transmission. Separation is now achieved by assigning each channel its own 'code', **guard spaces** are realized by using codes with the necessary 'distance' in code space, e.g., **orthogonal codes**. The technical realization of CDM is discussed in section 2.7 and chapter 3 together with the medium access mechanisms. An excellent book dealing with all aspects of CDM is Viterbi (1995).

The typical everyday example of CDM is a party with many participants from different countries around the world who establish communication channels, i.e., they talk to each other, using the same frequency range (approx. 300–6000 Hz depending on a person's voice) at the same time. If everybody speaks the same language, SDM is needed to be able to communicate (i.e., standing in groups,

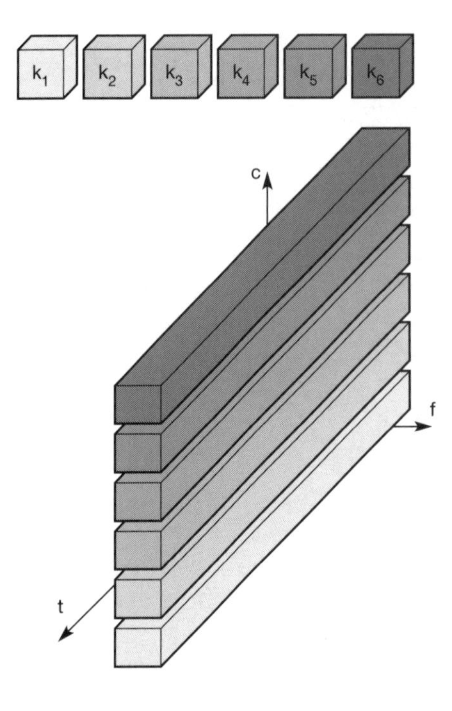

Figure 2.20 Code division multiplexing (CDM)

talking with limited transmit power). But as soon as another code, i.e., another language, is used, one can tune in to this language and clearly separate communication in this language from all the other languages. (The other languages appear as background noise.) This explains why CDM has built-in security: if the language is unknown, the signals can still be received, but they are useless. By using a secret code (or language), a secure channel can be established in a 'hostile' environment. (At parties this may cause some confusion.) Guard spaces are also of importance in this illustrative example. Using, e.g., Swedish and Norwegian does not really work; the languages are too close. But Swedish and Finnish are 'orthogonal' enough to separate the communication channels.

The main advantage of CDM for wireless transmission is that it gives good protection against interference and tapping. Different codes have to be assigned, but code space is huge compared to the frequency space. Assigning individual codes to each sender does not usually cause problems. The main disadvantage of this scheme is the relatively high complexity of the receiver (see section 3.5). A receiver has to know the code and must separate the channel with user data from the background noise composed of other signals and environmental noise. Additionally, a receiver must be precisely synchronized with the transmitter to apply the decoding correctly. The voice example also gives a hint to another problem of CDM receivers. All signals should reach a receiver with almost equal strength, otherwise some signals could drain others. If some people close to a receiver talk very loudly the language does not matter. The receiver cannot listen to any other person. To apply CDM, precise power control is required.

2.6 Modulation

Section 2.2 introduced the basic function of a sine wave which already indicates the three basic modulation schemes (typically, the cosine function is used for explanation):

$$g(t) = A_t \cos(2\pi f_t t + \varphi_t)$$

This function has three parameters: amplitude A_t, frequency f_t, and phase φ_t which may be varied in accordance with data or another modulating signal. For **digital modulation**, which is the main topic in this section, digital data (0 and 1) is translated into an analog signal (baseband signal). Digital modulation is required if digital data has to be transmitted over a medium that only allows for analog transmission. One example for wired networks is the old analog telephone system – to connect a computer to this system a modem is needed. The modem then performs the translation of digital data into analog signals and vice versa. Digital transmission is used, for example, in wired local area networks or within a computer (Halsall, 1996), (Stallings, 1997). In wireless networks, however, digital transmission cannot be used. Here, the binary bit-stream has to be translated into an analog signal first. The three basic methods for this translation are **amplitude shift keying (ASK)**, **frequency shift keying (FSK)**, and **phase shift keying (PSK)**. These are discussed in more detail in the following sections.

Apart from the translation of digital data into analog signals, wireless transmission requires an additional modulation, an **analog modulation** that shifts the center frequency of the baseband signal generated by the digital modulation up to the radio carrier. For example, digital modulation translates a 1 Mbit/s bit-stream into a baseband signal with a bandwidth of 1 MHz. There are several reasons why this baseband signal cannot be directly transmitted in a wireless system:

- **Antennas**: As shown in section 2.3, an antenna must be the order of magnitude of the signal's wavelength in size to be effective. For the 1 MHz signal in the example this would result in an antenna some hundred meters high, which is obviously not very practical for handheld devices. With 1 GHz, antennas a few centimeters in length can be used.
- **Frequency division multiplexing**: Using only baseband transmission, FDM could not be applied. Analog modulation shifts the baseband signals to different carrier frequencies as required in section 2.5.2. The higher the carrier frequency, the more bandwidth that is available for many baseband signals.
- **Medium characteristics**: Path-loss, penetration of obstacles, reflection, scattering, and diffraction – all the effects discussed in section 2.4 depend heavily on the wavelength of the signal. Depending on the application, the right carrier frequency with the desired characteristics has to be chosen: long waves for submarines, short waves for handheld devices, very short waves for directed microwave transmission etc.

As for digital modulation, three different basic schemes are known for analog modulation: **amplitude modulation (AM)**, **frequency modulation (FM)**, and **phase modulation (PM)**. The reader is referred to Halsall (1996) and Stallings (2002) for more details about these analog modulation schemes.

Figure 2.21 shows a (simplified) block diagram of a radio transmitter for digital data. The first step is the digital modulation of data into the analog baseband signal according to one of the schemes presented in the following sections. The analog modulation then shifts the center frequency of the analog signal up to the radio carrier. This signal is then transmitted via the antenna.

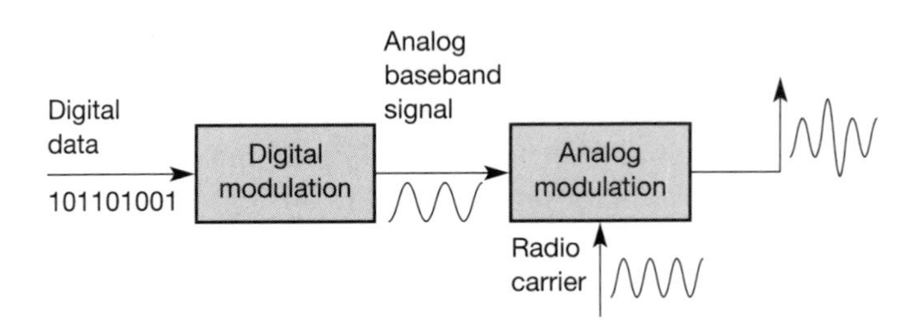

Figure 2.21 Modulation in a transmitter

The receiver (see Figure 2.22) receives the analog radio signal via its antenna and demodulates the signal into the analog baseband signal with the help of the known carrier. This would be all that is needed for an analog radio tuned in to a radio station. (The analog baseband signal would constitute the music.) For digital data, another step is needed. Bits or frames have to be detected, i.e., the receiver must synchronize with the sender. How synchronization is achieved, depends on the digital modulation scheme. After synchronization, the receiver has to decide if the signal represents a digital 1 or a 0, reconstructing the original data.

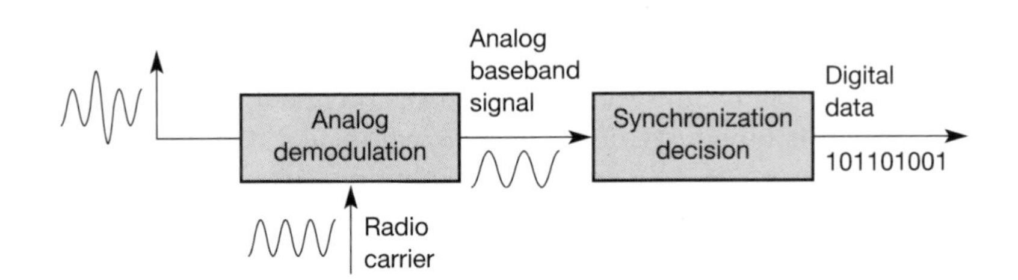

Figure 2.22 Demodulation and data reconstruction in a receiver

The digital modulation schemes presented in the following sections differ in many issues, such as **spectral efficiency** (i.e., how efficiently the modulation scheme utilizes the available frequency spectrum), **power efficiency** (i.e., how much power is needed to transfer bits – which is very important for portable devices that are battery dependent), and **robustness** to multi-path propagation, noise, and interference (Wesel, 1998).

2.6.1 Amplitude shift keying

Figure 2.23 illustrates **amplitude shift keying (ASK)**, the most simple digital modulation scheme. The two binary values, 1 and 0, are represented by two different amplitudes. In the example, one of the amplitudes is 0 (representing the binary 0). This simple scheme only requires low bandwidth, but is very susceptible to interference. Effects like multi-path propagation, noise, or path loss heavily influence the amplitude. In a wireless environment, a constant amplitude cannot be guaranteed, so ASK is typically not used for wireless radio transmission. However, the wired transmission scheme with the highest performance, namely optical transmission, uses ASK. Here, a light pulse may represent a 1, while the absence of light represents a 0. The carrier frequency in optical systems is some hundred THz. ASK can also be applied to wireless infra red transmission, using a directed beam or diffuse light (see chapter 7, Wireless LANs).

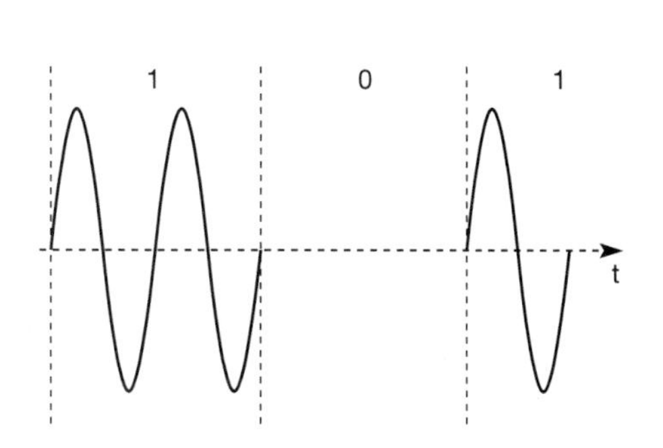

Figure 2.23 Amplitude shift keying (ASK)

2.6.2 Frequency shift keying

A modulation scheme often used for wireless transmission is **frequency shift keying (FSK)** (see Figure 2.24). The simplest form of FSK, also called **binary FSK (BFSK)**, assigns one frequency f_1 to the binary 1 and another frequency f_2 to the binary 0. A very simple way to implement FSK is to switch between two oscillators, one with the frequency f_1 and the other with f_2, depending on the input. To avoid sudden changes in phase, special frequency modulators with **continuous phase modulation, (CPM)** can be used. Sudden changes in phase cause high frequencies, which is an undesired side-effect.

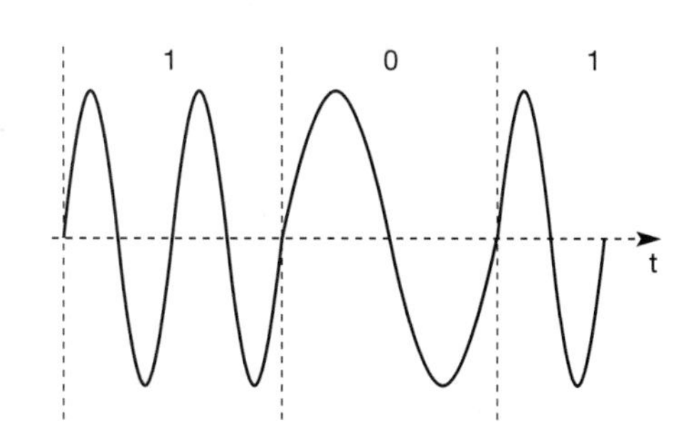

Figure 2.24 Frequency shift keying (FSK)

A simple way to implement demodulation is by using two bandpass filters, one for f_1 the other for f_2. A comparator can then compare the signal levels of the filter outputs to decide which of them is stronger. FSK needs a larger bandwidth compared to ASK but is much less susceptible to errors.

2.6.3 Phase shift keying

Finally, **phase shift keying (PSK)** uses shifts in the phase of a signal to represent data. Figure 2.25 shows a phase shift of 180° or π as the 0 follows the 1 (the same happens as the 1 follows the 0). This simple scheme, shifting the phase by 180° each time the value of data changes, is also called **binary PSK (BPSK)**. A simple

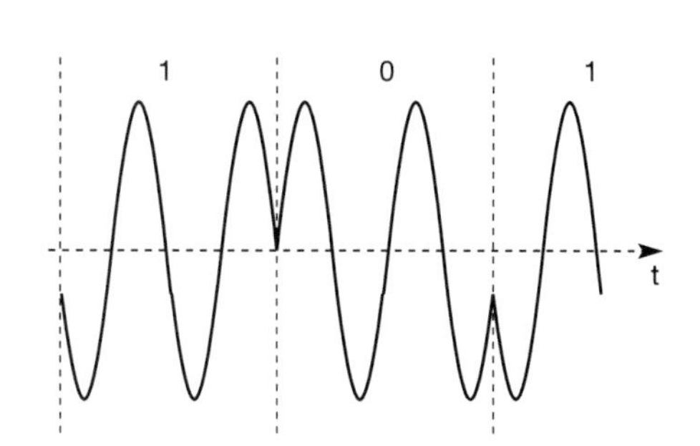

Figure 2.25 Phase shift keying (PSK)

implementation of a BPSK modulator could multiply a frequency f with +1 if the binary data is 1 and with –1 if the binary data is 0.

To receive the signal correctly, the receiver must synchronize in frequency and phase with the transmitter. This can be done using a **phase lock loop (PLL).** Compared to FSK, PSK is more resistant to interference, but receiver and transmitter are also more complex.

2.6.4 Advanced frequency shift keying

A famous FSK scheme used in many wireless systems is **minimum shift keying (MSK)**. MSK is basically BFSK without abrupt phase changes, i.e., it belongs to CPM schemes. Figure 2.26 shows an example for the implementation of MSK. In a first step, data bits are separated into even and odd bits, the duration of each bit being doubled. The scheme also uses two frequencies: f_1, the lower frequency, and f_2, the higher frequency, with $f_2 = 2f_1$.

Figure 2.26
Minimum shift keying (MSK)

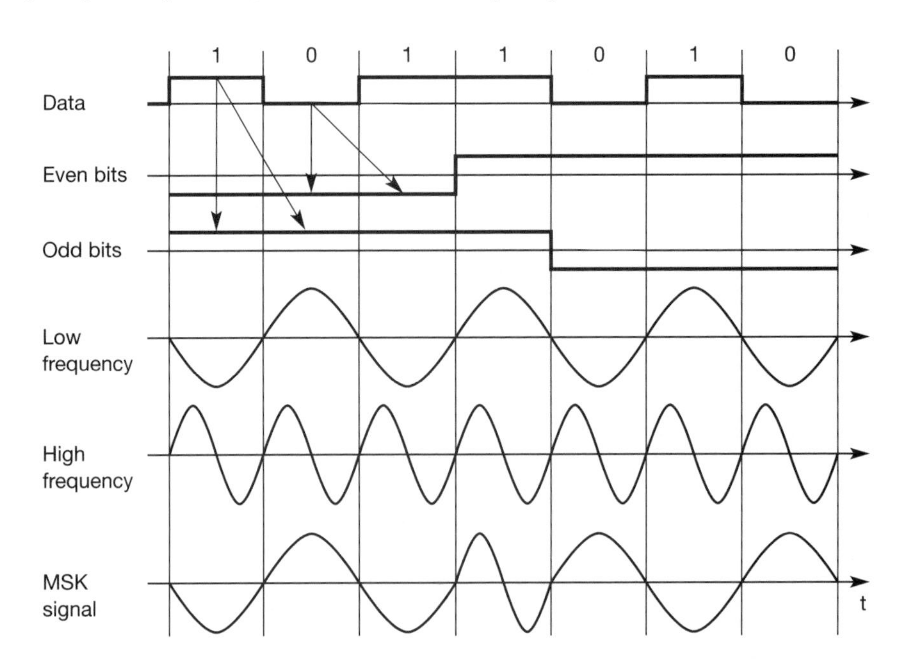

According to the following scheme, the lower or higher frequency is chosen (either inverted or non-inverted) to generate the MSK signal:

- if the even and the odd bit are both 0, then the higher frequency f_2 is inverted (i.e., f_2 is used with a phase shift of 180°);
- if the even bit is 1, the odd bit 0, then the lower frequency f_1 is inverted. This is the case, e.g., in the fifth to seventh columns of Figure 2.26,

- if the even bit is 0 and the odd bit is 1, as in columns 1 to 3, f_1 is taken without changing the phase,
- if both bits are 1 then the original f_2 is taken.

A high frequency is always chosen if even and odd bits are equal. The signal is inverted if the odd bit equals 0. This scheme avoids all phase shifts in the resulting MSK signal.

Adding a so-called Gaussian lowpass filter to the MSK scheme results in **Gaussian MSK (GMSK)**, which is the digital modulation scheme for many European wireless standards (see chapter 4 for GSM, DECT). The filter reduces the large spectrum needed by MSK.

2.6.5 Advanced phase shift keying

The simple PSK scheme can be improved in many ways. The basic BPSK scheme only uses one possible phase shift of 180°. The left side of Figure 2.27 shows BPSK in the phase domain (which is typically the better representation compared to the time domain in Figure 2.25). The right side of Figure 2.27 shows **quadrature PSK (QPSK)**, one of the most common PSK schemes (sometimes also called quaternary PSK). Here, higher bit rates can be achieved for the same bandwidth by coding two bits into one phase shift. Alternatively, one can reduce the bandwidth and still achieve the same bit rates as for BPSK.

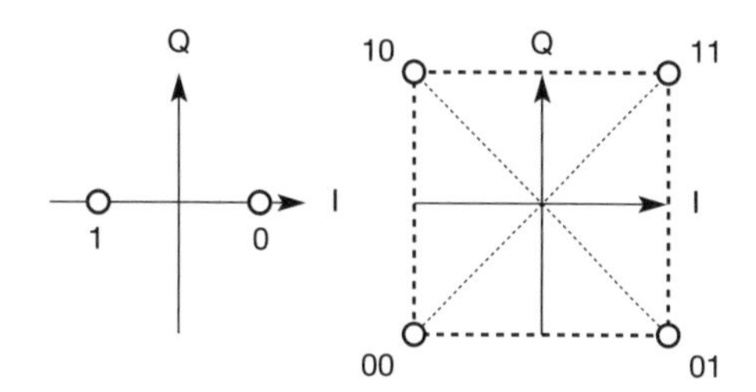

Figure 2.27 BPSK and QPSK in the phase domain

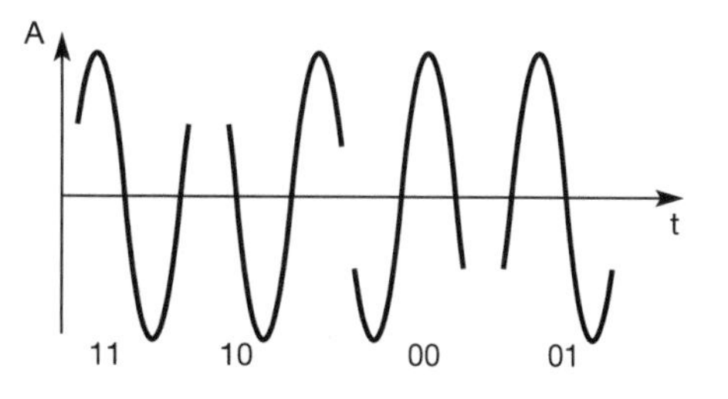

Figure 2.28 QPSK in the time domain

QPSK (and other PSK schemes) can be realized in two variants. The phase shift can always be relative to a **reference signal** (with the same frequency). If this scheme is used, a phase shift of 0 means that the signal is in phase with the reference signal. A QPSK signal will then exhibit a phase shift of 45° for the data 11, 135° for 10, 225° for 00, and 315° for 01 – with all phase shifts being relative to the reference signal. The transmitter 'selects' parts of the signal as shown in Figure 2.28 and concatenates them. To reconstruct data, the receiver has to compare the incoming signal with the reference signal. One problem of this scheme involves producing a reference signal at the receiver. Transmitter and receiver have to be synchronized very often, e.g., by using special synchronization patterns before user data arrives or via a pilot frequency as reference.

One way to avoid this problem is to use **differential QPSK (DQPSK)**. Here the phase shift is not relative to a reference signal but to the phase of the previous two bits. In this case, the receiver does not need the reference signal but only compares two signals to reconstruct data. DQPSK is used in US wireless technologies IS-136 and PACS and in Japanese PHS.

One could now think of extending the scheme to more and more angles for shifting the phase. For instance, one can think of coding 3 bits per phase shift using 8 angles. Additionally, the PSK scheme could be combined with ASK as is done for example in **quadrature amplitude modulation (QAM)** for standard 9,600 bit/s modems (left side of Figure 2.29). Here, three different amplitudes and 12 angles are combined coding 4 bits per phase/amplitude change. Problems occur for wireless communication in case of noise or ISI. The more 'points' used in the phase domain, the harder it is to separate them. DQPSK has been proven as one of the most efficient schemes under these considerations (Wesel, 1998).

Figure 2.29
16 quadrature amplitude modulation and hierarchical 64 QAM

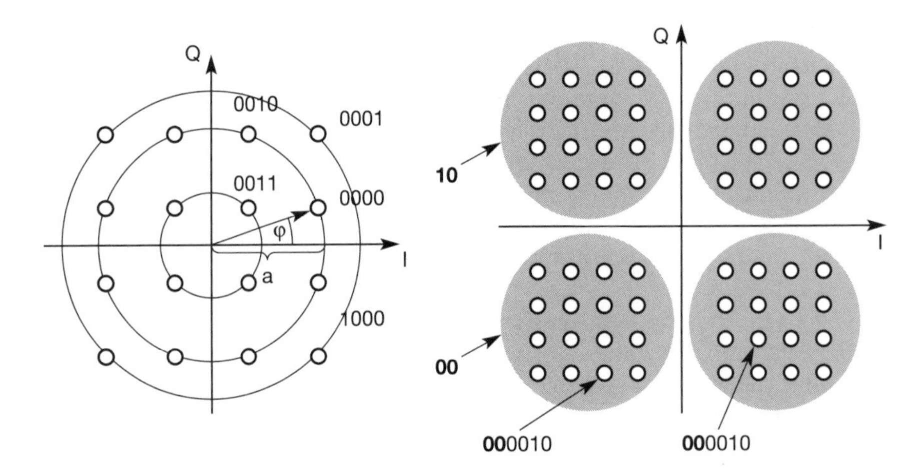

A more advanced scheme is a hierarchical modulation as used in the digital TV standard DVB-T. The right side of Figure 2.29 shows a 64 QAM that contains a QPSK modulation. A 64 QAM can code 6 bit per symbol. Here the two most significant bits are used for the QPSK signal embedded in the QAM signal. If the reception of the signal is good the entire QAM constellation can be resolved. Under poor reception conditions, e.g., with moving receivers, only the QPSK portion can be resolved. A high priority data stream in DVB-T is coded with QPSK using the two most significant bits. The remaining 4 bits represent low priority data. For TV this could mean that the standard resolution data stream is coded with high priority, the high resolution information with low priority. If the signal is distorted, at least the standard TV resolution can be received.

2.6.6 Multi-carrier modulation

Special modulation schemes that stand somewhat apart from the others are **multi-carrier modulation (MCM)**, **orthogonal frequency division multiplexing (OFDM)** or **coded OFDM (COFDM)** that are used in the context of the European digital radio system DAB (see section 6.3) and the WLAN standards IEEE 802.11a and HiperLAN2 (see chapter 7). The main attraction of MCM is its good ISI mitigation property. As explained in section 2.4.3, higher bit rates are more vulnerable to ISI. MCM splits the high bit rate stream into many lower bit rate streams (see Figure 2.30), each stream being sent using an independent carrier frequency. If, for example, n symbols/s have to be transmitted, each subcarrier transmits n/c symbols/s with c being the number of subcarriers. One symbol could, for example represent 2 bit as in QPSK. DAB, for example, uses between 192 and 1,536 of these subcarriers. The physical layer of HiperLAN2 and IEEE 802.11a uses 48 subcarriers for data.

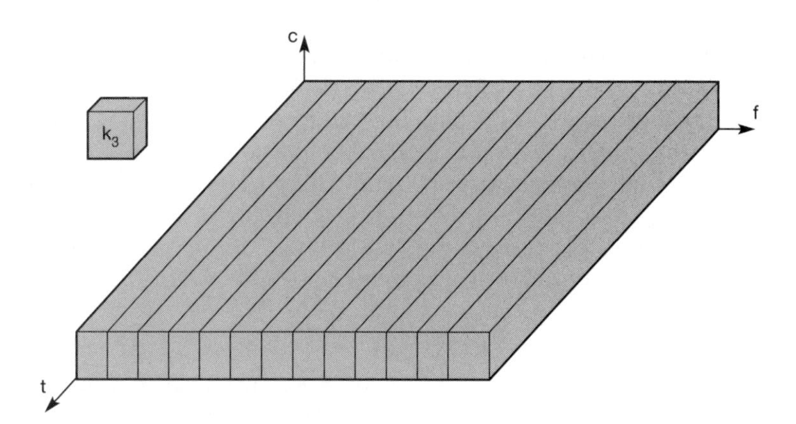

Figure 2.30 Parallel data transmission on several subcarriers with lower rate

Figure 2.31 shows the superposition of orthogonal frequencies. The maximum of one subcarrier frequency appears exactly at a frequency where all other subcarriers equal zero.

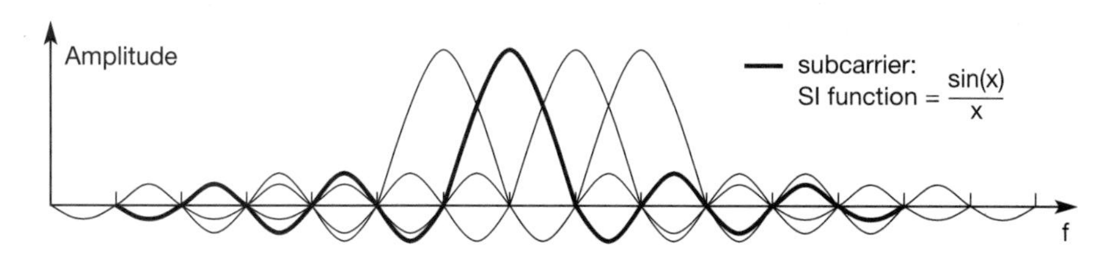

Figure 2.31 Superposition of orthogonal frequencies

Using this scheme, frequency selective fading only influences some subcarriers, and not the whole signal – an additional benefit of MCM. Typically, MCM transmits symbols with guard spaces between single symbols or groups of symbols. This helps the receiver to handle multi-path propagation. OFDM is a special method of implementing MCM using orthogonal carriers. Computationally, this is a very efficient algorithm based on fast Fourier transform (FFT) for modulation/demodulation. If additional error-control coding across the symbols in different subcarriers is applied, the system is referred to as COFDM. More details about the implementation of MCM, OFDM, and COFDM can be found in Wesel (1998), Pahlavan (2002), ETSI (1997) and in section 6.3 or chapter 7.

2.7 Spread spectrum

As the name implies, **spread spectrum** techniques involve spreading the bandwidth needed to transmit data – which does not make sense at first sight. Spreading the bandwidth has several advantages. The main advantage is the resistance to **narrowband interference**. In Figure 2.32, diagram i) shows an idealized narrowband signal from a sender of user data (here power density dP/df versus frequency f). The sender now spreads the signal in step ii), i.e., converts the narrowband signal into a broadband signal. The energy needed to transmit the signal (the area shown in the diagram) is the same, but it is now spread over a larger frequency range. The power level of the spread signal can be much lower than that of the original narrowband signal without losing data. Depending on the generation and reception of the spread signal, the power level of the user signal can even be as low as the background noise. This makes it difficult to distinguish the user signal from the background noise and thus hard to detect.

Figure 2.32
Spread spectrum: spreading and despreading

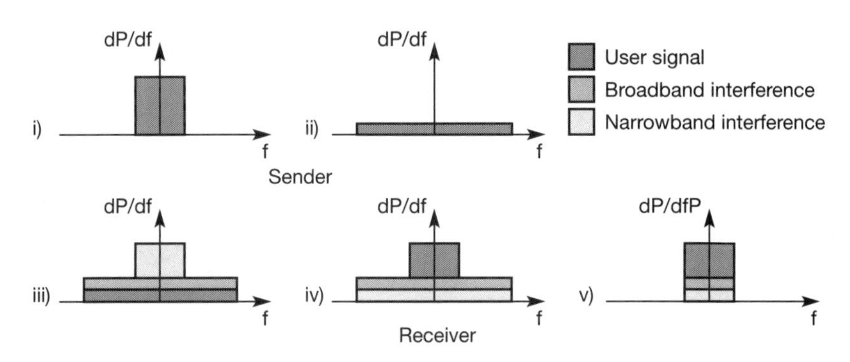

During transmission, narrowband and broadband interference add to the signal in step iii). The sum of interference and user signal is received. The receiver now knows how to despread the signal, converting the spread user signal into a narrowband signal again, while spreading the narrowband interference and leaving the broadband interference. In step v) the receiver applies a bandpass filter to cut off frequencies left and right of the narrowband signal. Finally, the receiver can reconstruct the original data because the power level of the user signal is high enough, i.e., the signal is much stronger than the remaining interference. The following sections show how spreading can be performed.

Just as spread spectrum helps to deal with narrowband interference for a single channel, it can be used for several channels. Consider the situation shown in Figure 2.33. Six different channels use FDM for multiplexing, which means that each channel has its own narrow frequency band for transmission. Between each frequency band a guard space is needed to avoid adjacent channel interference. As mentioned in section 2.5.2, this method requires careful frequency planning. Additionally, Figure 2.33 depicts a certain channel quality. This is frequency dependent and is a measure for interference at this frequency. Channel quality also changes over time – the diagram only shows a snapshot at one moment. Depending on receiver characteristics, channels 1, 2, 5, and 6 could be received while the quality of channels 3 and 4 is too bad to reconstruct transmitted data. Narrowband interference destroys the transmission of channels 3 and 4. This illustration only represents a snapshot and the situation could be completely different at the next moment. All in all, communication may be very difficult using such narrowband signals.

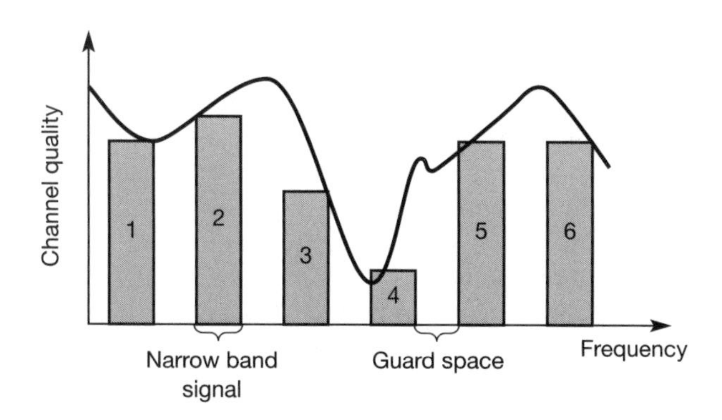

Figure 2.33
Narrowband interference without spread spectrum

How can spread spectrum help in such a situation? As already shown, spread spectrum can increase resistance to narrowband interference. The same technique is now applied to all narrowband signals. As shown in Figure 2.34, all narrowband signals are now spread into broadband signals using the same frequency range. No more frequency planning is needed (under these simplified assumptions), and all senders use the same frequency band. But how can receivers recover their signal?

To separate different channels, CDM is now used instead of FDM. This application shows the tight coupling of CDM and spread spectrum (explained in more detail in chapter 3). Spreading of a narrowband signal is achieved using a special code as shown in sections 2.7.1 and 2.7.2. Each channel is allotted its own code, which the receivers have to apply to recover the signal. Without knowing the code, the signal cannot be recovered and behaves like background noise. This is the security effect of spread spectrum if a secret code is used for spreading. Features that make spread spectrum and CDM very attractive for military applications are the coexistence of several signals without coordination (apart from the fact that the codes must have certain properties), robustness against narrowband interference, relative high security, and a characteristic like background noise. Only the appropriate (secret) codes have to be exchanged.

Figure 2.34 Spread spectrum to avoid narrowband interference

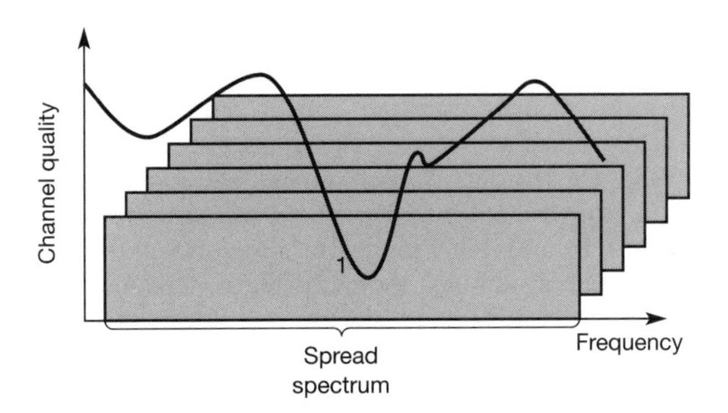

Apart from military uses, the combination of spread spectrum and CDM is becoming more and more attractive for everyday applications. As mentioned before, frequencies are a scarce resource around the world, particularly license-free bands. Spread spectrum now allows an overlay of new transmission technology at exactly the same frequency at which current narrowband systems are already operating. This is used by US mobile phone systems. While the frequency band around 850 MHz had already been in use for TDM and FDM systems (AMPS and IS-54), the introduction of a system using CDM (IS-95) was still possible.

Spread spectrum technologies also exhibit drawbacks. One disadvantage is the increased complexity of receivers that have to despread a signal. Today despreading can be performed up to high data rates thanks to digital signal processing. Another problem is the large frequency band that is needed due to the spreading of the signal. Although spread signals appear more like noise, they still raise the background noise level and may interfere with other transmissions if no special precautions are taken.

Spreading the spectrum can be achieved in two different ways as shown in the following two sections.

2.7.1 Direct sequence spread spectrum

Direct sequence spread spectrum (DSSS) systems take a user bit stream and perform an (XOR) with a so-called **chipping sequence** as shown in Figure 2.35. The example shows that the result is either the sequence 0110101 (if the user bit equals 0) or its complement 1001010 (if the user bit equals 1). While each user bit has a duration t_b, the chipping sequence consists of smaller pulses, called **chips**, with a duration t_c. If the chipping sequence is generated properly it appears as random noise: this sequence is also sometimes called **pseudo-noise** sequence. The **spreading factor** $s = t_b/t_c$ determines the bandwidth of the resulting signal. If the original signal needs a bandwidth w, the resulting signal needs s·w after spreading. While the spreading factor of the very simple example is only 7 (and the chipping sequence 0110101 is not very random), civil applications use spreading factors between 10 and 100, military applications use factors of up to 10,000. Wireless LANs complying with the standard IEEE 802.11 (see section 7.3) use, for example, the sequence 10110111000, a so-called Barker code, if implemented using DSSS. Barker codes exhibit a good robustness against interference and insensitivity to multi-path propagation. Other known Barker codes are 11, 110, 1110, 11101, 1110010, and 1111100110101 (Stallings, 2002).

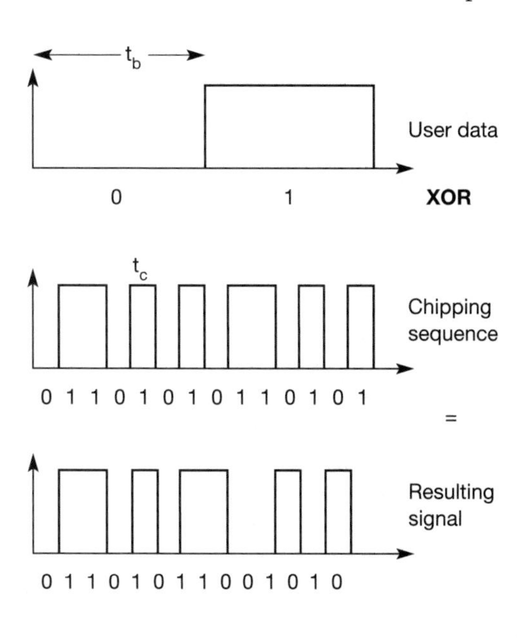

Figure 2.35
Spreading with DSSS

Up to now only the spreading has been explained. However, transmitters and receivers using DSSS need additional components as shown in the simplified block diagrams in Figure 2.36 and Figure 2.37. The first step in a DSSS transmitter, Figure 2.36 is the spreading of the user data with the chipping sequence (**digital modulation**). The spread signal is then modulated with a radio carrier as explained in section 2.6 (**radio modulation**). Assuming for example a user signal with a bandwidth of 1 MHz. Spreading with the above 11-chip Barker code would result in a signal with 11 MHz bandwidth. The radio carrier then shifts this signal to the carrier frequency (e.g., 2.4 GHz in the ISM band). This signal is then transmitted.

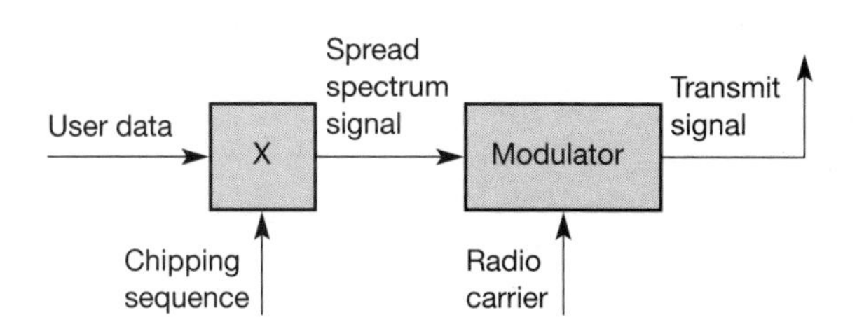

Figure 2.36
DSSS transmitter

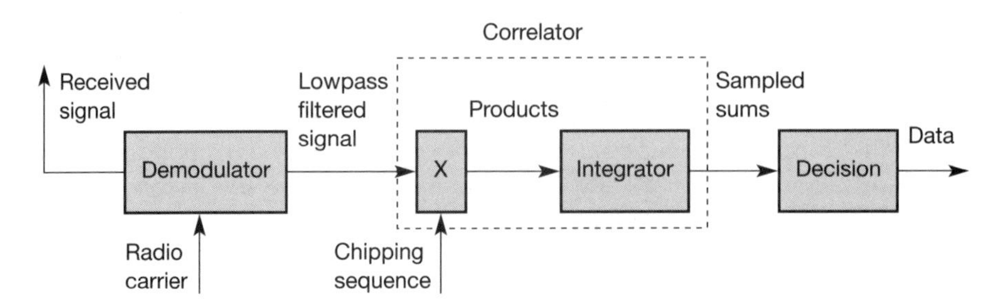

Figure 2.37
DSSS receiver

The DSSS receiver is more complex than the transmitter. The receiver only has to perform the inverse functions of the two transmitter modulation steps. However, noise and multi-path propagation require additional mechanisms to reconstruct the original data. The first step in the receiver involves demodulating the received signal. This is achieved using the same carrier as the transmitter reversing the modulation and results in a signal with approximately the same bandwidth as the original spread spectrum signal. Additional filtering can be applied to generate this signal.

While demodulation is well known from ordinary radio receivers, the next steps constitute a real challenge for DSSS receivers, contributing to the complexity of the system. The receiver has to know the original chipping sequence, i.e., the receiver basically generates the same pseudo random sequence as the transmitter. Sequences at the sender and receiver have to be precisely synchronized because the receiver calculates the product of a chip with the incoming signal. This comprises another XOR operation as explained in section 3.5, together with a medium access mechanism that relies on this scheme. During a bit period, which also has to be derived via synchronization, an **integrator** adds all these products. Calculating the products of chips and signal, and adding the products in an integrator is also called correlation, the device a **correlator**. Finally, in each bit period a **decision unit** samples the sums generated by the integrator and decides if this sum represents a binary 1 or a 0.

If transmitter and receiver are perfectly synchronized and the signal is not too distorted by noise or multi-path propagation, DSSS works perfectly well according to the simple scheme shown. Sending the user data 01 and applying the 11-chip Barker code 10110111000 results in the spread 'signal' 1011011100001001000111. On the receiver side, this 'signal' is XORed bit-wise after demodulation with the same Barker code as chipping sequence. This results in the sum of products equal to 0 for the first bit and to 11 for the second bit. The decision unit can now map the first sum (=0) to a binary 0, the second sum (=11) to a binary 1 – this constitutes the original user data.

In real life, however, the situation is somewhat more complex. Assume that the demodulated signal shows some distortion, e.g., 1010010100001101000111. The sum of products for the first bit would be 2, 10 for the second bit. Still, the decision unit can map, e.g., sums less than 4 to a binary 0 and sums larger than

7 to a binary 1. However, it is important to stay synchronized with the transmitter of a signal. But what happens in case of multi-path propagation? Then several paths with different delays exist between a transmitter and a receiver. Additionally, the different paths may have different path losses. In this case, using so-called rake receivers provides a possible solution. A **rake receiver** uses n correlators for the n strongest paths. Each correlator is synchronized to the transmitter plus the delay on that specific path. As soon as the receiver detects a new path which is stronger than the currently weakest path, it assigns this new path to the correlator with the weakest path. The output of the correlators are then combined and fed into the decision unit. Rake receivers can even take advantage of the multi-path propagation by combining the different paths in a constructive way (Viterbi, 1995).

2.7.2 Frequency hopping spread spectrum

For **frequency hopping spread spectrum (FHSS)** systems, the total available bandwidth is split into many channels of smaller bandwidth plus guard spaces between the channels. Transmitter and receiver stay on one of these channels for a certain time and then hop to another channel. This system implements FDM and TDM. The pattern of channel usage is called the **hopping sequence**, the time spend on a channel with a certain frequency is called the **dwell time**. FHSS comes in two variants, slow and fast hopping (see Figure 2.38).

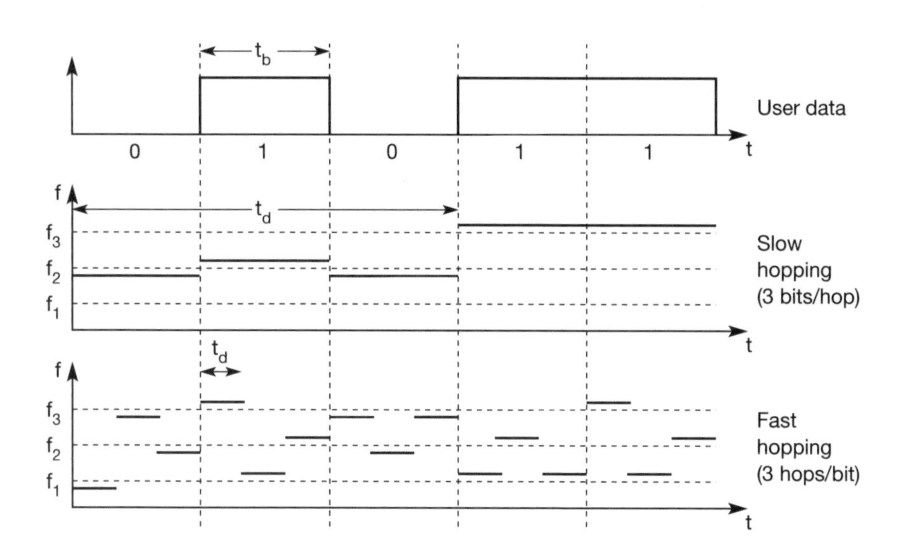

Figure 2.38
Slow and fast frequency hopping

In **slow hopping**, the transmitter uses one frequency for several bit periods.[3] Figure 2.38 shows five user bits with a bit period t_b. Performing slow hopping, the transmitter uses the frequency f_2 for transmitting the first three bits during the dwell time t_d. Then, the transmitter hops to the next frequency f_3. Slow hopping systems are typically cheaper and have relaxed tolerances, but they are not as immune to narrowband interference as fast hopping systems. Slow frequency hopping is an option for GSM (see section 4.1).

For **fast hopping** systems, the transmitter changes the frequency several times during the transmission of a single bit. In the example of Figure 2.38, the transmitter hops three times during a bit period. Fast hopping systems are more complex to implement because the transmitter and receiver have to stay synchronized within smaller tolerances to perform hopping at more or less the same points in time. However, these systems are much better at overcoming the effects of narrowband interference and frequency selective fading as they only stick to one frequency for a very short time.

Another example of an FHSS system is Bluetooth, which is presented in section 7.5. Bluetooth performs 1,600 hops per second and uses 79 hop carriers equally spaced with 1 MHz in the 2.4 GHz ISM band.

Figures 2.39 and 2.40 show simplified block diagrams of FHSS transmitters and receivers respectively. The first step in an FHSS transmitter is the modulation of user data according to one of the digital-to-analog modulation schemes, e.g., FSK or BPSK, as discussed in section 2.6. This results in a narrowband signal, if FSK is used with a frequency f_0 for a binary 0 and f_1 for a binary 1. In the next step, frequency hopping is performed, based on a hopping sequence. The hopping sequence is fed into a frequency synthesizer generating the carrier frequencies f_i. A second modulation uses the modulated narrowband signal and the carrier frequency to generate a new spread signal with frequency of f_i+f_0 for a 0 and f_i+f_1 for a 1 respectively. If different FHSS transmitters use hopping sequences that never overlap, i.e., if two transmitters never use the same frequency f_i at the same time, then these two transmissions do not interfere. This requires the coordination of all transmitters and their hopping sequences. As for DSSS systems, pseudo-random hopping sequences can also be used without coordination. These sequences only have to fulfill certain properties to keep interference minimal.[4] Two or more transmitters may choose the same frequency for a hop, but dwell time is short for fast hopping systems, so interference is minimal.

The receiver of an FHSS system has to know the hopping sequence and must stay synchronized. It then performs the inverse operations of the modulation to reconstruct user data. Several filters are also needed (these are not shown in the simplified diagram in Figure 2.40).

3 Another definition refers to the number of hops per signal element instead of bits.

4 These sequences should have a low cross-correlation. More details are given in section 3.5.

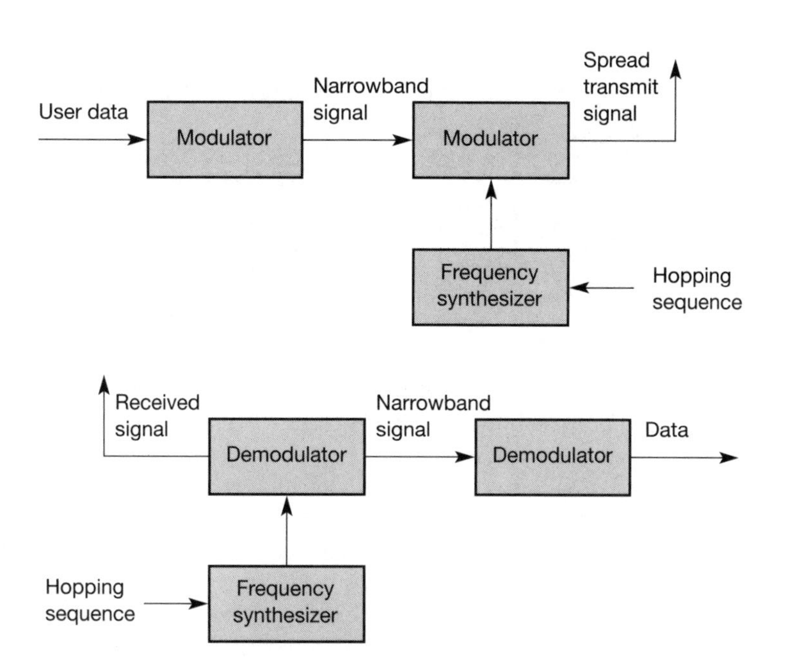

Figure 2.39
FHSS transmitter

Figure 2.40
FHSS receiver

Compared to DSSS, spreading is simpler using FHSS systems. FHSS systems only use a portion of the total band at any time, while DSSS systems always use the total bandwidth available. DSSS systems on the other hand are more resistant to fading and multi-path effects. DSSS signals are much harder to detect – without knowing the spreading code, detection is virtually impossible. If each sender has its own pseudo-random number sequence for spreading the signal (DSSS or FHSS), the system implements CDM. More details about spread spectrum applications and their theoretical background can be found in Viterbi (1995), Peterson (1995), Ojanperä (1998), and Dixon (1994).

2.8 Cellular systems

Cellular systems for mobile communications implement SDM. Each transmitter, typically called a **base station**, covers a certain area, a **cell**. Cell radii can vary from tens of meters in buildings, and hundreds of meters in cities, up to tens of kilometers in the countryside. The shape of cells are never perfect circles or hexagons (as shown in Figure 2.41), but depend on the environment (buildings, mountains, valleys etc.), on weather conditions, and sometimes even on system load. Typical systems using this approach are mobile telecommunication systems (see chapter 4), where a mobile station within the cell around a base station communicates with this base station and vice versa.

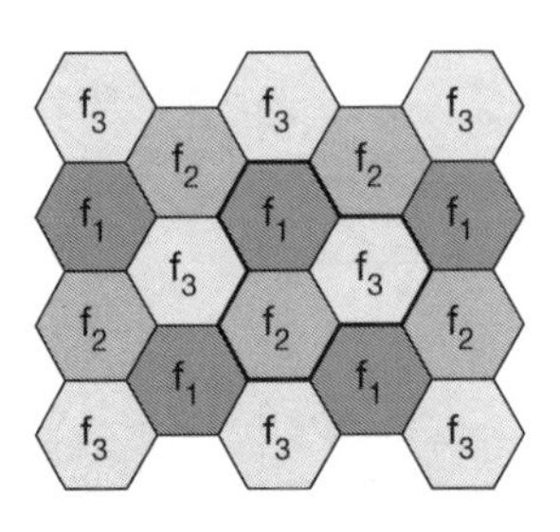

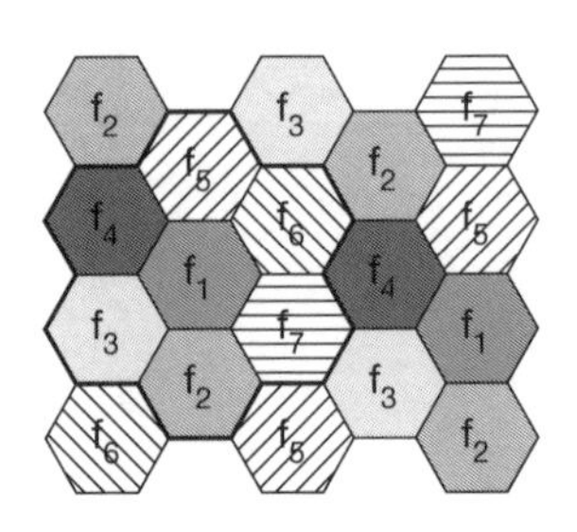

Figure 2.41
Cellular system with three and seven cell clusters

In this context, the question arises as to why mobile network providers install several thousands of base stations throughout a country (which is quite expensive) and do not use powerful transmitters with huge cells like, e.g., radio stations, use.

Advantages of cellular systems with small cells are the following:

- **Higher capacity**: Implementing SDM allows frequency reuse. If one transmitter is far away from another, i.e., outside the interference range, it can reuse the same frequencies. As most mobile phone systems assign frequencies to certain users (or certain hopping patterns), this frequency is blocked for other users. But frequencies are a scarce resource and, the number of concurrent users per cell is very limited. Huge cells do not allow for more users. On the contrary, they are limited to less possible users per km^2. This is also the reason for using very small cells in cities where many more people use mobile phones.
- **Less transmission power**: While power aspects are not a big problem for base stations, they are indeed problematic for mobile stations. A receiver far away from a base station would need much more transmit power than the current few Watts. But energy is a serious problem for mobile handheld devices.
- **Local interference only**: Having long distances between sender and receiver results in even more interference problems. With small cells, mobile stations and base stations only have to deal with 'local' interference.
- **Robustness**: Cellular systems are decentralized and so, more robust against the failure of single components. If one antenna fails, this only influences communication within a small area.

Small cells also have some **disadvantages**:

- **Infrastructure needed**: Cellular systems need a complex infrastructure to connect all base stations. This includes many antennas, switches for call forwarding, location registers to find a mobile station etc, which makes the whole system quite expensive.

- **Handover needed**: The mobile station has to perform a handover when changing from one cell to another. Depending on the cell size and the speed of movement, this can happen quite often.
- **Frequency planning**: To avoid interference between transmitters using the same frequencies, frequencies have to be distributed carefully. On the one hand, interference should be avoided, on the other, only a limited number of frequencies is available.

To avoid interference, different transmitters within each other's interference range use FDM. If FDM is combined with TDM (see Figure 2.19), the hopping pattern has to be coordinated. The general goal is never to use the same frequency at the same time within the interference range (if CDM is not applied). Two possible models to create cell patterns with minimal interference are shown in Figure 2.41. Cells are combined in **clusters** – on the left side three cells form a cluster, on the right side seven cells form a cluster. All cells within a cluster use disjointed sets of frequencies. On the left side, one cell in the cluster uses set f_1, another cell f_2, and the third cell f_3. In real-life transmission, the pattern will look somewhat different. The hexagonal pattern is chosen as a simple way of illustrating the model. This pattern also shows the repetition of the same frequency sets. The transmission power of a sender has to be limited to avoid interference with the next cell using the same frequencies.

To reduce interference even further (and under certain traffic conditions, i.e., number of users per km^2) **sectorized antennas** can be used. Figure 2.42 shows the use of three sectors per cell in a cluster with three cells. Typically, it makes sense to use sectorized antennas instead of omni-directional antennas for larger cell radii.

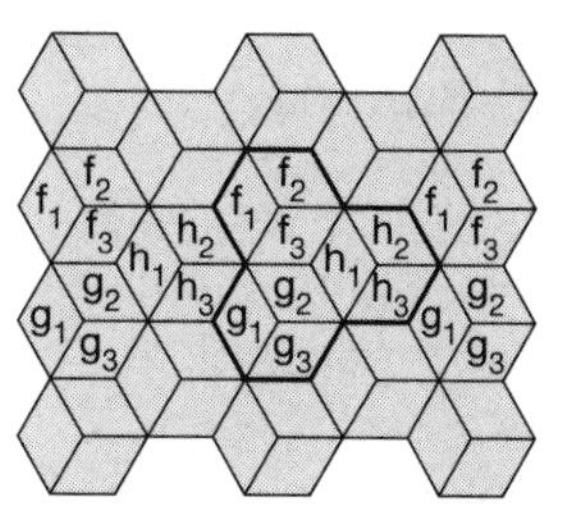

Figure 2.42 Cellular system with three cell clusters and three sectors per cell

The fixed assignment of frequencies to cell clusters and cells respectively, is not very efficient if traffic load varies. For instance, in the case of a heavy load in one cell and a light load in a neighboring cell, it could make sense to 'borrow' frequencies. Cells with more traffic are dynamically allotted more frequencies. This scheme is known as **borrowing channel allocation (BCA)**, while the first fixed scheme is called **fixed channel allocation (FCA)**. FCA is used in the GSM system as it is much simpler to use, but it requires careful traffic analysis before installation.

A **dynamic channel allocation (DCA)** scheme has been implemented in DECT (see section 4.2). In this scheme, frequencies can only be borrowed, but it is also possible to freely assign frequencies to cells. With dynamic assignment of frequencies to cells, the danger of interference with cells using the same frequency exists. The 'borrowed' frequency can be blocked in the surrounding cells.

Cellular systems using CDM instead of FDM do not need such elaborate channel allocation schemes and complex frequency planning. Here, users are separated through the code they use, not through the frequency. Cell planning faces another problem – the cell size depends on the current load. Accordingly, **CDM cells** are commonly said to **'breathe'**. While a cell can cover a larger area under a light load, it shrinks if the load increases. The reason for this is the growing noise level if more users are in a cell. (Remember, if you do not know the code, other signals appear as noise, i.e., more and more people join the party.) The higher the noise, the higher the path loss and the higher the transmission errors. Finally, mobile stations further away from the base station drop out of the cell. (This is similar to trying to talk to someone far away at a crowded party.) Figure 2.43 illustrates this phenomenon with a user transmitting a high bit rate stream within a CDM cell. This additional user lets the cell shrink with the result that two users drop out of the cell. In a real-life scenario this additional user could request a video stream (high bit rate) while the others use standard voice communication (low bit rate).

Figure 2.43
Cell breathing depending on the current load

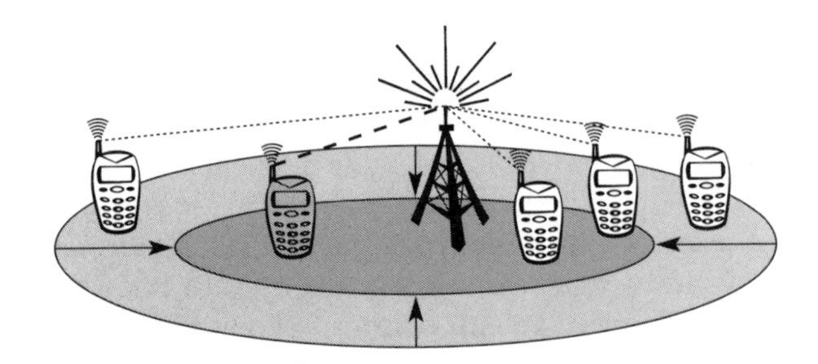

2.9 Summary

This chapter introduced the basics of wireless communications, leaving out most formulae found in books dedicated to wireless transmission and the effects of radio propagation. However, the examples, mechanisms, and problems discussed will hopefully give the reader a good idea as to why wireless communication is fundamentally different from wired communication and why protocols and applications on higher layers have to follow different principles to take the missing wire into account.

A topic of worldwide importance is the regulation and harmonization of frequencies used for radio transmission. The chapter showed many different systems using either different or the same frequencies. Hopefully, the future will bring more frequencies which are available worldwide to avoid more expensive multi-mode devices. At least some harmonization has taken and continues to take place in the area of WLANs (see chapter 7) and 3G mobile phone systems (see chapter 4).

As electromagnetic waves are the basis for wireless communication, antennas are needed for the transmission and reception of waves. While base stations of mobile phone systems often use directed antennas, omni-directional antennas are the choice for mobile devices. On the way from sender to receiver, many things can happen to electromagnetic waves. The standard effects, such as shadowing, fading, reflection, diffraction, and scattering have been presented. All these effects lead to one of the biggest problems in wireless communication: multi-path propagation. Multi-path propagation limits the bandwidth of a channel due to intersymbol interference, i.e., one symbol is 'smeared' into another symbol due to delay spread.

As we only have one 'medium' for wireless transmission, several multiplexing schemes can be applied to raise overall capacity. The standard schemes are SDM, FDM, TDM, and CDM. To achieve FDM, data has to be 'translated' into a signal with a certain carrier frequency. Therefore, two modulation steps can be applied. Digital modulation encodes data into a baseband signal, whereas analog modulation then shifts the centre frequency of the signal up to the radio carrier. Some advanced schemes have been presented that can code many bits into a single phase shift, raising the efficiency.

With the help of spread spectrum technology, several features can be implemented. One is (at least some) security – without knowing the spreading code, the signal appears as noise. As we will see in more detail in chapter 3, spread spectrum lays the basis for special medium access schemes using the code space. Spread spectrum also makes a transmission more robust against narrowband interference, as the signal is spread over a larger bandwidth so, the narrowband interference only influences a small fraction of the signal.

Finally, this chapter has presented the concept of cellular systems. Cellular systems implement SDM to raise the overall capacity of mobile phone systems. While these systems require detailed planning (i.e., matching the cell size with the traffic expected), it presents one of the basic solutions for using the scarce frequency resources efficiently.

2.10 Review exercises

1. Frequency regulations may differ between countries. Check out the regulations valid for your country (within Europe the European Radio Office may be able to help you, www.ero.dk, for the US try the FCC, www.fcc.gov, for Japan ARIB, www.arib.or.jp).
2. Why can waves with a very low frequency follow the earth's surface? Why are they not used for data transmission in computer networks?
3. Why does the ITU-R only regulate 'lower' frequencies (up to some hundred GHz) and not higher frequencies (in the THz range)?

4 What are the two different approaches in regulation regarding mobile phone systems in Europe and the US? What are the consequences?
5 Why is the international availability of the same ISM bands important?
6 Is it possible to transmit a digital signal, e.g., coded as square wave as used inside a computer, using radio transmission without any loss? Why?
7 Is a directional antenna useful for mobile phones? Why? How can the gain of an antenna be improved?
8 What are the main problems of signal propagation? Why do radio waves not always follow a straight line? Why is reflection both useful and harmful?
9 Name several methods for ISI mitigation. How does ISI depend on the carrier frequency, symbol rate, and movement of sender/receiver? What are the influences of ISI on TDM schemes?
10 What are the means to mitigate narrowband interference? What is the complexity of the different solutions?
11 Why, typically, is digital modulation not enough for radio transmission? What are general goals for digital modulation? What are typical schemes?
12 Think of a phase diagram and the points representing bit patterns for a PSK scheme (see Figure 2.29). How can a receiver decide which bit pattern was originally sent when a received 'point' lies somewhere in between other points in the diagram? Why is it difficult to code more and more bits per phase shift?
13 What are the main benefits of a spread spectrum system? How can spreading be achieved? What replaces the guard space in Figure 2.33 when compared to Figure 2.34? How can DSSS systems benefit from multi-path propagation?
14 What are the main reasons for using cellular systems? How is SDM typically realized and combined with FDM? How does DCA influence the frequencies available in other cells?
15 What limits the number of simultaneous users in a TDM/FDM system compared to a CDM system? What happens to the transmission quality of connections if the load gets higher in a cell, i.e., how does an additional user influence the other users in the cell?

2.11 References

Anderson, J.B., Rappaport, T.S., Yoshida, S. (1995) 'Propagation measurements and models for wireless communications channels,' *IEEE Communications Magazine*, 33, (1).

Dixon, R. (1994) *Spread spectrum systems with commercial applications*. John Wiley.

ETSI (1997) *Digital Audio Broadcasting (DAB) to mobile, portable, and fixed receivers*, European Telecommunications Standards Institute, ETS 300 401.

GSM World (2002), GSM Association, http://www.gsmworld.com/.

Halsall, F. (1996) *Data communications, computer networks and open systems*. Addison-Wesley Longman.

Ojanperä, T., Prasad, R. (1998) *Wideband CDMA for Third Generation Mobile Communications*. Artech House.

Pahlavan, K., Krishnamurthy, P. (2002) *Principles of Wireless Network*. Prentice Hall.

Peterson, R., Ziemer, R., Borth, D. (1995) *Introduction to spread spectrum communications*. Prentice Hall.

Stallings, W. (1997) *Data and computer communications*. Prentice Hall.

Stallings, W. (2002) *Wireless Communications and Networking*. Prentice Hall.

Viterbi, A. (1995) *CDMA: Principles of spread spectrum communication*. Addison-Wesley Longman.

Wesel, E. (1998) *Wireless multimedia communications: networking video, voice, and data*. Addison-Wesley Longman.

Medium access control 3

This chapter introduces several **medium access control (MAC)** algorithms which are specifically adapted to the wireless domain. Medium access control comprises all mechanisms that regulate user access to a medium using SDM, TDM, FDM, or CDM. MAC is thus similar to traffic regulations in the highway/multiplexing example introduced in chapter 2. The fact that several vehicles use the same street crossing in TDM, for example, requires rules to avoid collisions; one mechanism to enforce these rules is traffic lights. While the previous chapter mainly introduced mechanisms of the physical layer, layer 1, of the ISO/OSI reference model, MAC belongs to layer 2, the **data link control layer (DLC)**. Layer 2 is subdivided into the **logical link control (LLC)**, layer 2b, and the MAC, layer 2a (Halsall, 1996). The task of DLC is to establish a reliable point to point or point to multi-point connection between different devices over a wired or wireless medium. The basic MAC mechanisms are introduced in the following sections, whereas LLC and higher layers, as well as specific relevant technologies will be presented in later chapters together with mobile and wireless systems.

This chapter aims to explain why special MACs are needed in the wireless domain and why standard MAC schemes known from wired networks often fail. (In contrast to wired networks, hidden and exposed terminals or near and far terminals present serious problems here.) Then, several MAC mechanisms will be presented for the multiplexing schemes introduced in chapter 2. While SDM and FDM are typically used in a rather fixed manner, i.e., a certain space or frequency (or frequency hopping pattern) is assigned for a longer period of time; the main focus of this chapter is on TDM mechanisms. TDM can be used in a very flexible way, as tuning in to a certain frequency does not present a problem, but time can be allocated on demand and in a distributed fashion. Well-known algorithms are Aloha (in several versions), different reservation schemes, or simple polling.

Finally, the use of CDM is discussed again to show how a MAC scheme using CDM has to assign certain codes to allow the separation of different users in code space. This chapter also shows that one typically does not use a single scheme in its pure form but mixes schemes to benefit from the specific advantages. A comparison of the four basic schemes concludes the chapter.

3.1 Motivation for a specialized MAC

The main question in connection with MAC in the wireless is whether it is possible to use elaborated MAC schemes from wired networks, for example, CSMA/CD as used in the original specification of IEEE 802.3 networks (aka Ethernet).

So let us consider **carrier sense multiple access with collision detection, (CSMA/CD)** which works as follows. A sender senses the medium (a wire or coaxial cable) to see if it is free. If the medium is busy, the sender waits until it is free. If the medium is free, the sender starts transmitting data and continues to listen into the medium. If the sender detects a collision while sending, it stops at once and sends a jamming signal.

Why does this scheme fail in wireless networks? CSMA/CD is not really interested in collisions at the sender, but rather in those at the receiver. The signal should reach the receiver without collisions. But the sender is the one detecting collisions. This is not a problem using a wire, as more or less the same signal strength can be assumed all over the wire if the length of the wire stays within certain often standardized limits. If a collision occurs somewhere in the wire, everybody will notice it. It does not matter if a sender listens into the medium to detect a collision at its own location while in reality is waiting to detect a possible collision at the receiver.

The situation is different in wireless networks. As shown in chapter 2, the strength of a signal decreases proportionally to the square of the distance to the sender. Obstacles attenuate the signal even further. The sender may now apply carrier sense and detect an idle medium. The sender starts sending – but a collision happens at the receiver due to a second sender. Section 3.1.1 explains this hidden terminal problem. The same can happen to the collision detection. The sender detects no collision and assumes that the data has been transmitted without errors, but a collision might actually have destroyed the data at the receiver. Collision detection is very difficult in wireless scenarios as the transmission power in the area of the transmitting antenna is several magnitudes higher than the receiving power. So, this very common MAC scheme from wired network fails in a wireless scenario. The following sections show some more scenarios where schemes known from fixed networks fail.

3.1.1 Hidden and exposed terminals

Consider the scenario with three mobile phones as shown in Figure 3.1. The transmission range of A reaches B, but not C (the detection range does not reach C either). The transmission range of C reaches B, but not A. Finally, the transmission range of B reaches A and C, i.e., A cannot detect C and vice versa.

A starts sending to B, C does not receive this transmission. C also wants to send something to B and senses the medium. The medium appears to be free, the carrier sense fails. C also starts sending causing a collision at B. But A cannot detect this collision at B and continues with its transmission. A is **hidden** for C and vice versa.

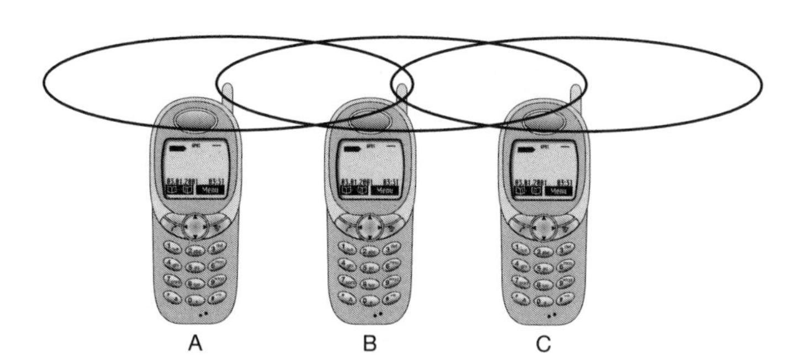

Figure 3.1 Hidden and exposed terminals

While hidden terminals may cause collisions, the next effect only causes unnecessary delay. Now consider the situation that B sends something to A and C wants to transmit data to some other mobile phone outside the interference ranges of A and B. C senses the carrier and detects that the carrier is busy (B's signal). C postpones its transmission until it detects the medium as being idle again. But as A is outside the interference range of C, waiting is not necessary. Causing a 'collision' at B does not matter because the collision is too weak to propagate to A. In this situation, C is **exposed** to B.

3.1.2 Near and far terminals

Consider the situation as shown in Figure 3.2. A and B are both sending with the same transmission power. As the signal strength decreases proportionally to the square of the distance, B's signal drowns out A's signal. As a result, C cannot receive A's transmission.

Now think of C as being an arbiter for sending rights (e.g., C acts as a base station coordinating media access). In this case, terminal B would already drown out terminal A on the physical layer. C in return would have no chance of applying a fair scheme as it would only hear B.

The **near/far effect** is a severe problem of wireless networks using CDM. All signals should arrive at the receiver with more or less the same strength. Otherwise (referring again to the party example of chapter 2) a person standing closer to somebody could always speak louder than a person further away. Even

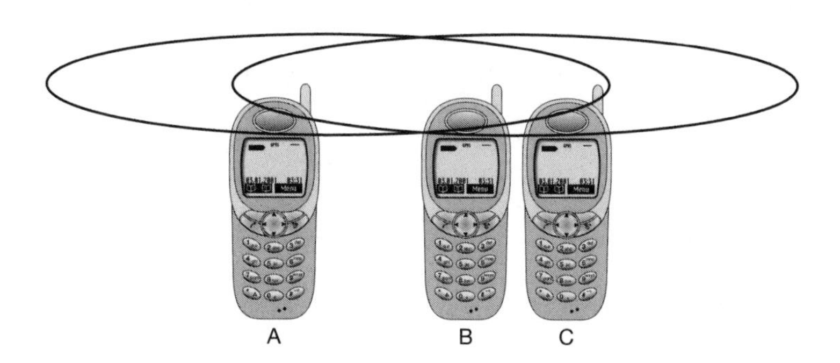

Figure 3.2 Near and far terminals

if the senders were separated by code, the closest one would simply drown out the others. Precise power control is needed to receive all senders with the same strength at a receiver. For example, the UMTS system presented in chapter 4 adapts power 1,500 times per second.

3.2 SDMA

Space Division Multiple Access (SDMA) is used for allocating a separated space to users in wireless networks. A typical application involves assigning an optimal base station to a mobile phone user. The mobile phone may receive several base stations with different quality. A MAC algorithm could now decide which base station is best, taking into account which frequencies (FDM), time slots (TDM) or code (CDM) are still available (depending on the technology). Typically, SDMA is never used in isolation but always in combination with one or more other schemes. The basis for the SDMA algorithm is formed by cells and sectorized antennas which constitute the infrastructure implementing **space division multiplexing (SDM)** (see section 2.5.1). A new application of SDMA comes up together with beam-forming antenna arrays as explained in chapter 2. Single users are separated in space by individual beams. This can improve the overall capacity of a cell (e.g., measured in bit/s/m^2 or voice calls/m^2) tremendously.

3.3 FDMA

Frequency division multiple access (FDMA) comprises all algorithms allocating frequencies to transmission channels according to the **frequency division multiplexing (FDM)** scheme as presented in section 2.5.2. Allocation can either be fixed (as for radio stations or the general planning and regulation of frequencies) or dynamic (i.e., demand driven).

Channels can be assigned to the same frequency at all times, i.e., pure FDMA, or change frequencies according to a certain pattern, i.e., FDMA combined with TDMA. The latter example is the common practice for many wireless systems to circumvent narrowband interference at certain frequencies, known as frequency hopping. Sender and receiver have to agree on a hopping pattern, otherwise the receiver could not tune to the right frequency. Hopping patterns are typically fixed, at least for a longer period. The fact that it is not possible to arbitrarily jump in the frequency space (i.e., the receiver must be able to tune to the right frequency) is one of the main differences between FDM schemes and TDM schemes.

Furthermore, FDM is often used for simultaneous access to the medium by base station and mobile station in cellular networks. Here the two partners typically establish a **duplex channel**, i.e., a channel that allows for simultaneous transmission in both directions. The two directions, mobile station to base station and vice versa are now separated using different frequencies. This scheme is then called **frequency division duplex (FDD)**. Again, both partners have to

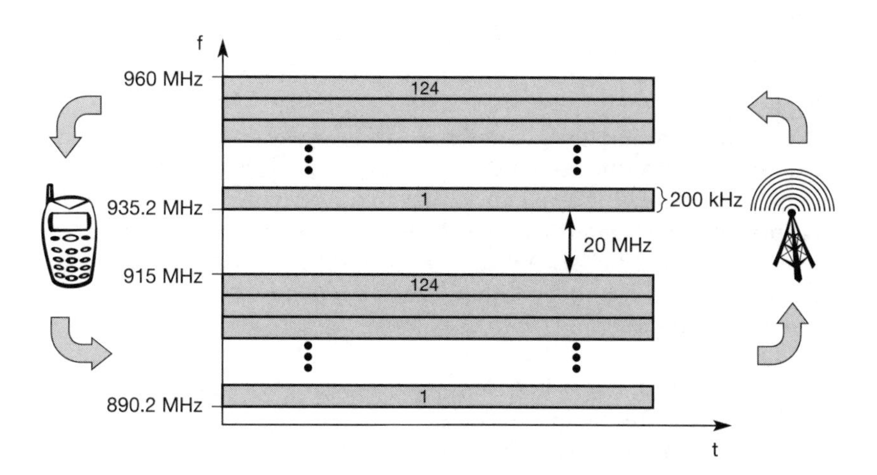

Figure 3.3 Frequency division multiplexing for multiple access and duplex

know the frequencies in advance; they cannot just listen into the medium. The two frequencies are also known as **uplink**, i.e., from mobile station to base station or from ground control to satellite, and as **downlink**, i.e., from base station to mobile station or from satellite to ground control.

As for example FDM and FDD, Figure 3.3 shows the situation in a mobile phone network based on the GSM standard for 900 MHz (see chapter 4). The basic frequency allocation scheme for GSM is fixed and regulated by national authorities. (Certain variations exist regarding the frequencies mentioned in the examples.) All uplinks use the band between 890.2 and 915 MHz, all downlinks use 935.2 to 960 MHz. According to FDMA, the base station, shown on the right side, allocates a certain frequency for up- and downlink to establish a duplex channel with a mobile phone. Up- and downlink have a fixed relation. If the uplink frequency is f_u = 890 MHz + n·0.2 MHz, the downlink frequency is $f_d = f_u$ + 45 MHz, i.e., f_d = 935 MHz + n·0.2 MHz for a certain channel n. The base station selects the channel. Each channel (uplink and downlink) has a bandwidth of 200 kHz. This illustrates the use of FDM for multiple access (124 channels per direction are available at 900 MHz) and duplex according to a predetermined scheme. Similar FDM schemes for FDD are implemented in AMPS, IS-54, IS-95, IS-136, PACS, and UMTS (FDD mode). Chapter 4 presents some more details regarding the combination of this scheme with TDM as implemented in GSM.

3.4 TDMA

Compared to FDMA, **time division multiple access (TDMA)** offers a much more flexible scheme, which comprises all technologies that allocate certain time slots for communication, i.e., controlling **TDM**. Now tuning in to a certain frequency is not necessary, i.e., the receiver can stay at the same frequency the whole time. Using only one frequency, and thus very simple receivers and transmitters, many different algorithms exist to control medium access. As already mentioned, listening to different frequencies at the same time is quite difficult, but listening to many channels separated in time at the same frequency is simple. Almost all MAC schemes for wired networks work according to this principle, e.g., Ethernet, Token Ring, ATM etc. (Halsall, 1996), (Stallings, 1997).

Now synchronization between sender and receiver has to be achieved in the time domain. Again this can be done by using a fixed pattern similar to FDMA techniques, i.e., allocating a certain time slot for a channel, or by using a dynamic allocation scheme. Dynamic allocation schemes require an identification for each transmission as this is the case for typical wired MAC schemes (e.g., sender address) or the transmission has to be announced beforehand. MAC addresses are quite often used as identification. This enables a receiver in a broadcast medium to recognize if it really is the intended receiver of a message. Fixed schemes do not need an identification, but are not as flexible considering varying bandwidth requirements. The following sections present several examples for fixed and dynamic schemes as used for wireless transmission. Typically, those schemes can be combined with FDMA to achieve even greater flexibility and transmission capacity.

3.4.1 Fixed TDM

The simplest algorithm for using TDM is allocating time slots for channels in a fixed pattern. This results in a fixed bandwidth and is the typical solution for wireless phone systems. MAC is quite simple, as the only crucial factor is accessing the reserved time slot at the right moment. If this synchronization is assured, each mobile station knows its turn and no interference will happen. The fixed pattern can be assigned by the base station, where competition between different mobile stations that want to access the medium is solved.

Fixed access patterns (at least fixed for some period in time) fit perfectly well for connections with a fixed bandwidth. Furthermore, these patterns guarantee a fixed delay – one can transmit, e.g., every 10 ms as this is the case for standard DECT systems. TDMA schemes with fixed access patterns are used for many digital mobile phone systems like IS-54, IS-136, GSM, DECT, PHS, and PACS.

Figure 3.4 shows how these fixed TDM patterns are used to implement multiple access and a duplex channel between a base station and mobile station. Assigning different slots for uplink and downlink using the same frequency is called **time division duplex (TDD)**. As shown in the figure, the base station uses one out of 12 slots for the downlink, whereas the mobile station uses one out of 12 different slots for the uplink. Uplink and downlink are separated in time. Up to 12 different mobile stations can use the same frequency without interference using this scheme. Each connection is allotted its own up- and downlink pair. In the example below, which is the standard case for the DECT cordless phone system, the pattern is repeated every 10 ms, i.e., each slot has a duration of 417 μs. This repetition guarantees access to the medium every 10 ms, independent of any other connections.

While the fixed access patterns, as shown for DECT, are perfectly apt for connections with a constant data rate (e.g., classical voice transmission with 32 or 64 kbit/s duplex), they are very inefficient for bursty data or asymmetric connections. If temporary bursts in data are sent from the base station to the

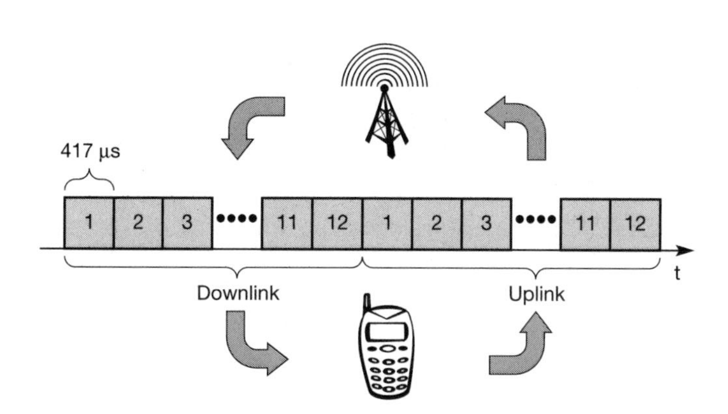

Figure 3.4
Time division multiplexing for multiple access and duplex

mobile station often or vice versa (as in the case of web browsing, where no data transmission occurs while reading a page, whereas clicking on a hyperlink triggers a data transfer from the mobile station, often to the base station, often followed by huge amounts of data returned from the web server). While DECT can at least allocate asymmetric bandwidth (see section 4.2), this general scheme still wastes a lot of bandwidth. It is too static, too inflexible for data communication. In this case, connectionless, demand-oriented TDMA schemes can be used, as the following sections show.

3.4.2 Classical Aloha

As mentioned above, TDMA comprises all mechanisms controlling medium access according to TDM. But what happens if TDM is applied without controlling access? This is exactly what the classical **Aloha** scheme does, a scheme which was invented at the University of Hawaii and was used in the ALOHANET for wireless connection of several stations. Aloha neither coordinates medium access nor does it resolve contention on the MAC layer. Instead, each station can access the medium at any time as shown in Figure 3.5. This is a random access scheme, without a central arbiter controlling access and without coordination among the stations. If two or more stations access the medium at the same time, a **collision** occurs and the transmitted data is destroyed. Resolving this problem is left to higher layers (e.g., retransmission of data).

The simple Aloha works fine for a light load and does not require any complicated access mechanisms. On the classical assumption[1] that data packet arrival follows a Poisson distribution, maximum throughput is achieved for an 18 per cent load (Abramson, 1977), (Halsall, 1996).

1 This assumption is often used for traffic in classical telephone networks but does not hold for today's Internet traffic. Internet traffic is considered as self-similar following – a so-called heavy-tail distribution. An important feature of this distribution is the existence of many values far away from the average. Self-similarity describes the independence of the observed event pattern from the duration of the observation. For example, the interarrival times of www sessions, TCP connection set-ups, IP packets or ATM cells all look similar within their respective timescale (Willinger, 1998a, b).

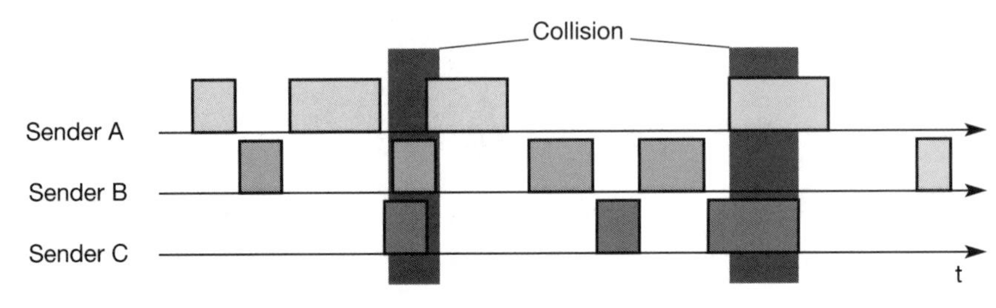

Figure 3.5 Classical Aloha multiple access

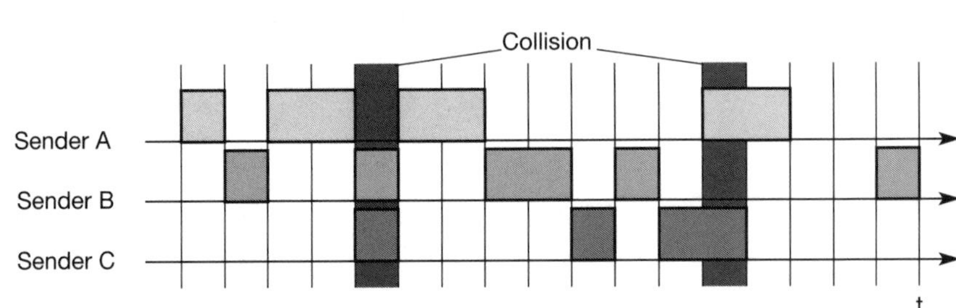

Figure 3.6 Slotted Aloha multiple access

3.4.3 Slotted Aloha

The first refinement of the classical Aloha scheme is provided by the introduction of time slots (**slotted Aloha**). In this case, all senders have to be **synchronized**, transmission can only start at the beginning of a **time slot** as shown in Figure 3.6. Still, access is not coordinated. Under the assumption stated above, the introduction of slots raises the throughput from 18 per cent to 36 per cent, i.e., slotting doubles the throughput.

As we will see in the following sections, both basic Aloha principles occur in many systems that implement distributed access to a medium. Aloha systems work perfectly well under a light load (as most schemes do), but they cannot give any hard transmission guarantees, such as maximum delay before accessing the medium, or minimum throughput. Here one needs additional mechanisms, e.g., combining fixed schemes and Aloha schemes. However, even new mobile communication systems like UMTS have to rely on slotted Aloha for medium access in certain situations (random access for initial connection set-up).

3.4.4 Carrier sense multiple access

One improvement to the basic Aloha is sensing the carrier before accessing the medium. This is what **carrier sense multiple access (CSMA)** schemes generally do (Kleinrock, 1975, Halsall, 1996). Sensing the carrier and accessing the medium only if the carrier is idle decreases the probability of a collision. But, as already mentioned in the introduction, hidden terminals cannot be detected, so, if a hidden terminal transmits at the same time as another sender, a collision might occur at the receiver. This basic scheme is still used in most wireless LANs (this will be explained in more detail in chapter 7).

Several versions of CSMA exist. In **non-persistent CSMA**, stations sense the carrier and start sending immediately if the medium is idle. If the medium is busy, the station pauses a random amount of time before sensing the medium again and repeating this pattern. In **p-persistent CSMA** systems nodes also sense the medium, but only transmit with a probability of p, with the station deferring to the next slot with the probability 1-p, i.e., access is slotted in addition. In **1-persistent CSMA systems**, all stations wishing to transmit access the medium at the same time, as soon as it becomes idle. This will cause many collisions if many stations wish to send and block each other. To create some fairness for stations waiting for a longer time, back-off algorithms can be introduced, which are sensitive to waiting time as this is done for standard Ethernet (Halsall, 1996).

CSMA with collision avoidance (**CSMA/CA**) is one of the access schemes used in wireless LANs following the standard IEEE 802.11. Here sensing the carrier is combined with a back-off scheme in case of a busy medium to achieve some fairness among competing stations. Another, very elaborate scheme is elimination yield – non-preemptive multiple access (**EY-NMPA**) used in the HIPERLAN 1 specification. Here several phases of sensing the medium and accessing the medium for contention resolution are interleaved before one "winner" can finally access the medium for data transmission. Here, priority schemes can be included to assure preference of certain stations with more important data.

3.4.5 Demand assigned multiple access

A general improvement of Aloha access systems can also be achieved by **reservation** mechanisms and combinations with some (fixed) TDM patterns. These schemes typically have a reservation period followed by a transmission period. During the reservation period, stations can reserve future slots in the transmission period. While, depending on the scheme, collisions may occur during the reservation period, the transmission period can then be accessed without collision. Alternatively, the transmission period can be split into periods with and without collision. In general, these schemes cause a higher delay under a light load (first the reservation has to take place), but allow higher throughput due to less collisions.

One basic scheme is **demand assigned multiple access (DAMA)** also called **reservation Aloha**, a scheme typical for satellite systems. DAMA, as shown in Figure 3.7 has two modes. During a contention phase following the slotted Aloha scheme, all stations can try to reserve future slots. For example, different stations on earth try to reserve access time for satellite transmission. Collisions during the reservation phase do not destroy data transmission, but only the short requests for data transmission. If successful, a time slot in the future is reserved, and no other station is allowed to transmit during this slot. Therefore, the satellite collects all successful requests (the others are destroyed) and sends back a reservation list indicating access rights for future slots. All ground stations have to obey this list. To maintain the fixed TDM pattern of reservation and transmission, the stations have to be synchronized from time to time. DAMA is an **explicit reservation** scheme. Each transmission slot has to be reserved explicitly.

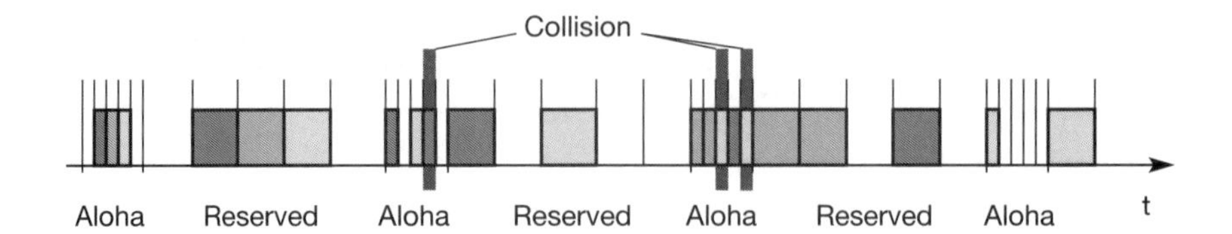

Figure 3.7 Demand assignment multiple access with explicit reservation

3.4.6 PRMA packet reservation multiple access

An example for an **implicit reservation** scheme is **packet reservation multiple access (PRMA)**. Here, slots can be reserved implicitly according to the following scheme. A certain number of slots forms a frame (Figure 3.8 shows eight slots in a frame). The frame is repeated in time (forming frames one to five in the example), i.e., a fixed TDM pattern is applied.

A base station, which could be a satellite, now broadcasts the status of each slot (as shown on the left side of the figure) to all mobile stations. All stations receiving this vector will then know which slot is occupied and which slot is currently free. In the illustration, a successful transmission of data is indicated by the station's name (A to F). In the example, the base station broadcasts the reservation status 'ACDABA-F' to all stations, here A to F. This means that slots one to six and eight are occupied, but slot seven is free in the following transmission. All stations wishing to transmit can now compete for this free slot in Aloha fashion. The already occupied slots are not touched. In the example shown, more than one station wants to access this slot, so a collision occurs. The base station returns the reservation status 'ACDABA-F', indicating that the reservation of slot seven failed (still indicated as free) and that nothing has changed for the other slots. Again, stations can compete for this slot. Additionally, station D has stopped sending in slot three and station F in slot eight. This is noticed by the base station after the second frame.

Before the third frame starts, the base station indicates that slots three and eight are now idle. Station F has succeeded in reserving slot seven as also indicated by the base station. PRMA constitutes yet another combination of fixed

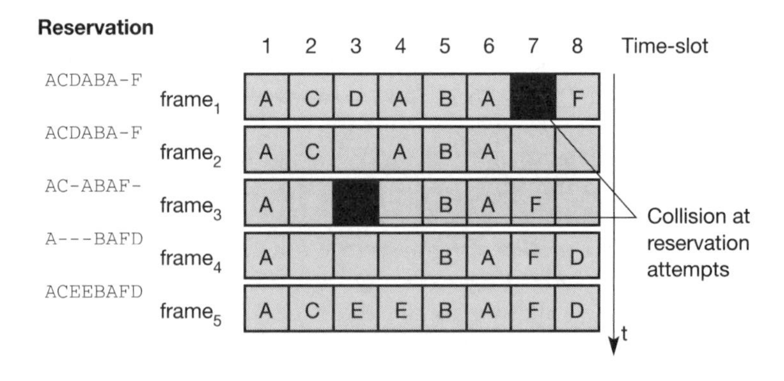

Figure 3.8 Demand assignment multiple access with implicit reservation

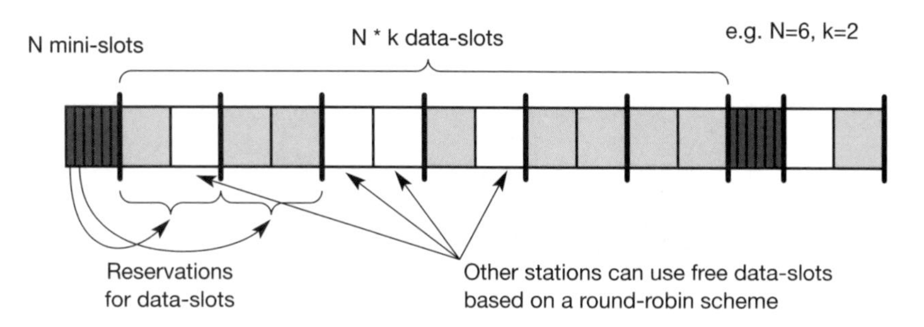

Figure 3.9
Reservation TDMA access scheme

and random TDM schemes with reservation compared to the previous schemes. As soon as a station has succeeded with a reservation, all future slots are implicitly reserved for this station. This ensures transmission with a guaranteed data rate. The slotted aloha scheme is used for idle slots only, data transmission is not destroyed by collision.

3.4.7 Reservation TDMA

An even more fixed pattern that still allows some random access is exhibited by **reservation TDMA** (see Figure 3.9). In a fixed TDM scheme N mini-slots followed by N·k data-slots form a frame that is repeated. Each station is allotted its own mini-slot and can use it to reserve up to k data-slots. This guarantees each station a certain bandwidth and a fixed delay. Other stations can now send data in unused data-slots as shown. Using these free slots can be based on a simple round-robin scheme or can be uncoordinated using an Aloha scheme. This scheme allows for the combination of, e.g., isochronous traffic with fixed bitrates and best-effort traffic without any guarantees.

3.4.8 Multiple access with collision avoidance

Let us go back to one of the initial problems: hidden terminals. How do the previous access schemes solve this? To all schemes with central base stations assigning TDM patterns, the problem of hidden terminals is unknown. If the terminal is hidden for the base station it cannot communicate anyway. But as mentioned above, more or less fixed access patterns are not as flexible as Aloha schemes. What happens when no base station exists at all? This is the case in so-called ad-hoc networks (presented in more detail in chapter 7).

Multiple access with collision avoidance (MACA) presents a simple scheme that solves the hidden terminal problem, does not need a base station, and is still a random access Aloha scheme – but with dynamic reservation. Figure 3.10 shows the same scenario as Figure 3.1 with the hidden terminals. Remember, A and C both want to send to B. A has already started the transmission, but is hidden for C, C also starts with its transmission, thereby causing a collision at B.

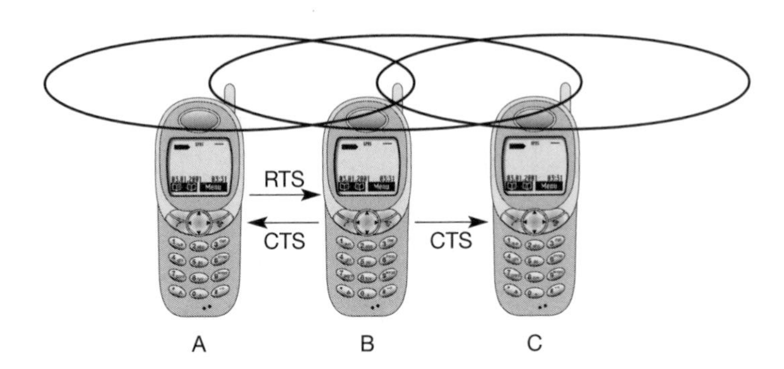

Figure 3.10
MACA can avoid hidden terminals

With MACA, A does not start its transmission at once, but sends a **request to send (RTS)** first. B receives the RTS that contains the name of sender and receiver, as well as the length of the future transmission. This RTS is not heard by C, but triggers an acknowledgement from B, called **clear to send (CTS)**. The CTS again contains the names of sender (A) and receiver (B) of the user data, and the length of the future transmission. This CTS is now heard by C and the medium for future use by A is now reserved for the duration of the transmission. After receiving a CTS, C is not allowed to send anything for the duration indicated in the CTS toward B. A collision cannot occur at B during data transmission, and the hidden terminal problem is solved – provided that the transmission conditions remain the same. (Another station could move into the transmission range of B after the transmission of CTS.)

Still, collisions can occur during the sending of an RTS. Both A and C could send an RTS that collides at B. RTS is very small compared to the data transmission, so the probability of a collision is much lower. B resolves this contention and acknowledges only one station in the CTS (if it was able to recover the RTS at all). No transmission is allowed without an appropriate CTS. This is one of the medium access schemes that is optionally used in the standard IEEE 802.11 (more details can be found in section 7.3).

Can MACA also help to solve the 'exposed terminal' problem? Remember, B wants to send data to A, C to someone else. But C is polite enough to sense the medium before transmitting, sensing a busy medium caused by the transmission from B. C defers, although C could never cause a collision at A.

With MACA, B has to transmit an RTS first (as shown in Figure 3.11) containing the name of the receiver (A) and the sender (B). C does not react to this message as it is not the receiver, but A acknowledges using a CTS which identifies B as the sender and A as the receiver of the following data transmission. C does not receive this CTS and concludes that A is outside the detection range. C can start its transmission assuming it will not cause a collision at A. The problem with exposed terminals is solved without fixed access patterns or a base station. One problem of MACA is clearly the overheads associated with the RTS and CTS transmissions – for short and time-critical data packets, this is

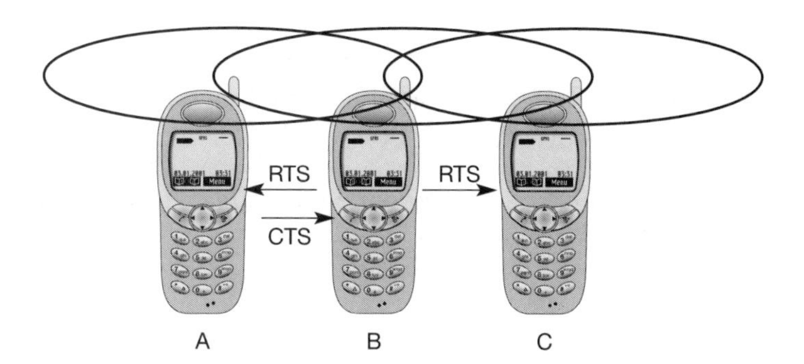

Figure 3.11
MACA can avoid exposed terminals

not negligible. MACA also assumes symmetrical transmission and reception conditions. Otherwise, a strong sender, directed antennas etc. could counteract the above scheme.

Figure 3.12 shows simplified state machines for a sender and receiver. The sender is idle until a user requests the transmission of a data packet. The sender then issues an RTS and waits for the right to send. If the receiver gets an RTS and is in an idle state, it sends back a CTS and waits for data. The sender receives the CTS and sends the data. Otherwise, the sender would send an RTS again after a time-out (e.g., the RTS could be lost or collided). After transmission of the data, the sender waits for a positive acknowledgement to return into an idle state. The receiver sends back a positive acknowledgement if the received data was correct. If not, or if the waiting time for data is too long, the receiver returns into idle state. If the sender does not receive any acknowledgement or a negative acknowledgement, it sends an RTS and again waits for the right to send. Alternatively, a receiver could indicate that it is currently busy via a separate RxBusy. Real implementations have to add more states and transitions, e.g., to limit the number of retries.

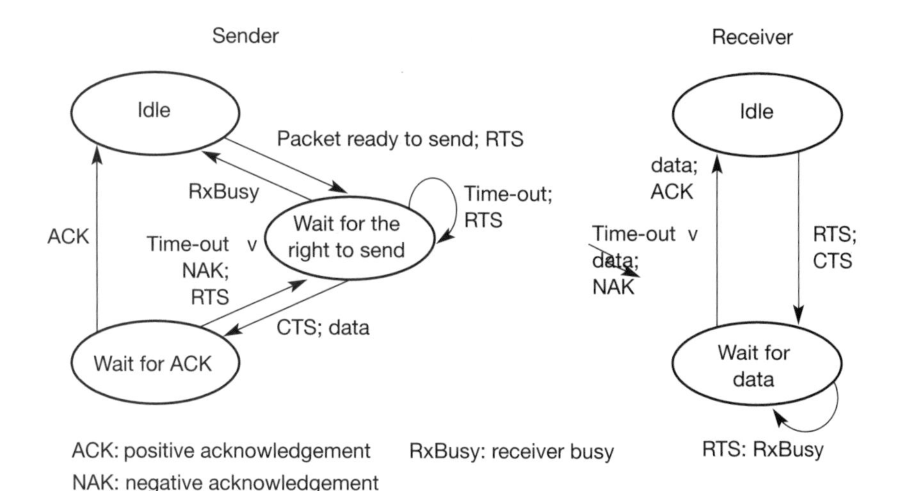

Figure 3.12
Protocol machines for multiple access with collision avoidance

3.4.9 Polling

Where one station is to be heard by all others (e.g., the base station of a mobile phone network or any other dedicated station), **polling** schemes (known from the mainframe/terminal world) can be applied. Polling is a strictly centralized scheme with one master station and several slave stations. The master can poll the slaves according to many schemes: round robin (only efficient if traffic patterns are similar over all stations), randomly, according to reservations (the classroom example with polite students) etc. The master could also establish a list of stations wishing to transmit during a contention phase. After this phase, the station polls each station on the list. Similar schemes are used, e.g., in the Bluetooth wireless LAN and as one possible access function in IEEE 802.11 systems as described in chapter 7.

3.4.10 Inhibit sense multiple access

Another combination of different schemes is represented by **inhibit sense multiple access (ISMA)**. This scheme, which is used for the packet data transmission service Cellular Digital Packet Data (CDPD) in the AMPS mobile phone system, is also known as **digital sense multiple access (DSMA)**. Here, the base station only signals a busy medium via a busy tone (called BUSY/IDLE indicator) on the downlink (see Figure 3.13). After the busy tone stops, accessing the uplink is not coordinated any further. The base station acknowledges successful transmissions, a mobile station detects a collision only via the missing positive acknowledgement. In case of collisions, additional back-off and retransmission mechanisms are implemented. (Salkintzis, 1999)

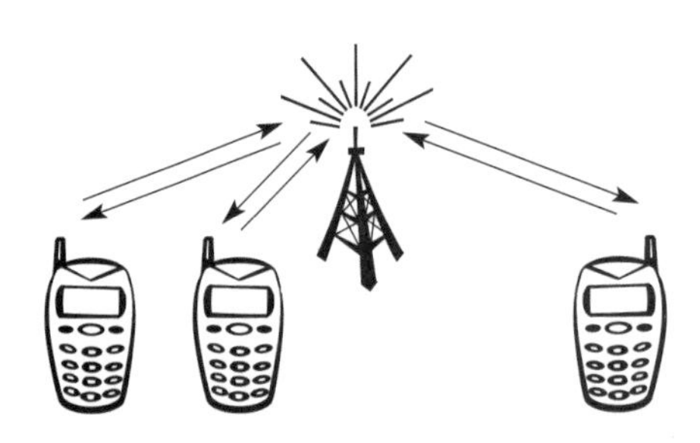

Figure 3.13 Inhibit sense multiple access using a busy tone

3.5 CDMA

Finally, codes with certain characteristics can be applied to the transmission to enable the use of **code division multiplexing (CDM)**. **Code division multiple access (CDMA)** systems use exactly these codes to separate different users in code space and to enable access to a shared medium without interference. The main problem is how to find "good" codes and how to separate the signal from noise generated by other signals and the environment.

Chapter 2 demonstrated how the codes for spreading a signal (e.g., using DSSS) could be used. The code directly controls the chipping sequence. But what is a good code for CDMA? A code for a certain user should have a good autocorrelation[2] and should be **orthogonal** to other codes. Orthogonal in code space has the same meaning as in standard space (i.e., the three dimensional space). Think of a system of coordinates and vectors starting at the origin, i.e., in (0, 0, 0).[3] Two vectors are called orthogonal if their inner product is 0, as is the case for the two vectors (2, 5, 0) and (0, 0, 17): (2, 5, 0)*(0, 0, 17) = 0 + 0 + 0 = 0. But also vectors like (3, –2, 4) and (–2, 3, 3) are orthogonal: (3, –2, 4)*(–2, 3, 3) = –6 – 6 + 12 = 0. By contrast, the vectors (1,2,3) and (4,2, –6) are not orthogonal (the inner product is –10), and (1, 2, 3) and (4, 2, –3) are "almost" orthogonal, with their inner product being –1 (which is "close" to zero). This description is not precise in a mathematical sense. However, it is useful to remember these simplified definitions when looking at the following examples where the original code sequences may be distorted due to noise. Orthogonality cannot be guaranteed for initially orthogonal codes.

Now let us translate this into code space and explain what we mean by a good **autocorrelation**. The Barker code (+1, –1, +1, +1, –1, +1, +1, +1, –1, –1, –1), for example, has a good autocorrelation, i.e., the inner product with itself is large, the result is 11. This code is used for ISDN and IEEE 802.11. But as soon as this Barker code is shifted 1 chip further (think of shifting the 11 chip Barker code over itself concatenated several times), the correlation drops to an absolute value of 1. It stays at this low value until the code matches itself again perfectly. This helps, for example, to synchronize a receiver with the incoming data stream. The peak in the matching process helps the receiver to reconstruct the original data precisely, even if noise distorts the original signal up to a certain level.

After this quick introduction to orthogonality and autocorrelation, the following (theoretical) example explains the basic function of CDMA before it is applied to signals:

- Two senders, A and B, want to send data. CDMA assigns the following unique and orthogonal key sequences: key A_k = 010011 for sender A, key B_K = 110101 for sender B. Sender A wants to send the bit $A_d = 1$, sender B sends $B_d = 0$. To illustrate this example, let us assume that we code a binary 0 as –1, a binary 1 as +1. We can then apply the standard addition and multiplication rules.
- Both senders spread their signal using their key as chipping sequence (the term 'spreading' here refers to the simple multiplication of the data bit with the whole chipping sequence). In reality, parts of a much longer chipping sequence are applied to single bits for spreading. Sender A then sends the signal $A_s = A_d*A_k = +1*(-1, +1, -1, -1, +1, +1) = (-1, +1, -1, -1, +1, +1)$. Sender B does the same with its data to spread the signal with the code: $B_s = B_d*B_k = -1*(+1, +1, -1, +1, -1, +1) = (-1, -1, +1, -1, +1, -1)$.

---

2 The absolute value of the inner product of a vector multiplied with itself should be large. The inner product of two vectors a and b with $a = (a_1, a_2, \ldots, a_n)$ and $b = (b_1, b_2, \ldots, b_n)$ is defined as $a * b = \sum_{i=1}^{n} a_i b_i$.

3 82This example could also be n dimensional.

- Both signals are then transmitted at the same time using the same frequency, so, the signals superimpose in space (analog modulation is neglected in this example). Discounting interference from other senders and environmental noise from this simple example, and assuming that the signals have the same strength at the receiver, the following signal C is received at a receiver: $C = A_s + B_s = (-2, 0, 0, -2, +2, 0)$.
- The receiver now wants to receive data from sender A and, therefore, tunes in to the code of A, i.e., applies A's code for despreading: $C^*A_k = (-2, 0, 0, -2, +2, 0)^*(-1, +1, -1, -1, +1, +1) = 2 + 0 + 0 + 2 + 2 + 0 = 6$. As the result is much larger than 0, the receiver detects a binary 1. Tuning in to sender B, i.e., applying B's code gives $C^*B_k = (-2, 0, 0, -2, +2, 0)^*$ $(+1, +1, -1, +1, -1, +1) = -2 + 0 + 0 - 2 - 2 + 0 = -6$. The result is negative, so a 0 has been detected.

This example involved several simplifications. The codes were extremely simple, but at least orthogonal. More importantly, noise was neglected. Noise would add to the transmitted signal C, the results would not be as even with –6 and +6, but would maybe be close to 0, making it harder to decide if this is still a valid 0 or 1. Additionally, both spread bits were precisely superimposed and both signals are equally strong when they reach the receiver. What would happen if, for example, B was much stronger? Assume that B's strength is five times A's strength. Then, $C' = A_s + 5^*B_s = (-1, +1, -1, -1, +1, +1) + (-5, -5, +5, -5, +5, -5) = (-6, -4, +4, -6, +6, -4)$. Again, a receiver wants to receive B: $C'^*B_k = -6 - 4 - 4 - 6 - 6 - 4 = -30$. It is easy to detect the binary 0 sent by B. Now the receiver wants to receive A: $C'^*A_k = 6 - 4 - 4 + 6 + 6 - 4 = 6$. Clearly, the (absolute) value for the much stronger signal is higher (30 compared to 6). While –30 might still be detected as 0, this is not so easy for the 6 because compared to 30, 6 is quite close to zero and could be interpreted as noise. Remember the party example. If one person speaks in one language very loudly, it is of no more use to have another language as orthogonal code – no one can understand you, your voice will only add to the noise. Although simplified, this example shows that power control is essential for CDMA systems. This is one of the biggest problems CDMA systems face as the power has to be adjusted over one thousand times per second in some systems – this consumes a lot of energy.

The following examples summarize the behaviour of CDMA together with the DSSS spreading using orthogonal codes. The examples now use longer codes or key sequences (i.e., longer as a single bit). Code sequences in IS-95, for example, (a mobile phone system that uses CDMA) are $2^{42} - 1$ chips long, the chipping rate is 1228800 chips/s (i.e., the code repeats after 41.425 days). More details about CDMA can be found in Viterbi (1995).

Figure 3.14 shows a sender A that wants to transmit the bits 101. The key of A is shown as signal and binary key sequence A_k. In this example, the binary "0" is assigned a positive signal value, the binary "1" a negative signal value. After spreading, i.e., XORing A_d and A_k, the resulting signal is A_s.

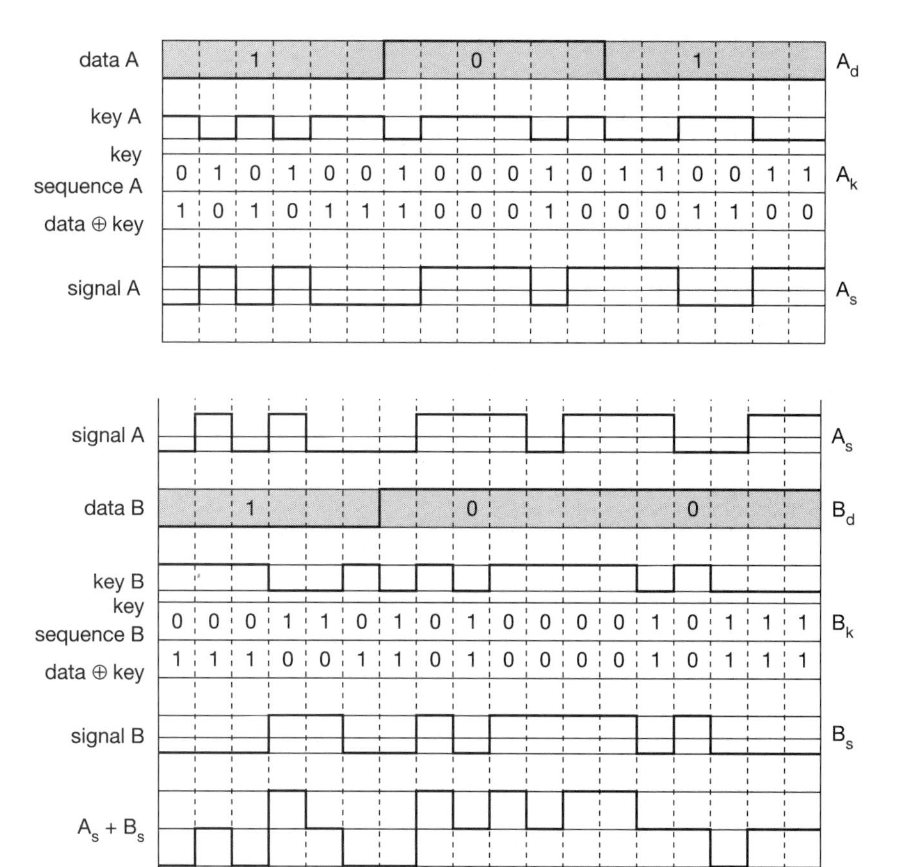

Figure 3.14
Coding and spreading of data from sender A

Figure 3.15
Coding and spreading of data from sender B

The same happens with data from sender B, here the bits are 100. The result of spreading with the code is the signal B_s. A_s and B_s now superimpose during transmission (again without noise and both signals having the same strength). The resulting signal is simply the sum $A_s + B_s$ as shown in Figure 3.15.

A receiver now tries to reconstruct the original data from A, A_d. Therefore the receiver applies A's key, A_k, to the received signal and feeds the result into an integrator (see section 2.7.1). The integrator adds the products (i.e., calculates the inner product), a comparator then has to decide if the result is a 0 or a 1 as shown in Figure 3.16. As we can see, although the original signal form is distorted by B's signal, the result is still quite clear.

The same happens if a receiver wants to receive B's data (see Figure 3.17). The comparator can easily detect the original data. Looking at $(A_s + B_s)*B_k$ one can also imagine what could happen if A's signal was much stronger and noise distorted the signal. The little peaks which are now caused by A's signal would

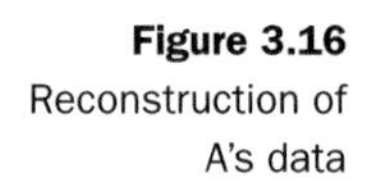

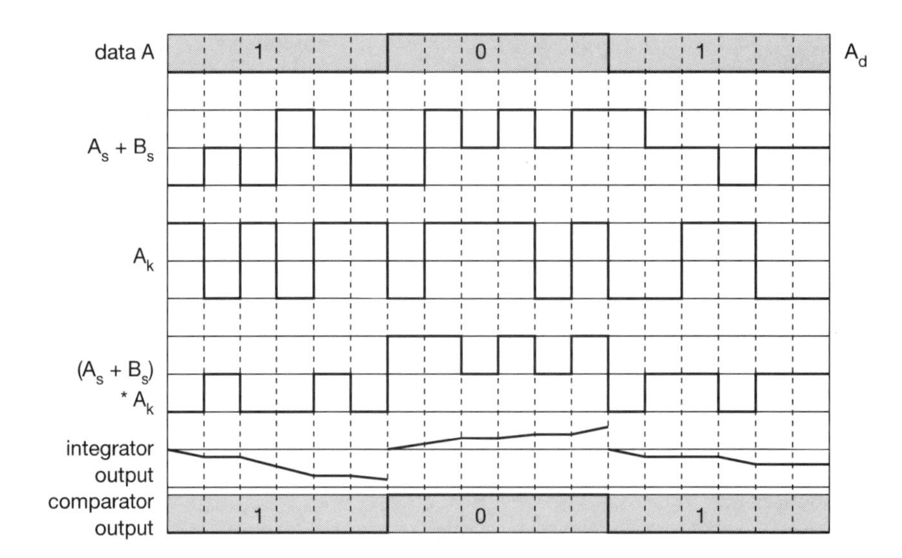

Figure 3.16 Reconstruction of A's data

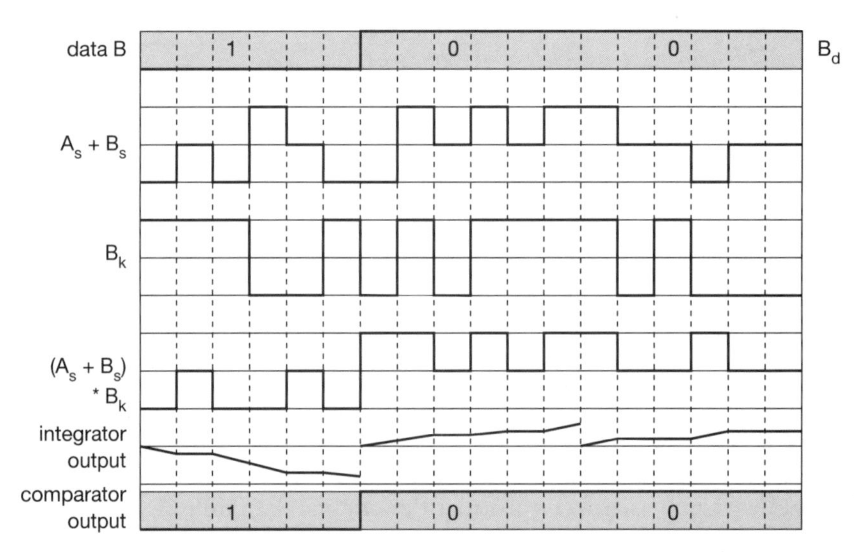

Figure 3.17 Reconstruction of B's data

be much higher, and the result of the integrator would be wrong. If A_k and B_k are perfectly orthogonal and no noise disturbs the transmission, the method works (in theory) for arbitrarily different signal strengths.

Finally, Figure 3.18 shows what happens if a receiver has the wrong key or is not synchronized with the chipping sequence of the transmitter. The integrator still presents a value after each bit period, but now it is not always possible for the comparator to decide for a 1 or a 0, as the signal rather resembles noise. Integrating over noise results in values close to zero. Even if the comparator

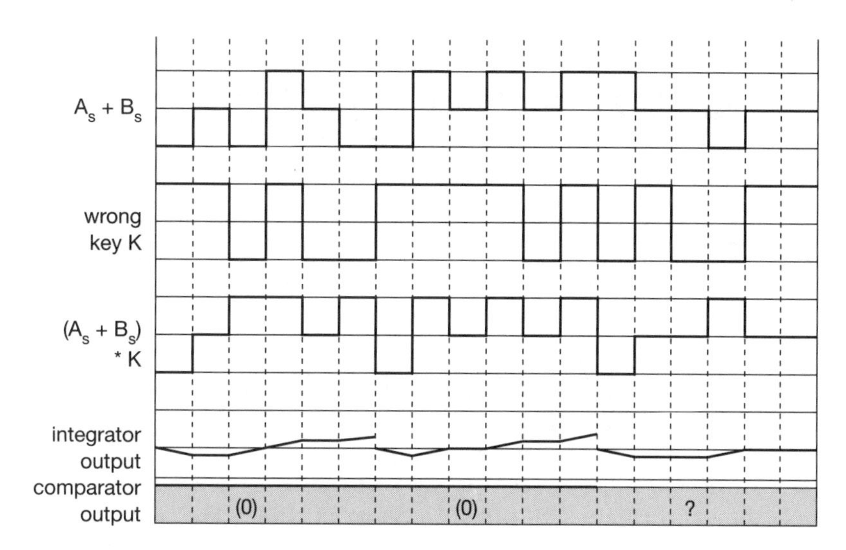

Figure 3.18
Receiving a signal with the wrong key

could detect a clear 1, this could still not reconstruct the whole bit sequence transmitted by a sender. A checksum on layer 2 would detect the erroneous packet. This illustrates CDMA's inherent protection against tapping. It is also the reason for calling the spreading code a key, as it is simultaneously used for encryption on the physical layer.

3.5.1 Spread Aloha multiple access

As shown in the previous section, using different codes with certain properties for spreading data results in a nice and powerful multiple access scheme – namely CDMA. But CDMA senders and receivers are not really simple devices. Communicating with *n* devices requires programming of the receiver to be able to decode *n* different codes (and probably sending with *n* codes, too). For mobile phone systems, a lot of the complexity needed for CDMA is integrated in the base stations. The wireless and mobile devices communicate with the base station only. However, if spontaneous, bursty traffic has to be supported between an arbitrary number of devices, the CDMA technique seems to pose too much overhead. No one wants to program many different spreading codes for, e.g., ad-hoc networks. On the other hand, Aloha was a very simple scheme, but could only provide a relatively low bandwidth due to collisions.

What happens if we combine the spreading of CDMA and the medium access of Aloha or, in other words, what if we use CDMA with only a single code, i.e., without CD? The resulting scheme is called **spread Aloha multiple access (SAMA)** and is a combination of CDMA and TDMA (Abramson, 1996).

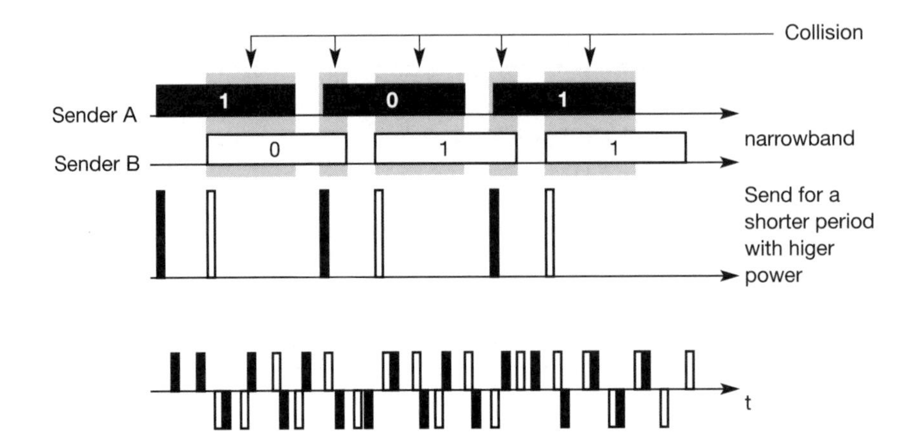

Figure 3.19 Spread Aloha multiple access

SAMA works as follows: each sender uses the same spreading code (in the example shown in Figure 3.19 this is the code 110101).[4] The standard case for Aloha access is shown in the upper part of the figure. Sender A and sender B access the medium at the same time in their narrowband spectrum, so that all three bits shown cause a collision.

The same data could also be sent with higher power for a shorter period as shown in the middle, but now spread spectrum is used to spread the shorter signals, i.e., to increase the bandwidth (spreading factor $s = 6$ in the example). Both signals are spread, but the chipping phase differs slightly. Separation of the two signals is still possible if one receiver is synchronized to sender A and another one to sender B. The signal of an unsynchronized sender appears as noise. The probability of a 'collision' is quite low if the number of simultaneous transmitters stays below 0.1–0.2s (Abramson, 1996). This also depends on the noise level of the environment. The main problem in using this approach is finding good chipping sequences. Clearly, the code is not orthogonal to itself – it should have a good autocorrelation but, at the same time, correlation should be low if the phase differs slightly. The maximum throughput is about 18 per cent, which is very similar to Aloha, but the approach benefits from the advantages of spread spectrum techniques: robustness against narrowband interference and simple coexistence with other systems in the same frequency bands.

4 Clearly, this is not a good code, for it is much too short. Here, coding is only done per bit, a much longer code could also stretch over many bits.

3.6 Comparison of S/T/F/CDMA

To conclude the chapter, a comparison of the four basic multiple access versions is given in Table 3.1. The table shows the MAC schemes without combination with other schemes. However, in real systems, the MAC schemes always occur in combinations. A very typical combination is constituted by SDMA/TDMA/FDMA as used in IS-54, GSM, DECT, PHS, and PACS phone systems, or the Iridium and ICO satellite systems. CDMA together with SDMA is used in the IS-95 mobile phone system and the Globalstar satellite system (see chapters 4 and 5).

Although many network providers and manufacturers have lowered their expectations regarding the performance of CDMA compared to the early 1980s (due to experiences with the IS-95 mobile phone system) CDMA is integrated into almost all third generation mobile phone systems either as W-CDMA (FOMA, UMTS) or cdma2000 (see chapter 4). CDMA can be used in combination with FDMA/TDMA access schemes to increase the capacity of a cell. In contrast to other schemes, CDMA has the advantage of a soft handover and soft capacity. Handover, explained in more detail in chapter 4, describes the switching from one cell to another, i.e., changing the base station that a mobile station is connected to. Soft handover means that a mobile station can smoothly switch cells. This is achieved by communicating with two base stations at the same time. CDMA does this using the same code and the receiver even benefits from both signals. TDMA/FDMA systems perform a hard handover, i.e., they switch base station and hopping sequences (time/frequency) precisely at the moment of handover. Handover decision is based on the signal strength, and oscillations between base stations are possible.

Soft capacity in CDMA systems describes the fact that CDMA systems can add more and more users to a cell, i.e., there is no hard limit. For TDMA/FDMA systems, a hard upper limit exists – if no more free time/frequency slots are available, the system rejects new users. If a new user is added to a CDMA cell, the noise level rises and the cell shrinks, but the user can still communicate. However, the shrinking of a cell can cause problems, as other users could now drop out of it. Cell planning is more difficult in CDMA systems compared to the more fixed TDMA/FDMA schemes (see chapter 2).

While mobile phone systems using SDMA/TDMA/FDMA or SDMA/CDMA are centralized systems – a base station controls many mobile stations – arbitrary wireless communication systems need different MAC algorithms. Most distributed systems use some version of the basic Aloha. Typically, Aloha is slotted and some reservation mechanisms are applied to guarantee access delay and bandwidth. Each of the schemes has advantages and disadvantages. Simple CSMA is very efficient at low load, MACA can overcome the problem of hidden or exposed terminals, and polling guarantees bandwidth. No single scheme combines all benefits, which is why, for example, the wireless LAN standard IEEE 802.11 combines all three schemes (see section 7.3). Polling is used to set up a time structure via a base station. A CSMA version is used to access the medium during uncoordinated periods, and additionally, MACA can be used to avoid hidden terminals or in cases where no base station exists.

Table 3.1 Comparison of SDMA, TDMA, FDMA, and CDMA mechanisms

Approach	SDMA	TDMA	FDMA	CDMA
Idea	Segment space into cells/sectors	Segment sending time into disjoint time-slots, demand driven or fixed patterns	Segment the frequency band into disjoint sub-bands	Spread the spectrum using orthogonal codes
Terminals	Only one terminal can be active in one cell/one sector	All terminals are active for short periods of time on the same frequency	Every terminal has its own frequency, uninterrupted	All terminals can be active at the same place at the same moment, uninterrupted
Signal separation	Cell structure directed antennas	Synchronization in the time domain	Filtering in the frequency domain	Code plus special receivers
Advantages	Very simple, increases capacity per km^2	Established, fully digital, very flexible	Simple, established, robust	Flexible, less planning needed, soft handover
Disadvantages	Inflexible, antennas typically fixed	Guard space needed (multi-path propagation), synchronization difficult	Inflexible, frequencies are a scarce resource	Complex receivers, needs more complicated power control for senders
Comment	Only in combination with TDMA, FDMA or CDMA useful	Standard in fixed networks, together with FDMA/SDMA used in many mobile networks	Typically combined with TDMA (frequency hopping patterns) and SDMA (frequency reuse)	Used in many 3G systems, higher complexity, lowered expectations; integrated with TDMA/FDMA

3.7 Review exercises

1. What is the main physical reason for the failure of many MAC schemes known from wired networks? What is done in wired networks to avoid this effect?
2. Recall the problem of hidden and exposed terminals. What happens in the case of such terminals if Aloha, slotted Aloha, reservation Aloha, or MACA is used?
3. How does the near/far effect influence TDMA systems? What happens in CDMA systems? What are countermeasures in TDMA systems, what about CDMA systems?
4. Who performs the MAC algorithm for SDMA? What could be possible roles of mobile stations, base stations, and planning from the network provider?
5. What is the basic prerequisite for applying FDMA? How does this factor increase complexity compared to TDMA systems? How is MAC distributed if we consider the whole frequency space as presented in chapter 1?
6. Considering duplex channels, what are alternatives for implementation in wireless networks? What about typical wired networks?
7. What are the advantages of a fixed TDM pattern compared to random, demand driven TDM? Compare the efficiency in the case of several connections with fixed data rates or in the case of varying data rates. Now explain why traditional mobile phone systems use fixed patterns, while computer networks generally use random patterns. In the future, the main data being transmitted will be computer-generated data. How will this fact change mobile phone systems?
8. Explain the term interference in the space, time, frequency, and code domain. What are countermeasures in SDMA, TDMA, FDMA, and CDMA systems?
9. Assume all stations can hear all other stations. One station wants to transmit and senses the carrier idle. Why can a collision still occur after the start of transmission?
10. What are benefits of reservation schemes? How are collisions avoided during data transmission, why is the probability of collisions lower compared to classical Aloha? What are disadvantages of reservation schemes?
11. How can MACA still fail in case of hidden/exposed terminals? Think of mobile stations and changing transmission characteristics.
12. Which of the MAC schemes can give hard guarantees related to bandwidth and access delay?
13. How are guard spaces realized between users in CDMA?
14. Redo the simple CDMA example of section 3.5, but now add random 'noise' to the transmitted signal (–2,0,0,–2,+2,0). Add, for example, (1,–1,0,1,0,–1). In this case, what can the receiver detect for sender A and B respectively? Now include the near/far problem. How does this complicate the situation? What would be possible countermeasures?

3.8 References

Abramson, N. (1977) 'The throughput of packet broadcasting channels,' *IEEE Transactions on Communication*, COM-25(1).

Abramson, N. (1996) 'Wideband random access for the last mile,' *IEEE Personal Communications*, 3(6).

Halsall, F. (1996) *Data communications, computer networks and open systems*. Addison-Wesley Longman.

Kleinrock, L., Tobagi, F. (1975) 'Packet switching in radio channels: part 1 – carrier sense multiple-access modes and their throughput-delay characteristics,' *IEEE Transactions on Communications*, COM-23(12).

Salkintzis, A. (1999) 'Packet data over cellular networks: The CDPD approach,' *IEEE Communications Magazine*, 37(6).

Stallings, W. (1997) *Data and computer communications*. Prentice Hall.

Viterbi, A. (1995) *CDMA: principles of spread spectrum communication*. Addison-Wesley Longman.

Willinger, W., Paxson, V. (1998a) 'Where Mathematics meets the Internet,' *Notices of the American Mathematical Society*, 45(8).

Willinger, W, Paxson, V., Taqqu, M. (1998b) 'Self-similarity and Heavy Tails: Structural Modeling of Network Traffic,' *A Practical Guide to Heavy Tails: Statistical Techniques and Applications*. Adler, Taqqu (eds.), Birkhäuser-Verlag, Boston.

4 Telecommunication systems

Digital cellular networks are the segment of the market for mobile and wireless devices which are growing most rapidly. They are the wireless extensions of traditional PSTN or ISDN networks and allow for seamless roaming with the same mobile phone nation or even worldwide. Today, these systems are mainly used for voice traffic. However, data traffic is continuously growing and, therefore, this chapter presents several technologies for wireless data transmission using cellular systems.[1]

The systems presented fit into the traditional telephony architecture and do not originate from computer networks. The basic versions typically implement a circuit-switched service, focused on voice, and only offer data rates of up to, e.g., 9.6 kbit/s. However, service is provided up to a speed of 250 km/h (e.g., using GSM in a car) where most other wireless systems fail.

The **worldwide market** figures for cellular networks are as follows (GSM Association, 2002). The most popular digital system is GSM, with approximately 70 per cent market share. (This system will be presented in section 4.1.) The analog AMPS system still holds three per cent, whereas the Japanese PDC holds five per cent (60 million users). The remainder is split between CDMA (12 per cent) and TDMA (10 per cent) systems, and other technologies. In **Europe** almost everyone uses the digital GSM system (over 370 million) with almost no analog systems left. The situation is different in the **US** and some other countries that have adopted US technology (e.g., South Korea, Canada). Here, the digital market is split into TDMA, CDMA, and GSM systems with 107 million TDMA, 135 million CDMA, and only 16 million GSM users (North America only). While only one digital system exists in Europe, the US market is divided into several systems. This leads to severe problems regarding coverage and service availability, and is one of the examples where market forces did not ensure improved services (compared to the common standard in Europe).

Figure 4.1 shows the worldwide number of subscribers to different mobile phone technologies (GSM Association, 2002). The figure combines different versions of the same technology (e.g., GSM working on 900, 1,800, and 1,900 MHz).

1 All systems presented here are digital, for older analog systems such as the US AMPS (advanced mobile phone system) the reader is referred to, e.g., Goodman (1997).

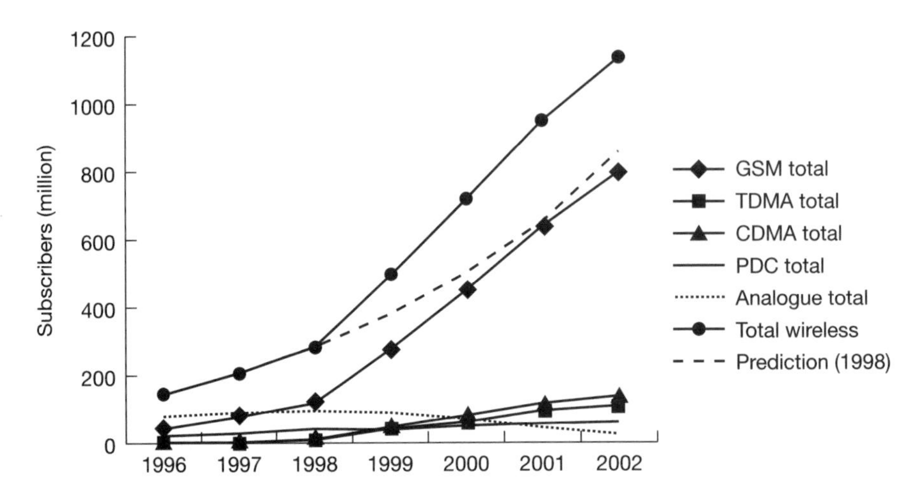

Figure 4.1
Worldwide subscribers of different mobile phone technologies

The two upper lines in the graph show the total number of users and the predictions from 1998. It is interesting that no one foresaw the tremendous success of the mobile communication technology. The graph shows, too, that the time for analog systems is over and GSM is heavily dominating the current market. GSM, TDMA, CDMA, and PDC are all second generation systems. It is important to note that today more people use mobile phone systems than fixed telephones! The graphs of mobile and fixed users crossed in March 2002.

The following sections present prominent examples for second generation (2G) mobile phone networks, cordless telephones, trunked radio systems, and third generation (3G) mobile phone networks. This chapter uses GSM as the main example for a 2G fully digital mobile phone system, not only because of market success, but also due to the system architecture that served many other systems as an early example. Other systems adopted mobility management, mobile assisted handover and other basic ideas (Goodman, 1997), (Stallings, 2002), (Pahlavan, 2002). While US systems typically focus on the air interface for their specification, a system like GSM has many open interfaces and network entities defined in the specification. While the first approach enables companies to have their own, proprietary and possibly better solutions, the latter enables network providers to choose between many different products from different vendors. DECT and TERTA are used, respectively as examples for cordless telephony and trunked radio systems. One reason for this is their system architecture which is similar to GSM. This is not very surprising as all three systems have been standardized by ETSI. The main focus is always on data service, so the evolution of GSM offering higher data rates and packet-oriented transfer is also presented. The chapter concludes with UMTS as a prominent example for 3G mobile telecommunication networks. UMTS is Europe's and Japan's proposal

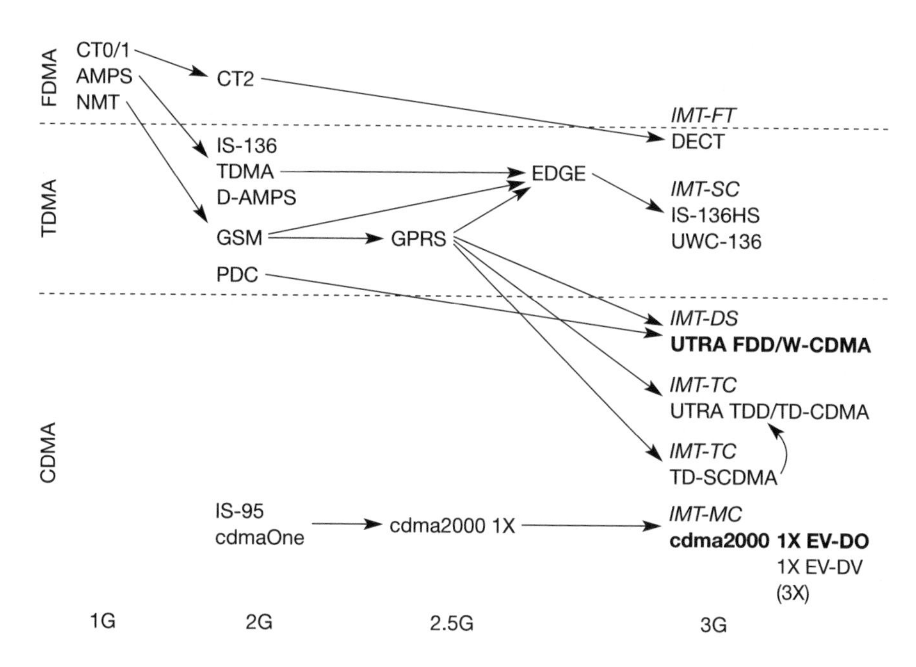

Figure 4.2
Development of different generations of mobile telecommunication systems

for the next generation mobile and wireless system within the ITU IMT-2000 framework. The early phases of UMTS show the evolutionary path from GSM via GSM with higher data rates to UMTS, which allows for saving a lot of investment into the infrastructure. Later phases of UMTS development show more and more the integration of Internet technology that simplifies service creation and offers a migration path to more advanced networks.

Figure 4.2 shows several development and migration paths for different mobile telecommunication systems presented in this chapter. The diagram is divided into the three main multiplexing schemes, FDMA, TDMA, and CDMA. The figure classifies the technologies into three generations. The first generation comprises analog systems, which typically rely on FDMA. The first 2G systems hit the market in the early nineties. In the US **D-AMPS** was a digital successor of **AMPS**, in Europe **GSM** was developed as a replacement for several versions of **NMT**, and **PDC** was introduced in Japan. All these 2G systems introduced a TDMA mechanism in addition to FDMA, which is still used for channel separation. With cdmaOne the first CDMA technology was available in the US as a competitor to the TDMA technologies. Between the second and third generation there is no real revolutionary step. The systems evolved over time: **GPRS** introduced a packet-oriented service and higher data rates to GSM (but can also be used for TDMA systems in general), **EDGE** proposes a new modulation scheme, and **cdmaOne** was enhanced to **cdma2000 1x** offering higher data rates. These systems are often called 2.5G systems.[2] Most, but not all, systems

2 Note that cdma2000 1x, the first version of cdma2000, was not accepted as 3G system by the ITU.

added CDMA technology to become 3G systems. Cordless telephone systems started with CT0 and CT1, became digital with CT2, and ended in Europe in the fully digital standard **DECT**. This standard has even been chosen as one of the candidates for a 3G system (IMT-FT).

While the number of different systems might be confusing, there are some "natural" development paths. Most network providers offering GSM service today will deploy UMTS, while cdmaOne users will choose cdma2000 for simpler migration. The reasons for this are quite simple. With the introduction of GPRS in GSM networks, the core of the network was already enhanced in a way that it can be directly used for UMTS with the radio technologies **UTRA FDD** and **UTRA TDD**. A similar path for evolution exists for **TD-SCDMA**, the Chinese proposal for a 3G system (which has been integrated into UTRA TDD). With some simplification it can be said that UMTS mainly adds a new radio interface but relies in its initial phase on the same core network as GSM/GPRS. Also for cdmaOne the evolution to cdma2000 technologies is quite natural, as the new standard is backward compatible and can reuse frequencies. Cdma2000 1x still uses the same 1.25 MHz channels as cdmaOne does, but offers data rates of up to 153 kbit/s. The **cdma2000 3x** standard uses three 1.25 MHz channels to fit into ITU's frequency scheme for 3G. However, this standard is not pushed as much as the following enhancements of cdma2000 1x. These enhancements are:

- **cdma2000 1x EV-DO** (evolution-data optimized, also known as high data rate (HDR), some call it data only) promising peak data rates of 2.4 Mbit/s using a second 1.25 MHz channel; and
- **cdma2000 1x EV-DV** (evolution-data and voice) aiming at 1.2 Mbit/s for mobile and 5.2 Mbit/s for stationary users.

Cdma2000 1x EV-DO was the first version of cdma2000 accepted by the ITU as 3G system. More information about the technologies and acronyms used in the diagram is provided in the following sections.

4.1 GSM

GSM is the most successful digital mobile telecommunication system in the world today. It is used by over 800 million people in more than 190 countries. In the early 1980s, Europe had numerous coexisting analog mobile phone systems, which were often based on similar standards (e.g., NMT 450), but ran on slightly different carrier frequencies. To avoid this situation for a second generation fully digital system, the **groupe spéciale mobile (GSM)** was founded in 1982. This system was soon named the **global system for mobile communications (GSM)**, with the specification process lying in the hands of ETSI (ETSI, 2002), (GSM Association, 2002). In the context of UMTS and the creation of 3GPP (Third generation partnership project, 3GPP, 2002a) the whole development process of GSM was transferred to 3GPP and further development is combined with 3G development. 3GPP assigned new numbers to all GSM standards. However, to remain consistent with most of the GSM literature, this GSM section stays with the original numbering (see 3GPP, 2002a, for conversion). Section 4.4 will present the ongoing joint specification process in more detail.

The primary goal of GSM was to provide a mobile phone system that allows users to roam throughout Europe and provides voice services compatible to ISDN and other PSTN systems. The specification for the initial system already covers more than 5,000 pages; new services, in particular data services, now add even more specification details. Readers familiar with the ISDN reference model will recognize many similar acronyms, reference points, and interfaces. GSM standardization aims at adopting as much as possible.

GSM is a typical second generation system, replacing the first generation analog systems, but not offering the high worldwide data rates that the third generation systems, such as UMTS, are promising. GSM has initially been deployed in Europe using 890–915 MHz for uplinks and 935–960 MHz for downlinks – this system is now also called **GSM 900** to distinguish it from the later versions. These versions comprise GSM at 1800 MHz (1710–1785 MHz uplink, 1805–1880 MHz downlink), also called **DCS (digital cellular system) 1800**, and the GSM system mainly used in the US at 1900 MHz (1850–1910 MHz uplink, 1930–1990 MHz downlink), also called **PCS (personal communications service) 1900**. Two more versions of GSM exist. **GSM 400** is a proposal to deploy GSM at 450.4–457.6/478.8–486 MHz for uplinks and 460.4–467.6/488.8–496 MHz for downlinks. This system could replace analog systems in sparsely populated areas.

A GSM system that has been introduced in several European countries for railroad systems is **GSM-Rail** (GSM-R, 2002), (ETSI, 2002). This system does not only use separate frequencies but offers many additional services which are unavailable using the public GSM system. GSM-R offers 19 exclusive channels for railroad operators for voice and data traffic (see section 4.1.3 for more information about channels). Special features of this system are, e.g., emergency calls with acknowledgements, voice group call service (VGCS), voice broadcast service (VBS). These so-called advanced speech call items (ASCI) resemble features typically available in trunked radio systems only (see section 4.3). Calls are prioritized: high priority calls pre-empt low priority calls. Calls have very short set-up times: emergency calls less than 2 s, group calls less than 5 s. Calls can be directed for example, to all users at a certain location, all users with a certain function, or all users within a certain number space. However, the most sophisticated use of GSM-R is the control of trains, switches, gates, and signals. Trains going not faster than 160 km/h can control all gates, switches, and signals themselves. If the train goes faster than 160 km/h (many trains are already capable of going faster than 300 km/h) GSM-R can still be used to maintain control.

The following section describes the architecture, services, and protocols of GSM that are common to all three major solutions, **GSM 900**, **GSM 1800**, and **GSM 1900**. GSM has mainly been designed for this and voice services and this still constitutes the main use of GSM systems. However, one can foresee that many future applications for mobile communications will be data driven. The relationship of data to voice traffic will shift more and more towards data.

4.1.1 Mobile services

GSM permits the integration of different voice and data services and the interworking with existing networks. Services make a network interesting for customers. GSM has defined three different categories of services: bearer, tele, and supplementary services. These are described in the following subsections. Figure 4.3 shows a reference model for GSM services. A **mobile station MS** is connected to the **GSM public land mobile network (PLMN)** via the U_m interface. (GSM-PLMN is the infrastructure needed for the GSM network.) This network is connected to transit networks, e.g., **integrated services digital network (ISDN)** or traditional **public switched telephone network (PSTN)**. There might be an additional network, the source/destination network, before another **terminal TE** is connected. **Bearer services** now comprise all services that enable the transparent transmission of data between the interfaces to the network, i.e., S in case of the mobile station, and a similar interface for the other terminal (e.g., S_0 for ISDN terminals). Interfaces like U, S, and R in case of ISDN have not been defined for all networks, so it depends on the specific network which interface is used as a reference for the transparent transmission of data. In the classical GSM model, bearer services are connection-oriented and circuit- or packet-switched. These services only need the lower three layers of the ISO/OSI reference model.

Within the mobile station MS, the **mobile termination (MT)** performs all network specific tasks (TDMA, FDMA, coding etc.) and offers an interface for data transmission (S) to the terminal TE which can then be network independent. Depending on the capabilities of TE, further interfaces may be needed, such as R, according to the ISDN reference model (Halsall, 1996). **Tele services** are application specific and may thus need all seven layers of the ISO/OSI reference model. These services are specified end-to-end, i.e., from one terminal TE to another.

4.1.1.1 Bearer services

GSM specifies different mechanisms for data transmission, the original GSM allowing for data rates of up to 9600 bit/s for non-voice services. Bearer services permit transparent and non-transparent, synchronous or asynchronous data transmission. **Transparent bearer services** only use the functions of the physical layer (layer 1) to transmit data. Data transmission has a constant delay and throughput if no transmission errors occur. The only mechanism to increase

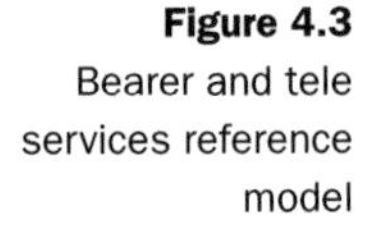
Figure 4.3 Bearer and tele services reference model

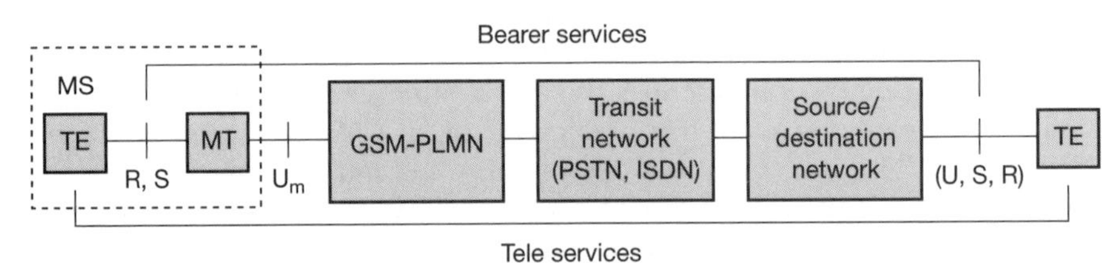

transmission quality is the use of **forward error correction (FEC)**, which codes redundancy into the data stream and helps to reconstruct the original data in case of transmission errors. Depending on the FEC, data rates of 2.4, 4.8, or 9.6 kbit/s are possible. Transparent bearer services do not try to recover lost data in case of, for example, shadowing or interruptions due to handover.

Non-transparent bearer services use protocols of layers two and three to implement error correction and flow control. These services use the transparent bearer services, adding a **radio link protocol (RLP)**. This protocol comprises mechanisms of **high-level data link control (HDLC)**, (Halsall, 1996) and special selective-reject mechanisms to trigger retransmission of erroneous data. The achieved bit error rate is less than 10^{-7}, but now throughput and delay may vary depending on transmission quality.

Using transparent and non-transparent services, GSM specifies several bearer services for interworking with PSTN, ISDN, and packet switched public data networks (PSPDN) like X.25, which is available worldwide. Data transmission can be full-duplex, synchronous with data rates of 1.2, 2.4, 4.8, and 9.6 kbit/s or full-duplex, asynchronous from 300 to 9,600 bit/s (ETSI, 1991a). Clearly, these relatively low data rates reflect the assumption that data services will only constitute some small percentage of the overall traffic. While this is still true of GSM networks today, the relation of data and voice services is changing, with data becoming more and more important. This development is also reflected in the new data services (see section 4.1.8).

4.1.1.2 Tele services

GSM mainly focuses on voice-oriented tele services. These comprise encrypted voice transmission, message services, and basic data communication with terminals as known from the PSTN or ISDN (e.g., fax). However, as the main service is **telephony**, the primary goal of GSM was the provision of high-quality digital voice transmission, offering at least the typical bandwidth of 3.1 kHz of analog phone systems. Special codecs (coder/decoder) are used for voice transmission, while other codecs are used for the transmission of analog data for communication with traditional computer modems used in, e.g., fax machines.

Another service offered by GSM is the **emergency number**. The same number can be used throughout Europe. This service is mandatory for all providers and free of charge. This connection also has the highest priority, possibly pre-empting other connections, and will automatically be set up with the closest emergency center.

A useful service for very simple message transfer is the **short message service (SMS)**, which offers transmission of messages of up to 160 characters. SMS messages do not use the standard data channels of GSM but exploit unused capacity in the signalling channels (see section 4.1.3.1). Sending and receiving of SMS is possible during data or voice transmission. SMS was in the GSM standard from the beginning; however, almost no one used it until millions of young people discovered this service in the mid-nineties as a fun service. SMS can be used for "serious" applications such as displaying road conditions, e-mail headers or stock quotes, but it can also transferr logos, ring tones, horoscopes and love letters. Today more than 30 billion short messages are transferred worldwide per month! SMS is big business today, not only for the network operators, but also for many content providers. It should be noted that SMS is typically the only way to reach a mobile phone from within the network. Thus, SMS is used for updating mobile phone software or for implementing so-called push services (see chapter 10).

The successor of SMS, the **enhanced message service (EMS)**, offers a larger message size (e.g., 760 characters, concatenating several SMs), formatted text, and the transmission of animated pictures, small images and ring tones in a standardized way (some vendors offered similar proprietary features before). EMS never really took off as the **multimedia message service (MMS)** was available. (Nokia never liked EMS but pushed Smart Messaging, a proprietary system.) MMS offers the transmission of larger pictures (GIF, JPG, WBMP), short video clips etc. and comes with mobile phones that integrate small cameras. MMS is further discussed in the context of WAP in chapter 10.

Another non-voice tele service is **group 3 fax**, which is available worldwide. In this service, fax data is transmitted as digital data over the analog telephone network according to the ITU-T standards T.4 and T.30 using modems. Typically, a transparent fax service is used, i.e., fax data and fax signaling is transmitted using a transparent bearer service. Lower transmission quality causes an automatic adaptation of the bearer service to lower data rates and higher redundancy for better FEC.

4.1.1.3 Supplementary services

In addition to tele and bearer services, GSM providers can offer **supplementary services**. Similar to ISDN networks, these services offer various enhancements for the standard telephony service, and may vary from provider to provider. Typical services are user **identification**, call **redirection**, or **forwarding** of ongoing calls. Standard ISDN features such as **closed user groups** and **multi-party** communication may be available. Closed user groups are of special interest to companies because they allow, for example, a company-specific GSM sub-network, to which only members of the group have access.

4.1.2 System architecture

As with all systems in the telecommunication area, GSM comes with a hierarchical, complex system architecture comprising many entities, interfaces, and acronyms. Figure 4.4 gives a simplified overview of the GSM system as specified in ETSI (1991b). A GSM system consists of three subsystems, the **radio sub system (RSS)**, the **network and switching subsystem (NSS)**, and the **operation subsystem (OSS)**. Each subsystem will be discussed in more detail in the following sections. Generally, a GSM customer only notices a very small fraction of the whole network – the mobile stations (MS) and some antenna masts of the base transceiver stations (BTS).

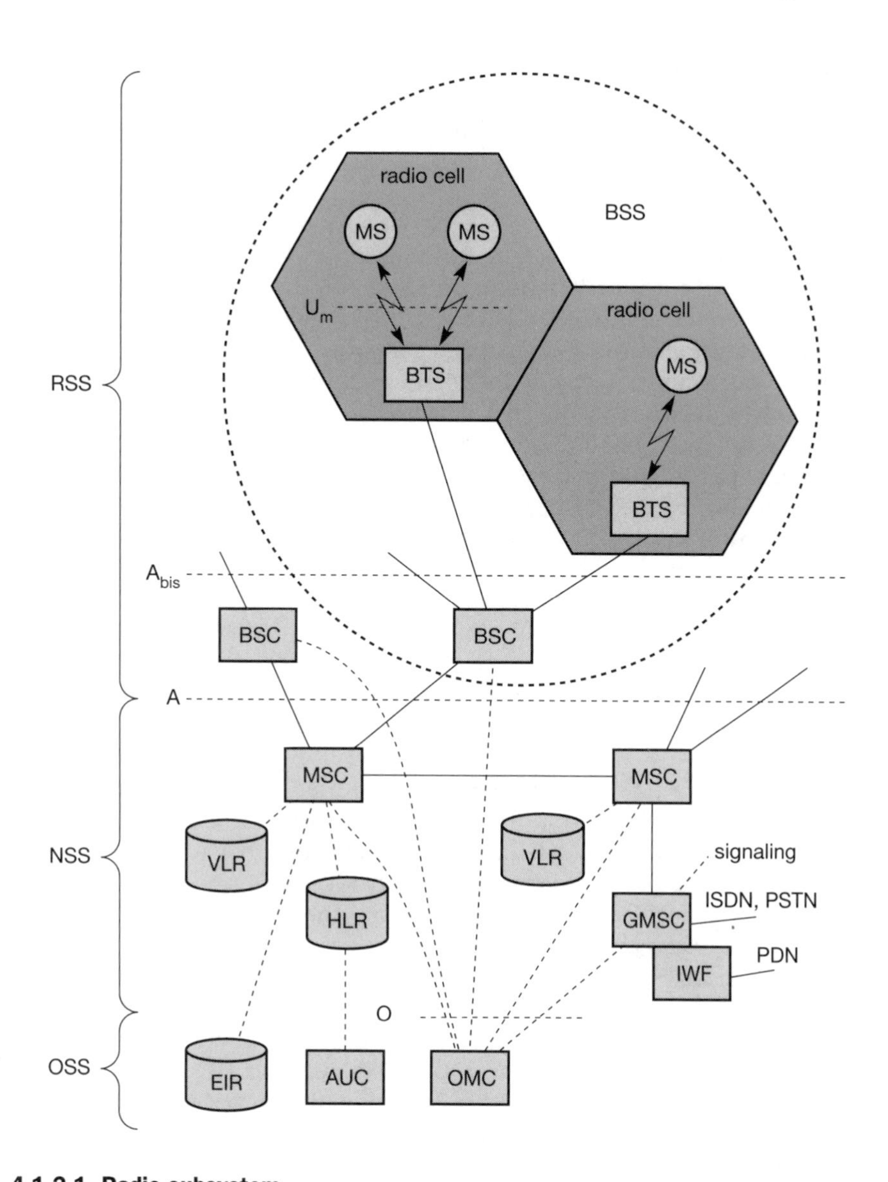

Figure 4.4
Functional architecture of a GSM system

4.1.2.1 Radio subsystem

As the name implies, the **radio subsystem (RSS)** comprises all radio specific entities, i.e., the **mobile stations (MS)** and the **base station subsystem (BSS)**. Figure 4.4 shows the connection between the RSS and the NSS via the **A interface** (solid lines) and the connection to the OSS via the **O interface** (dashed lines). The A interface is typically based on circuit-switched PCM-30 systems (2.048 Mbit/s), carrying up to 30 64 kbit/s connections, whereas the O interface uses the Signalling System No. 7 (SS7) based on X.25 carrying management data to/from the RSS.

- **Base station subsystem (BSS)**: A GSM network comprises many BSSs, each controlled by a base station controller (BSC). The BSS performs all functions necessary to maintain radio connections to an MS, coding/decoding of voice, and rate adaptation to/from the wireless network part. Besides a BSC, the BSS contains several BTSs.
- **Base transceiver station (BTS)**: A BTS comprises all radio equipment, i.e., antennas, signal processing, amplifiers necessary for radio transmission. A BTS can form a radio cell or, using sectorized antennas, several cells (see section 2.8), and is connected to MS via the **U_m interface** (ISDN U interface for mobile use), and to the BSC via the **A_{bis} interface**. The U_m interface contains all the mechanisms necessary for wireless transmission (TDMA, FDMA etc.) and will be discussed in more detail below. The A_{bis} interface consists of 16 or 64 kbit/s connections. A GSM cell can measure between some 100 m and 35 km depending on the environment (buildings, open space, mountains etc.) but also expected traffic.
- **Base station controller (BSC)**: The BSC basically manages the BTSs. It reserves radio frequencies, handles the handover from one BTS to another within the BSS, and performs paging of the MS. The BSC also multiplexes the radio channels onto the fixed network connections at the A interface.

Table 4.1 gives an overview of the tasks assigned to the BSC and BTS or of tasks in which these entities support other entities in the network.

- **Mobile station (MS)**: The MS comprises all user equipment and software needed for communication with a GSM network. An MS consists of user independent hard- and software and of the **subscriber identity module (SIM)**, which stores all user-specific data that is relevant to GSM.[3] While an MS can be identified via the **international mobile equipment identity (IMEI)**, a user can personalize any MS using his or her SIM, i.e., user-specific mechanisms like charging and authentication are based on the SIM, not on the device itself. Device-specific mechanisms, e.g., theft protection, use the device specific IMEI. Without the SIM, only emergency calls are possible. The SIM card contains many identifiers and tables, such as card-type, serial number, a list of subscribed services, a **personal identity number (PIN)**, a **PIN unblocking key (PUK)**, an **authentication key K_i**, and the **international mobile subscriber identity (IMSI)** (ETSI, 1991c). The PIN is used to unlock the MS. Using the wrong PIN three times will lock the SIM. In such cases, the PUK is needed to unlock the SIM. The MS stores dynamic information while logged onto the GSM system, such as, e.g., the **cipher key K_c** and the location information consisting of a **temporary mobile subscriber identity (TMSI)** and the **location area identification (LAI)**. Typical MSs for GSM 900 have a transmit power of up to 2 W, whereas for GSM 1800 1 W is enough due to the smaller cell size. Apart from the telephone interface, an

---

3 Many additional items can be stored on the mobile device. However, this is irrelevant to GSM.

Function	BTS	BSC
Management of radio channels		X
Frequency hopping	X	X
Management of terrestrial channels		X
Mapping of terrestrial onto radio channels		X
Channel coding and decoding	X	
Rate adaptation	X	
Encryption and decryption	X	X
Paging	X	X
Uplink signal measurement	X	
Traffic measurement		X
Authentication		X
Location registry, location update		X
Handover management		X

Table 4.1 Tasks of the BTS and BSC within a BSS

MS can also offer other types of interfaces to users with display, loudspeaker, microphone, and programmable soft keys. Further interfaces comprise computer modems, IrDA, or Bluetooth. Typical MSs, e.g., mobile phones, comprise many more vendor-specific functions and components, such as cameras, fingerprint sensors, calendars, address books, games, and Internet browsers. Personal digital assistants (PDA) with mobile phone functions are also available. The reader should be aware that an MS could also be integrated into a car or be used for location tracking of a container.

4.1.2.2 Network and switching subsystem

The "heart" of the GSM system is formed by the **network and switching subsystem (NSS)**. The NSS connects the wireless network with standard public networks, performs handovers between different BSSs, comprises functions for worldwide localization of users and supports charging, accounting, and roaming of users between different providers in different countries. The NSS consists of the following switches and databases:

- **Mobile services switching center (MSC)**: MSCs are high-performance digital ISDN switches. They set up connections to other MSCs and to the BSCs via the A interface, and form the fixed backbone network of a GSM system. Typically, an MSC manages several BSCs in a geographical region. A **gateway MSC (GMSC)** has additional connections to other fixed networks, such as **PSTN** and **ISDN**. Using additional **interworking functions (IWF)**, an MSC can also connect to **public data networks (PDN)** such as X.25. An MSC handles all signaling needed for connection setup, connection release and handover of connections to other MSCs. The **standard signaling system No. 7 (SS7)** is used for this purpose. SS7 covers all aspects of control signaling for digital networks (reliable routing and delivery of control messages, establishing and monitoring of calls). Features of SS7 are number portability, free phone/toll/collect/credit calls, call forwarding, three-way calling etc. An MSC also performs all functions needed for supplementary services such as call forwarding, multi-party calls, reverse charging etc.

- **Home location register (HLR)**: The HLR is the most important database in a GSM system as it stores all user-relevant information. This comprises static information, such as the **mobile subscriber ISDN number (MSISDN)**, subscribed services (e.g., call forwarding, roaming restrictions, GPRS), and the **international mobile subscriber identity (IMSI)**. Dynamic information is also needed, e.g., the current **location area (LA)** of the MS, the **mobile subscriber roaming number (MSRN)**, the current VLR and MSC. As soon as an MS leaves its current LA, the information in the HLR is updated. This information is necessary to localize a user in the worldwide GSM network. All these user-specific information elements only exist once for each user in a single HLR, which also supports charging and accounting. The parameters will be explained in more detail in section 4.1.5. HLRs can manage data for several million customers and contain highly specialized data bases which must fulfill certain real-time requirements to answer requests within certain time-bounds.
- **Visitor location register (VLR)**: The VLR associated to each MSC is a dynamic database which stores all important information needed for the MS users currently in the LA that is associated to the MSC (e.g., IMSI, MSISDN, HLR address). If a new MS comes into an LA the VLR is responsible for, it copies all relevant information for this user from the HLR. This hierarchy of VLR and HLR avoids frequent HLR updates and long-distance signaling of user information. The typical use of HLR and VLR for user localization will be described in section 4.1.5. Some VLRs in existence, are capable of managing up to one million customers.

4.1.2.3 Operation subsystem

The third part of a GSM system, the **operation subsystem (OSS)**, contains the necessary functions for network operation and maintenance. The OSS possesses network entities of its own and accesses other entities via SS7 signaling (see Figure 4.4). The following entities have been defined:

- **Operation and maintenance center (OMC)**: The OMC monitors and controls all other network entities via the O interface (SS7 with X.25). Typical OMC management functions are traffic monitoring, status reports of network entities, subscriber and security management, or accounting and billing. OMCs use the concept of **telecommunication management network (TMN)** as standardized by the ITU-T.

- **Authentication centre (AuC)**: As the radio interface and mobile stations are particularly vulnerable, a separate AuC has been defined to protect user identity and data transmission. The AuC contains the algorithms for authentication as well as the keys for encryption and generates the values needed for user authentication in the HLR. The AuC may, in fact, be situated in a special protected part of the HLR.
- **Equipment identity register (EIR)**: The EIR is a database for all IMEIs, i.e., it stores all device identifications registered for this network. As MSs are mobile, they can be easily stolen. With a valid SIM, anyone could use the stolen MS. The EIR has a blacklist of stolen (or locked) devices. In theory an MS is useless as soon as the owner has reported a theft. Unfortunately, the blacklists of different providers are not usually synchronized and the illegal use of a device in another operator's network is possible (the reader may speculate as to why this is the case). The EIR also contains a list of valid IMEIs (white list), and a list of malfunctioning devices (gray list).

4.1.3 Radio interface

The most interesting interface in a GSM system is U_m, the radio interface, as it comprises many mechanisms presented in chapters 2 and 3 for multiplexing and media access. GSM implements SDMA using cells with BTS and assigns an MS to a BTS. Furthermore, FDD is used to separate downlink and uplink as shown in Figures 3.3 and 4.5. Media access combines TDMA and FDMA. In GSM 900, 124 channels, each 200 kHz wide, are used for FDMA, whereas GSM 1800 uses, 374 channels. Due to technical reasons, channels 1 and 124 are not used for transmission in GSM 900. Typically, 32 channels are reserved for organizational data; the remaining 90 are used for customers. Each BTS then manages a single channel for organizational data and, e.g., up to 10 channels for user data. The following example is based on the GSM 900 system, but GSM works in a similar way at 1800 and 1900 MHz.

While Figure 3.3 in chapter 3 has already shown the FDM in GSM, Figure 4.5 also shows the TDM used. Each of the 248 channels is additionally separated in time via a **GSM TDMA frame**, i.e., each 200 kHz carrier is subdivided into frames that are repeated continuously. The duration of a frame is 4.615 ms. A frame is again subdivided into 8 **GSM time slots**, where each slot represents a physical TDM channel and lasts for 577 μs. Each TDM channel occupies the 200 kHz carrier for 577 μs every 4.615 ms.

Data is transmitted in small portions, called **bursts**. Figure 4.5 shows a so-called **normal burst** as used for data transmission inside a time slot (user and signaling data). In the diagram, the burst is only 546.5 μs long and contains 148 bits. The remaining 30.5 μs are used as **guard space** to avoid overlapping with other bursts due to different path delays and to give the transmitter time to turn on and off. Filling the whole slot with data allows for the transmission of

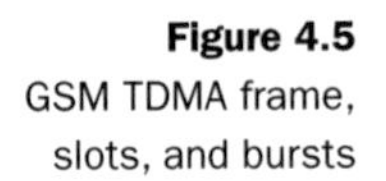

Figure 4.5
GSM TDMA frame, slots, and bursts

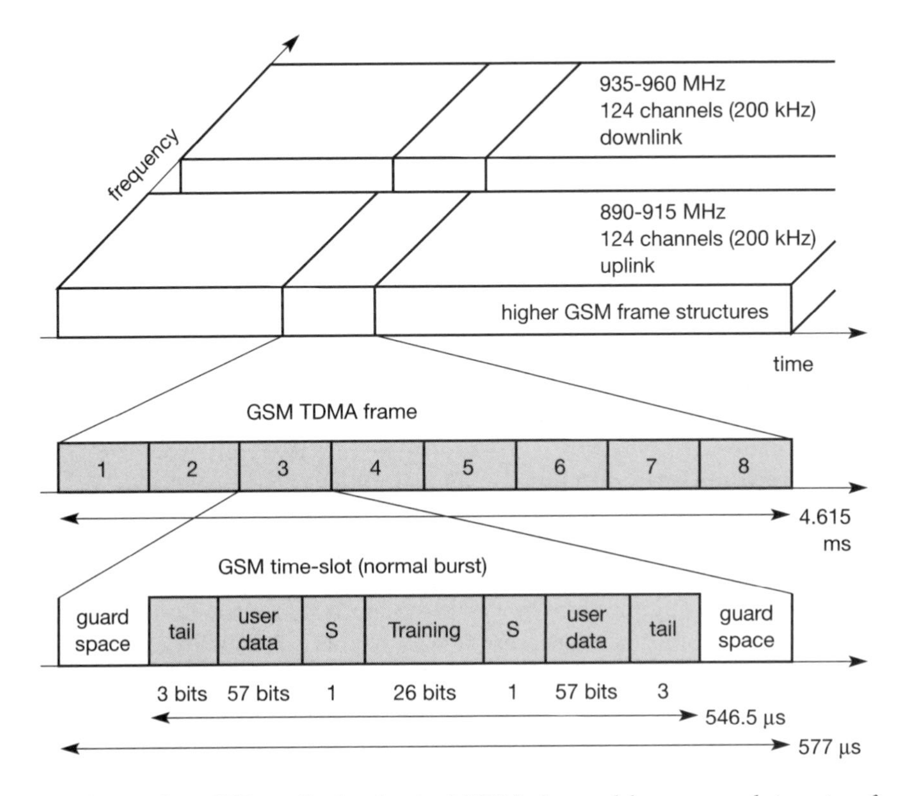

156.25 bit within 577 μs. Each physical TDM channel has a raw data rate of about 33.8 kbit/s, each radio carrier transmits approximately 270 kbit/s over the U_m interface.

The first and last three bits of a normal burst (**tail**) are all set to 0 and can be used to enhance the receiver performance. The **training** sequence in the middle of a slot is used to adapt the parameters of the receiver to the current path propagation characteristics and to select the strongest signal in case of multi-path propagation. A flag **S** indicates whether the **data** field contains user or network control data. Apart from the normal burst, ETSI (1993a) defines four more bursts for data transmission: a **frequency correction** burst allows the MS to correct the local oscillator to avoid interference with neighboring channels, a **synchronization burst** with an extended training sequence synchronizes the MS with the BTS in time, an **access burst** is used for the initial connection setup between MS and BTS, and finally a **dummy burst** is used if no data is available for a slot.

Two factors allow for the use of simple transmitter hardware: on the one hand, the slots for uplink and downlink of a physical TDM channel are separated in frequency (45 MHz for GSM 900, 95 MHz for GSM 1800 using FDD). On the other hand, the TDMA frames are shifted in time for three slots, i.e., if the BTS sends data at time t_0 in slot one on the downlink, the MS accesses slot one on the uplink at time t_0+3·577 μs. An MS does not need a full-duplex transmitter, a simpler half-duplex transmitter switching between receiving and sending is enough.

To avoid frequency selective fading, GSM specifies an optional **slow frequency hopping** mechanism. MS and BTS may change the carrier frequency after each frame based on a common hopping sequence. An MS changes its frequency between up and downlink slots respectively.

4.1.3.1 Logical channels and frame hierarchy

While the previous section showed the physical separation of the medium into 8*124 duplex channels, this section presents logical channels and a hierarchy of frames based on the combination of these physical channels. A physical channel consists of a slot, repeated every 4.615 ms. Think of a logical channel C_1 that only takes up every fourth slot and another logical channel C_2 that uses every other slot. Both logical channels could use the same physical channel with the pattern $C_1C_2xC_2C_1C_2xC_2C_1$ etc. (The x indicates that the physical channel still has some capacity left.)

GSM specifies two basic groups of logical channels, i.e., traffic channels and control channels:[4]

- **Traffic channels (TCH)**: GSM uses a TCH to transmit user data (e.g., voice, fax). Two basic categories of TCHs have been defined, i.e., **full-rate TCH (TCH/F)** and **half-rate TCH (TCH/H)**. A TCH/F has a data rate of 22.8 kbit/s, whereas TCH/H only has 11.4 kbit/s. With the voice codecs available at the beginning of the GSM standardization, 13 kbit/s were required, whereas the remaining capacity of the TCH/F (22.8 kbit/s) was used for error correction (**TCH/FS**). Improved codes allow for better voice coding and can use a TCH/H. Using these TCH/HSs doubles the capacity of the GSM system for voice transmission. However, speech quality decreases with the use of TCH/HS and many providers try to avoid using them. The standard codecs for voice are called **full rate** (FR, 13 kbit/s) and **half rate** (HR, 5.6 kbit/s). A newer codec, **enhanced full rate** (EFR), provides better voice quality than FR as long as the transmission error rate is low. The generated data rate is only 12.2 kbit/s. New codecs, which automatically choose the best mode of operation depending on the error rate (AMR, adaptive multi-rate), will be used together with 3G systems. An additional increase in voice quality is provided by the so-called **tandem free operation (TFO)**. This mode can be used if two MSs exchange voice data. In this case, coding to and from PCM encoded voice (standard in ISDN) can be skipped and the GSM encoded voice data is directly exchanged. Data transmission in GSM is possible at many different data rates, e.g., **TCH/F4.8** for 4.8 kbit/s, **TCH/F9.6** for 9.6 kbit/s, and, as a newer specification, **TCH/F14.4** for 14.4 kbit/s. These logical channels differ in terms of their coding schemes and error correction capabilities.

---

4 More information about channels can be found in Goodman (1997) and ETSI (1993a).

- **Control channels (CCH)**: Many different CCHs are used in a GSM system to control medium access, allocation of traffic channels or mobility management. Three groups of control channels have been defined, each again with subchannels (maybe you can imagine why the initial specification already needed over 5,000 pages):
  - **Broadcast control channel (BCCH)**: A BTS uses this channel to signal information to all MSs within a cell. Information transmitted in this channel is, e.g., the cell identifier, options available within this cell (frequency hopping), and frequencies available inside the cell and in neighboring cells. The BTS sends information for frequency correction via the **frequency correction channel (FCCH)** and information about time synchronization via the **synchronization channel (SCH)**, where both channels are subchannels of the BCCH.
  - **Common control channel (CCCH)**: All information regarding connection setup between MS and BS is exchanged via the CCCH. For calls toward an MS, the BTS uses the **paging channel (PCH)** for paging the appropriate MS. If an MS wants to set up a call, it uses the **random access channel (RACH)** to send data to the BTS. The RACH implements multiple access (all MSs within a cell may access this channel) using slotted Aloha. This is where a collision may occur with other MSs in a GSM system. The BTS uses the **access grant channel (AGCH)** to signal an MS that it can use a TCH or SDCCH for further connection setup.
  - **Dedicated control channel (DCCH)**: While the previous channels have all been unidirectional, the following channels are bidirectional. As long as an MS has not established a TCH with the BTS, it uses the **stand-alone dedicated control channel (SDCCH)** with a low data rate (782 bit/s) for signaling. This can comprise authentication, registration or other data needed for setting up a TCH. Each TCH and SDCCH has a **slow associated dedicated control channel (SACCH)** associated with it, which is used to exchange system information, such as the channel quality and signal power level. Finally, if more signaling information needs to be transmitted and a TCH already exists, GSM uses a **fast associated dedicated control channel (FACCH)**. The FACCH uses the time slots which are otherwise used by the TCH. This is necessary in the case of handovers where BTS and MS have to exchange larger amounts of data in less time.

However, these channels cannot use time slots arbitrarily – GSM specifies a very elaborate multiplexing scheme that integrates several hierarchies of frames. If we take a simple TCH/F for user data transmission, each TCH/F will have an associated SACCH for slow signaling. If fast signaling is required, the FACCH uses the time slots for the TCH/F. A typical usage pattern of a physical channel for data transmission now looks like this (with T indicating the user traffic in the TCH/F and S indicating the signalling traffic in the SACCH):

TTTTTTTTTTTTSTTTTTTTTTTTTx

TTTTTTTTTTTTSTTTTTTTTTTTTx

Twelve slots with user data are followed by a signalling slot. Again 12 slots with user data follow, then an unused slot. This pattern of 26 slots is repeated over and over again. In this case, only 24 out of 26 physical slots are used for the TCH/F. Now recall that each normal burst used for data transmission carries 114 bit user data and is repeated every 4.615 ms. This results in a data rate of 24.7 kbit/s. As the TCH/F only uses 24/26 of the slots, the final data rate is 22.8 kbit/s as specified for the TCH/F. The SACCH thus has a capacity of 950 bit/s.

This periodic pattern of 26 slots occurs in all TDMA frames with a TCH. The combination of these frames is called **traffic multiframe**. Figure 4.6 shows the logical combination of 26 frames (TDMA frames with a duration of 4.615 ms) to a multiframe with a duration of 120 ms. This type of multiframe is used for TCHs, SACCHs for TCHs, and FACCHs. As these logical channels are all associated with user traffic, the multiframe is called traffic multiframe. TDMA frames containing (signaling) data for the other logical channels are combined to a **control multiframe**. Control multiframes consist of 51 TDMA frames and have a duration of 235.4 ms.

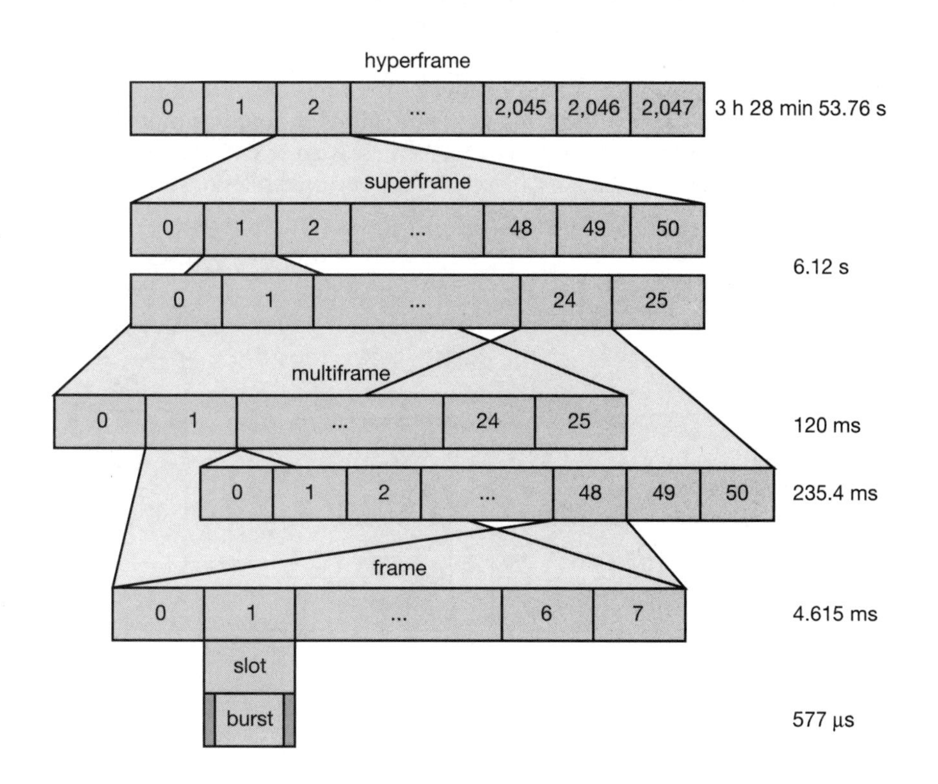

Figure 4.6
GSM structuring of time using a frame hierarchy

This logical frame hierarchy continues, combining 26 multiframes with 51 frames or 51 multiframes with 26 frames to form a **superframe**. 2,048 superframes build a **hyperframe** with a duration of almost 3.5 hours. Altogether, 2,715,648 TDMA frames form a hyperframe. This large logical structure is needed for encryption – GSM counts each TDMA frame, with the frame number forming input for the encryption algorithm. The frame number plus the slot number uniquely identify each time slot in GSM.

4.1.4 Protocols

Figure 4.7 shows the protocol architecture of GSM with signaling protocols, interfaces, as well as the entities already shown in Figure 4.4. The main interest lies in the U_m interface, as the other interfaces occur between entities in a fixed network. **Layer 1**, the physical layer, handles all **radio**-specific functions. This includes the creation of bursts according to the five different formats, **multiplexing** of bursts into a TDMA frame, **synchronization** with the BTS, detection of idle channels, and measurement of the **channel quality** on the downlink. The physical layer at U_m uses GMSK for digital **modulation** and performs **encryption/decryption** of data, i.e., encryption is not performed end-to-end, but only between MS and BSS over the air interface.

Synchronization also includes the correction of the individual path delay between an MS and the BTS. All MSs within a cell use the same BTS and thus must be synchronized to this BTS. The BTS generates the time-structure of frames, slots etc. A problematic aspect in this context are the different round trip times (RTT). An MS close to the BTS has a very short RTT, whereas an MS 35 km away already exhibits an RTT of around 0.23 ms. If the MS far away used the slot structure with-

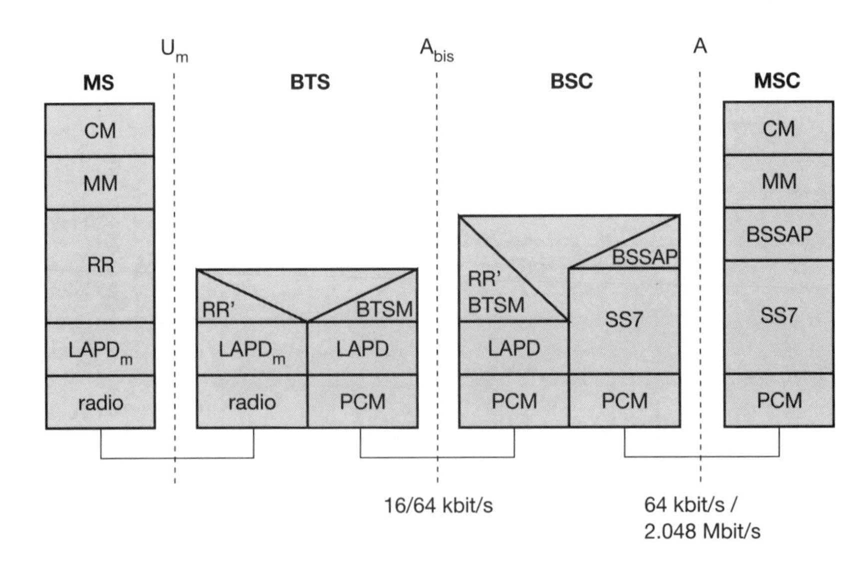

Figure 4.7
Protocol architecture for signaling

out correction, large guard spaces would be required, as 0.23 ms are already 40 per cent of the 0.577 ms available for each slot. Therefore, the BTS sends the current RTT to the MS, which then adjusts its access time so that all bursts reach the BTS within their limits. This mechanism reduces the guard space to only 30.5 μs or five per cent (see Figure 4.5). Adjusting the access is controlled via the variable **timing advance**, where a burst can be shifted up to 63 bit times earlier, with each bit having a duration of 3.69 μs (which results in the 0.23 ms needed). As the variable timing advance cannot be extended a burst cannot be shifted earlier than 63 bit times. This results in the 35 km maximum distance between an MS and a BTS. It might be possible to receive the signals over longer distances; to avoid collisions at the BTS, access cannot be allowed.[5]

The main tasks of the physical layer comprise **channel coding** and **error detection/correction**, which is directly combined with the coding mechanisms. Channel coding makes extensive use of different **forward error correction (FEC)** schemes. FEC adds redundancy to user data, allowing for the detection and correction of selected errors. The power of an FEC scheme depends on the amount of redundancy, coding algorithm and further interleaving of data to minimize the effects of burst errors. The FEC is also the reason why error detection and correction occurs in layer one and not in layer two as in the ISO/OSI reference model. The GSM physical layer tries to correct errors, but it does not deliver erroneous data to the higher layer.

Different logical channels of GSM use different coding schemes with different correction capabilities. Speech channels need additional coding of voice data after analog to digital conversion, to achieve a data rate of 22.8 kbit/s (using the 13 kbit/s from the voice codec plus redundancy, CRC bits, and interleaving (Goodman, 1997). As voice was assumed to be the main service in GSM, the physical layer also contains special functions, such as **voice activity detection (VAD)**, which transmits voice data only when there is a voice signal. This mechanism helps to decrease interference as a channel might be silent approximately 60 per cent of the time (under the assumption that only one person speaks at the same time and some extra time is needed to switch between the speakers). During periods of silence (e.g., if a user needs time to think before talking), the physical layer generates a **comfort noise** to fake a connection (complete silence would probably confuse a user), but no actual transmission takes place. The noise is even adapted to the current background noise at the communication partner's location.

All this interleaving of data for a channel to minimize interference due to burst errors and the recurrence pattern of a logical channel generates a **delay** for transmission. The delay is about 60 ms for a TCH/FS and 100 ms for a TCH/F9.6

5 A special trick allows for larger cells. If the timing advance for MSs that are further away than 35 km is set to zero, the bursts arriving from these MSs will fall into the following time slot. Reception of data is simply shifted one time slot and again the timing advance may be used up to a distance of 70 km (under simplified assumptions). Using this special trick, the capacity of a cell is decreased (near and far MSs cannot be mixed arbitrarily), but coverage of GSM is extended. Network operators may choose this approach, e.g., in coastal regions.

(within 100 ms signals in fixed networks easily travel around the globe). These times have to be added to the transmission delay if communicating with an MS instead of a standard fixed station (telephone, computer etc.) and may influence the performance of any higher layer protocols, e.g., for computer data transmission (see chapter 9).

Signaling between entities in a GSM network requires higher layers (see Figure 4.7). For this purpose, the **LAPD$_m$** protocol has been defined at the U_m interface for **layer two**. LAPD$_m$, as the name already implies, has been derived from link access procedure for the D-channel (**LAPD**) in ISDN systems, which is a version of HDLC (Goodman, 1997), (Halsall, 1996). LAPD$_m$ is a lightweight LAPD because it does not need synchronization flags or checksumming for error detection. (The GSM physical layer already performs these tasks.) LAPD$_m$ offers reliable data transfer over connections, re-sequencing of data frames, and flow control (ETSI, 1993b), (ETSI, 1993c). As there is no buffering between layer one and two, LAPD$_m$ has to obey the frame structures, recurrence patterns etc. defined for the U_m interface. Further services provided by LAPD$_m$ include segmentation and reassembly of data and acknowledged/unacknowledged data transfer.

The network layer in GSM, **layer three**, comprises several sublayers as Figure 4.7 shows. The lowest sublayer is the **radio resource management (RR)**. Only a part of this layer, **RR'**, is implemented in the BTS, the remainder is situated in the BSC. The functions of RR' are supported by the BSC via the **BTS management (BTSM)**. The main tasks of RR are setup, maintenance, and release of radio channels. RR also directly accesses the physical layer for radio information and offers a reliable connection to the next higher layer.

Mobility management (MM) contains functions for registration, authentication, identification, location updating, and the provision of a **temporary mobile subscriber identity (TMSI)** that replaces the **international mobile subscriber identity (IMSI)** and which hides the real identity of an MS user over the air interface. While the IMSI identifies a user, the TMSI is valid only in the current location area of a VLR. MM offers a reliable connection to the next higher layer.

Finally, the **call management (CM)** layer contains three entities: **call control (CC)**, **short message service (SMS)**, and **supplementary service (SS)**. SMS allows for message transfer using the control channels SDCCH and SACCH (if no signaling data is sent), while SS offers the services described in section 4.1.1.3. CC provides a point-to-point connection between two terminals and is used by higher layers for call establishment, call clearing and change of call parameters. This layer also provides functions to send in-band tones, called **dual tone multiple frequency (DTMF)**, over the GSM network. These tones are used, e.g., for the remote control of answering machines or the entry of PINs in electronic banking and are, also used for dialing in traditional analog telephone systems. These tones cannot be sent directly over the voice codec of a GSM MS, as the codec would distort the tones. They are transferred as signals and then converted into tones in the fixed network part of the GSM system.

Additional protocols are used at the A_{bis} and A interfaces (the internal interfaces of a GSM system not presented here). Data transmission at the physical layer typically uses **pulse code modulation (PCM)** systems. While PCM systems offer transparent 64 kbit/s channels, GSM also allows for the submultiplexing of four 16 kbit/s channels into a single 64 kbit/s channel (16 kbit/s are enough for user data from an MS). The physical layer at the A interface typically includes leased lines with 2.048 Mbit/s capacity. LAPD is used for layer two at A_{bis}, BTSM for BTS management.

Signaling system No. 7 (SS7) is used for signaling between an MSC and a BSC. This protocol also transfers all management information between MSCs, HLR, VLRs, AuC, EIR, and OMC. An MSC can also control a BSS via a **BSS application part (BSSAP)**.

4.1.5 Localization and calling

One fundamental feature of the GSM system is the automatic, worldwide localization of users. The system always knows where a user currently is, and the same phone number is valid worldwide. To provide this service, GSM performs periodic location updates even if a user does not use the mobile station (provided that the MS is still logged into the GSM network and is not completely switched off). The HLR always contains information about the current location (only the location area, not the precise geographical location), and the VLR currently responsible for the MS informs the HLR about location changes. As soon as an MS moves into the range of a new VLR (a new location area), the HLR sends all user data needed to the new VLR. Changing VLRs with uninterrupted availability of all services is also called **roaming**. Roaming can take place within the network of one provider, between two providers in one country (national roaming is, often not supported due to competition between operators), but also between different providers in different countries (international roaming). Typically, people associate international roaming with the term roaming as it is this type of roaming that makes GSM very attractive: one device, over 190 countries!

To locate an MS and to address the MS, several numbers are needed:

- **Mobile station international ISDN number (MSISDN):**[6] The only important number for a user of GSM is the phone number. Remember that the phone number is not associated with a certain device but with the SIM, which is personalized for a user. The MSISDN follows the ITU-T standard E.164 for addresses as it is also used in fixed ISDN networks. This number consists of the **country code (CC)** (e.g., +49 179 1234567 with 49 for Germany), the **national destination code (NDC)** (i.e., the address of the network provider, e.g., 179), and the **subscriber number (SN)**.

6 In other types of documentation, this number is also called 'Mobile Subscriber ISDN Number' or 'Mobile Station ISDN Number'. Even the original ETSI standards use different wordings for the same acronym. However, the term 'subscriber' is much better suited as it expresses the independence of the user related number from the device (station).

- **International mobile subscriber identity (IMSI):** GSM uses the IMSI for internal unique identification of a subscriber. IMSI consists of a **mobile country code (MCC)** (e.g., 240 for Sweden, 208 for France), the **mobile network code (MNC)** (i.e., the code of the network provider), and finally the **mobile subscriber identification number (MSIN)**.
- **Temporary mobile subscriber identity (TMSI):** To hide the IMSI, which would give away the exact identity of the user signaling over the air interface, GSM uses the 4 byte TMSI for local subscriber identification. TMSI is selected by the current VLR and is only valid temporarily and within the location area of the VLR (for an ongoing communication TMSI and LAI are sufficient to identify a user; the IMSI is not needed). Additionally, a VLR may change the TMSI periodically.
- **Mobile station[7] roaming number (MSRN):** Another temporary address that hides the identity and location of a subscriber is MSRN. The VLR generates this address on request from the MSC, and the address is also stored in the HLR. MSRN contains the current **visitor country code (VCC)**, the **visitor national destination code (VNDC)**, the identification of the current MSC together with the subscriber number. The MSRN helps the HLR to find a subscriber for an incoming call.

All these numbers are needed to find a subscriber and to maintain the connection with a mobile station. The interesting case is the **mobile terminated call (MTC)**, i.e., a situation in which a station calls a mobile station (the calling station could be outside the GSM network or another mobile station). Figure 4.8 shows the basic steps needed to connect the calling station with the mobile user. In step 1, a user dials the phone number of a GSM subscriber. The fixed network (PSTN) notices (looking at the destination code) that the number belongs to a user in the GSM network and forwards the call setup to the Gateway MSC (2). The GMSC identifies the HLR for the subscriber (which is coded in the phone number) and signals the call setup to the HLR (3). The HLR now checks whether the number exists and whether the user has subscribed to the requested services, and requests an MSRN from the current VLR (4). After receiving the MSRN (5), the HLR can determine the MSC responsible for the MS and forwards this information to the GMSC (6). The GMSC can now forward the call setup request to the MSC indicated (7).

From this point on, the MSC is responsible for all further steps. First, it requests the current status of the MS from the VLR (8). If the MS is available, the MSC initiates paging in all cells it is responsible for (i.e. the location area, LA, 10), as searching for the right cell would be too time consuming (but this approach puts some load on the signaling channels so optimizations exist). The

7 Here, a discrepancy exists between ITU-T standards and ETSI's GSM. MS can denote mobile station or mobile subscriber. Typically, almost all MS in GSM refer to subscribers, as identifiers are not dependent on the station, but on the subscriber identity (stored in the SIM).

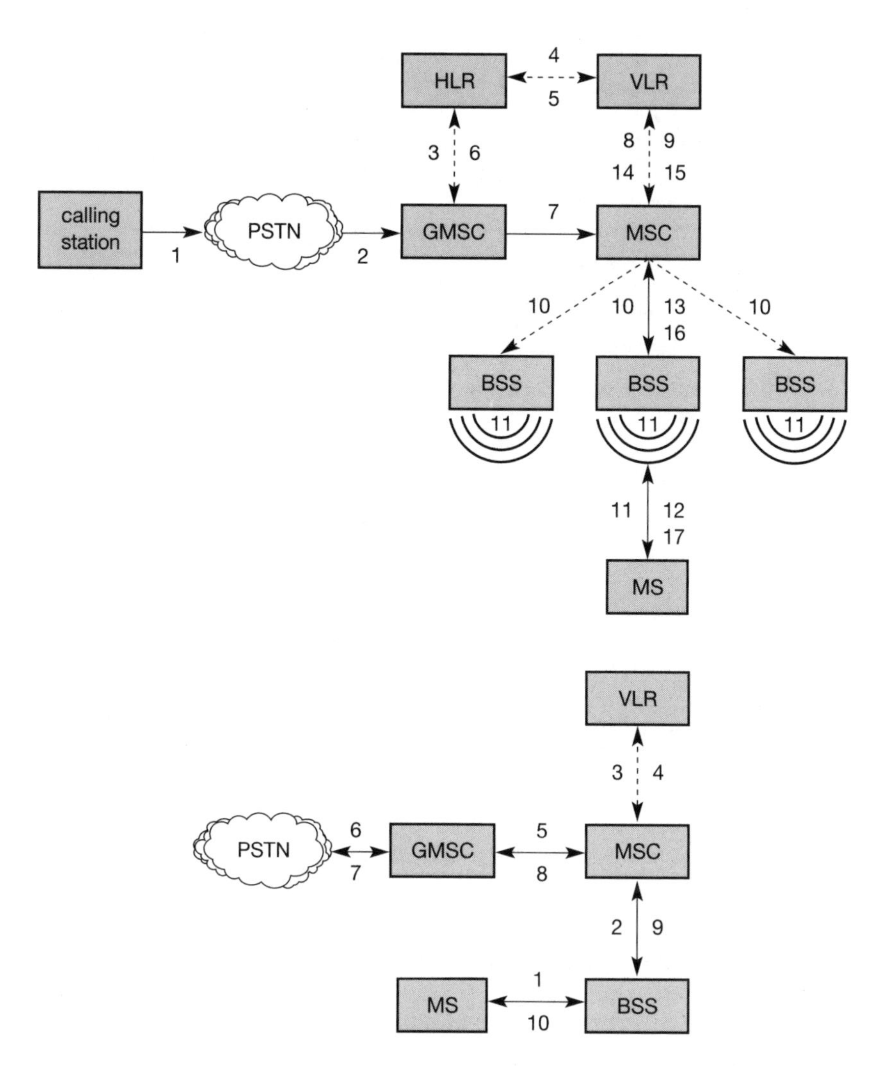

Figure 4.8
Mobile terminated call (MTC)

Figure 4.9
Mobile originated call (MOC)

BTSs of all BSSs transmit this paging signal to the MS (11). If the MS answers (12 and 13), the VLR has to perform security checks (set up encryption etc.). The VLR then signals to the MSC to set up a connection to the MS (steps 15 to 17).

It is much simpler to perform a **mobile originated call (MOC)** compared to a MTC (see Figure 4.9). The MS transmits a request for a new connection (1), the BSS forwards this request to the MSC (2). The MSC then checks if this user is allowed to set up a call with the requested service (3 and 4) and checks the availability of resources through the GSM network and into the PSTN. If all resources are available, the MSC sets up a connection between the MS and the fixed network.

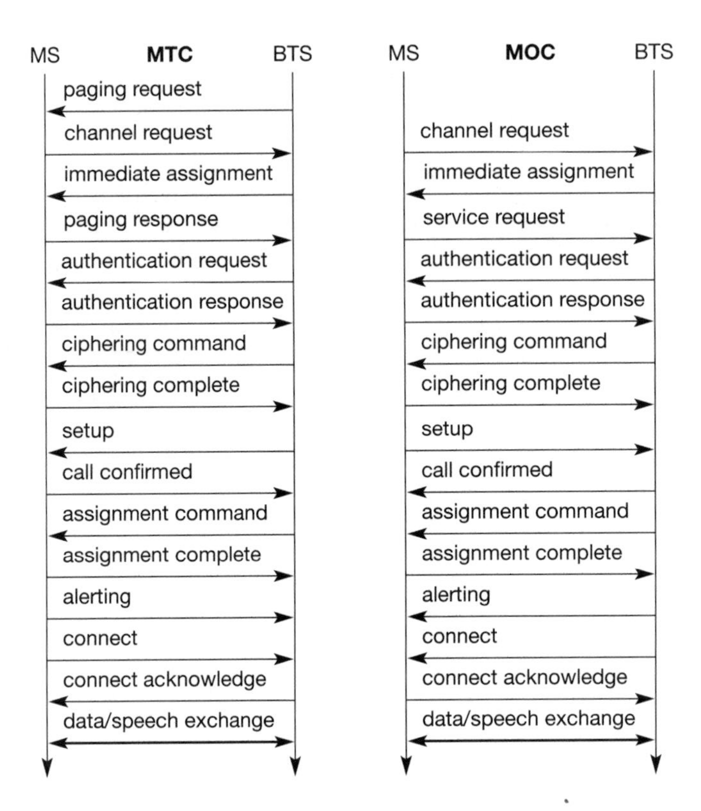

Figure 4.10 Message flow for MTC and MOC

In addition to the steps mentioned above, other messages are exchanged between an MS and BTS during connection setup (in either direction). These messages can be quite often heard in radios or badly shielded loudspeakers as crackling noise before the phone rings. Figure 4.10 shows the messages for an MTC and MOC. Paging is only necessary for an MTC, then similar message exchanges follow. The first step in this context is the channel access via the random access channel (RACH) with consecutive channel assignment; the channel assigned could be a traffic channel (TCH) or a slower signalling channel SDCCH.

The next steps, which are needed for communication security, comprise the authentication of the MS and the switching to encrypted communication. The system now assigns a TCH (if this has not been done). This has the advantage of only having to use an SDCCH during the first setup steps. If the setup fails, no TCH has been blocked. However, using a TCH from the beginning has a speed advantage.

The following steps depend on the use of MTC or MOC. If someone is calling the MS, it answers now with 'alerting' that the MS is ringing and with 'connect' that the user has pressed the connect button. The same actions happen the other way round if the MS has initiated the call. After connection acknowledgement, both parties can exchange data.

Closing the connection comprises a user-initiated disconnect message (both sides can do this), followed by releasing the connection and the radio channel.

4.1.6 Handover

Cellular systems require **handover** procedures, as single cells do not cover the whole service area, but, e.g., only up to 35 km around each antenna on the countryside and some hundred meters in cities (Tripathi, 1998). The smaller the cell size and the faster the movement of a mobile station through the cells (up to 250 km/h for GSM), the more handovers of ongoing calls are required. However, a handover should not cause a cut-off, also called **call drop**. GSM aims at maximum handover duration of 60 ms.

There are two basic reasons for a handover (about 40 have been identified in the standard):

- The mobile station **moves out of the range** of a BTS or a certain antenna of a BTS respectively. The received **signal level** decreases continuously until it falls below the minimal requirements for communication. The **error rate** may grow due to interference, the distance to the BTS may be too high (max. 35 km) etc. – all these effects may diminish the **quality of the radio link** and make radio transmission impossible in the near future.
- The wired infrastructure (MSC, BSC) may decide that the **traffic in one cell is too high** and shift some MS to other cells with a lower load (if possible). Handover may be due to **load balancing**.

Figure 4.11 shows four possible handover scenarios in GSM:

- **Intra-cell handover:** Within a cell, narrow-band interference could make transmission at a certain frequency impossible. The BSC could then decide to change the carrier frequency (scenario 1).
- **Inter-cell, intra-BSC handover:** This is a typical handover scenario. The mobile station moves from one cell to another, but stays within the control of the same BSC. The BSC then performs a handover, assigns a new radio channel in the new cell and releases the old one (scenario 2).
- **Inter-BSC, intra-MSC handover:** As a BSC only controls a limited number of cells; GSM also has to perform handovers between cells controlled by different BSCs. This handover then has to be controlled by the MSC (scenario 3). This situation is also shown in Figure 4.13.
- **Inter MSC handover:** A handover could be required between two cells belonging to different MSCs. Now both MSCs perform the handover together (scenario 4).

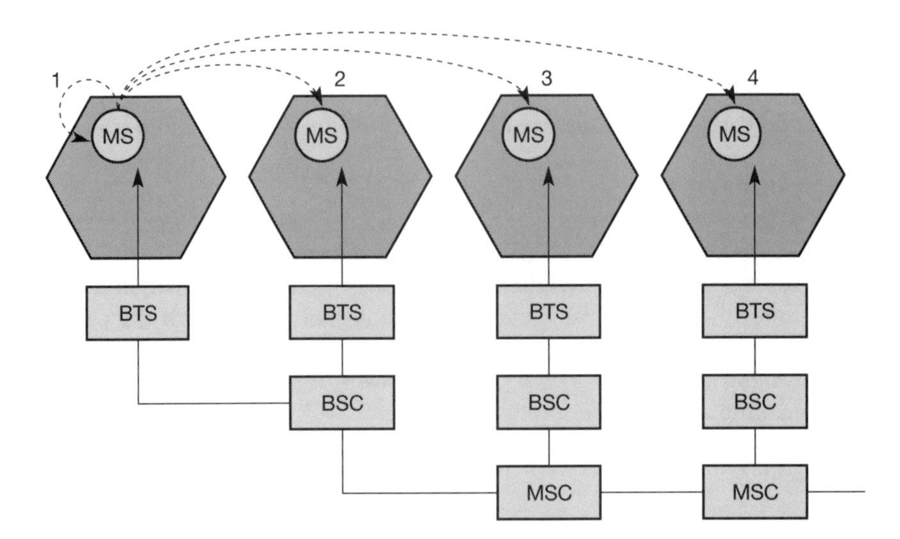

Figure 4.11 Types of handover in GSM

To provide all the necessary information for a handover due to a weak link, MS and BTS both perform periodic measurements of the downlink and uplink quality respectively. (Link quality comprises signal level and bit error rate.) Measurement reports are sent by the MS about every half-second and contain the quality of the current link used for transmission as well as the quality of certain channels in neighboring cells (the BCCHs).

Figure 4.12 shows the typical behavior of the received signal level while an MS moves away from one BTS (BTS_{old}) closer to another one (BTS_{new}). In this case, the handover decision does not depend on the actual value of the received signal level, but on the average value. Therefore, the BSC collects all values (bit error rate and signal levels from uplink and downlink) from BTS and MS and calculates average values. These values are then compared to thresholds, i.e., the handover margin (HO_MARGIN), which includes some hysteresis to avoid a ping-pong effect (Wong, 1997). (Without hysteresis, even short-term interference, e.g., shadowing due to a building, could cause a handover.) Still, even with the HO_MARGIN, the ping-pong effect may occur in GSM – a value which is too high could cause a cut-off, and a value which is too low could cause too many handovers.

Figure 4.13 shows the typical signal flow during an inter-BSC, intra-MSC handover. The MS sends its periodic measurements reports, the BTS_{old} forwards these reports to the BSC_{old} together with its own measurements. Based on these values and, e.g., on current traffic conditions, the BSC_{old} may decide to perform a handover and sends the message HO_required to the MSC. The task of the MSC then comprises the request of the resources needed for the handover from the new BSC, BSC_{new}. This BSC checks if enough resources (typically frequencies or time slots) are available and activates a physical channel at the BTS_{new} to prepare for the arrival of the MS.

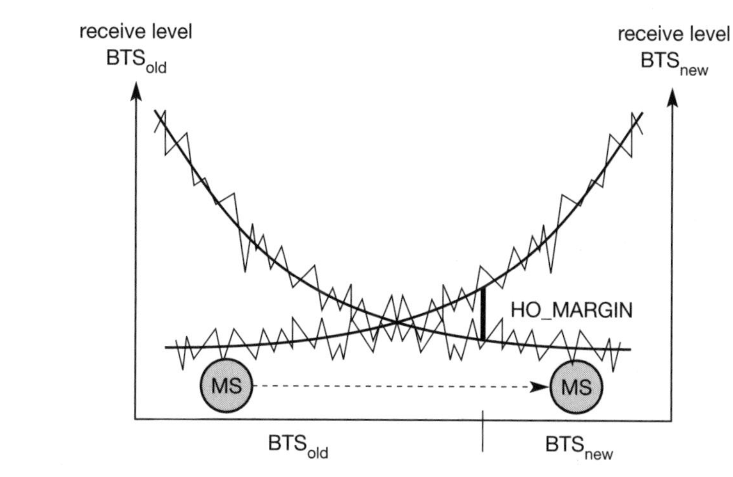

Figure 4.12
Handover decision depending on receive level

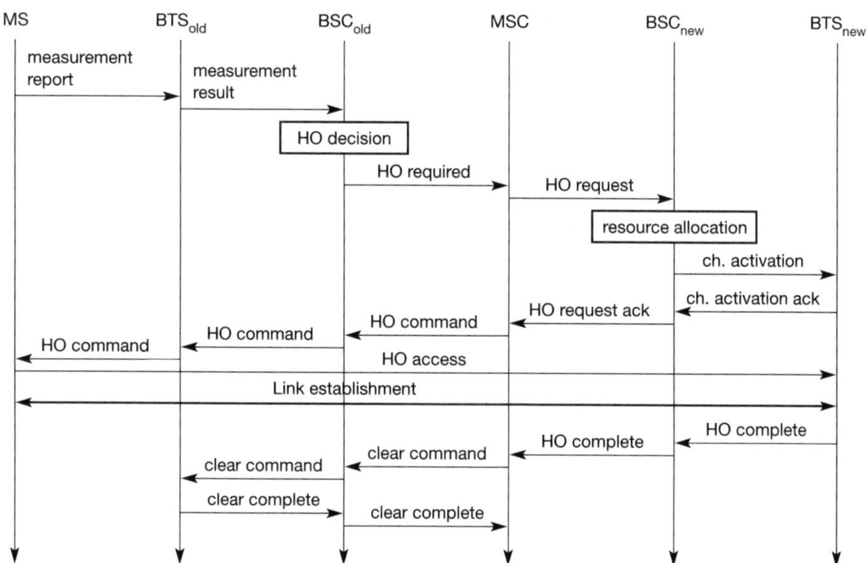

Figure 4.13
Intra-MSC handover

The BTS_{new} acknowledges the successful channel activation, BSC_{new} acknowledges the handover request. The MSC then issues a handover command that is forwarded to the MS. The MS now breaks its old radio link and accesses the new BTS. The next steps include the establishment of the link (this includes layer two link establishment and handover complete messages from the MS). Basically, the MS has then finished the handover, but it is important to release the resources at the old BSC and BTS and to signal the successful handover using the handover and clear complete messages as shown.

More sophisticated handover mechanisms are needed for seamless handovers between different systems. For example, future 3G networks will not cover whole countries but focus on cities and highways. Handover from, e.g., UMTS to GSM without service interruption must be possible. Even more challenging is the seamless handover between wireless LANs (see chapter 7) and 2G/3G networks. This can be done using multimode mobile stations and a more sophisticated roaming infrastructure. However, it is still not obvious how these systems may scale for a large number of users and many handovers, and what handover quality guarantees they can give.

4.1.7 Security

GSM offers several security services using confidential information stored in the AuC and in the individual SIM (which is plugged into an arbitrary MS). The SIM stores personal, secret data and is protected with a PIN against unauthorized use. (For example, the secret key K_i used for authentication and encryption procedures is stored in the SIM.) The security services offered by GSM are explained below:

- **Access control and authentication:** The first step includes the authentication of a valid user for the SIM. The user needs a secret PIN to access the SIM. The next step is the subscriber authentication (see Figure 4.10). This step is based on a challenge-response scheme as presented in section 4.1.7.1.
- **Confidentiality:** All user-related data is encrypted. After authentication, BTS and MS apply encryption to voice, data, and signaling as shown in section 4.1.7.2. This confidentiality exists only between MS and BTS, but it does not exist end-to-end or within the whole fixed GSM/telephone network.
- **Anonymity:** To provide user anonymity, all data is encrypted before transmission, and user identifiers (which would reveal an identity) are not used over the air. Instead, GSM transmits a temporary identifier (TMSI), which is newly assigned by the VLR after each location update. Additionally, the VLR can change the TMSI at any time.

Three algorithms have been specified to provide security services in GSM. **Algorithm A3** is used for **authentication**, **A5** for **encryption**, and **A8** for the **generation of a cipher key**. In the GSM standard only algorithm A5 was publicly available, whereas A3 and A8 were secret, but standardized with open interfaces. Both A3 and A8 are no longer secret, but were published on the internet in 1998. This demonstrates that security by obscurity does not really work. As it turned out, the algorithms are not very strong. However, network providers can use stronger algorithms for authentication – or users can apply stronger end-to-end encryption. Algorithms A3 and A8 (or their replacements) are located on the SIM and in the AuC and can be proprietary. Only A5 which is implemented in the devices has to be identical for all providers.

4.1.7.1 Authentication

Before a subscriber can use any service from the GSM network, he or she must be authenticated. Authentication is based on the SIM, which stores the **individual authentication key K_i**, the **user identification IMSI**, and the algorithm used for authentication A3. Authentication uses a challenge-response method: the access

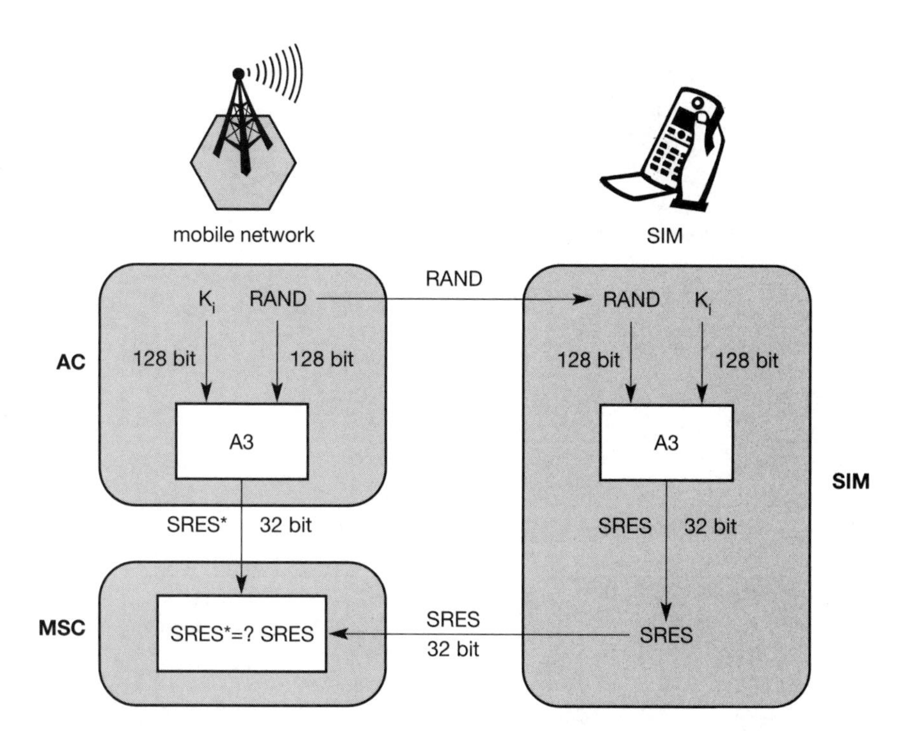

Figure 4.14 Subscriber authentication

control AC generates a random number **RAND** as challenge, and the SIM within the MS answers with **SRES** (signed response) as response (see Figure 4.14). The AuC performs the basic generation of random values RAND, signed responses SRES, and cipher keys K_c for each IMSI, and then forwards this information to the HLR. The current VLR requests the appropriate values for RAND, SRES, and K_c from the HLR.

For authentication, the VLR sends the random value RAND to the SIM. Both sides, network and subscriber module, perform the same operation with RAND and the key K_i, called A3. The MS sends back the SRES generated by the SIM; the VLR can now compare both values. If they are the same, the VLR accepts the subscriber, otherwise the subscriber is rejected.

4.1.7.2 Encryption

To ensure privacy, all messages containing user-related information are encrypted in GSM over the air interface. After authentication, MS and BSS can start using encryption by applying the cipher key K_c (the precise location of security functions for encryption, BTS and/or BSC are vendor dependent). K_c is generated using the individual key K_i and a random value by applying the algorithm A8. Note that the SIM in the MS and the network both calculate the same K_c based on the random value RAND. The key K_c itself is not transmitted over the air interface.

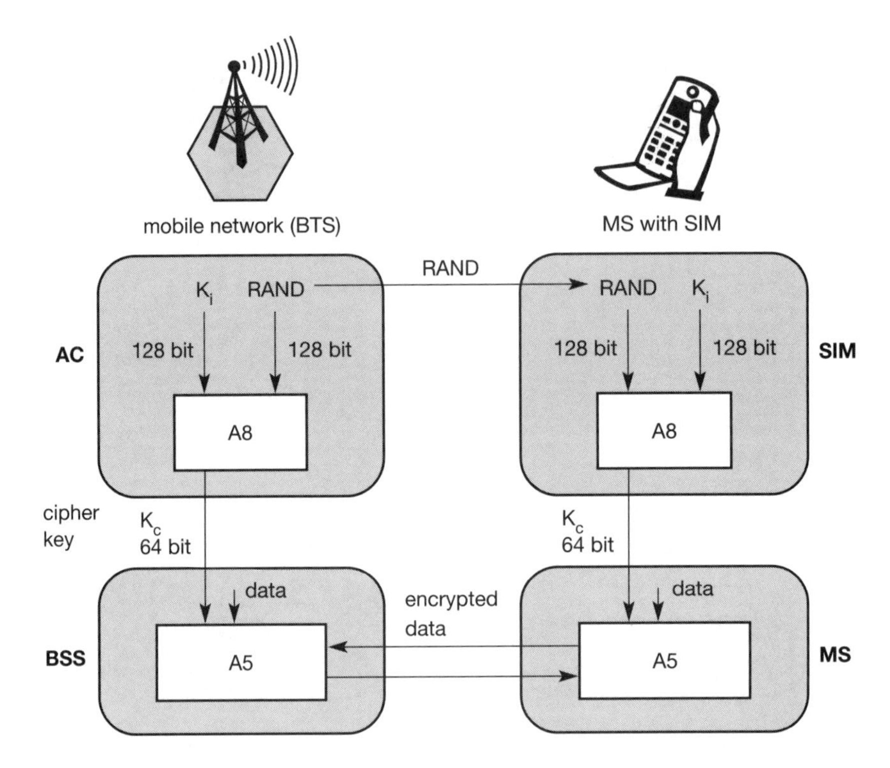

Figure 4.15 Data encryption

MS and BTS can now encrypt and decrypt data using the algorithm A5 and the cipher key K_c. As Figure 4.15 shows, K_c should be a 64 bit key – which is not very strong, but is at least a good protection against simple eavesdropping. However, the publication of A3 and A8 on the internet showed that in certain implementations 10 of the 64 bits are always set to 0, so that the real length of the key is thus only 54 consequently, the encryption is much weaker.

4.1.8 New data services

As mentioned above, the standard bandwidth of 9.6 kbit/s (14.4 kbit/s with some providers) available for data transmission is not sufficient for the requirements of today's computers. When GSM was developed, not many people anticipated the tremendous growth of data communication compared to voice communication. At that time, 9.6 kbit/s was a lot, or at least enough for standard group 3 fax machines. But with the requirements of, e.g., web browsing, file download, or even intensive e-mail exchange with attachments, this is not enough.

To enhance the data transmission capabilities of GSM, two basic approaches are possible. As the basic GSM is based on connection-oriented traffic channels, e.g., with 9.6 kbit/s each, several channels could be combined to increase bandwidth. This system is called HSCSD and is presented in the following section. A more progressive step is the introduction of packet-oriented traffic in GSM, i.e., shifting the paradigm from connections/telephone thinking to packets/internet thinking. The system, called GPRS, is presented in section 4.1.8.2.

4.1.8.1 HSCSD

A straightforward improvement of GSM's data transmission capabilities is **high speed circuit switched data (HSCSD)**, which is available with some providers. In this system, higher data rates are achieved by bundling several TCHs. An MS requests one or more TCHs from the GSM network, i.e., it allocates several TDMA slots within a TDMA frame. This allocation can be asymmetrical, i.e., more slots can be allocated on the downlink than on the uplink, which fits the typical user behavior of downloading more data compared to uploading. Basically, HSCSD only requires software upgrades in an MS and MSC (both have to be able to split a traffic stream into several streams, using a separate TCH each, and to combine these streams again).

In theory, an MS could use all eight slots within a TDMA frame to achieve an **air interface user rate (AIUR)** of, e.g., 8 TCH/F14.4 channels or 115.2 kbit/s (ETSI, 1998e). One problem of this configuration is that the MS is required to send and receive at the same time. Standard GSM does not require this capability – uplink and downlink slots are always shifted for three slots. ETSI (1997a) specifies the AIUR available at 57.6 kbit/s (duplex) using four slots in the uplink and downlink (Table 4.2 shows the permitted combinations of traffic channels and allocated slots for non-transparent services).

Although it appears attractive at first glance, HSCSD exhibits some major disadvantages. It still uses the connection-oriented mechanisms of GSM. These are not at all efficient for computer data traffic, which is typically bursty and asymmetrical. While downloading a larger file may require all channels reserved, typical web browsing would leave the channels idle most of the time. Allocating channels is reflected directly in the service costs, as, once the channels have been reserved, other users cannot use them.

Table 4.2 Available data rates for HSCSD in GSM

AIUR	TCH / F4.8	TCH / F9.6	TCH / F14.4
4.8 kbit/s	1	–	–
9.6 kbit/s	2	1	–
14.4 kbit/s	3	–	1
19.2 kbit/s	4	2	–
28.8 kbit/s	–	3	2
38.4 kbit/s	–	4	–
43.2 kbit/s	–	–	3
57.6 kbit/s	–	–	4

For *n* channels, HSCSD requires *n* times signaling during handover, connection setup and release. Each channel is treated separately. The probability of blocking or service degradation increases during handover, as in this case a BSC has to check resources for *n* channels, not just one. All in all, HSCSD may be an attractive interim solution for higher bandwidth and rather constant traffic (e.g., file download). However, it does not make much sense for bursty internet traffic as long as a user is charged for each channel allocated for communication.

4.1.8.2 GPRS

The next step toward more flexible and powerful data transmission avoids the problems of HSCSD by being fully packet-oriented. The **general packet radio service (GPRS)** provides packet mode transfer for applications that exhibit traffic patterns such as frequent transmission of small volumes (e.g., typical web requests) or infrequent transmissions of small or medium volumes (e.g., typical web responses) according to the requirement specification (ETSI, 1998a). Compared to existing data transfer services, GPRS should use the existing network resources more efficiently for packet mode applications, and should provide a selection of QoS parameters for the service requesters. GPRS should also allow for broadcast, multicast, and unicast service. The overall goal in this context is the provision of a more efficient and, thus, cheaper packet transfer service for typical internet applications that usually rely solely on packet transfer. Network providers typically support this model by charging on volume and not on connection time as is usual for traditional GSM data services and for HSCSD. The main benefit for users of GPRS is the 'always on' characteristic – no connection has to be set up prior to data transfer. Clearly, GPRS was driven by the tremendous success of the packet-oriented internet, and by the new traffic models and applications. However, GPRS, as shown in the following sections, needs additional network elements, i.e., software and hardware. Unlike HSCSD, GPRS does not only represent a software update to allow for the bundling of channels, it also represents a big step towards UMTS as the main internal infrastructure needed for UMTS (in its initial release) is exactly what GPRS uses (see section 4.4).

The main concepts of GPRS are as follows (ETSI, 1998b). For the new GPRS radio channels, the GSM system can allocate between one and eight time slots within a TDMA frame. Time slots are not allocated in a fixed, pre-determined manner but on demand. All time slots can be shared by the active users; up- and downlink are allocated separately. Allocation of the slots is based on current load and operator preferences. Depending on the coding, a transfer rate of up to 170 kbit/s is possible. For GPRS, operators often reserve at least a time slot per cell to guarantee a minimum data rate. The GPRS concept is independent of channel characteristics and of the type of channel (traditional GSM traffic or control channel), and does not limit the maximum data rate (only the GSM transport system limits the rate). All GPRS services can be used in parallel to conventional services. Table 4.3 shows the typical data rates available with GPRS if it is used together with GSM (GPRS can also be used for other TDMA systems).

Coding scheme	1 slot	2 slots	3 slots	4 slots	5 slots	6 slots	7 slots	8 slots
CS-1	9.05	18.2	27.15	36.2	45.25	54.3	63.35	72.4
CS-2	13.4	26.8	40.2	53.6	67	80.4	93.8	107.2
CS-3	15.6	31.2	46.8	62.4	78	93.6	109.2	124.8
CS-4	21.4	42.8	64.2	85.6	107	128.4	149.8	171.2

Table 4.3 GPRS data rates in kbit/s

In the beginning, only coding schemes CS-1 and CS-2 are available. The system chooses a coding scheme depending on the current error rate (CS-4 provides no error correction capabilities).

It should be noted that the real available data rate heavily depends on the current load of the cell as GPRS typically only uses idle time slots. The transfer rate depends on the capabilities of the MS as not all devices are able to send and receive at the same time. Table 4.4 gives examples for device classes together with their ability to use time slots for sending and receiving data. The maximum possible number of slots limits the transfer rate even more. For example, a class 12 device may receive data using 4 slots within a GSM time frame or it may send data using 4 slots. However, a maximum number of 5 slots may be used altogether. Using all 8 slots for data encoded using CS-4 yields the maximum rate of 171.2 kbit/s. Today, a typical MS is a class 10 device using CS-2, which results in a receiving rate of 53.6 kbit/s and a sending rate of 26.8 kbit/s.

In phase 1, GPRS offers a **point-to-point (PTP)** packet transfer service (ETSI, 1998c). One of the PTP versions offered is the **PTP connection oriented network service (PTP-CONS)**, which includes the ability of GPRS to maintain a virtual circuit upon change of the cell within the GSM network. This type of

Class	Receiving slots	Sending slots	Maximum number of slots
1	1	1	2
2	2	1	3
3	2	2	3
5	2	2	4
8	4	1	5
10	4	2	5
12	4	4	5

Table 4.4 Examples for GPRS device classes

Table 4.5 Reliability classes in GPRS according to ETSI (1998c)

Reliability class	Lost SDU probability	Duplicate SDU probability	Out of sequence SDU probability	Corrupt SDU probability
1	10^{-9}	10^{-9}	10^{-9}	10^{-9}
2	10^{-4}	10^{-5}	10^{-5}	10^{-6}
3	10^{-2}	10^{-5}	10^{-5}	10^{-2}

service corresponds to **X.25**, the typical circuit-switched packet-oriented transfer protocol available worldwide. The other PTP version offered is the **PTP connectionless network service (PTP-CLNS)**, which supports applications that are based on the Internet Protocol **IP**. Multicasting, called **point-to-multipoint (PTM)** service, is left for GPRS phase 2.

Users of GPRS can specify a **QoS-profile**. This determines the **service precedence** (high, normal, low), **reliability class** and **delay class** of the transmission, and **user data throughput**. GPRS should adaptively allocate radio resources to fulfill these user specifications. Table 4.5 shows the three reliability classes together with the maximum probabilities for a lost service data unit (SDU), a duplicated SDU, an SDU out of the original sequence, and the probability of delivering a corrupt SDU to the higher layer. Reliability class 1 could be used for very error-sensitive applications that cannot perform error corrections themselves. If applications exhibit greater error tolerance, class 2 could be appropriate. Finally, class 3 is the choice for error-insensitive applications or applications that can handle error corrections themselves.

Delay within a GPRS network is incurred by channel access delay, coding for error correction, and transfer delays in the fixed and wireless part of the GPRS network. The delay introduced by external fixed networks is out of scope. However, GPRS does not produce additional delay by buffering packets as store-and-forward networks do. If possible, GPRS tries to forward packets as fast as possible. Table 4.6 shows the specified maximum mean and 95 percentile delay values for packet sizes of 128 and 1,024 byte. As we can clearly see, no matter which class, all delays are orders of magnitude higher than fixed network delays. This is a very important characteristic that has to be taken into account when implementing higher layer protocols such as TCP on top of GPRS networks (see chapter 9). Typical round trip times (RTT) in fixed networks are in the order of 10 to 100 ms. Using real unloaded GPRS networks round trip times of well above 1 s for even small packets (128–512 byte) are common. Additionally, GPRS exhibits a large jitter compared to fixed networks (several 100 ms are not uncommon). This characteristic has a strong impact on user experience when, e.g., interactive Internet applications are used on top of GPRS.

	SDU size 128 byte		SDU size 1,024 byte	
Delay Class	Mean	95 percentile	Mean	95 percentile
1	<0.5 s	<1.5 s	<2 s	<7 s
2	<5 s	<25 s	<15 s	<75 s
3	<50 s	<250 s	<75 s	<375 s
4		Unspecified		

Table 4.6 Delay classes in GRPS according to ETSI (1998c)

Finally, GPRS includes several **security services** such as authentication, access control, user identity confidentiality, and user information confidentiality. Even a completely **anonymous service** is possible, as, e.g., applied for road toll systems that only charge a user via the MS independent of the user's identity.

The **GPRS architecture** introduces two new network elements, which are called **GPRS support nodes (GSN)** and are in fact routers. All GSNs are integrated into the standard GSM architecture, and many new interfaces have been defined (see Figure 4.16). The **gateway GPRS support node (GGSN)** is the interworking unit between the GPRS network and external **packet data networks (PDN)**. This node contains routing information for GPRS users, performs address conversion, and tunnels data to a user via encapsulation. The GGSN is connected to external networks (e.g., IP or X.25) via the G_i interface and transfers packets to the SGSN via an IP-based GPRS backbone network (G_n interface).

The other new element is the **serving GPRS support node (SGSN)** which supports the MS via the G_b interface. The SGSN, for example, requests user addresses from the **GPRS register (GR)**, keeps track of the individual MSs' location, is responsible for collecting billing information (e.g., counting bytes), and performs several security functions such as access control. The SGSN is connected to a BSC via frame relay and is basically on the same hierarchy level as an MSC. The GR, which is typically a part of the HLR, stores all GPRS-relevant data. GGSNs and SGSNs can be compared with home and foreign agents, respectively, in a mobile IP network (see chapter 8).

As shown in Figure 4.16, packet data is transmitted from a PDN, via the GGSN and SGSN directly to the BSS and finally to the MS. The MSC, which is responsible for data transport in the traditional circuit-switched GSM, is only used for signaling in the GPRS scenario. Additional interfaces to further network elements and other PLMNs can be found in ETSI (1998b).

Before sending any data over the GPRS network, an MS must attach to it, following the procedures of the **mobility management**. The attachment procedure includes assigning a temporal identifier, called a **temporary logical link identity (TLLI)**, and a **ciphering key sequence number (CKSN)** for data encryption. For each MS, a **GPRS context** is set up and stored in the MS and in

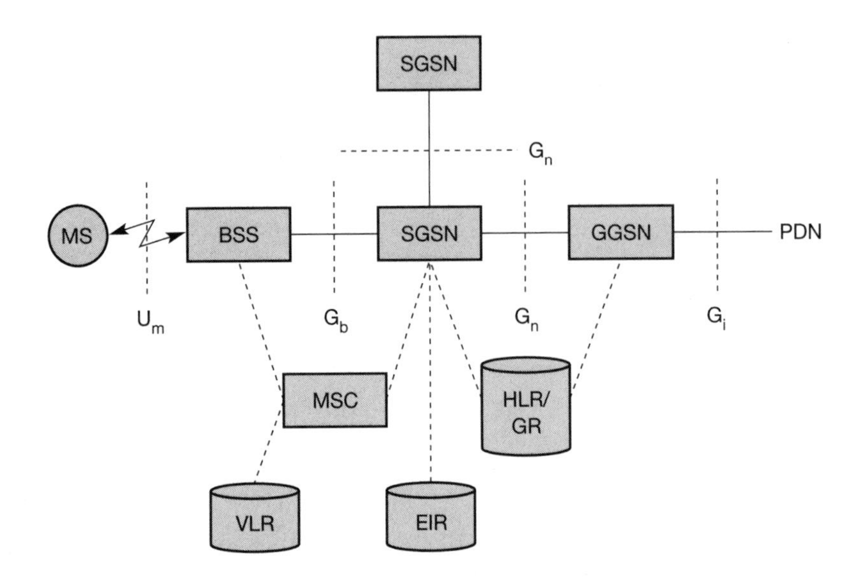

Figure 4.16
GPRS architecture reference model

the corresponding SGSN. This context comprises the status of the MS (which can be ready, idle, or standby; ETSI, 1998b), the CKSN, a flag indicating if compression is used, and routing data (TLLI, the routing area RA, a cell identifier, and a packet data channel, PDCH, identifier). Besides attaching and detaching, mobility management also comprises functions for authentication, location management, and ciphering (here, the scope of ciphering lies between MS and SGSN, which is more than in standard GSM). In **idle** mode an MS is not reachable and all context is deleted. In the **standby** state only movement across routing areas is updated to the SGSN but not changes of the cell. Permanent updating would waste battery power, no updating would require system-wide paging. The update procedure in standby mode is a compromise. Only in the **ready** state every movement of the MS is indicated to the SGSN.

Figure 4.17 shows the protocol architecture of the transmission plane for GPRS. Architectures for the signaling planes can be found in ETSI (1998b). All data within the GPRS backbone, i.e., between the GSNs, is transferred using the **GPRS tunnelling protocol (GTP)**. GTP can use two different transport protocols, either the reliable **TCP** (needed for reliable transfer of X.25 packets) or the non-reliable **UDP** (used for IP packets). The network protocol for the GPRS backbone is **IP** (using any lower layers). To adapt to the different characteristics of the underlying networks, the **subnetwork dependent convergence protocol (SNDCP)** is used between an SGSN and the MS. On top of SNDCP and GTP, user packet data is tunneled from the MS to the GGSN and vice versa. To achieve a high reliability of packet transfer between SGSN and MS, a special LLC is used, which comprises ARQ and FEC mechanisms for PTP (and later PTM) services.

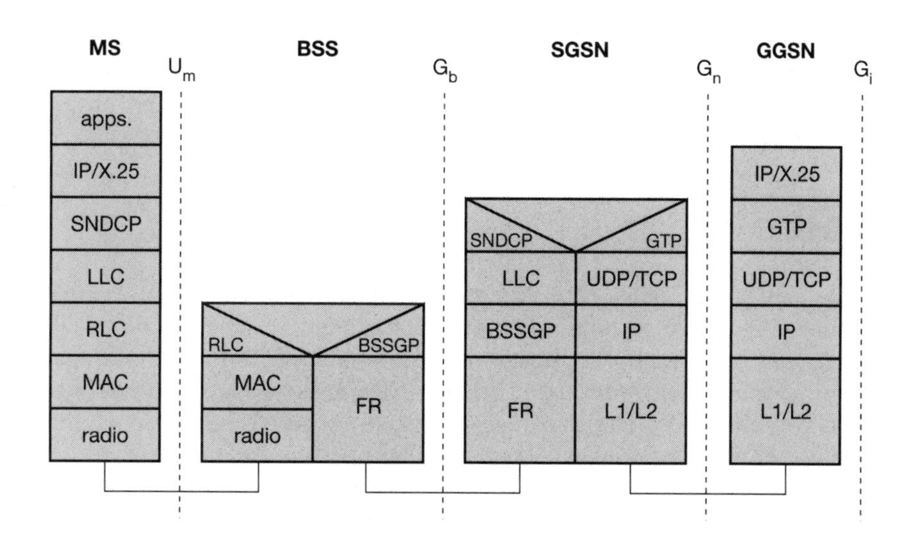

Figure 4.17 GPRS transmission plane protocol reference model

A **base station subsystem GPRS protocol (BSSGP)** is used to convey routing and QoS-related information between the BSS and SGSN. BSSGP does not perform error correction and works on top of a **frame relay (FR)** network. Finally, radio link dependent protocols are needed to transfer data over the U_m interface. The **radio link protocol (RLC)** provides a reliable link, while the **MAC** controls access with signaling procedures for the radio channel and the mapping of LLC frames onto the GSM physical channels. The **radio interface** at U_m needed for GPRS does not require fundamental changes compared to standard GSM (Brasche, 1997), (ETSI, 1998d). However, several new logical channels and their mapping onto physical resources have been defined. For example, one MS can allocate up to eight **packet data traffic channels (PDTCHs)**. Capacity can be allocated on demand and shared between circuit-switched channels and GPRS. This allocation can be done dynamically with load supervision or alternatively, capacity can be pre-allocated.

A very important factor for any application working end-to-end is that it does not 'notice' any details from the GSM/GPRS-related infrastructure. The application uses, e.g., TCP on top of IP, IP packets are tunneled to the GGSN, which forwards them into the PDN. All PDNs forward their packets for a GPRS user to the GGSN, the GGSN asks the current SGSN for tunnel parameters, and forwards the packets via SGSN to the MS. Although MSs using GPRS may be considered as part of the internet, one should know that operators typically perform an address translation in the GGSN using NAT. All MSs are assigned private IP addresses which are then translated into global addresses at the GGSN. The advantage of this approach is the inherent protection of MSs from attacks (the subscriber typically has to pay for traffic even if it originates from an attack!) – private addresses are not routed through the internet so it is not possible to reach an MS from the internet. This is also a disadvantage if an MS wants to offer a service using a fixed, globally visible IP address. This is difficult with IPv4 and NAT and it will be interesting to see how IPv6 is used for this purpose (while still protecting the MSs from outside attacks as air traffic is expensive).

4.2 DECT

Another fully digital cellular network is the **digital enhanced cordless telecommunications (DECT)** system specified by ETSI (2002, 1998j, k), (DECT Forum, 2002). Formerly also called **digital European cordless telephone and digital European cordless telecommunications**, DECT replaces older analog cordless phone systems such as CT1 and CT1+. These analog systems only ensured security to a limited extent as they did not use encryption for data transmission and only offered a relatively low capacity. DECT is also a more powerful alternative to the digital system CT2, which is mainly used in the UK (the DECT standard works throughout Europe), and has even been selected as one of the 3G candidates in the IMT-2000 family (see section 4.4). DECT is mainly used in offices, on campus, at trade shows, or in the home. Furthermore, access points to the PSTN can be established within, e.g., railway stations, large government buildings and hospitals, offering a much cheaper telephone service compared to a GSM system. DECT could also be used to bridge the last few hundred meters between a new network operator and customers. Using this 'small range' local loop, new companies can offer their service without having their own lines installed in the streets. DECT systems offer many different interworking units, e.g., with GSM, ISDN, or data networks. Currently, over 100 million DECT units are in use (DECT, 2002).

A big difference between DECT and GSM exists in terms of cell diameter and cell capacity. While GSM is designed for outdoor use with a cell diameter of up to 70 km, the range of DECT is limited to about 300 m from the base station (only around 50 m are feasible inside buildings depending on the walls). Due to this limited range and additional multiplexing techniques, DECT can offer its service to some 10,000 people within one km^2. This is a typical scenario within a big city, where thousands of offices are located in skyscrapers close together. DECT also uses base stations, but these base stations together with a mobile station are in a price range of €100 compared to several €10,000 for a GSM base station. GSM base stations can typically not be used by individuals for private networks. One reason is licensing as all GSM frequencies have been licensed to network operators. DECT can also handle handover, but it was not designed to work at a higher speed (e.g., up to 250 km/h like GSM systems). Devices handling GSM and DECT exist but have never been a commercial success.

DECT works at a frequency range of 1880–1990 MHz offering 120 full duplex channels. Time division duplex (TDD) is applied using 10 ms frames. The frequency range is subdivided into 10 carrier frequencies using FDMA, each frame being divided into 24 slots using TDMA. For the TDD mechanism,

12 slots are used as uplink, 12 slots as downlink (see Figure 3.4). The digital modulation scheme is GMSK – each station has an average transmission power of only 10 mW with a maximum of 250 mW.

4.2.1 System architecture

A DECT system, may have various different physical implementation depending on its actual use. Different DECT entities can be integrated into one physical unit; entities can be distributed, replicated etc. However, all implementations are based on the same logical reference model of the system architecture as shown in Figure 4.18. A **global network** connects the local communication structure to the outside world and offers its services via the interface D_1. Global networks could be integrated services digital networks (ISDN), public switched telephone networks (PSTN), public land mobile networks (PLMN), e.g., GSM, or packet switched public data network (PSPDN). The services offered by these networks include transportation of data and the translation of addresses and routing of data between the local networks.

Local networks in the DECT context offer local telecommunication services that can include everything from simple switching to intelligent call forwarding, address translation etc. Examples for such networks are analog or digital private branch exchanges (PBXs) or LANs, e.g., those following the IEEE 802.x family of LANs. As the core of the DECT system itself is quite simple, all typical network functions have to be integrated in the local or global network, where the databases **home data base (HDB)** and **visitor data base (VDB)** are also located. Both databases support mobility with functions that are similar to those in the HLR and VLR in GSM systems. Incoming calls are automatically forwarded to the current subsystem responsible for the DECT user, and the current VDB informs the HDB about changes in location.

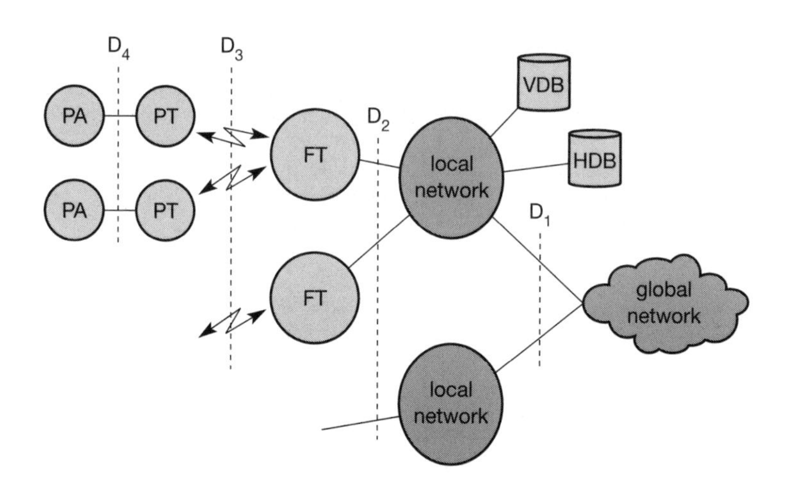

Figure 4.18 DECT system architecture reference model

The DECT core network consists of the **fixed radio termination (FT)** and the **portable radio termination (PT)**, and basically only provides a multiplexing service. FT and PT cover layers one to three at the fixed network side and mobile network side respectively. Additionally, several portable applications (PA) can be implemented on a device.

4.2.2 Protocol architecture

The DECT protocol reference architecture follows the OSI reference model. Figure 4.19 shows the layers covered by the standard: the physical layer, medium access control, and data link control[8] for both the **control plane (C-Plane)** and the **user plane (U-Plane)**. An additional network layer has been specified for the C-Plane, so that user data from layer two is directly forwarded to the U-Plane. A management plane vertically covers all lower layers of a DECT system.

4.2.2.1 Physical layer

As in all wireless networks, the **physical layer** comprises all functions for modulation/demodulation, incoming signal detection, sender/receiver synchronization, and collection of status information for the management plane. This layer generates the physical channel structure with a certain, guaranteed throughput. On request from the MAC layer, the physical layer assigns a channel for data transmission.

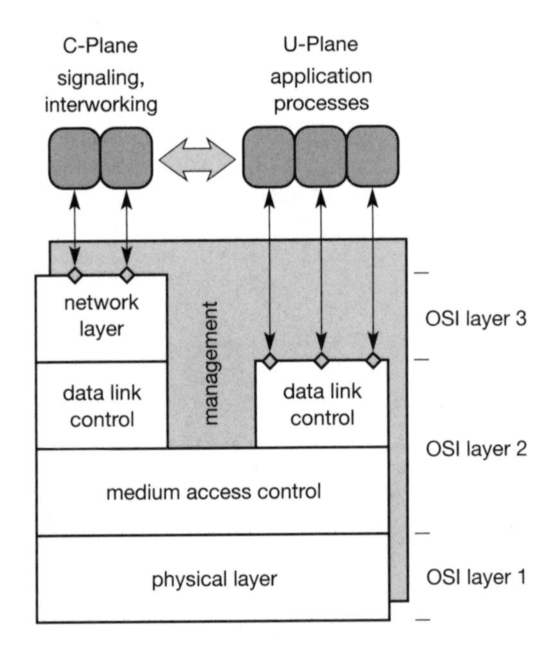

Figure 4.19 DECT protocol layers

8 Strictly speaking, the name "data link control" for the upper part of layer two is wrong in this architecture. According to the OSI reference model, the data link control (layer two) comprises the logical link control (layer 2b) and the medium access control (layer 2a).

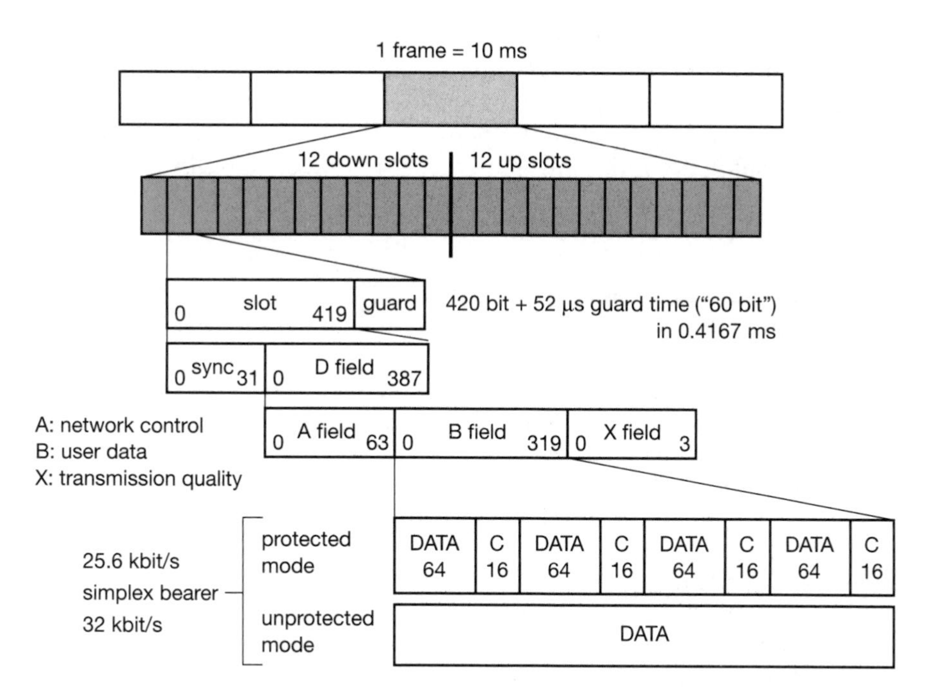

Figure 4.20
DECT multiplex and frame structure

Figure 4.20 shows the standard TDMA frame structure used in DECT and some typical data packets. Each frame has a duration of 10 ms and contains 12 slots for the downlink and 12 slots for the uplink in the **basic connection** mode. If a mobile node receives data in slot s, it returns data in slot s+12. An **advanced connection** mode allows different allocation schemes. Each slot has a duration of 0.4167 ms and can contain several different physical packets. Typically, 420 bits are used for data; the remaining 52 μs are left as **guard space**. The 420 data bits are again divided into a 32 bit **synchronization pattern** followed by the **data** field D.

The fields for data transmission now use these remaining 388 bits for **network control** (A field), **user data** (B field), and the transfer of the **transmission quality** (X field). While network control is transmitted with a data rate of 6.4 kbit/s (64 bit each 10 ms), the user data rate depends on additional error correction mechanisms. The **simplex bearer** provides a data rate of 32 kbit/s in an **unprotected mode**, while using a 16 bit CRC **checksum** C for a data block of 64 bit in the **protected mode** reduces the data rate to 25.6 kbit/s. A **duplex bearer** service is produced by combining two simplex bearers. DECT also defines bearer types with higher throughputs by combining slots, e.g., the **double duplex bearer** offers 80 kbit/s full-duplex.

4.2.2.2 Medium access control layer

The **medium access control** (MAC) layer establishes, maintains, and releases channels for higher layers by activating and deactivating physical channels. MAC multiplexes several logical channels onto physical channels. Logical channels exist for signaling network control, user data transmission, paging, or sending broadcast messages. Additional services offered include segmentation/reassembly of packets and error control/error correction.

4.2.2.3 Data link control layer

The **data link control (DLC)** layer creates and maintains reliable connections between the mobile terminal and the base station. Two services have been defined for the **C-Plane**: a **connectionless broadcast** service for paging (called **Lb**) and a **point-to-point** protocol similar to LAPD in ISDN, but adapted to the underlying MAC (called **LAPC+Lc**).

Several services exist for the **U-Plane**, e.g., a transparent unprotected service (basically a null service), a forward error correction service, rate adaptation services, and services for future enhancements. If services are used, e.g., to transfer ISDN data at 64 kbit/s, then DECT also tries to transfer 64 kbit/s. However, in case of errors, DECT raises the transfer rate to 72 kbit/s, and includes FEC and a buffer for up to eight blocks to perform ARQ. This buffer then introduces an additional delay of up to 80 ms.

4.2.2.4 Network layer

The **network layer** of DECT is similar to those in ISDN and GSM and only exists for the **C-Plane**. This layer provides services to request, check, reserve, control, and release resources at the fixed station (connection to the fixed network, wireless connection) and the mobile terminal (wireless connection). The **mobility management (MM)** within the network layer is responsible for identity management, authentication, and the management of the location data bases. **Call control** (CC) handles connection setup, release, and negotiation. Two message services, the **connection oriented message service (COMS)** and the **connectionless message service (CLMS)** transfer data to and from the interworking unit that connects the DECT system with the outside world.

4.3 TETRA

Trunked radio systems constitute another method of wireless data transmission. These systems use many different radio carriers but only assign a specific carrier to a certain user for a short period of time according to demand. While, for example, taxi services, transport companies with fleet management systems and rescue teams all have their own unique carrier frequency in traditional systems, they can share a whole group of frequencies in trunked radio systems for better frequency reuse via FDM and TDM techniques. These types of radio systems typically offer interfaces to the fixed telephone network, i.e., voice and data services, but are not publicly accessible. These systems are not only simpler than most other networks, they are also reliable and relatively cheap to set up and operate, as they only have to cover the region where the local users operate, e.g., a city taxi service.

To allow a common system throughout Europe, ETSI standardized the **TETRA** system (**terrestrial trunked radio**)[9] in 1991 (ETSI, 2002), (TETRA MoU, 2002). This system should replace national systems, such as MODACOM, MOBITEX and COGNITO in Europe that typically connect to an X.25 packet network. (An example system from the US is ARDIS.) TETRA offers two standards: the **Voice+Data (V+D)** service (ETSI, 1998l) and the **packet data optimized (PDO)** service (ETSI, 1998m). While V+D offers circuit-switched voice and data transmission, PDO only offers packet data transmission, either connection-oriented to connect to X.25 or connectionless for the ISO CLNS (connectionless network service). The latter service can be point-to-point or point-to-multipoint, the typical delay for a short message (128 byte) being less than 100 ms. V+D connection modes comprise unicast and broadcast connections, group communication within a certain protected group, and a direct ad hoc mode without a base station. However, delays for short messages can be up to 500 ms or higher depending on the priority.

TETRA also offers bearer services of up to 28.8 kbit/s for unprotected data transmission and 9.6 kbit/s for protected transmission. Examples for end-to-end services are call forwarding, call barring, identification, call hold, call priorities, emergency calls and group joins. The system architecture of TETRA is very similar to GSM. Via the radio interface U_m, the **mobile station (MS)** connects to the **switching and management infrastructure (SwMI)**, which contains the user data bases (HDB, VDB), the base station, and interfaces to PSTN, ISDN, or PDN. The system itself, however, is much simpler in real implementation compared to GSM, as typically no handover is needed. Taxis usually remain within a certain area which can be covered by one TETRA cell.

Several frequencies have been specified for TETRA which uses FDD (e.g., 380–390 MHz uplink/390–400 MHz downlink, 410–420 MHz uplink/420–430 MHz downlink). Each channel has a bandwidth of 25 kHz and can carry 36 kbit/s. Modulation is DQPSK. While V+D uses up to four TDMA voice or data channels per carrier, PDO performs statistical multiplexing. For accessing a channel, slotted Aloha is used.

Figure 4.21 shows the typical **TDMA frame structure** of TETRA. Each **frame** consists of four slots (four channels in the V+D service per carrier), with a frame duration of 56.67 ms. Each **slot** carries 510 bits within 14.17 ms, i.e., 36 kbit/s. 16 frames together with one **control frame** (CF) form a **multiframe**, and finally, a **hyperframe** contains 60 multiframes. To avoid sending and receiving at the same time, TETRA shifts the uplink for a period of two slots compared to the downlink.

9 Formerly known as trans-European trunked radio, but worldwide marketing is better without "Europe" in the name (see DECT).

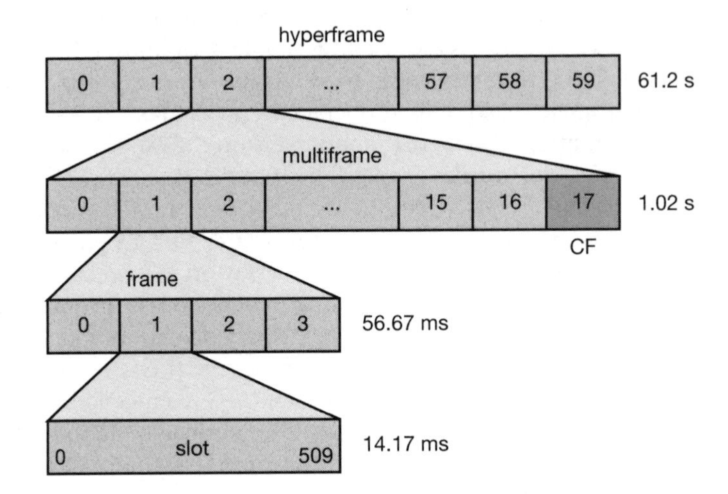

Figure 4.21 TETRA frame structure

TETRA offers **traffic channels (TCH)** and **control channels (CCH)** similar to GSM. Typical TCHs are TCH/S for voice transmission, and TCH/7.2, TCH/4.8, TCH/2.4 for data transmission (depending on the FEC mechanisms required).

However, in contrast to GSM, TETRA offers additional services like group call, acknowledged group call, broadcast call, and discreet listening. Emergency services need a sub-second group-call setup in harsh environments which possibly lack all infrastructure. These features are currently not available in GSM or other typical mobile telephone networks, so TETRA is complementary to other systems. TETRA has been chosen by many government organizations in Europe and China.

4.4 UMTS and IMT-2000

A lot has been written about third generation (or 3G) networks in the last few years. After a lot of hype and frustration these networks are currently deployed in many countries around the world. But how did it all start? First of all, the International Telecommunication Union (ITU) made a request for proposals for radio transmission technologies (RTT) for the **international mobile telecommunications (IMT) 2000** program (ITU, 2002), (Callendar, 1997), (Shafi, 1998). IMT-2000, formerly called future public land mobile telecommunication system (FPLMTS), tried to establish a common worldwide communication system that allowed for terminal and user mobility, supporting the idea of universal personal telecommunication (UPT). Within this context, ITU has created several recommendations for FPLMTS systems, e.g., network architectures for FPLMTS (M.817), Requirements for the Radio Interface(s) for FPLMTS (M.1034), or Framework for Services Supported by FPLMTS (M.816). The number 2000 in IMT-2000 should indicate the start of the system (year 2000+x) and the spectrum used (around 2000 MHz). IMT-2000 includes different environments such as indoor use, vehicles, satellites and pedestrians. The world radio conference (WRC) 1992 identified 1885–2025 and 2110–2200 MHz as the frequency bands that should be available worldwide for the new IMT-2000 systems (Recommendation ITU-R M.1036). Within these bands, two times 30 MHz have been reserved for mobile satellite services (MSS).

Figure 4.22 shows the ITU frequency allocation (from the world administrative radio conference, 1992) together with examples from several regions that already indicate the problem of worldwide common frequency bands. In Europe, some parts of the ITU's frequency bands for IMT-2000 are already allocated for DECT (see section 4.2). The remaining frequencies have been split into bands for UTRA-FDD (uplink: 1920–1980 MHz, downlink: 2110–2170 MHz) and UTRA-TDD (1900–1920 MHz and 2010–2025 MHz). The technology behind UTRA-FDD and –TDD will subsequently be explained in more detail as they form the basis of UMTS. Currently, no other system is planned for IMT-2000 in Europe. More bandwidth is available in China for the Chinese 3G system TD-SCDMA or possibly other 3G technologies (such as W-CDMA or cdma2000 – it is still open which system will dominate the Chinese market; Chen, 2002). Again slightly different frequencies are used by the 3G services in Japan, which are based on W-CDMA (like UTRA-FDD) or cdma2000. An open question is the future of 3G in the US as the ITU's frequency bands have already been allocated for 2G networks or are reserved for other use. In addition to the original frequency allocations, the world radio conference (WRC) allocated new terrestrial IMT-2000 bands in the range of 800–1000 MHz, 1700–1900 MHz and 2500–2700 MHz in 2000. This approach includes the reuse of 2G spectrum (Evci, 2001).

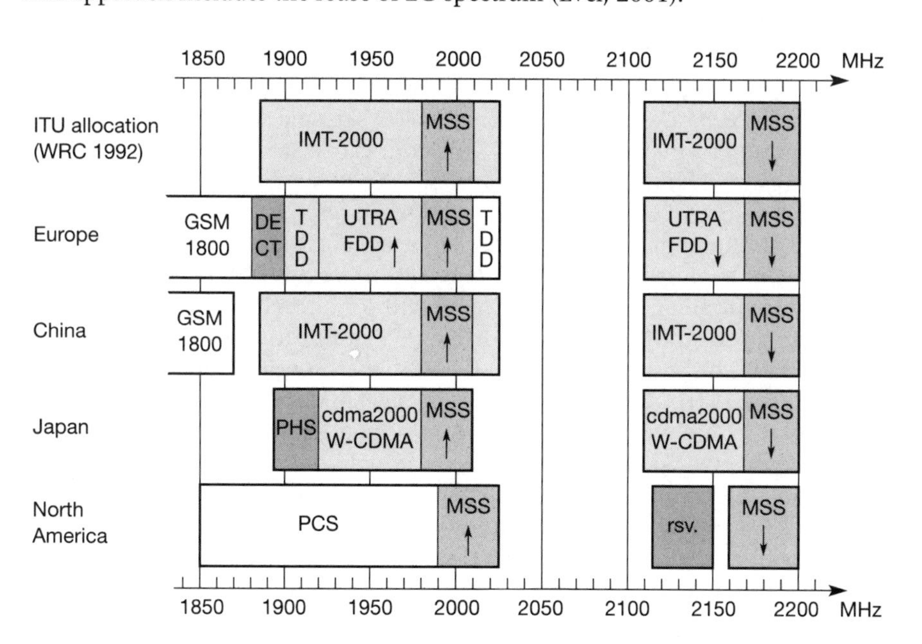

Figure 4.22
IMT-2000 frequencies

Now the reader might be confused by all the different technologies mentioned in the context of IMT-2000. Wasn't the plan to have a common global system? This was the original plan, but after many political discussions and fights about patents this idea was dropped and a so-called family of 3G standards was adopted.

For the RTT, several proposals were received in 1998 for indoor, pedestrian, vehicular, and satellite environments. These came from many different organizations, e.g., **UWC-136** from the Universal Wireless Communications Consortium (US) that extends the IS-136 standard into the third generation systems, **cdma2000** that is based on the IS-95 system (US), and wideband packet CDMA (WP-CDMA) which tries to align to the European UTRA proposal. Basically, three big regions were submitting proposals to the ITU: ETSI for Europe, ARIB (Association of Radio Industries and Broadcasting) and TTC (Telecommunications Technology Council) for Japan, and ANSI (American National Standards Institute) for the US.

The European proposal for IMT-2000 prepared by ETSI is called **universal mobile telecommunications system (UMTS)** (Dasilva, 1997), (Ojanperä, 1998), the specific proposal for the radio interface RTT is **UMTS (now: universal) terrestrial radio access (UTRA)** (ETSI, 1998n), (UMTS Forum, 2002). UMTS as initially proposed by ETSI rather represents an evolution from the second generation GSM system to the third generation than a completely new system. In this way, many solutions have been proposed for a smooth transition from GSM to UMTS, saving money by extending the current system rather than introducing a new one (GSMMoU, 1998).

One initial enhancement of GSM toward UMTS was **enhanced data rates for global (or: GSM) evolution (EDGE)**, which uses enhanced modulation schemes (8 PSK instead of GSM's GMSK, see chapter 2) and other techniques for data rates of up to 384 kbit/s using the same 200 kHz wide carrier and the same frequencies as GSM (i.e., a data rate of 48 kbit/s per time slot is available). EDGE can be introduced incrementally offering some channels with EDGE enhancement that can switch between EDGE and GSM/GPRS. In Europe, EDGE was never used as a step toward UMTS but operators directly jumped onto UMTS. However, EDGE can also be applied to the US IS-136 system and may be a choice for operators that want to enhance their 2G systems (3G Americas, 2002).

Besides enhancing data rates, new additions to GSM, like **customized application for mobile enhanced logic (CAMEL)** introduce intelligent network support. This system supports, for example, the creation of a **virtual home environment (VHE)** for visiting subscribers. GSMMoU (1998) provides many proposals covering QoS aspects, roaming, services, billing, accounting, radio aspects, core networks, access networks, terminal requirements, security, application domains, operation and maintenance, and several migration aspects.

UMTS fits into a bigger framework developed in the mid-nineties by ETSI, called **global multimedia mobility (GMM)**. GMM provides an architecture to integrate mobile and fixed **terminals**, many different **access networks** (GSM BSS, DECT, ISDN, UMTS, LAN, WAN, CATV, MBS), and several **core transport**

networks (GSM NSS+IN, ISDN+IN, B-ISDN+TINA, TCP/IP) (ETSI, 2002). Within this framework, ETSI developed **basic requirements** for UMTS and for UTRA, the radio interface (ETSI, 1998h). Key requirements are minimum data rates of 144 kbit/s for rural outdoor access (with the goal of 384 kbit/s) at a maximum speed of 500 km/h.[10] For suburban outdoor use a minimum of 384 kbit/s should be achieved with the goal of 512 kbit/s at 120 km/h. For indoor or city use with relatively short ranges, up to 2 Mbit/s are required at 10 km/h (walking).

UMTS should also provide several bearer services, real-time and non real-time services, circuit and packet switched transmission, and many different data rates. Handover should be possible between UMTS cells, but also between UMTS and GSM or satellite networks. The system should be compatible with GSM, ATM, IP, and ISDN-based networks. To reflect the asymmetric bandwidth needs of typical users, UMTS should provide a variable division of uplink and downlink data rates. Finally, UMTS has to fit into the IMT-2000 framework (this is probably the decisive factor for its success). As the global UMTS approach is rather ambitious, a more realistic alternative for the initial stages would be UMTS cells in cities providing a subset of services.

Several companies and interest groups have handed in proposals for UTRA (ETSI, 1998i), of which ETSI selected two for UMTS in January 1998. For the **paired band** (using FDD as a duplex mechanism), ETSI adopted the **W-CDMA** (Wideband CDMA) proposal, for the **unpaired band** (using TDD as duplex mechanism) the **TD-CDMA** (Time Division CDMA) proposal is used (Adachi, 1998), (Dahlman, 1998), (ETSI, 1998n). The paired band is typically used for public mobile network providers (wide area, see GSM), while the unpaired band is often used for local and indoor communication (see DECT). The following sections will present key properties of the initial UMTS system.

What happened to the IMT-2000 family? Figure 4.23 gives an overview. As a single standard could not be found, the ITU standardized five groups of 3G radio access technologies.

- **IMT-DS:** The **direct spread** technology comprises wideband CDMA (**W-CDMA**) systems. This is the technology specified for UTRA-FDD and used by all European providers and the Japanese NTT DoCoMo for 3G wide area services. To avoid complete confusion ITU's name for the technology is IMT-DS, ETSI called it UTRA-FDD in the UMTS context, and technology used is called W-CDMA (in Japan this is promoted as FOMA, freedom of mobile multimedia access). Today, standardization of this technology takes place in 3GPP (Third generation partnership project, 3GPP, 2002a). Section 4.4.1 provides more detail about the standardization process.
- **IMT-TC:** Initially, this family member, called **time code**, contained only the UTRA-TDD system which uses time-division CDMA (**TD-CDMA**). Later on, the Chinese proposal, TD-synchronous CDMA (**TD-SCDMA**) was added.

10 This speed is a problem as currently, only DAB can provide higher bit rates at high speeds.

Both standards have been combined and 3GPP fosters the development of this technology. It is unclear when and to what extent this technology will be introduced. The initial UMTS installations are based on W-CDMA.

- **IMT-MC:** cdma2000 is a **multi-carrier** technology standardized by 3GPP2 (Third generation partnership project 2, 3GPP2, 2002), which was formed shortly after 3GPP to represent the second main stream in 3G technology. Version cdma2000 EV-DO has been accepted as the 3G standard.
- **IMT-SC:** The enhancement of the US TDMA systems, UWC-136, is a **single carrier** technology originally promoted by the Universal Wireless Communications Consortium (UWCC). It is now integrated into the 3GPP efforts. This technology applies EDGE, among others, to enhance the 2G IS-136 standard.
- **IMT-FT**: As **frequency time** technology, an enhanced version of the cordless telephone standard DECT has also been selected for applications that do not require high mobility. ETSI is responsible for the standardization of DECT.

The main driving forces in the standardization process are 3GPP and 3GPP2. ETSI has moved its GSM standardization process to 3GPP and plays a major role there. 3GPP tends to be dominated by European and Japanese manufacturers and standardization bodies, while 3GPP2 is dominated by the company Qualcomm and CDMA network operators. The quarrels between Qualcomm and European manufacturers (e.g., Nokia, Ericsson) regarding CDMA patents (UMTS and cdma2000 use CDMA) even escalated into the political arena back in 1998 (US vs EU). Everything cooled down when hundreds of patents had been exchanged and the systems had been harmonized (e.g., CDMA chipping rates).

Figure 4.23
The IMT-2000 family

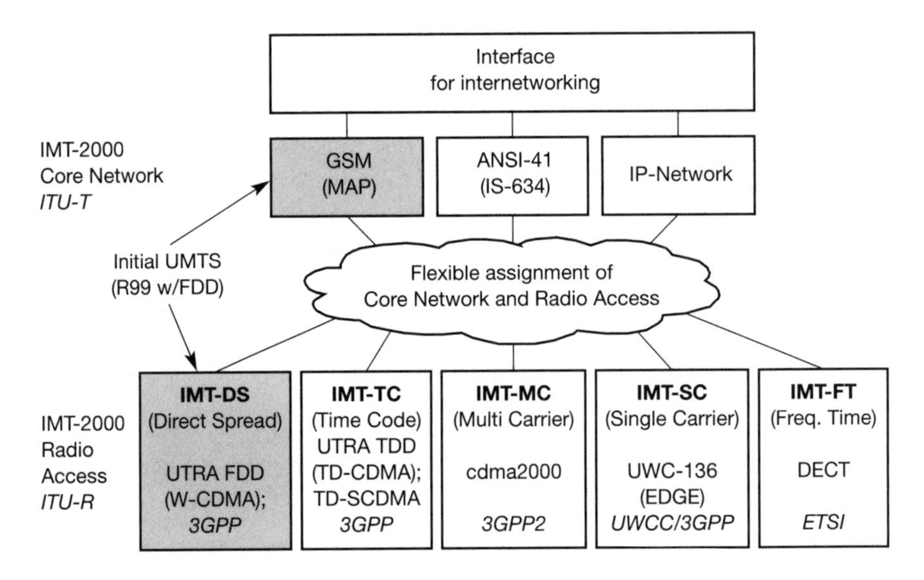

Figure 4.23 shows more than just the radio access technologies. One idea of IMT-2000 is the flexible assignment of a core network to a radio access system. The classical core network uses SS7 for signaling which is enhanced by ANSI-41 (cdmaOne, cdma2000, TDMA) or MAP (GSM) to enable roaming between different operators. The evolution toward 4G systems is indicated by the use of all-IP core networks (see Chapter 11). Obviously, internet-working functions have to be provided to enable cross-system data transfer, roaming, billing etc.

4.4.1 UMTS releases and standardization

UMTS as discussed today and introduced in many countries relies on the initial release of the UMTS standard called **release 99** or **R99** for short. This release of the specification describes the new radio access technologies UTRA FDD and UTRA TDD, and standardizes the use of a GSM/GPRS network as core within 440 separate specifications. This enables a cost effective migration from GSM to UMTS. The initial installations will even offer the FDD mode only as indicated in Figure 4.23. This release was (almost) finalized in 1999 – hence the name R99. The following sections will focus on this release as it is unclear when, and to what extent, the following releases will be realized.

After R99 the release 2000 or R00 followed. However, in September 2000 3GPP realized that it would be impossible to finalize the standard within the year 2000. 3GPP decided to split R2000 into two standards and call them release 4 (Rel-4) and release 5 (Rel-5). The version of all standards finalized for R99 start with 3.x.y (a reason for renaming R99 into Rel-3), Rel-4 and Rel-5 versions start with 4.x.y and 5.x.y, respectively. The standards are grouped into series. For example, radio aspects are specified in series 25, technical realization in series 23, and codecs in series 26. The complete standard number (e.g., TS 25.401 V3.10.0) then identifies the series (25), the standard itself (401), the release (3), and the version within the release (10.0). All standards can be downloaded from `www.3gpp.org` (the example given is the UTRAN overall description, release 99, from June 2002).

Release 4 introduces quality of service in the fixed network plus several execution environments (e.g., MExE, mobile execution environment, see chapter 10) and new service architectures. Furthermore, the Chinese proposal, TD-SCDMA was added as low chiprate option to UTRA-TDD (only 1.28 Mchip/s occupying only 1.6 MHz bandwidth). This release already consists of over 500 specifications and was frozen in March 2001.

Release 5 specifies a radically different core network. The GSM/GPRS based network will be replaced by an almost all-IP-core. While the radio interfaces remain the same, the changes in the core are tremendous for telecommunication network operators who have used traditional telephone technologies for many years. The content of this specification was frozen March 2002. This standard integrates IP-based multimedia services (IMS) controlled by the IETF's session initiation protocol (SIP, RFC 3261; Rosenberg, 2002; SIP Forum, 2002). A high speed downlink packet access (HSDPA) with speeds in the order of

8–10 Mbit/s was added as well as a wideband 16 kHz AMR codec for better audio quality. Additional features are end-to-end QoS messaging and several data compression mechanisms.

3GPP is currently working on **release 6** (and thinking of release 7) which is expected to be frozen in March 2003. This release comprises the use of multiple input multiple output (MIMO) antennas, enhanced MMS, security enhancements, WLAN/UMTS interworking, broadcast/multicast services, enhanced IMS, IP emergency calls, and many more management features (3GPP, 2002a).

The reader should not forget that many companies still have to make any money from, release 99, so it is not clear at what time and to what extent the new releases will be implemented. The following describes the initial UMTS standard, release 99, which is currently deployed.

4.4.2 UMTS system architecture

Figure 4.24 shows the very simplified UMTS reference architecture which applies to both UTRA solutions (3GPP, 2000). The **UTRA network (UTRAN)** handles cell level mobility and comprises several **radio network subsystems (RNS)**. The functions of the RNS include radio channel ciphering and deciphering, handover control, radio resource management etc. The UTRAN is connected to the **user equipment (UE)** via the radio interface U_u (which is comparable to the U_m interface in GSM). Via the $\mathbf{I_u}$ interface (which is similar to the A interface in GSM), UTRAN communicates with the **core network (CN)**. The CN contains functions for inter-system handover, gateways to other networks (fixed or wireless), and performs location management if there is no dedicated connection between UE and UTRAN.

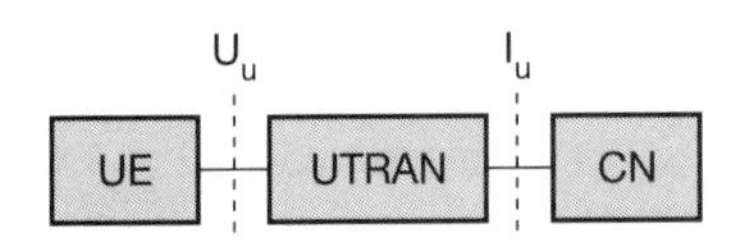

Figure 4.24
Main components of the UMTS reference architecture

UMTS further subdivides the above simplified architecture into so-called **domains** (see Figure 4.25). The **user equipment** domain is assigned to a single user and comprises all the functions that are needed to access UMTS services. Within this domain are the USIM domain and the mobile equipment domain. The **USIM** domain contains the SIM for UMTS which performs functions for encryption and authentication of users, and stores all the necessary user-related data for UMTS. Typically, this USIM belongs to a service provider and contains a micro processor for an enhanced program execution environment (USAT, UMTS SIM application toolkit). The end device itself is in the **mobile equipment** domain. All functions for radio transmission as well as user interfaces are located here.

The **infrastructure** domain is shared among all users and offers UMTS services to all accepted users. This domain consists of the **access network** domain, which contains the radio access networks (RAN), and the core network domain, which contains access network independent functions. The **core network** domain can be separated into three domains with specific tasks. The **serving network** domain comprises all functions currently used by a user for accessing

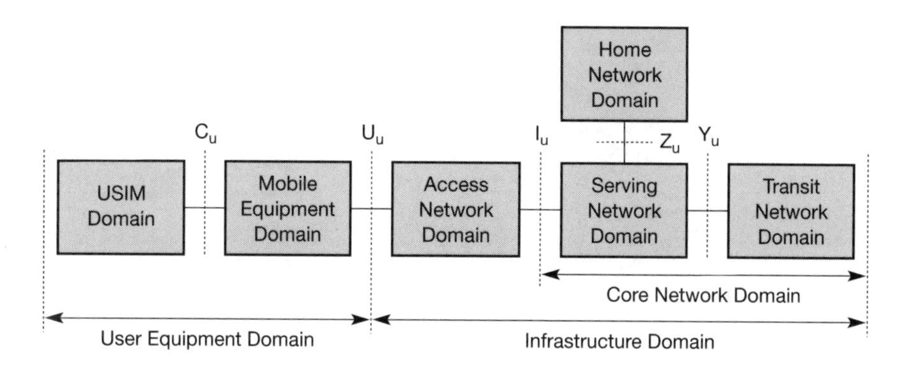

Figure 4.25
UMTS domains and interfaces

UMTS services. All functions related to the home network of a user, e.g., user data look-up, fall into the **home network** domain. Finally, the **transit network** domain may be necessary if, for example, the serving network cannot directly contact the home network. All three domains within the core network may be in fact the same physical network. These domains only describe functionalities.

4.4.3 UMTS radio interface

The biggest difference between UMTS and GSM comes with the new radio interface (U_u). The duplex mechanisms are already well known from GSM (FDD) and DECT (TDD). However, the direct sequence (DS) CDMA used in UMTS is new (for European standards, not in the US where CDMA technology has been available since the early nineties). DS-CDMA was introduced in chapters 2 and 3. This technology multiplies a stream of bits with a chipping sequence. This spreads the signal and, if the chipping sequence is unique, can separate different users. All signals use the same frequency band (in UMTS/IMT-2000 5 MHz-wide bands have been specified and licensed to network operators). To separate different users, the codes used for spreading should be (quasi) orthogonal, i.e., their cross-correlation should be (almost) zero.

UMTS uses a constant **chipping rate** of 3.84 Mchip/s. Different user data rates can be supported using different spreading factors (i.e., the number of chips per bit). Figure 4.26 shows the basic ideas of spreading and separation of different senders in UMTS. The first step in a sender is spreading of user data ($data_i$) using orthogonal **spreading** codes. Using orthogonal codes separates the different data streams of a sender. UMTS uses so-called **orthogonal variable spreading factor (OVSF)** codes. Figure 4.27 shows the basic idea of OVSF. Orthogonal codes are generated by doubling a chipping sequence X with and without flipping the sign of the chips. This results in X and –X, respectively. Doubling the chipping sequence also results in spreading a bit twice as much as before. The spreading factor SF=n becomes 2n. Starting with a spreading factor of 1, Figure 4.27 shows the generation of orthogonal codes with different spreading factors. Two codes are orthogonal as long as one code is never a part of the other code. Looking at the coding tree in Figure 4.27 and considering the construction of the codes, orthogonality is guaranteed if one code has not been generated based on another. For example, if a sender uses the code (1,–1)

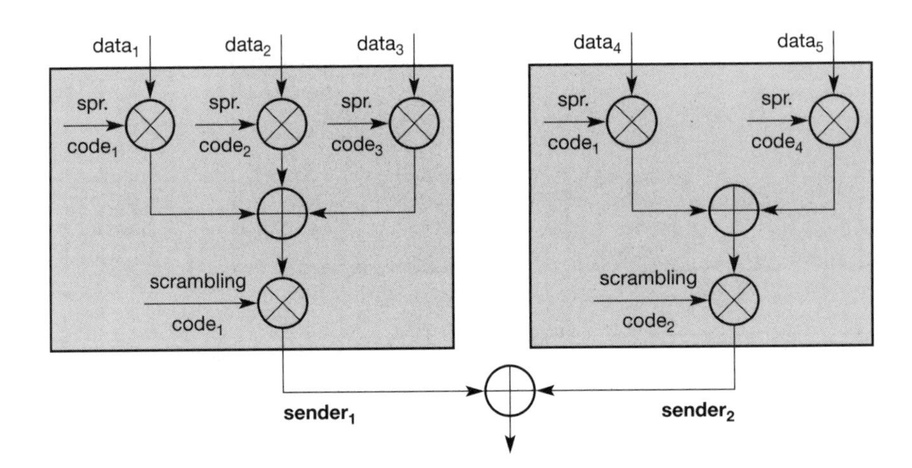

Figure 4.26 Spreading and scrambling of user date

with spreading factor 2, it is not allowed to use any of the codes located in the subtrees generated out of (1,–1). This means that, e.g., (1,–1,1,–1), (1,–1,–1,1,–1,1,1,–1), or (1,–1,–1,1,–1,1,1,–1,–1,1,1,–1,1,–1,–1,1) cannot be used anymore. However, it is no problem to use codes with different spreading factors if one code has not been generated using the other. Thus, (1,–1) block only the lower subtree in Figure 4.27, many other codes from the upper part can still be used. An example for a valid combination in OVSF is (1,–1), (1,1,–1,–1), (1,1,1,1,1,1,1,1), (1,1,1,1,–1,–1,–1,–1, 1,1,1,1,–1,–1,–1,–1), (1,1,1,1,–1,–1,–1,–1,–1,–1,–1,–1,1,1,1,1). This combination occupies the whole code spaces and allows for the transmission of data with different spreading factors (2, 4, 8, and 2*16). This example shows the tight coupling of available spreading factors and orthogonal codes.

Now remember that UMTS uses a constant chipping rate (3.84 Mchip/s). Using different spreading factors this directly translates into the support of different data rates. If the chipping rate is constant, doubling the spreading factor means dividing the data rate by two. But this also means that UMTS can only support a single data stream with SF=1 as then no other code may be used. Using the example combination above, a stream with half the maximum data rate, one with a fourth, one with an eighth, and two with a sixteenth are supported at the same time.

Each sender uses OVSF to spread its data streams as Figure 4.26 shows. The spreading codes chosen in the senders can be the same. Using different spreading codes in all senders within a cell would require a lot of management and would increase the complexity. After spreading all chip streams are added and scrambled. **Scrambling** does not spread the chip sequence any further but XORs chips based on a code. In the **FDD** mode, this scrambling code is unique for each sender and separates all senders (UE and base station) in a cell. After scrambling, the signals of different senders are quasi orthogonal. Quasi-orthogonal signals have the nice feature that they stay quasi-orthogonal even if they are not synchronized. Using orthogonal codes would require chip-synchronous reception and tight synchronization (this is done in other CDMA networks). For **TDD** the scrambling code is cell specific, i.e., all stations in a cell use the same scrambling code and cells are separated using different codes. The scrambled chips are **QPSK** modulated and transmitted.

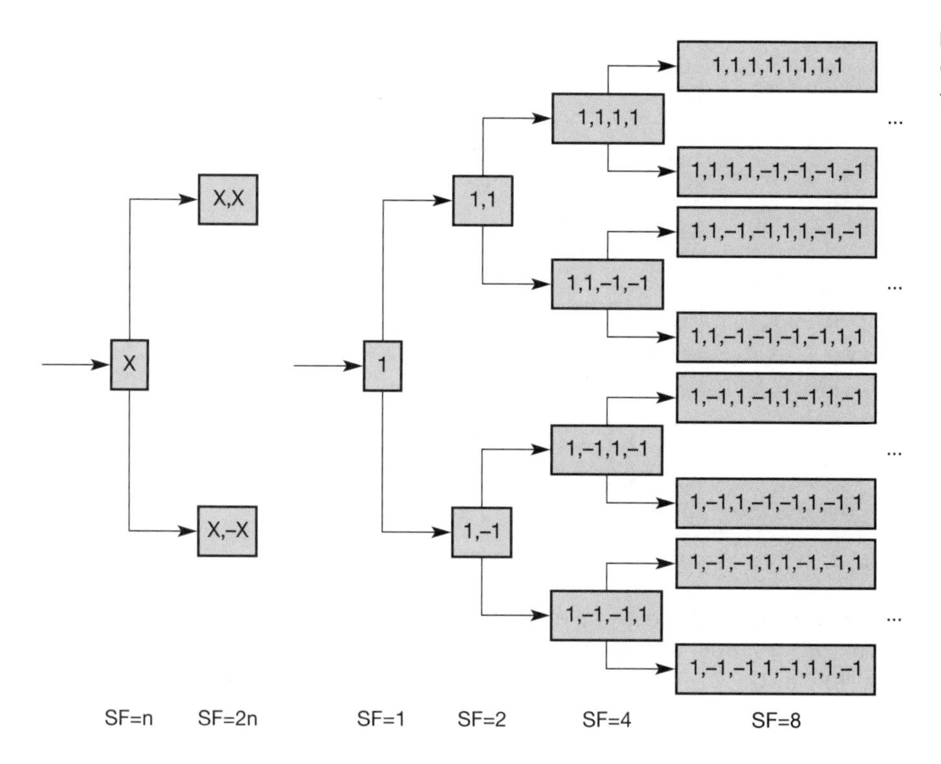

Figure 4.27
OVSF code tree used for orthogonal spreading

4.4.3.1 UTRA-FDD (W-CDMA)

The FDD mode for UTRA uses **wideband CDMA (W-CDMA)** with direct sequence spreading. As implied by FDD, uplink and downlink use different frequencies. A mobile station in Europe sends via the uplink using a carrier between 1920 and 1980 MHz, the base station uses 2110 to 2170 MHz for the downlink (see Figure 4.22). Figure 4.28 shows a radio frame comprising 15 time slots. Time slots in W-CDMA are not used for user separation but to support periodic functions (note that this is in contrast to, e.g., GSM, where time slots are used to separate users!). A radio frame consists of 38,400 chips and has a duration of 10 ms. Each time slot consists of 2,560 chips, which roughly equals 666.6 μs.[11] The occupied bandwidth per W-CDMA channel is 4.4 to 5 MHz (channel spacing can be varied to avoid interference between channels of different operators). These 5 MHz bands of the spectrum have been sold in many countries using an auction or a beauty contest. In Germany, the FDD spectrum was sold for over 50 billion Euros during an auction! But that was at a time when marketing people tried to convince everyone that UMTS would bring

11 Early version of W-CDMA specified a chipping rate of 4.096 M chip/s and 16 time slots per frame. This was changed during the harmonization process which was necessary to avoid patent conflicts and to enable devices that can handle different CDMA standards. The harmonization process is fostered by the operators harmonization group (OHG), which is an informal steering group of wireless operator companies promoting 3G harmonization. The OHG was founded in 1999.

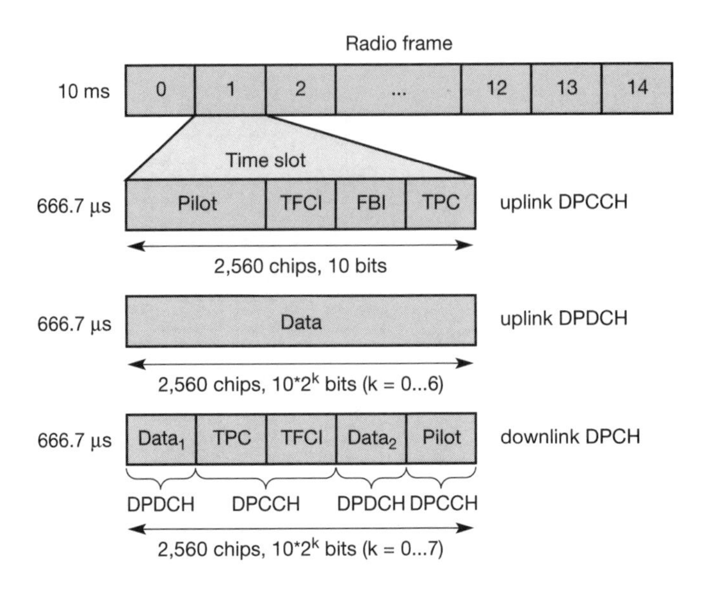

Figure 4.28
UTRA FDD (W-CDMA) frame structure

high-bandwidth applications to any mobile device with high profits for all. Today, most people are much more realistic and know that data rates will be quite low in the beginning (150 kbit/s per user are realistic, 2 Mbit/s are not). The capacity of a cell under realistic assumptions (interference etc.), i.e., the sum of all data rates, will rather be 2 Mbit/s. To provide high data rates a lot of money has to be invested in the infrastructure: UTRA FDD requires at least twice as many base stations as GSM; cell diameters of 500 m will be commonplace. This shows clearly that this technology will not cover whole countries in the near future but cities and highways only. People in the countryside will have to rely on GSM/GPRS for many more years to come.

Back to the frame structure shown in Figure 4.28. Similar to GSM, UMTS defines many logical and physical channels, and their mapping. The figure shows three examples of physical channels as they are used for data transmission. Two physical channels are shown for the uplink.

- **Dedicated physical data channel (DPDCH):** This channel conveys user or signaling data. The spreading factor of this channel can vary between 4 and 256. This directly translates into the data rates this channel can offer: 960 kbit/s (spreading factor 4, 640 bits per slot, 15 slots per frame, 100 frames per second), 480, 240, 120, 60, 30, and 15 kbit/s (spreading factor 256). This also shows one of the problems of using OVSF for spreading: only certain multiples of the basic data rate of 15 kbit/s can be used. If, for example, 250 kbit/s are needed the device has to choose 480 kbit/s, which wastes bandwidth. In each connection in layer 1 it can have between zero and six DPDCHs. This results in a theoretical maximum data rate of 5,740 kbit/s (UMTS describes UEs with a maximum of 1,920 kbit/s only). Table 4.7 shows typical user data rates together with the required data rates on the physical channels.

User data rate [kbit/s]	12.2 (voice)	64	144	384
DPDCH [kbit/s]	60	240	480	960
DPCCH [kbit/s]	15	15	15	15
Spreading	64	16	8	4

Table 4.7 Typical UTRA–FDD uplink data rates

- **Dedicated physical control channel (DPCCH):** In each connection layer 1 needs exactly one DPCCH. This channel conveys control data for the physical layer only and uses the constant spreading factor 256. The **pilot** is used for channel estimation. The **transport format combination identifier (TFCI)** specifies the channels transported within the DPDCHs. Signaling for a soft handover is supported by the **feedback information field (FBI)**. The last field, **transmit power control (TPC)** is used for controlling the transmission power of a sender. Power control is performed in each slot, thus 1,500 power control cycles are available per second. Tight power control is necessary to mitigate near-far-effects as explained in chapter 2. Six different DPCCH bursts have been defined which differ in the size of the fields.
- **Dedicated physical channel (DPCH):** The downlink time multiplexes control and user data. Spreading factors between 4 and 512 are available. Again, many different burst formats (17 altogether) have been defined which differ in the size of the field shown in Figure 4.28. The available data rates for data channels (DPDCH) within a DPCH are 6 (SF=512), 24, 51, 90, 210, 432, 912, and 1,872 kbit/s (SF=4).

While no collisions can occur on the downlink (only the base station sends on the downlink), medium access on the uplink has to be coordinated. A **physical random access channel (PRACH)** is used for this purpose. UTRA-FDD defines 15 random access slots within 20 ms; within each access slot 16 different access preambles can be used for random access. Using slotted Aloha, a UE can access an access slot by sending a preamble. The UE starts with the lowest available transmission power to avoid interfering with other stations. If no positive acknowledgement is received, the UE tries another slot and another preamble with the next higher power level (power ramping). The number of available access slots can be defined per cell and is transmitted via a broadcast channel to all UEs.

A UE has to perform the following steps during the **search for a cell** after power on:

- **Primary synchronization:** A UE has to synchronize with the help of a 256 chip primary synchronization code. This code is the same for all cells and helps to synchronize with the time slot structure.

- **Secondary synchronization:** During this second phase the UE receives a secondary synchronization code which defines the group of scrambling codes used in this cell. The UE is now synchronized with the frame structure.
- **Identification of the scrambling code:** The UE tries all scrambling codes within the group of codes to find the right code with the help of a correlator. After these three steps the UE can receive all further data over a broadcast channel.

4.4.3.2 UTRA-TDD (TD-CDMA)

The second UTRA mode, UTRA-TDD, separates up and downlink in time using a radio frame structure similar to FDD. 15 slots with 2,560 chips per slot form a radio frame with a duration of 10 ms. The chipping rate is also 3.84 Mchip/s. To reflect different user needs in terms of data rates, the TDD frame can be **symmetrical** or **asymmetrical**, i.e., the frame can contain the same number of uplink and downlink slots or any arbitrary combination. The frame can have only one **switching point** from uplink to downlink or several switching points. However, at least one slot must be allocated for the uplink and downlink respectively.

The system can change the spreading factor (1, 2, 4, 8, 16) as a function of the desired data rate. Using the burst type shown in Figure 4.29 results in data rates of 6,624, 3,312, 1,656, 828, and 414 kbit/s respectively (if all slots are used for data transmission). The figure shows a burst of type 2 which comprises two **data** fields of 1,104 chips each. Spreading is applied to these data fields only. Additionally, a **midample** is used for training and channel estimation. As TDD uses the same scrambling codes for all stations, the stations must be tightly synchronized and the spreading codes are available only once per slot. This results in a maximum number of 16 simultaneous sending stations. To loosen the tight synchronization a little bit, a **guard period (GP)** has been introduced at the end of each slot. Due to the tight synchronization and the use of orthogonal codes, a simpler power control scheme with less power control cycles (e.g., 100 per second) is sufficient.

Figure 4.29
UTRA TDD (TD–CDMA) frame structure

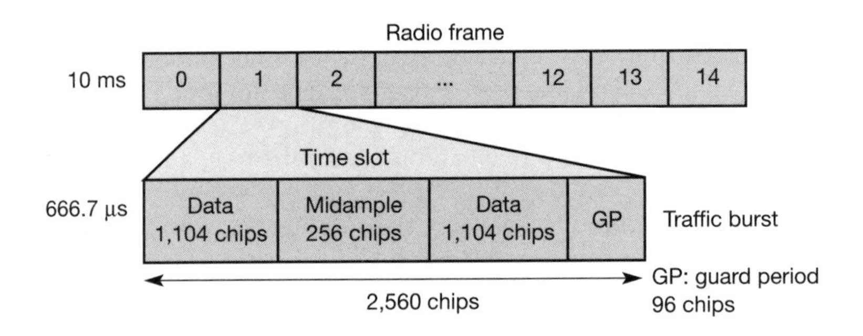

UTRA TDD occupies 5 MHz bandwidth per channel as UTRA FDD does per direction (FDD needs 2x 5 MHz). Compared to the license for FDD, TDD was quite cheap. Germany paid less than €300 million. Figure 4.22 shows the location of the spectrum for this UMTS mode, but it is unclear to what extend this system will be deployed. The coverage per cell is even less than using FDD, UEs must not move too fast – this sounds like the characteristics of WLANs which are currently deployed in many places.

4.4.4 UTRAN

Figure 4.30 shows the basic architecture of the UTRA network (UTRAN; 3GPP, 2002b). This consists of several **radio network subsystems (RNS)**. Each RNS is controlled by a **radio network controller (RNC)** and comprises several components that are called node B. An RNC in UMTS can be compared with the BSC; a node B is similar to a BTS. Each **node B** can control several antennas which make a radio cell. The mobile device, UE, can be connected to one or more antennas as will subsequently be explained in the context of handover. Each RNC is connected with the core network (CN) over the interface $\mathbf{I_u}$ (similar to the role of the A interface in GSM) and with a node B over the interface $\mathbf{I_{ub}}$. A new interface, which has no counterpart in GSM, is the interface $\mathbf{I_{ur}}$ connecting two RNCs with each other. The use of this interface is explained together with the UMTS handover mechanisms.

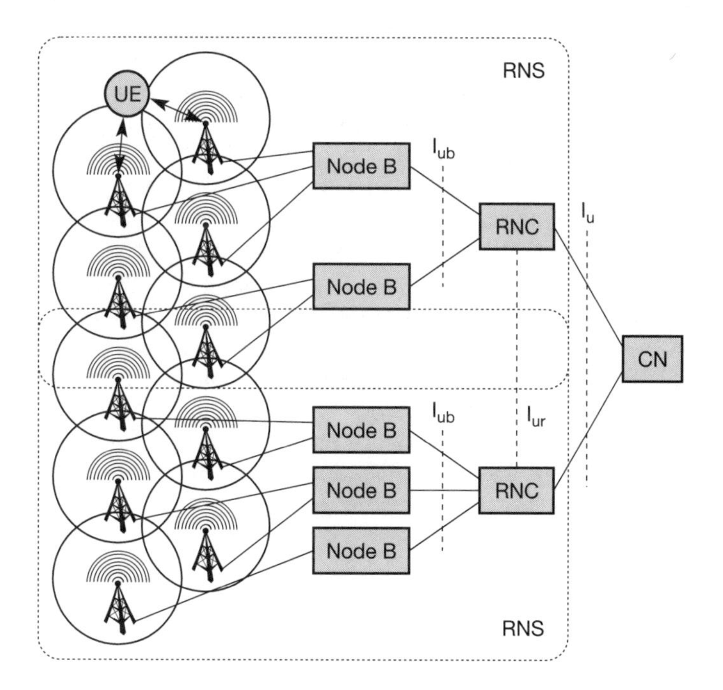

Figure 4.30
Basic architecture of the UTRA network

4.4.4.1 Radio network controller

An **RNC** in UMTS has a broad spectrum of tasks as listed in the following:

- **Call admission control:** It is very important for CDMA systems to keep the interference below a certain level. The RNC calculates the traffic within each cell and decides, if additional transmissions are acceptable or not.
- **Congestion control:** During packet-oriented data transmission, several stations share the available radio resources. The RNC allocates bandwidth to each station in a cyclic fashion and must consider the QoS requirements.
- **Encryption/decryption:** The RNC encrypts all data arriving from the fixed network before transmission over the wireless link (and vice versa).
- **ATM switching and multiplexing, protocol conversion:** Typically, the connections between RNCs, node Bs, and the CN are based on ATM. An RNC has to switch the connections to multiplex different data streams. Several protocols have to be converted – this is explained later.
- **Radio resource control:** The RNC controls all radio resources of the cells connected to it via a node B. This task includes interference and load measurements. The priorities of different connections have to be obeyed.
- **Radio bearer setup and release:** An RNC has to set-up, maintain, and release a logical data connection to a UE (the so-called UMTS radio bearer).
- **Code allocation:** The CDMA codes used by a UE are selected by the RNC. These codes may vary during a transmission.
- **Power control:** The RNC only performs a relatively loose power control (the outer loop). This means that the RNC influences transmission power based on interference values from other cells or even other RNCs. But this is not the tight and fast power control performed 1,500 times per second. This is carried out by a node B. This outer loop of power control helps to minimize interference between neighbouring cells or controls the size of a cell.
- **Handover control and RNS relocation:** Depending on the signal strengths received by UEs and node Bs, an RNC can decide if another cell would be better suited for a certain connection. If the RNC decides for handover it informs the new cell and the UE as explained in subsection 4.4.6. If a UE moves further out of the range of one RNC, a new RNC responsible for the UE has to be chosen. This is called RNS relocation.
- **Management:** Finally, the network operator needs a lot of information regarding the current load, current traffic, error states etc. to manage its network. The RNC provides interfaces for this task, too.

4.4.4.2 Node B

The name node B was chosen during standardization until a new and better name was found. However, no one came up with anything better so it remained. A node B connects to one or more antennas creating one or more cells (or sectors in GSM speak), respectively. The cells can either use FDD or TDD or both. An important task of a node B is the inner loop power control to mitigate near-far effects. This node also measures connection qualities and signal strengths. A node B can even support a special case of handover, a so-called softer handover which takes place between different antennas of the same node B (see section 4.4.6).

4.4.4.3 User equipment

The UE shown in Figure 4.30 is the counterpart of several nodes of the architecture.

- As the counterpart of a node B, the UE performs signal quality measurements, inner loop power control, spreading and modulation, and rate matching.
- As a counterpart of the RNC, the UE has to cooperate during handover and cell selection, performs encryption and decryption, and participates in the radio resource allocation process.
- As a counterpart of the CN, the UE has to implement mobility management functions, performs bearer negotiation, or requests certain services from the network.

This list of tasks of a UE, which is not at all exhaustive, already shows the complexity such a device has to handle. Additionally, users also want to have organizers, games, cameras, operating systems etc. and the stand-by time should be high.

4.4.5 Core network

Figure 4.31 shows a high-level view of the UMTS release 99 core network architecture together with a UTRAN RNS and a GSM BSS (see section 4.1). This shows the evolution from GSM/GPRS to UMTS. The core network (CN) shown here is basically the same as already explained in the context of GSM (see Figure 4.4) and GPRS (see Figure 4.16). The **circuit switched domain (CSD)** comprises the classical circuit switched services including signaling. Resources are reserved at connection setup and the GSM components MSC, GMSC, and VLR are used. The CSD connects to the RNS via a part of the I_u interface called **I_uCS**. The CSD components can still be part of a classical GSM network connected to a BSS but need additional functionalities (new protocols etc.).

The **packet switched domain (PSD)** uses the GPRS components SGSN and GGSN and connects to the RNS via the **I_uPS** part of the I_u interface. Both domains need the data-bases EIR for equipment identification and HLR for location management (including the AuC for authentication and GR for user specific GPRS data).

Reusing the existing infrastructure helps to save a lot of money and may convince many operators to use UMTS if they already use GSM. The UMTS industry pushes their technology with the help of the market dominance of GSM. This is basically the same as cdma2000, which is a evolution of cdmaOne. The real flexible core network comes with releases 5 and 6, where the GSM

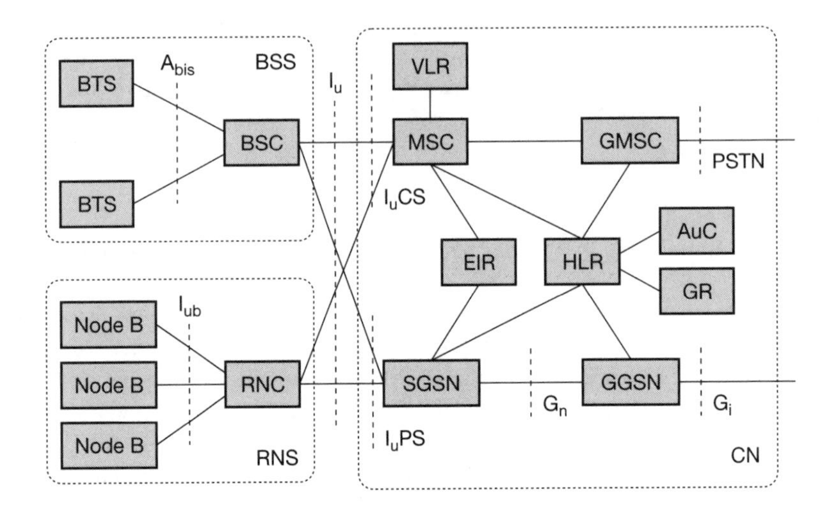

Figure 4.31 UMTS core network together with a 3G RNS and a 2G BSS

circuit switched part is being replaced by an all-IP core. Chapter 11 presents this idea in the context of 4G networks. It is not yet clear when this replacement of GSM will take place as many questions are still open (quality of service and security being the most important).

Figure 4.32 shows the protocol stacks of the users planes of the circuit switched and packet switched domains, respectively. The **CSD** uses the **ATM adaptation layer 2** (AAL2) for user data transmission on top of ATM as transport technology. The RNC in the UTRAN implements the radio link control (RLC) and the MAC layer, while the physical layer is located in the node B. The AAL2 **segmentation and reassembly** layer (SAR) is, for example, used to segment data packets received from the RLC into small chunks which can be transported in ATM. AAL2 and ATM has been chosen, too, because these protocols can transport and multiplex low bit rate voice data streams with low jitter and latency (compared to the protocols used in the PSD).

In the **PSD** several more protocols are needed. Basic data transport is performed by different lower layers (e.g., ATM with AAL5, frame relay). On top of these lower layers UDP/IP is used to create a UMTS internal IP network. All packets (e.g., IP, PPP) destined for the UE are encapsulated using the **GPRS tunneling protocol (GTP)**. The RNC performs protocol conversion from the combination GTP/UDP/IP into the **packet data convergence protocol (PDCP)**. This protocol performs header compression to avoid redundant data transmission using scarce radio resources. Comparing Figure 4.32 with Figure 4.17 (GPRS protocol reference model) shows a difference with respect to the tunnel. In UMTS the RNC handles the tunneling protocol GTP, while in GSM/GPRS GTP is used between an SGSN and GGSN only. The BSC in GSM is not involved in IP protocol processing.

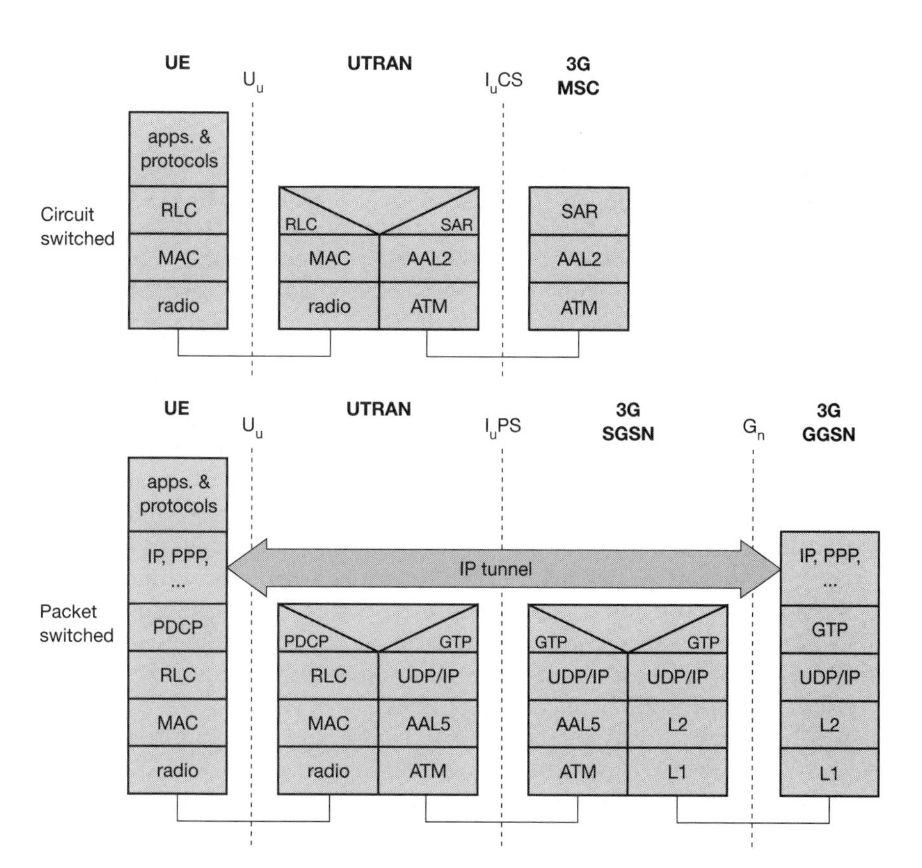

Figure 4.32 User plane protocol stacks (circuit and packet switched)

The **radio layer** (physical layer) depends on the UTRA mode (see sections 4.4.3.1 and 4.4.3.2). The **medium access control (MAC)** layer coordinates medium access and multiplexes logical channels onto transport channels. The MAC layers also help to identify mobile devices and may encrypt data. The **radio link control (RLC)** layer offers three different transport modes. The **acknowledged mode** transfer uses ARQ for error correction and guarantees one-time in-order delivery of data packets. The **unacknowledged mode** transfer does not perform ARQ but guarantees at least one-time delivery of packets with the help of sequence numbers. The **transparent mode** transfer simply forwards MAC data without any further processing. The system then has to rely on the FEC which is always used in the radio layer. The RLC also performs segmentation and reassembly and flow control. For certain services the RLC also encrypts.

4.4.6 Handover

UMTS knows two basic classes of handovers:

- **Hard handover:** This handover type is already known from GSM and other TDMA/FDMA systems. Switching between different antennas or different systems is performed at a certain point in time. **UTRA TDD** can only use this type. Switching between TDD cells is done between the slots of different frames. **Inter frequency handover**, i.e., changing the carrier frequency, is a hard handover. Receiving data at different frequencies at the same time requires a more complex receiver compared to receiving data from different sources at the same carrier frequency. Typically, all **inter system handovers** are hard handovers in UMTS. This includes handovers to and from GSM or other IMT-2000 systems. A special type of handover is the handover to a satellite system (inter-segment handover), which is also a hard handover, as different frequencies are used. However, it is unclear what technology will be used for satellite links if it will ever come. To enable a UE to listen into GSM or other frequency bands, UMTS specifies a **compressed mode** transmission for UTRA FDD. During this mode a UE stops all transmission. To avoid data loss, either the spreading factor can be lowered before and after the break in transmission (i.e., more data can be sent in shorter time) or less data is sent using different coding schemes.
- **Soft handover:** This is the real new mechanism in UMTS compared to GSM and is only available in the FDD mode. Soft handovers are well known from traditional CDMA networks as they use **macro diversity**, a basic property of CDMA. As shown in Figure 4.33, a UE can receive signals from up to three different antennas, which may belong to different node Bs. Towards the UE the RNC splits the data stream and forwards it to the node Bs. The UE combines the received data again. In the other direction, the UE simply sends its data which is then received by all node Bs involved. The RNC combines the data streams received from the node Bs. The fact that a UE receives data from different antennas at the same time makes a handover soft. Moving from one cell to another is a smooth, not an abrupt process.

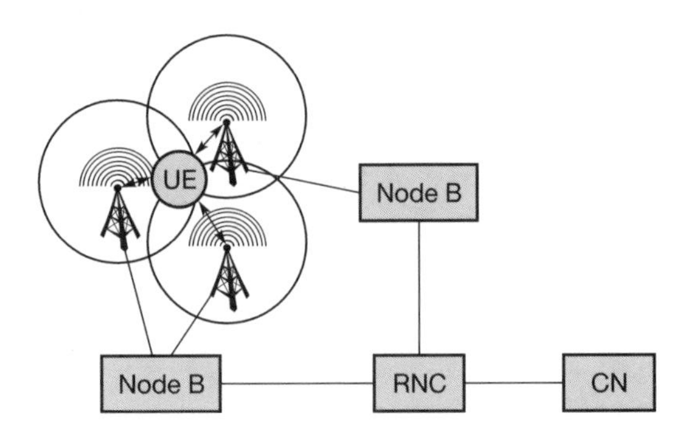

Figure 4.33 Marco-diversity supporting soft handovers

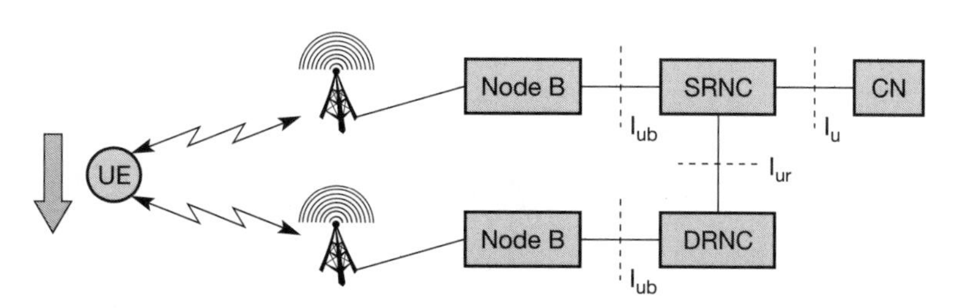

Figure 4.34
Serving RNC and drift RNC

Macro-diversity makes the transmission more robust with respect to fast fading, multi-path propagation, and shading. If one path is blocked by an obstacle the chances are good that data can still be received using another antenna. During a soft handover a UE receives power control commands from all involved node Bs. The UE then lowers transmission power as long as it still receives a command to lower the power. This avoids interference if, for example, the UE is in the transmission area of two antennas, one close, one further away. Without the above mechanism the UE's signal may be too strong when listening to the antenna further away. The lower the interference a UE introduces into a cell, the higher the capacity. Without this control, cell breathing would be even more problematic than it already is in CDMA networks.

As soft handover is not supported by the CN, all mechanisms related to soft handover must be located within UTRAN. Figure 4.34 shows a situation where a soft handover is performed between two node Bs that do not belong to the same RNC. In this case one RNC controls the connection and forwards all data to and from the CN. If the UE moves in the example from the upper cell to the lower cell, the upper RNC acts as a **serving RNC (SRNC)** while the other is the **drift RNC (DRNC)**. (If the whole RNS is considered, the terms are serving RNS and drift RNS, respectively.) The SRNC forwards data received from the CN to its node B and to the DRNC via the I_{ur} interface (splitting). This mechanism does not exist in, e.g., GSM. Data received by the lower node B is forwarded by the DRNC to the SRNC. The SRNC combines both data streams and forwards a single stream of data to the CN. The CN does not notice anything from the simultaneous reception. If the UE moves further down and drops out of the transmission area of the upper node B, two RNCs reserve resources for data transmission, SRNC and DRNC, although none of SRNC's node Bs transmit data for this UE. To avoid wasting resources, SRNC relocation can be performed. This involves the CN so is a hard handover.

Figure 4.35 gives an overview of several common handover types in a combined UMTS/GSM network (UMTS specifies ten different types which include soft and hard handover). The combination of a UTRA-FDD/GSM device will be the most common case in the beginning as coverage of 3G networks will be poor.

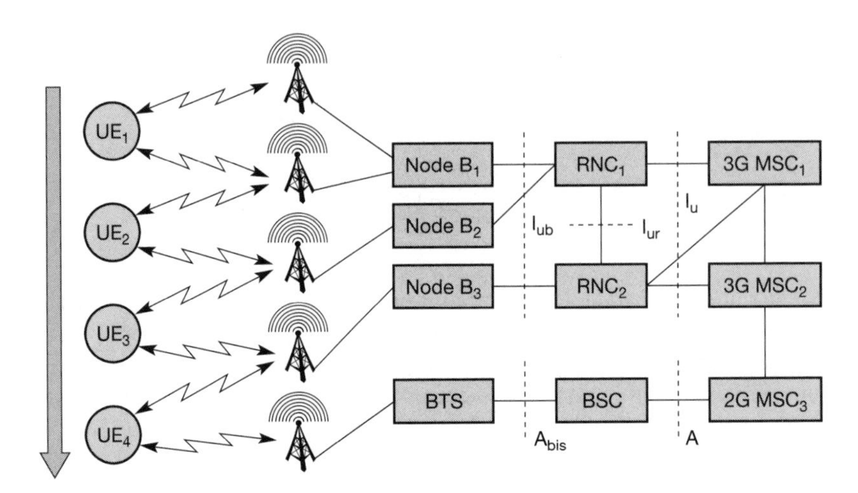

Figure 4.35 Overview of different handover types

- **Intra-node B, intra-RNC:** UE_1 moves from one antenna of node B_1 to another antenna. This type of handover is called **softer handover**. In this case node B_1 performs combining and splitting of the data streams.
- **Inter-node B, intra-RNC:** UE_2 moves from node B_1 to node B_2. In this case RNC_1 supports the soft handover by combining and splitting data.
- **Inter-RNC:** When UE_3 moves from node B_2 to node B_3 two different types of handover can take place. The **internal inter-RNC** handover is not visible for the CN, as described in Figure 4.34. RNC_1 can act as SRNC, RNC_2 will be the DRNC. The CN will communicate via the same interface I_u all the time. As soon as a relocation of the interface I_u takes place (relocation of the controlling RNC), the handover is called an **external inter-RNC** handover. Communication is still handled by the same MSC_1, but the external handover is now a hard handover.
- **Inter-MSC:** It could be also the case that MSC_2 takes over and performs a hard handover of the connection.
- **Inter-system:** UE_4 moves from a 3G UMTS network into a 2G GSM network. This hard handover is important for real life usability of the system due to the limited 3G coverage in the beginning.

4.5 Summary

This chapter has, for the most part, presented GSM as the most successful second generation digital cellular network. Although GSM was primarily designed for voice transmission, the chapter showed the evolution toward a more data-oriented transfer via HSCSD and GPRS. This evolution also includes the transition from a circuit-switched network to a packet-switched system that comes closer to the internet model. Other systems presented include DECT, the digital standard for cordless phones, and TETRA, a trunked radio system. DECT can be used for wireless data transmission on a campus or indoors, but also for wireless local loops (WLL). For special scenarios, e.g., emergencies, trunked radio systems such as TETRA can be the best choice. They offer a fast connection setup (even within communication groups) and can work in an ad hoc network, i.e., without a base station.

The situation in the US is different from Europe. Based on the analog AMPS system, the US industry developed the TDMA system IS-54 that adds digital traffic channels. IS-54 uses dual mode mobile phones and incorporates several GSM ideas, such as, associated control channels, authentication procedures using encryption, and mobile assisted handover (called handoff). The Japanese PDC system was designed using many ideas in IS-54.

The next step, IS-136, includes digital control channels (IS-54 uses analog AMPS control channels) and is more efficient. Now fully digital phones can be used, several additional services are offered, e.g., voice mail, call waiting, identification, group calling, or SMS. IS-136 is also called North American TDMA (NA-TDMA) or Digital AMPS (D-AMPS) and operates at 800 and 1,900 MHz. Enhancements of D-AMPS/IS-136 toward IMT-2000 include advanced modulation techniques for the 30 kHz radio carrier, shifting data rates up to 64 kbit/s (first phase, called 136+). The second phase, called 136HS (High Speed) comprises a new air interface specification based on the EDGE technology.

IS-95 (promoted as cdmaOne) is based on CDMA, which is a completely different medium access method. Before deployment, the system was proclaimed as having many advantages over TDMA systems, such as its much higher capacity of users per cell, e.g., 20 times the capacity of AMPS. Today, CDMA providers are making more realistic estimates of around five times as many users. IS-95 offers soft handover, avoiding the GSM ping-pong effect (Wong, 1997). However, IS-95 needs precise synchronization of all base stations (using GPS satellites which are military satellites, so are not under control of the network provider), frequent power control, and typically, dual mode mobile phones due to the limited coverage. The basic ideas of CDMA have been integrated into most 3G systems.

This chapter also presented an overview of current and future third generation systems. UMTS, a proposal of operators and companies involved in the GSM business, was discussed in more detail. This standard is more an evolutionary approach than a revolution. To avoid even higher implementation costs, UMTS tries to reuse as much infrastructure as possible while introducing new services and higher data rates based on CDMA technology. The initial installations will basically use the GSM/GPRS infrastructure and offer only moderate data rates. The initial capacity of a UMTS cell is approximately 2 Mbit/s; cell diameters are in the order of 500 m. UMTS will be used to offload GSM networks and to offer enhanced data rates in cities as a first step. Future releases aim to replace the infrastructure by an (almost) all-IP network. These ideas will be presented together with a look at fourth generation systems in chapter 11. It is quite clear that it will take a long time before 3G services are available in many places. It took GSM 10 years to become the most successful 2G mobile communication system. A similar period of time will be needed for 3G systems to succeed. Meanwhile, customers will need multiple mode phones offering, e.g., GSM 900/1800/1900 and UMTS UTRA-FDD services. It is not clear if and when UTRA-TDD will succeed. Providers already using cdmaOne will take the evolutionary path via cdma2000 1x toward the 3G system cdma2000 1x EV-DO. Several tests have already been conducted for 3G satellite services in the MSS spectrum (e.g., satellite based multicast, Nussli, 2002). However, right now many companies will wait before investing money in satellite services (see chapter 5). The main problem of multi-mode systems is the inter-system handover. While this chapter introduces handover scenarios within UMTS and GSM, and between GSM and UMTS, even more complex scenarios could comprise wireless LANs (see chapter 7) or other packet-oriented networks (Pahlavan, 2000).

4.6 Review exercises

1. Name some key features of the GSM, DECT, TETRA, and UMTS systems. Which features do the systems have in common? Why have the three older different systems been specified? In what scenarios could one system replace another? What are the specific advantages of each system?
2. What are the main problems when transmitting data using wireless systems that were made for voice transmission? What are the possible steps to mitigate the problems and to raise efficiency? How can this be supported by billing?
3. Which types of different services does GSM offer? Give some examples and reasons why these services have been separated.
4. Compared to the TCHs offered, standard GSM could provide a much higher data rate (33.8 kbit/s) when looking at the air interface. What lowers the data rates available to a user?
5. Name the main elements of the GSM system architecture and describe their functions. What are the advantages of specifying not only the radio interface but also all internal interfaces of the GSM system?
6. Describe the functions of the MS and SIM. Why does GSM separate the MS and SIM? How and where is user-related data represented/stored in the GSM system? How is user data protected from unauthorized access, especially over the air interface? How could the position of an MS (not only the current BTS) be localized? Think of the MS reports regarding signal quality.
7. Looking at the HLR/VLR database approach used in GSM – how does this architecture limit the scalability in terms of users, especially moving users?
8. Why is a new infrastructure needed for GPRS, but not for HSCSD? Which components are new and what is their purpose?

9 What are the limitations of a GSM cell in terms of diameter and capacity (voice, data) for the traditional GSM, HSCSD, GPRS? How can the capacity be increased?

10 What multiplexing schemes are used in GSM and for what purpose? Think of other layers apart from the physical layer.

11 How is synchronization achieved in GSM? Who is responsible for synchronization and why is it so important?

12 What are the reasons for the delays in a GSM system for packet data traffic? Distinguish between circuit-switched and packet-oriented transmission.

13 Where and when can collisions occur while accessing the GSM system? Compare possible collisions caused by data transmission in standard GSM, HSCSD, and GPRS.

14 Why and when are different signaling channels needed? What are the differences?

15 How is localization, location update, roaming, etc. done in GSM and reflected in the data bases? What are typical roaming scenarios?

16 Why are so many different identifiers/addresses (e.g., MSISDN, TMSI, IMSI) needed in GSM? Give reasons and distinguish between user-related and system-related identifiers.

17 Give reasons for a handover in GSM and the problems associated with it. What are the typical steps for handover, what types of handover can occur? Which resources need to be allocated during handover for data transmission using HSCSD or GPRS respectively? What about QoS guarantees?

18 What are the functions of authentication and encryption in GSM? How is system security maintained?

19 How can higher data rates be achieved in standard GSM, how is this possible with the additional schemes HSCSD, GPRS, EDGE? What are the main differences of the approaches, also in terms of complexity? What problems remain even if the data rate is increased?

20 What limits the data rates that can be achieved with GPRS and HSCSD using real devices (compared to the theoretical limit in a GSM system)?

21 Using the best delay class in GPRS and a data rate of 115.2 kbit/s – how many bytes are in transit before a first acknowledgement from the receiver could reach the sender (neglect further delays in the fixed network and receiver system)? Now think of typical web transfer with 10 kbyte average transmission size – how would a standard TCP behave on top of GPRS (see chapters 9 and 10)? Think of congestion avoidance and its relation to the round-trip time. What changes are needed?

22 How much of the original GSM network does GPRS need? Which elements of the network perform the data transfer?

23 What are typical data rates in DECT? How are they achieved considering the TDMA frames? What multiplexing schemes are applied in DECT and for what purposes? Compare the complexity of DECT with that of GSM.

24 Who would be the typical users of a trunked radio system? What makes trunked radio systems particularly attractive for these user groups? What are the main differences to existing systems for that purpose? Why are trunked radio systems cheaper compared to, e.g., GSM systems for their main purposes?

25 Summarize the main features of third generation mobile phone systems. How do they achieve higher capacities and higher data rates? How does UMTS implement asymmetrical communication and different data rates?

26 Compare the current situation of mobile phone networks in Europe, Japan, China, and North America. What are the main differences, what are efforts to find a common system or at least interoperable systems?

27 What disadvantage does OVSF have with respect to flexible data rates? How does UMTS offer different data rates (distinguish between FDD and TDD mode)?

28 How are different DPDCHs from different UEs within one cell distinguished in UTRA FDD?

29 Which components can perform combining/splitting at what handover situation? What is the role of the interface I_{ur}? Why can CDMA systems offer soft handover?

30 How does UTRA-FDD counteract the near-far effect? Why is this not a problem in GSM?

4.7 References

3G Americas (2002) http://www.3gamericas.org/.

3GPP (2000) *General UMTS architecture*, 3rd Generation Partnership Project, 3G TS 23.101 3.1.0 (2000-12).

3GPP (2002a) 3rd Generation Partnership Project, http://www.3gpp.org/.

3GPP (2002b) *UTRAN overall description*, 3rd Generation Partnership Project, 3GPP TS 25.401 V3.10.0 (2002-06).

3GPP2 (2002) 3rd Generation Partnership Project 2, http://www.3gpp2.org/.

Adachi, F., Sawahashi, M., Suda, H. (1998) 'Wideband DS-CDMA for next-generation mobile communications systems,' *IEEE Communications Magazine*, 36(9).

Brasche, G., Walke, B. (1997) 'Concepts, services, and protocols of the new GSM phase 2+ General Packet Radio Service,' *IEEE Communications Magazine*, 35(8).

Callendar, M. (1997) 'International Mobile Telecommunications-2000 standards efforts of the ITU,' collection of articles in *IEEE Personal Communications*, 4(4).

Chen, H.-H., Fan, C.-X., Lu, W. (2002) 'China's Perspectives on 3G Mobile Communications and Beyond: TD-SCDMA Technology,' *IEEE Wireless Communications*, 9(2).

Dahlman, E., Gudmundson, B., Nilsson, M., Sköld, J. (1998) 'UMTS/IMT-2000 based on wideband CDMA,' *IEEE Communications Magazine*, 36(9).

Dasilva, J., Ikonomou, D., Erben, H. (1997) 'European R&D programs on third-generation mobile communication systems,' *IEEE Personal Communications*, 4(1).

DECT (2002), http://www.dect.ch/, http://www.dectweb.com/.

ETSI (1991a) *Bearer services supported by a GSM PLMN*, European Telecommunications Standards Institute, GSM recommendations 02.02.

ETSI (1991b) *General description of a GSM PLMN*, European Telecommunications Standards Institute, GSM recommendations 01.02.

ETSI (1991c) *Subscriber Identity Modules, Functional Characteristics*, European Telecommunications Standards Institute, GSM recommendations 02.17.

ETSI (1993a) *Multiplexing and multiple access on the radio path*, European Telecommunications Standards Institute, GSM recommendations 05.02.

ETSI (1993b) *MS-BSS data link layer – general aspects*, European Telecommunications Standards Institute, GSM recommendations 04.05.

ETSI (1993c) *MS-BSS data link layer specification*, European Telecommunications Standards Institute, GSM recommendations 04.06.

ETSI (1997a) *High Speed Circuit Switched Data (HSCSD)*, Stage 1, European Telecommunications Standards Institute, GSM 02.34, V5.2.1.

ETSI (1998a) *General Packet Radio Service (GPRS); Requirements specification of GPRS*, European Telecommunications Standards Institute, TR 101 186, V6.0.0 (1998–04).

ETSI (1998b) *General Packet Radio Service (GPRS); Service description; Stage 2*, European Telecommunications Standards Institute, EN 301 344, V6.1.1 (1998–08).

ETSI (1998c) *General Packet Radio Service (GPRS); Service description; Stage 1*, European Telecommunications Standards Institute, EN 301 113, V6.1.1 (1998–11).

ETSI (1998d) *General Packet Radio Service (GPRS); Overall description of the GPRS radio interface; Stage 2*, European Telecommunications Standards Institute, TS 101 350, V6.0.1 (1998–08).

ETSI (1998e) *High Speed Circuit Switched Data (HSCSD); Stage 2*, European Telecommunications Standards Institute, TS 101 038, V5.1.0 (1998–07).

ETSI (1998f) *Universal Mobile Telecommunications System (UMTS); Concept groups for the definition of the UMTS Terrestrial Radio Access (UTRA)*, European Telecommunications Standards Institute, TR 101 397, V3.0.1 (1998–10).

ETSI (1998h) *High level requirements relevant for the definition of the UMTS Terrestrial Radio Access (UTRA) concept*, European Telecommunications Standards Institute, TR 101 398, V3.0.1 (1998–10).

ETSI (1998i) *Concept groups for the definition of the UMTS Terrestrial Radio Access (UTRA)*, European Telecommunications Standards Institute, TR 101 397, V3.0.1 (1998–10).

ETSI (1998j) *Digital Enhanced Cordless Telecommunications (DECT), Generic Access Profile (GAP)*, European Telecommunications Standards Institute, EN 300 444, V1.3.2 (1998–03).

ETSI (1998k) *Digital Enhanced Cordless Telecommunications (DECT), Common Interface (CI)*, European Telecommunications Standards Institute, EN 300 175, V1.4.1 (1998–02).

ETSI (1998l) *Terrestrial Trunked Radio (TETRA), Voice plus Data (V+D)*, European Telecommunications Standards Institute, ETS 300 392 series of standards.

ETSI (1998m) *Terrestrial Trunked Radio (TETRA), Packet Data Optimized (PDO)*, European Telecommunications Standards Institute, ETS 300 393 series of standards.

ETSI (1998n) *The ETSI UMTS Terrestrial Radio Access (UTRA) ITU-R Radio Transmission Technologies (RTT) Candidate Submission*, European Telecommunications Standards Institute.

ETSI (2002) European Telecommunications Standards Institute, http://www.etsi.org/.

Evci, C. (2001) 'Optimizing and licensing the radio frequency spectrum for terrestrial 3G users,' *Alcatel Telecommunications Review*, 1/2001.

Goodman, D. (1997) *Wireless Personal Communications Systems*. Addison-Wesley Longman.

GSM Association (2002), http://www.gsmworld.com/.

GSMMoU (1998) *Vision for the evolution from GSM to UMTS*, GSM MoU Association, Permanent Reference Document, V 3.0.0.

GSM-R (2002), The GSM-R Industry Group, http://www.gsm-rail.com/.

Halsall, F. (1996) *Data communications, computer networks and open systems*. Addison-Wesley Longman.

ITU (2002) *International Mobile Telecommunications*, International Telecommunication Union, http://www.itu.int/imt/.

Nussli, C; Bertout, A. (2002) 'Satellite-based multicast architecture for multimedia services in 3G mobile networks,' *Alcatel Telecommunications Review*, 2/2002.

Ojanperä, T.; Prasad, R. (1998) 'An overview of third-generation wireless personal communications: A European perspective,' *IEEE Personal Communications*, 5(6).

Pahlavan, K., Krishnamurthy, P., Hatami, A., Ylianttila, M., Makela, J.-P., Pichna, R., Vallström, J. (2000) 'Handoff in Hybrid Mobile Data Networks,' *IEEE Personal Communications*, 7(2).

Pahlavan, K., Krishnamurthy, P. (2002) *Principles of Wireless Networks*. Prentice Hall.

Rosenberg, J., Schulzrinne, H., Camarillo, G., Johnston, A., Peterson, J., Sparks, R., Handley, M., Schooler, E. (2002) SIP: *Session Initiation Protocol*, RFC 3261, updated by RFC 3265.

SIP Forum (2002) http://www.sipforum.com/.

Shafi, M., Sasaki, A., Jeong, D. (1998) 'IMT-2000 developments in the Asia Pacific region,' collection of articles, *IEEE Communications Magazine*, 36(9).

Stallings, W. (2002) *Wireless Communications and Networks*. Prentice Hall.

TETRA MoU (2002) TETRA Memorandum of Understanding, http://www.tetramou.com/.

Tripathi, N.D., Reed, J.H., VanLandingham, H.F. (1998) 'Handoffs in cellular systems,' *IEEE Personal Communications*, 5(6).

UMTS Forum (2002) http://www.umts-forum.org/.

Wong, D., Lim, T. (1997) 'Soft handoffs in CDMA mobile systems,' *IEEE Personal Communications*, 4(6).

Satellite systems 5

Satellite communication introduces another system supporting mobile communications. Satellites offer global coverage without wiring costs for base stations and are almost independent of varying population densities. After a short history of satellite development and presentation of different areas of application, this chapter introduces the basics of satellite systems. Orbit, visibility, transmission quality, and other system characteristics are all closely linked. Several restrictions and application requirements result in three major classes of satellites, GEO, MEO, and LEO, as discussed later in this chapter. The high speed of satellites with a low altitude raises new problems for routing, localization of mobile users, and handover of communication links. Several aspects of these topics are therefore presented in separate sections. Finally, the chapter deals with four examples of global satellite communication systems that are currently planned or already installed. Following the ups and downs of satellite systems over the last years (bankruptcy of Iridium, over-ambitious systems etc.) it is clear that the future of these systems is uncertain. However, they may prove useful as an addition for existing systems (e.g., UMTS satellite segment as enhancement of the terrestrial service for multimedia broadcasting).

5.1 History

Satellite communication began after the Second World War. Scientists knew that it was possible to build rockets that would carry radio transmitters into space. In 1945, Arthur C. Clarke published his essay on 'Extra Terrestrial Relays'. But it was not until 1957, in the middle of the cold war, that the sudden launching of the first satellite SPUTNIK by the Soviet Union shocked the Western world. SPUTNIK is not at all comparable to a satellite today, it was basically a small sender transmitting a periodic 'beep'. But this was enough for the US to put all its effort into developing its first satellite. Only three years later, in 1960, the first reflecting communication satellite ECHO was in space. ECHO was basically a mirror in the sky enabling communication by reflecting signals. Three years further on, the first geostationary (or geosynchronous) satellite SYNCOM followed. Even today, geostationary satellites are the backbone of news broadcasting in the sky. Their great advantage, is their fixed position in the sky (see section 5.3.1). Their rotation is synchronous to the rotation of the earth, so they appear to be pinned to a certain location.

The first commercial geostationary communication satellite INTELSAT 1 (also known as 'Early Bird') went into operation in 1965. It was in service for one-and-a-half years, weighed 68 kg and offered 240 duplex telephone channels or, alternatively, a single TV channel. INTELSAT 2 followed in 1967, INTELSAT 3 in 1969 already offered 1,200 telephone channels. While communication on land always provides the alternative of using wires, this is not the case for ships at sea. Three MARISAT satellites went into operation in 1976 which offered worldwide maritime communication. However, Sender and receiver still had to be installed on the ships with large antennas (1.2 m antenna, 40 W transmit power). The first mobile satellite telephone system, INMARSAT-A, was introduced in 1982. Six years later, INMARSAT-C became the first satellite system to offer mobile phone and data services. (Data rates of about 600 bit/s, interfaces to the X.25 packet data network exist.) In 1993, satellite telephone systems finally became fully digital with INMARSAT-M. The actual mobility, however, was relative from a user's point of view, as the devices needed for communication via geostationary satellites were heavy (several kilograms) and needed a lot of transmit power to achieve decent data rates. Nineteen ninety-eight marked the beginning of a new age of satellite data communication with the introduction of global satellite systems for small mobile phones, such as, e.g., Iridium and Globalstar (see section 5.7). There are currently almost 200 geostationary satellites in commercial use which shows the impressive growth of satellite communication over the last 30 years (Miller, 1998), (Maral, 1998), (Pascall, 1997). However, satellite networks are currently facing heavy competition from terrestrial networks with nationwide coverage or at least enough coverage to support most applications and users.

5.2 Applications

Traditionally, satellites have been used in the following areas:

- **Weather forecasting:** Several satellites deliver pictures of the earth using, e.g., infra red or visible light. Without the help of satellites, the forecasting of hurricanes would be impossible.
- **Radio and TV broadcast satellites:** Hundreds of radio and TV programs are available via satellite. This technology competes with cable in many places, as it is cheaper to install and, in most cases, no extra fees have to be paid for this service. Today's satellite dishes have diameters of 30–40 cm in central Europe, (the diameters in northern countries are slightly larger).
- **Military satellites:** One of the earliest applications of satellites was their use for carrying out espionage. Many communication links are managed via satellite because they are much safer from attack by enemies.

- **Satellites for navigation:** Even though it was only used for military purposes in the beginning, the global positioning system (GPS) is nowadays well-known and available for everyone. The system allows for precise localization worldwide, and with some additional techniques, the precision is in the range of some metres. Almost all ships and aircraft rely on GPS as an addition to traditional navigation systems. Many trucks and cars come with installed GPS receivers. This system is also used, e.g., for fleet management of trucks or for vehicle localization in case of theft.

In the context of mobile communication, the capabilities of satellites to transmit data is of particular interest.

- **Global telephone backbones:** One of the first applications of satellites for communication was the establishment of international telephone backbones. Instead of using cables it was sometimes faster to launch a new satellite (aka 'big cable in the sky'). However, while some applications still use them, these, satellites are increasingly being replaced by fiber optical cables crossing the oceans. The main reason for this is the tremendous capacity of fiber optical links (commercially some 10 Gbit/s using wavelength division multiplexing, several Tbit/s in labs) and, in particular, the much lower delay compared to satellites. While the signal to a geostationary satellite has to travel about 72,000 km from a sender via the satellite to the receiver, the distance is typically less than 10,000 km if a fiber-optical link crossing the Pacific or Atlantic Ocean is used. Unfortunately, the speed of light is limited, resulting in a one-way, single-hop time delay of 0.25 s for geostationary satellites. Using satellites for telephone conversation is sometimes annoying and requires particular discipline in discussions.
- **Connections for remote or developing areas:** Due to their geographical location many places all over the world do not have direct wired connection to the telephone network or the internet (e.g., researchers on Antarctica) or because of the current state of the infrastructure of a country. Satellites now offer a simple and quick connection to global networks (Schwartz, 1996).
- **Global mobile communication:** The latest trend for satellites is the support of global mobile data communication. Due to the high latency, geostationary satellites are not ideal for this task; therefore, satellites using lower orbits are needed (see section 5.3). The basic purpose of satellites for mobile communication is not to replace the existing mobile phone networks, but to extend the area of coverage. Cellular phone systems, such as AMPS and GSM (and their successors) do not cover all parts of a country. Areas that are not covered usually have low population where it is too expensive to instal a base station. With the integration of satellite communication, however, the mobile phone can switch to satellites offering worldwide connectivity to a customer (Jamalipour, 1998). For the UMTS system (see chapter 4) frequency bands directly adjacent to the terrestrial bands have been allocated for the satellite segment (S-Band: 1980–2010 MHz uplink, 2170–2200 MHz downlink).

While in the beginning satellites were simple transponders, today's satellites rather resemble flying routers. Transponders basically receive a signal on one frequency, amplify the signal and transmit it on another frequency. While in the beginning only analog amplification was possible, the use of digital signals also allows for signal regeneration. The satellite decodes the signal into a bit-stream, and codes it again into a signal. The advantage of digital regeneration compared to pure analog amplification is the higher quality of the received signal on the earth. Today's communication satellites provide many functions of higher communication layers, e.g., inter-satellite routing, error correction etc.

Figure 5.1 shows a classical scenario for satellite systems supporting global mobile communication (Lutz, 1998). Depending on its type, each satellite can cover a certain area on the earth with its beam (the so-called 'footprint' (see section 5.3)). Within the footprint, communication with the satellite is possible for mobile users via a **mobile user link (MUL)** and for the base station controlling the satellite and acting as gateway to other networks via the **gateway link (GWL)**. Satellites may be able to communicate directly with each other via **inter-satellite links (ISL)**. This facilitates direct communication between users within different footprints without using base stations or other networks on earth. Saving extra links from satellite to earth can reduce latency for data packets and voice data. Some satellites have special antennas to create smaller cells using spot beams (e.g., 163 spot beams per satellite in the ICO system (ICO, 2002)). The required terrestrial service infrastructure for satellite control and the control links between Earth control stations and satellites not shown in Figure 5.1.

Figure 5.1
Typical satellite system for global mobile telecommunications

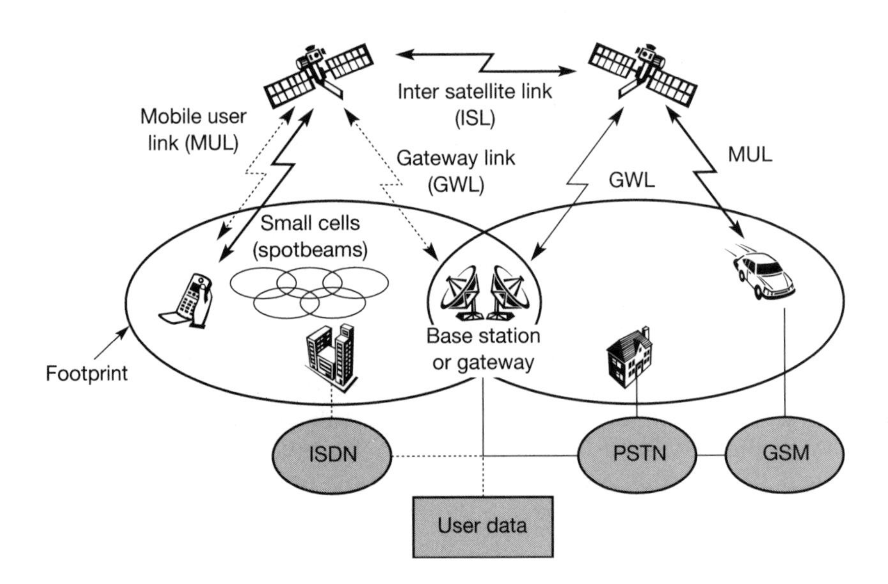

Satellite systems are, and will continue to be, a valuable addition to the many networks already in existance on earth. Users might communicate using ISDN or other PSTN, even cellular networks such as GSM and UMTS. Many gateways provide seamless communication between these different networks. A real challenge, for example, is the smooth, seamless handover between a cellular network and a satellite system (vertical handover) as it is already well known from within cellular networks (horizontal handover). Users should not notice the switching from, e.g., GSM, to a satellite network during conversation.

5.3 Basics

Satellites orbit around the earth. Depending on the application, these orbits can be circular or elliptical. Satellites in circular orbits always keep the same distance to the earth's surface following a simple law:

- The attractive force F_g of the earth due to gravity equals $m \cdot g \cdot (R/r)^2$.
- The centrifugal force F_c trying to pull the satellite away equals $m \cdot r \cdot \omega^2$.

The variables have the following meaning:

- m is the mass of the satellite;
- R is the radius of earth with $R = 6{,}370$ km;
- r is the distance of the satellite to the centre of the earth;
- g is the acceleration of gravity with $g = 9.81$ m/s^2;
- and ω is the angular velocity with $\omega = 2 \cdot \pi \cdot f$, f is the frequency of the rotation.

To keep the satellite in a stable circular orbit, the following equation must hold:

- $F_g = F_c$, i.e., both forces must be equal. Looking at this equation the first thing to notice is that the mass m of a satellite is irrelevant (it appears on both sides of the equation).
- Solving the equation for the distance r of the satellite to the center of the earth results in the following equation:
  The distance $r = (g \cdot R^2/(2 \cdot \pi \cdot f)^2)^{1/3}$

From the last equation it can be concluded that the distance of a satellite to the earth's surface depends on its rotation frequency. Figure 5.2 shows this dependency in addition to the relative velocity of a satellite. The interesting point in the diagram is when the satellite period equals 24 hours. This is exactly the case for a distance of 35,786 km. Having an orbiting time of 24 hours implies a geostationary satellite if it is additionally placed above the equator. (Satellites of this type will be discussed in a later section.)

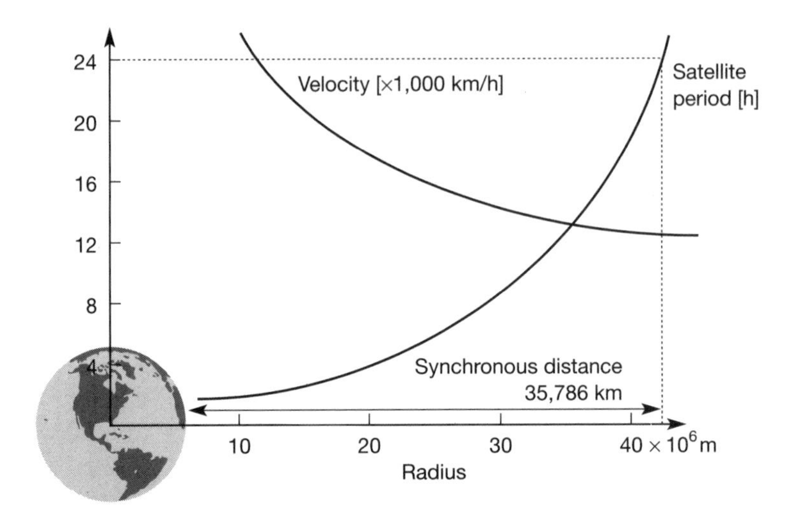

Figure 5.2 Dependency of satellite period and distance to earth

Important parameters in satellite communication are the inclination and elevation angles. The **inclination angle** δ (see Figure 5.3) is defined as the angle between the equatorial plane and the plane described by the satellite orbit. An inclination angle of 0 degrees means that the satellite is exactly above the equator. If the satellite does not have a circular orbit, the closest point to the earth is called the perigee.

The **elevation angle** ε (see Figure 5.4) is defined as the angle between the center of the satellite beam and the plane tangential to the earth's surface. A so-called **footprint** can be defined as the area on earth where the signals of the satellite can be received.

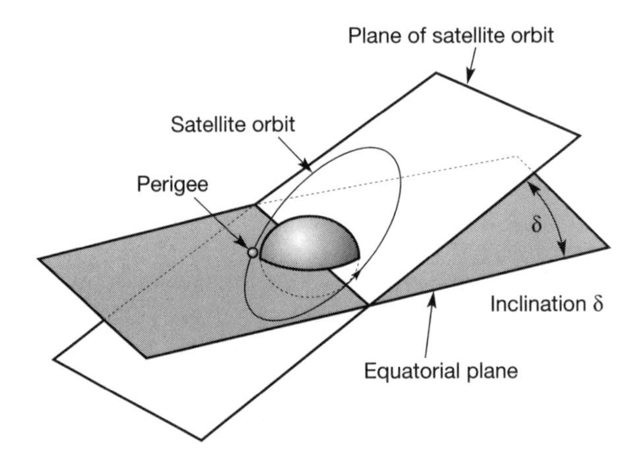

Figure 5.3 Inclination angle of a satellite

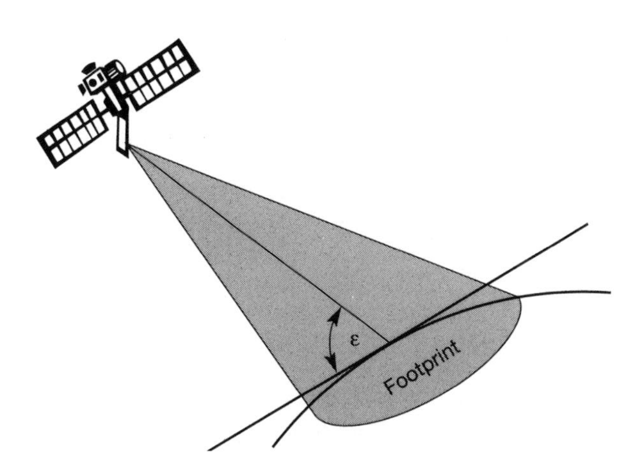

Figure 5.4
Elevation angle of a satellite

Another effect of satellite communication is the propagation loss of the signals. This attenuation of the signal power depends on the distance between a receiver on earth and the satellite, and, additionally, on satellite elevation and atmospheric conditions. The loss L depending on the distance r between sender and receiver can be calculated as:

$$L = (4 \cdot \pi \cdot r \cdot f \,/\, c)^2,$$

with f being the carrier frequency and c the speed of light.

This means that the power of the received signal decreases with the square of the distance. This also directly influences the maximum data rates achievable under certain assumptions (transmit power, antenna diameter, operating frequency etc.) as shown in Comparetto (1997). While with antennas used for mobile phones a data rate of 10 kbit/s is achievable with a 2 GHz carrier for satellites in some 100 km distance as discussed in section 5.3.2, only some 10 bit/s are possible with geostationary satellites in a distance of 36,000 km.

The attenuation of the signal due to certain atmospheric conditions is more complex (see Figure 5.5). Depending on the elevation, the signal has to penetrate a smaller or larger percentage of the atmosphere. Generally, an elevation less than 10 degrees is considered useless for communication. Especially rain absorption can be quite strong in tropical areas (here, the error rates increase dramatically during the afternoon rainfall).

Four different types of orbits can be identified as shown in Figure 5.6:

- **Geostationary (or geosynchronous) earth orbit (GEO):** GEO satellites have a distance of almost 36,000 km to the earth. Examples are almost all TV and radio broadcast satellites, many weather satellites and satellites operating as backbones for the telephone network (see section 5.3.1).

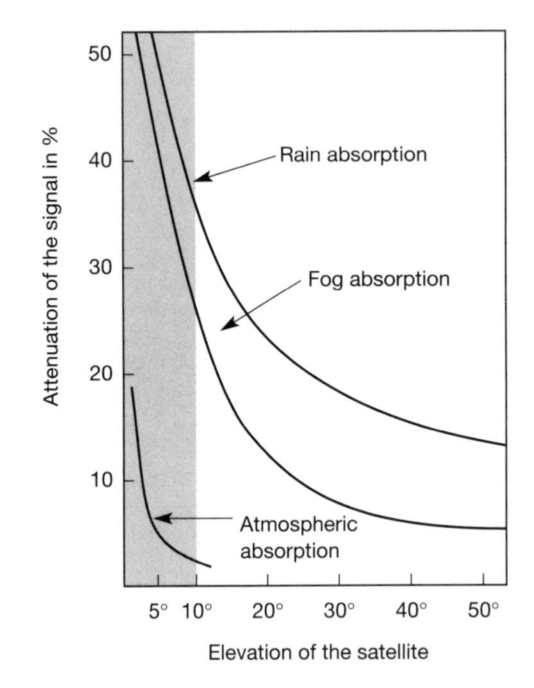

Figure 5.5 Signal attenuation due to atmospheric absorption

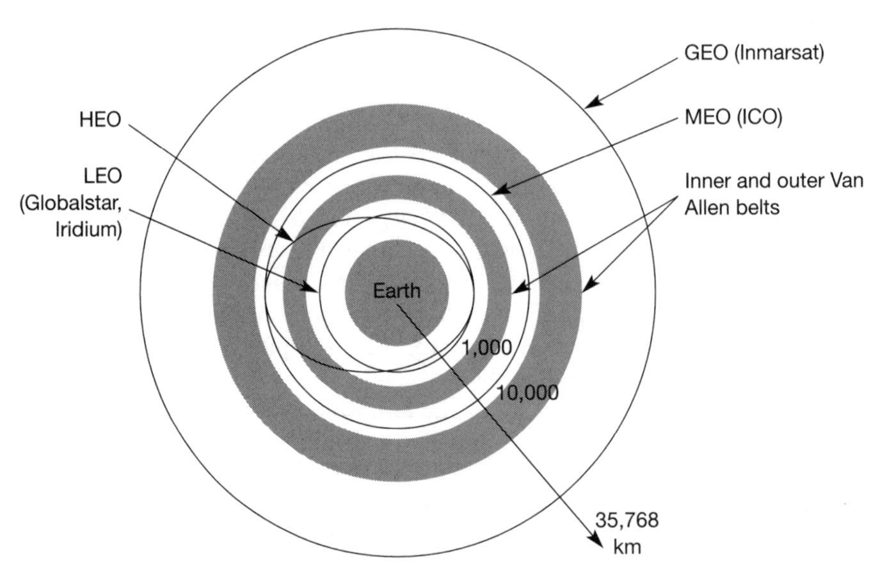

Figure 5.6 Different types of satellite orbits

- **Medium earth orbit (MEO):** MEOs operate at a distance of about 5,000–12,000 km. Up to now there have not been many satellites in this class, but some upcoming systems (e.g., ICO) use this class for various reasons (see section 5.3.3).
- **Low earth orbit (LEO):** While some time ago LEO satellites were mainly used for espionage, several of the new satellite systems now rely on this class using altitudes of 500–1,500 km (see section 5.3.2).
- **Highly elliptical orbit (HEO):** This class comprises all satellites with non-circular orbits. Currently, only a few commercial communication systems using satellites with elliptical orbits are planned. These systems have their perigee over large cities to improve communication quality.

The Van Allen radiation belts, belts consisting of ionized particles, at heights of about 2,000–6,000 km (inner Van Allen belt) and about 15,000–30,000 km (outer Van Allen belt) respectively make satellite communication very difficult in these orbits.

5.3.1 GEO

If a satellite should appear fixed in the sky, it requires a period of 24 hours. Using the equation for the distance between earth and satellite $r = (g \cdot R^2/(2 \cdot \pi \cdot f)^2)^{1/3}$ and the period of 24 hours $f = 1/24h$, the resulting distance is 35,786 km. The orbit must have an inclination of 0 degrees.

- **Advantages:** Three GEO satellites are enough for a complete coverage of almost any spot on earth. Senders and receivers can use fixed antenna positions, no adjusting is needed. GEOs are ideal for TV and radio broadcasting. Lifetime expectations for GEOs are rather high, at about 15 years. GEOs typically do not need a handover due to the large footprint. GEOs do not exhibit any Doppler shift because the relative movement is zero.
- **Disadvantages:** Northern or southern regions of the earth have more problems receiving these satellites due to the low elevation above a latitude of 60°, i.e., larger antennas are needed in this case. Shading of the signals in cities due to high buildings and the low elevation further away from the equator limit transmission quality. The transmit power needed is relatively high (some 10 W) which causes problems for battery powered devices. These satellites cannot be used for small mobile phones. The biggest problem for voice and also data communication is the high latency of over 0.25 s one-way – many retransmission schemes which are known from fixed networks fail. Due to the large footprint, either frequencies cannot be reused or the GEO satellite needs special antennas focusing on a smaller footprint. Transferring a GEO into orbit is very expensive.

5.3.2 LEO

As LEOs circulate on a lower orbit, it is obvious that they exhibit a much shorter period (the typical duration of LEO periods are 95 to 120 minutes). Additionally, LEO systems try to ensure a high elevation for every spot on earth to provide a high quality communication link. Each LEO satellite will only be visible from the earth for around ten minutes. A further classification of LEOs into little LEOs with low bandwidth services (some 100 bit/s), big LEOs (some 1,000 bit/s) and broadband LEOs with plans reaching into the Mbit/s range can be found in Comparetto (1997).

- **Advantages:** Using advanced compression schemes, transmission rates of about 2,400 bit/s can be enough for voice communication. LEOs even provide this bandwidth for mobile terminals with omni-directional antennas using low transmit power in the range of 1W. The delay for packets delivered via a LEO is relatively low (approx 10 ms). The delay is comparable to long-distance wired connections (about 5–10 ms). Smaller footprints of LEOs allow for better frequency reuse, similar to the concepts used for cellular networks (Gavish, 1998). LEOs can provide a much higher elevation in polar regions and so better global coverage.
- **Disadvantages:** The biggest problem of the LEO concept is the need for many satellites if global coverage is to be reached. Several concepts involve 50–200 or even more satellites in orbit. The short time of visibility with a high elevation requires additional mechanisms for connection handover between different satellites. (Different cases for handover are explained in section 5.4.) The high number of satellites combined with the fast movements results in a high complexity of the whole satellite system. One general problem of LEOs is the short lifetime of about five to eight years due to atmospheric drag and radiation from the inner Van Allen belt[1]. Assuming 48 satellites and a lifetime of eight years (as expected for the system Globalstar), a new satellite would be needed every two months. The low latency via a single LEO is only half of the story. Other factors are the need for routing of data packets from satellite to satellite (or several times from base stations to satellites and back) if a user wants to communicate around the world. Due to the large footprint, a GEO typically does not need this type of routing, as senders and receivers are most likely in the same footprint.

---

1 the life-time of satellites is typically limited by the limited amount of fuel on-board which is needed for correcting the orbit from time to time.

5.3.3 MEO

MEOs can be positioned somewhere between LEOs and GEOs, both in terms of their orbit and due to their advantages and disadvantages.

- **Advantages:** Using orbits around 10,000 km, the system only requires a dozen satellites which is more than a GEO system, but much less than a LEO system. These satellites move more slowly relative to the earth's rotation allowing a simpler system design (satellite periods are about six hours). Depending on the inclination, a MEO can cover larger populations, so requiring fewer handovers.
- **Disadvantages:** Again, due to the larger distance to the earth, delay increases to about 70–80 ms. The satellites need higher transmit power and special antennas for smaller footprints.

5.4 Routing

A satellite system together with gateways and fixed terrestrial networks as shown in Figure 5.1 has to route data transmissions from one user to another as any other network does. Routing in the fixed segment (on earth) is achieved as usual, while two different solutions exist for the satellite network in space. If satellites offer ISLs, traffic can be routed between the satellites. If not, all traffic is relayed to earth, routed there, and relayed back to a satellite.

Assume two users of a satellite network exchange data. If the satellite system supports ISLs, one user sends data up to a satellite and the satellite forwards it to the one responsible for the receiver via other satellites. This last satellite now sends the data down to the earth. This means that only one uplink and one downlink per direction is needed. The ability of routing within the satellite network reduces the number of gateways needed on earth.

If a satellite system does not offer ISLs, the user also sends data up to a satellite, but now this satellite forwards the data to a gateway on earth. Routing takes place in fixed networks as usual until another gateway is reached which is responsible for the satellite above the receiver. Again data is sent up to the satellite which forwards it down to the receiver. This solution requires two uplinks and two downlinks. Depending on the orbit and the speed of routing in the satellite network compared to the terrestrial network, the solution with ISLs might offer lower latency. The drawbacks of ISLs are higher system complexity due to additional antennas and routing hard- and software for the satellites.

5.5 Localization

Localization of users in satellite networks is similar to that of terrestrial cellular networks. One additional problem arises from the fact that now the 'base stations', i.e., the satellites, move as well. The gateways of a satellite network maintain several registers. A **home location register (HLR)** stores all static information about a user as well as his or her current location. The last known location of a mobile user is stored in the **visitor location register (VLR)**. Functions of the VLR and HLR are similar to those of the registers in, e.g., GSM (see chapter 4). A particularly important register in satellite networks is the **satellite user mapping register (SUMR)**. This stores the current position of satellites and a mapping of each user to the current satellite through which communication with a user is possible.

Registration of a mobile station is achieved as follows. The mobile station initially sends a signal which one or several satellites can receive. Satellites receiving such a signal report this event to a gateway. The gateway can now determine the location of the user via the location of the satellites. User data is requested from the user's HLR, VLR and SUMR are updated.

Calling a mobile station is again similar to GSM. The call is forwarded to a gateway which localizes the mobile station using HLR and VLR. With the help of the SUMR, the appropriate satellite for communication can be found and the connection can be set up.

5.6 Handover

An important topic in satellite systems using MEOs and in particular LEOs is handover. Imagine a cellular mobile phone network with fast moving base stations. This is exactly what such satellite systems are – each satellite represents a base station for a mobile phone. Compared to terrestrial mobile phone networks, additional instances of handover can be necessary due to the movement of the satellites.

- **Intra-satellite handover:** A user might move from one spot beam of a satellite to another spot beam of the same satellite. Using special antennas, a satellite can create several spot beams within its footprint. The same effect might be caused by the movement of the satellite.
- **Inter-satellite handover:** If a user leaves the footprint of a satellite or if the satellite moves away, a handover to the next satellite takes place. This might be a hard handover switching at one moment or a soft handover using both satellites (or even more) at the same time (as this is possible with CDMA systems). Inter-satellite handover can also take place between satellites if they support ISLs. The satellite system can trade high transmission

quality for handover frequency. The higher the transmission quality should be, the higher the elevation angles that are needed. High elevation angles imply frequent handovers which in turn, make the system more complex.

- **Gateway handover:** While the mobile user and satellite might still have good contact, the satellite might move away from the current gateway. The satellite has to connect to another gateway.
- **Inter-system handover:** While the three types of handover mentioned above take place within the satellite-based communication system, this type of handover concerns different systems. Typically, satellite systems are used in remote areas if no other network is available. As soon as traditional cellular networks are available, users might switch to this type usually because it is cheaper and offers lower latency. Current systems allow for the use of dual-mode (or even more) mobile phones but unfortunately, seamless handover between satellite systems and terrestrial systems or vice versa has not been possible up to now.

5.7 Examples

Table 5.1 shows four examples (two in operation, two planned) of MEO/LEO satellite networks (see also Miller, 1998 and Lutz, 1998). One system, which is in operation, is the **Iridium** system. This was originally targeted for 77 satellites (hence the name Iridium with its 77 electrons) and now runs with 66 satellites plus seven spare satellites (was six, Iridium, 2002). It is the first commercial LEO system to cover the whole world. Satellites orbit at an altitude of 780 km, the weight of a single satellite is about 700 kg. The fact that the satellites are heavier than, e.g., the competitor Globalstar results from their capability to route data between Iridium satellites by using ISLs, so a satellite needs more memory, processing power etc. Mobile stations (MS in Table 5.1) operate at 1.6138–1.6265 GHz according to an FDMA/TDMA scheme with TDD, feeder links to the satellites at 29.1–29.3 GHz for the uplink and 19.4–19.6 GHz for the downlink. ISLs use 23.18–23.38 GHz. The infrastructure of Iridium is GSM-based.

A direct competitor of Iridium is **Globalstar** (Globalstar, 2002). This system, which is also operational, uses a lower number of satellites with fewer capabilities per satellite. This makes the satellites lighter (about 450 kg weight) and the overall system cheaper. Globalstar does not provide ISLs and global coverage, but higher bandwidth is granted to the customers. Using CDMA and utilizing path diversity, Globalstar can provide soft handovers between different satellites by receiving signals from several satellites simultaneously. Globalstar uses 1.61–1.6265 GHz for uplinks from mobile stations to the satellites and 2.4835–2.5 GHz for the downlink. Feeder links for the satellites are at 5.091–5.250 GHz gateway to satellite and 6.875–7.055 GHz satellite to gateway.

Table 5.1 Example MEO and LEO systems

	Irdium (orbiting)	Globalstar (orbiting)	ICO (planned)	Teledesic (planned)
No. of satellites	66 + 7	48 + 4	10(?) + 2	288(?)
Altitude [km] coverage	780 global	1,414 ±70° latitude	10,390 global	Approx. 700 global
No. of planes	6	8	2	12
Inclination	86°	52°	45°	40°
Minimum elevation	8°	20°	20°	40°
Frequencies [GHz (circa)]	1.6 MS 29.2 ↗ 19.5 ↙ 23.3 ISL	1.6 MS ↗ 2.5 MS ↙ 5.1 ↗ 6.9 ↙	2 MS ↗ 2.2 MS ↙ 5.2 ↗ 7 ↙	19 ↙ 28.8 ↗ 62 ISL
Access method	FDMA/TDMA	CDMA	FDMA/TDMA	FDMA/TDMA
ISL	Yes	No	No	Yes
Bit rate	2.4 kbit/s	9.6 kbit/s	4.8 kbit/s (144 kbit/s planned)	64 Mbit/s ↙ 2/64 Mbit/s ↗
No. of channels	4,000	2,700	4,500	2,500
Lifetime [years]	5–8	7.5	12	10
Initial cost estimate	$4.4 bn	$2.9 bn	$4.5 bn	$9 bn

While the other three systems presented in Table 5.1 are LEOs, **Intermediate Circular Orbit**, (ICO) (ICO, 2002) represents a MEO system as the name indicates. ICO needs less satellites, 10 plus two spare are planned, to reach global coverage. Each satellite covers about 30 per cent of earth's surface, but the system works with an average elevation of 40°. Due to the higher complexity within the satellites (i.e., larger antennas and larger solar paddles to generate enough power for transmission), these satellites weigh about 2,600 kg. While launching ICO satellites is more expensive due to weight and higher orbit, their expected lifetime is higher with 12 years compared to Globalstar and Iridium with eight years and less. ICO satellites need fewer replacements making the whole system cheaper in return. The start of ICO has been delayed several times. The ICO consortium went through bankruptcy and several joint ventures with other satellite organizations, but still plans to start operation of the system within the next few years. The exact number of satellites is currently unclear, however, the system is shifted towards IP traffic with up to 144 kbit/s.

A very ambitious and maybe never realized LEO project is **Teledesic** which plans to provide high bandwidth satellite connections worldwide with high quality of service (Teledesic, 2002). In contrast to the other systems, this satellite network is not primarily planned for access using mobile phones, but to enable worldwide access to the internet via satellite. Primary customers are businesses, schools etc. in remote places. Teledesic wants to offer 64 Mbit/s downlinks and 2 Mbit/s uplinks. With special terminals even 64 Mbit/s uplinks should be possible. Receivers will be, e.g., roof-mounted laptop-sized terminals that connect to local networks in the building. Service start was targeted for 2003, however, currently only the web pages remained from the system and the start was shifted to 2005. The initial plans of 840 satellites plus 84 spares were dropped, then 288 plus spares were planned, divided into 12 planes with 24 satellites each. Considering an expected lifetime of ten years per satellite, this still means a new satellite will have to be launched at least every other week. Due to the high bandwidth, higher frequencies are needed, so Teledesic operates in the Ka-band with 28.6–29.1 GHz for the uplink and 18.8–19.3 GHz for the downlink. At these high frequencies, communication links can easily be blocked by rain or other obstacles. A high elevation of at least 40° is needed. Teledesic uses ISL for routing between the satellites and implements fast packet switching on the satellites.

Only Globalstar uses CDMA as access method, while the other systems rely on different TDMA/FDMA schemes. The cost estimates in Table 5.1 are just rough figures to compare the systems. They directly reflect system complexity. ICO satellites for example are more complicated compared to Iridium, so the ICO system has similar initial costs. Smaller and simpler Globalstar satellites make the system cheaper than Iridium.

5.8 Summary

Satellite systems evolved quickly from the early stages of GEOs in the late 1960s to many systems in different orbits of today. The trend for communication satellites is moving away from big GEOs, toward the smaller MEOs and LEOs mainly for the reason of lower delay which is essential for voice communication. Different systems will offer global coverage with services ranging from simple voice and low bit rate data up to high bandwidth communications with quality of service. However, satellite systems are not aimed at replacing terrestrial mobile communication systems but at complementing them. Up to now it has not been clear how high the costs for operation and maintenance of satellite systems are and how much data transmission via satellites really costs for a customer. Special problems for LEOs in this context are the high system complexity and the relatively short lifetime of the satellites. Initial system costs only constitute part of the overall costs. Before it is possible to offer any service to customers the whole satellite system has to be set up. An incremental growth as it is done for terrestrial networks is not possible in LEO systems. Operators instal new terrestrial networks in densely populated areas first to get a quick return on investment. This money can then be used for further extension of the system. Most LEO satellites fly over non- or sparsely populated areas (sea, deserts, polar regions etc.) and can not generate any revenue. In the end it turned out that there are too few customers for satellite systems such as Iridium leading to the bankruptcy of the operator in March 2000. Only the intervention of the US DoD could prevent a deorbiting of the satellites in August 2000 (which would have been the world's most expensive firework). Now the US DoD is the main customer of Iridium with its own gateway in Hawaii but service is still offered to everyone.

A new application for satellite systems is their use as an addition to terrestrial mobile communication systems in the following way. Point-to-point communication services are handled by, e.g., UMTS, additional multicast or broadcast delivery of multimedia content is performed by a satellite system. In this scenario the role of a satellite is similar to terrestrial broadcasters as explained in more detail in chapter 6. A project evaluating the possibilities of such a scenario is **satellite-digital multi-media broadcasting** (S-DMB). This project combines satellite and terrestrial transmission using the wideband CDMA terrestrial UMTS standard to achieve a multicast layer over 2.5/3G networks (Courseille, 2001), (Nussli, 2002).

Yet another market for new systems might appear between the low orbiting LEOs and terrestrial antennas. Several companies are planning to use high-altitude aircraft or Zeppelins for carrying base stations, so-called **high-altitude platforms** (HAP, Djuknic, 1997). These base stations could be placed high above large cities offering high-quality transmission at lower costs compared to satellite systems. A big advantage compared to LEO systems is the possibility for incremental growth and the requirement for only the number of systems that are actually needed to satisfy the bandwidth demand at certain locations. However, although high-altitude aircraft are feasible in principle it is unclear if it is cost-effective. The ITU has endorsed various frequency ranges for HAP applications opening the way for the provision of, e.g., UMTS services by means of HAPs (Avagnina, 2002).

5.9 Review exercises

1. Name basic applications for satellite communication and describe the trends.
2. Why are GEO systems for telecommunications currently being replaced by fiber optics?
3. How do inclination and elevation determine the use of a satellite?
4. What characteristics do the different orbits have? What are their pros and cons?
5. What are the general problems of satellite signals travelling from a satellite to a receiver?

6 Considered as an interworking unit in a communication network, what function can a satellite have?

7 What special problems do customers of a satellite system with mobile phones face if they are using it in big cities? Think of in-building use and skyscrapers.

8 Why is there hardly any space in space for GEOs?

5.10 References

Avagnina, D., Dovis, F., Ghiglione, A., Mulassano, P. (2002) 'Wireless Networks Based on the High-Altitude Platforms for the Provision of Integrated Navigation/Communication Services,' *IEEE Communications Magazine*, 40(2).

Comparetto, G., Ramirez, R. (1997) 'Trends in mobile satellite technology,' *IEEE Computer*, 30(2).

Courseille, O. (2001) 'Role of satellites in mobile systems,' *Alcatel Telecommunications Review*, 4/2001.

Djuknic, G., Freidenfels, J., Okunev, Y. (1997) 'Establishing Wireless Communications Services via High-Altitude Platforms: A Concept Whose Time has Come?,' *IEEE Communications Magazine*, 35(9).

Gavish, B., Kalvenes, J. (1998) 'The impact of satellite altitude on the performance of LEO based communication systems,' *Wireless Networks*, J.C. Baltzer, 4(2)

Globalstar (2002) Globalstar L.P., San Jose, CA, USA, http://www.globalstar.com/.

ICO (2002) ICO Global Communications London, UK, http://www.ico.com/.

Iridium (2002) Iridium Satellite LLC, Leesburg, VA, USA, http://www.iridium.com/.

Jamalipour, A. (1998) *Low earth orbital satellites for personal communication networks*. Artech House.

Lutz, E. (1998) 'Issues in satellite personal communication systems,' *Wireless Networks*, J.C. Baltzer, 4(2).

Maral, G., Bousquet, M. (1998) *Satellite communications systems: Systems, techniques and technology*. John Wiley & Sons.

Miller, B. (1998) 'Satellites free the mobile phone,' *IEEE Spectrum*, March.

Nussli, C.; Bertout, A. (2002) 'Satellite-based multicast architecture for multimedia services in 3G mobile networks,' *Alcatel Telecommunications Review*, 2/2002.

Pascall, S.C., Withers, D.J. (1997) *Commercial satellite communication*. Focal Press, 1997.

Schwartz, R. (1996) *Wireless communications in developing countries: cellular and satellite systems*. Artech House.

Teledesic (2002) Teledesic Corp., Bellevue, WA, USA, http://www.teledesic.com/.

6 Broadcast systems

Although this book mostly deals with different communication technologies allowing individual two-way communication, it is important to understand the role of unidirectional broadcast systems within future mobile communication scenarios. Typical broadcast systems, such as radio and television, distribute information regardless of the needs of individual users. As an addition to two-way communication technologies, broadcasting information can be very cost effective. Just imagine the distribution of a movie trailer to millions of potential customers and compare it with the abilities of 3G base stations to provide 10–20 simultaneous users with a 128 kbit/s video stream. The distribution of the trailer would block the whole mobile network for a long time even if tens of thousand base stations are assumed.

In the future, television and radio transmissions will be fully digital. Already several radio stations produce and transmit their programmes digitally via the internet or digital radio (see later sections in this chapter). Digital television is on its way. Besides transmitting video and audio, digital transmission allows for the distribution of arbitrary digital data, i.e., multimedia information can accompany radio and TV programmes at very low cost compared to individual wireless connections.

The following sections give a general introduction into asymmetric communication up to the extreme case of unidirectional broadcasting. One important issue is the cyclic repetition of data (as discussed in the sections about broadcast disks). Broadcasting systems which will be explained in detail are digital audio broadcasting (DAB) and digital video broadcasting (DVB). One interesting feature about data communication is the ability of DAB and DVB to carry multi-media information. In combination with satellite or terrestrial transmission and the use of the internet, these systems are able to deliver high bandwidth to individual customers at low cost (ETSI, 2002).

6.1 Overview

Unidirectional distribution systems or broadcast systems are an extreme version of asymmetric communication systems. Quite often, bandwidth limitations, differences in transmission power, or cost factors prevent a communication system from being symmetrical. **Symmetrical communication systems** offer the same transmission capabilities in both communication directions, i.e., the channel characteristics from A to B are the same as from B to A (e.g., bandwidth, delay, costs).

Examples of symmetrical communication services are the plain old telephone service (POTS) or GSM, if end-to-end communication is considered. In this case, it does not matter if one mobile station calls the other or the other way round, bandwidth and delay are the same in both scenarios.

This symmetry is necessary for a telephone service, but many other applications do not require the same characteristics for both directions of information transfer. Consider a typical client/server environment. Typically, the client needs much more data from the server than the server needs from the client. Today's most prominent example of this is the world wide web. Millions of users download data using their browsers (clients) from web servers. A user only returns information to the server from time to time. Single requests for new pages with a typical size of several hundred bytes result in responses of up to some 10 kbytes on average.

A television with a set-top box represents a more extreme scenario. While a high-resolution video stream requires several Mbit/s, a typical user returns some bytes from time to time to switch between channels or return some information for TV shopping.

Finally, today's pagers and radios work completely one-way. These devices can only receive information, and a user needs additional communication technology to send any information back to, e.g., the radio station. Typically, the telephone system is used for this purpose.

A special case of **asymmetrical communication systems** are **unidirectional broadcast systems** where typically a high bandwidth data stream exists from one sender to many receivers. The problem arising from this is that the sender can only optimize transmitted data for the whole group of receivers and not for an individual user. Figure 6.1 shows a simple broadcast scenario. A sender tries to optimize the transmitted packet stream for the access patterns of all receivers without knowing their exact requirements. All packets are then transmitted via a broadcast to all receivers. Each receiver now picks up the packets needed and drops the others or stores them for future use respectively.

These additional functions are needed to personalize distributed data depending on individual requirements and applications. A very simple example of this process could be a user-defined filter function that filters out all information which is not of interest to the user. A radio in a car, for example, could only present traffic information for the local environment, a set-top box could only store the starting times of movies and drop all information about sports.

However, the problem concerning which information to send at what time still remains for a sender. The following section shows several solutions to this.

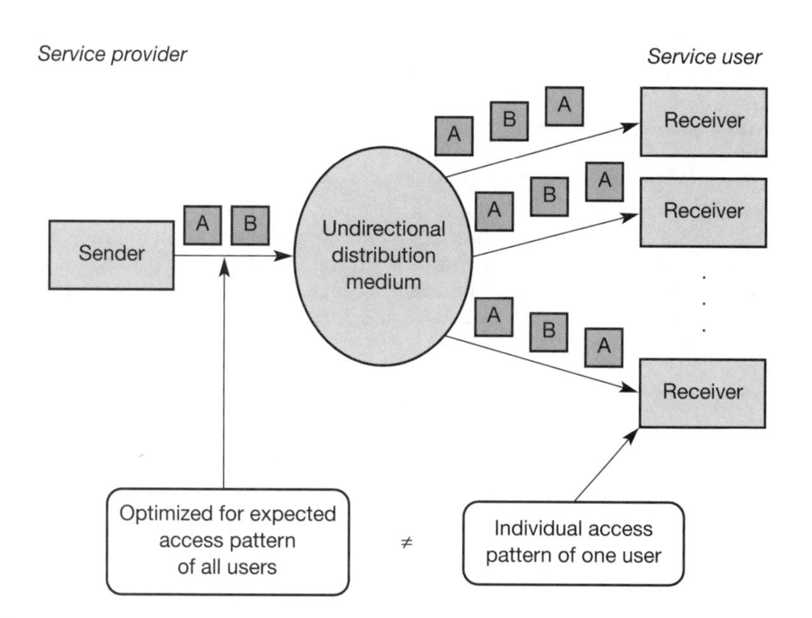

Figure 6.1 Broadcast transmission

6.2 Cyclical repetition of data

A broadcast sender of data does not know when a receiver starts to listen to the transmission. While for radio or television this is no problem (if you do not listen you will not get the message), transmission of other important information, such as traffic or weather conditions, has to be repeated to give receivers a chance to receive this information after having listened for a certain amount of time (like the news every full hour).

The cyclical repetition of data blocks sent via broadcast is often called a **broadcast disk** according to the project in Acharya (1995) or data carousel, e.g., according to the DAB/DVB standards (ETSI, 2002). Different patterns are possible (Figure 6.2 shows three examples). The sender repeats the three data blocks A, B, and C in a cycle. Using a **flat disk**, all blocks are repeated one after another. Every block is transmitted for an equal amount of time, the average waiting time for receiving a block is the same for A, B, and C. **Skewed disks** favor one or more data blocks by repeating them once or several times. This raises the probability of receiving a repeated block (here A) if the block was corrupted the first time. Finally, **multi-disks** distribute blocks that are repeated more often than others evenly over the cyclic pattern. This minimizes the delay if a user wants to access, e.g., block A.

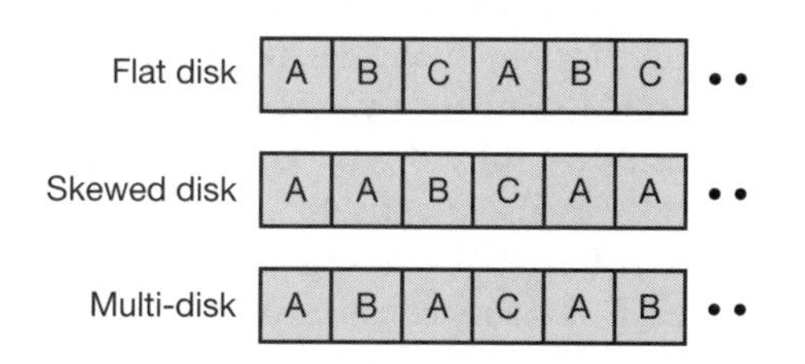

Figure 6.2 Different broadcast patterns

It is only possible to optimize these patterns if the sender knows something about the content of the data blocks and the access patterns of all users.

EXAMPLE BROADCAST DISK

Let us assume that the broadcast sender is a radio station transmitting information about road conditions (block A), the weather report (block B), the latest events in town (block C) and a menu to access these and other topics (block D) in addition to music. The sender can now assume, knowing something about the importance of the data blocks, that block D is the most important to enable access to the other information. The second important block is A, then B and finally C. A possible broadcast disk for this scenario could now look as follows:

DADBDADCDADBDADC ...

It is now the receiver's task to cache data blocks to minimize access delay as soon as a user needs a specific type of information. Again, the receiver can only optimize caching if it knows something about the content of the data blocks. The receiver can store typical access patterns of a user to be able to guess which blocks the user will access with a higher probability. **Caching** generally follows a cost-based strategy: what are the costs for a user (caused by the waiting time) if a data block has been requested but is currently not available in the cache?

Considering the above example, the mobile device of the future (e.g., a radio in a car, an enhanced mobile phone) might remember that a user always checks the latest events in town in the evening, but the road conditions in the morning. The device will cache block A in the morning and block C in the evening. This procedure will generally reduce the waiting time for a user if he or she stays with this access pattern.

6.3 Digital audio broadcasting

Today's analog radio system still follows the basic principle of frequency modulation invented back in 1933. In addition to audio transmission, very limited information such as the station identification can accompany the program. Transmission quality varies greatly depending on multi-path effects and interference. The fully digital **DAB** system does not only offer sound in a CD-like quality, it is also practically immune to interference and multi-path propagation effects (ETSI, 2001a), (DAB, 2002).

DAB systems can use **single frequency networks (SFN)**, i.e., all senders transmitting the same radio program operate at the same frequency. Today, different senders have to use different frequencies to avoid interference although they are transmitting the same radio program. Using an SFN is very frequency efficient, as a single radio station only needs one frequency throughout the whole country. Additionally, DAB transmission power per antenna is orders of magnitude lower compared to traditional FM stations. DAB uses VHF and UHF frequency bands (depending on national regulations), e.g., the terrestrial TV channels 5 to 12 (174–230 MHz) or the L-band (1452–1492 MHz). The modulation scheme used is **DQPSK**. DAB is one of the systems using **COFDM** (see chapter 2) with 192 to 1536 carriers (the so-called **ensemble**) within a DAB channel of 1.5 MHz. Additionally, DAB uses FEC to reduce the error rate and introduces **guard spaces** between single symbols during transmission. COFDM and the use of guard spaces reduce ISI to a minimum. DAB can even benefit from multipath propagation by recombining the signals from different paths.

EXAMPLE DAB ENSEMBLE

The following is an ensemble transmitted at 225.648 MHz in southern Germany. The ensemble contains six radio programs and two data channels.

- SWR 1 BW 192 kbit/s, stereo
- SWR 2 192 kbit/s, stereo
- SWR 3 192 kbit/s, stereo
- Hit Radio Antenne 1 192 kbit/s, stereo
- DAS DING 160 kbit/s, stereo
- SWR traffic information 16 kbit/s, data
- SWR service information 16 kbit/s, data

Within every frequency block of 1.5 MHz, DAB can transmit up to six stereo audio programmes with a data rate of 192 kbit/s each. Depending on the redundancy coding, a data service with rates up to 1.5 Mbit/s is available as an alternative. For the DAB transmission system, audio is just another type of data (besides different coding schemes). DAB uses two basic transport mechanisms:

- **Main service channel (MSC):** The MSC carries all user data, e.g. audio, multimedia data. The MSC consists of **common interleaved frames (CIF)**, i.e., data fields of 55,296 bits that are sent every 24 ms (this interval depends on the transmission mode (ETSI, 2001a)). This results in a data rate of 2.304 Mbit/s. A CIF consists of **capacity units (CU)** with a size of 64 bits, which form the smallest addressable unit within a DAB system.
- **Fast information channel (FIC):** The FIC contains **fast information blocks (FIB)** with 256 bits each (16 bit checksum). An FIC carries all control information which is required for interpreting the configuration and content of the MSC.

Two transport modes have been defined for the MSC. The **stream mode** offers a transparent data transmission from the source to the destination with a fixed bit rate in a sub channel. A **sub channel** is a part of the MSC and comprises several CUs within a CIF. The fixed data rate can be multiples of 8 kbit/s. The **packet mode** transfers data in addressable blocks (packets). These blocks are used to convey MSC data within a sub channel.

DAB defines many service information structures accompanying an audio stream. This **program associated data (PAD)** can contain program information, control information, still pictures for display on a small LCD, title display etc. Audio coding uses PCM with a sampling rate of 48 kHz and MPEG audio compression. The compressed audio stream can have bit rates ranging from 8 kbit/s to 384 kbit/s. Audio data is interleaved for better burst tolerance.[1]

Figure 6.3 shows the general frame structure of DAB. Each frame has a duration T_F of 24, 48, or 96 ms depending on the transmission mode. DAB defines four different transmission modes, each of which has certain strengths that make it more efficient for either cable, terrestrial, or satellite transmission (ETSI, 2001a). Within each frame, 76 or 153 symbols are transmitted using 192, 384, 768, or 1,536 different carriers for COFDM. The guard intervals T_d protecting each symbol can be 31, 62, 123, or 246 μs.

Each frame consists of three parts. The **synchronization channel (SC)** marks the start of a frame. It consists of a null symbol and a phase reference symbol to synchronize the receiver. The **fast information channel (FIC)** follows, containing control data in the FIBs. Finally, the **main service channel (MSC)** carries audio and data service components.

Figure 6.4 gives a simplified overview of a DAB sender. Audio services are encoded (MPEG compression) and coded for transmission (FEC). All data services are multiplexed and also coded with redundancy. The MSC multiplexer combines all user data streams and forwards them to the transmission multiplexer. This unit creates the frame structure by interleaving the FIC. Finally, OFDM coding is applied and the DAB signal is transmitted.

Figure 6.3
DAB frame structure

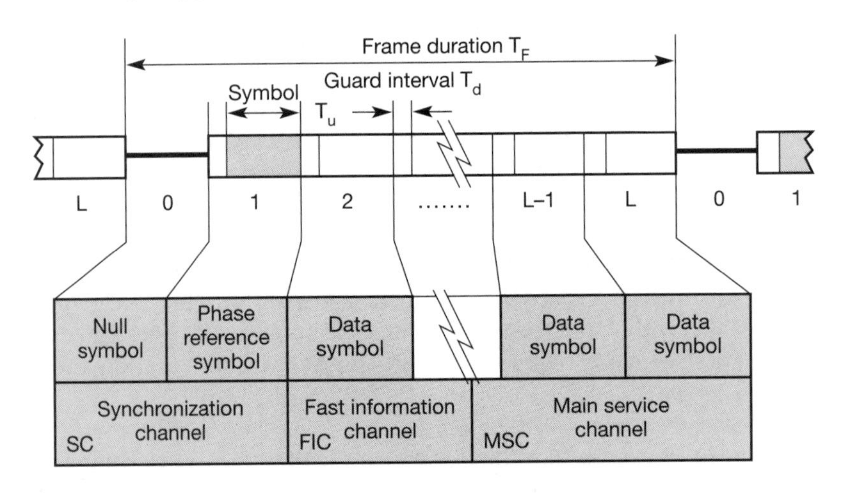

1 As the focus of this book is in data transmission, the reader is referred to ETSI (2001a) for more details about audio coding, audio transmission, multiplexing etc.

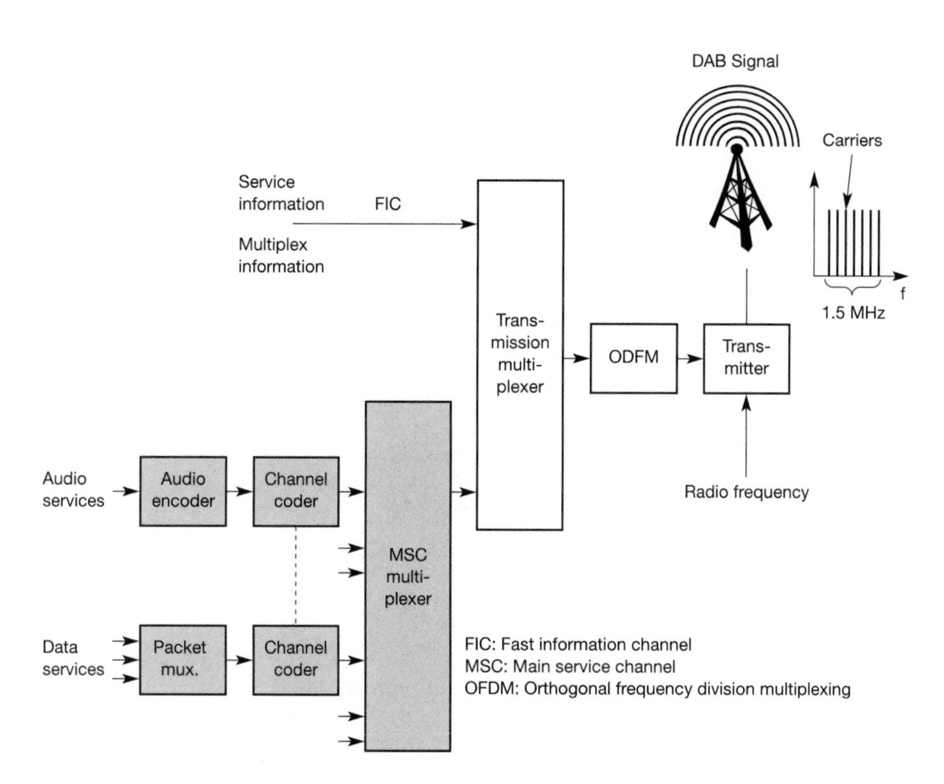

Figure 6.4
Components of a DAB sender (simplified)

DAB does not require fixed, pre-determined allocation of channels with certain properties to services. Figure 6.5 shows the possibilities of dynamic reconfiguration during transmission. Initially, DAB transmits six audio programmes of different quality together with nine data services. Each audio program has its PAD. In the example, audio 1, 2, and 3 have high quality, 4 and 5 lower quality, while 6 has the lowest quality. Programmes 1 to 3 could, e.g., be higher quality classic transmissions, while program 6 could be voice transmissions (news etc.). The radio station could now decide that for audio 3 128 kbit/s are enough when, for example, the news program starts. News may be in mono or stereo with lower quality but additional data (here D10 and D11 – headlines, pictures etc.). The DAB multiplexer dynamically interleaves data from all different sources. To inform the receiver about the current configuration of the MSC carrying the different data streams, the FIC sends **multiplex configuration information (MCI)**.

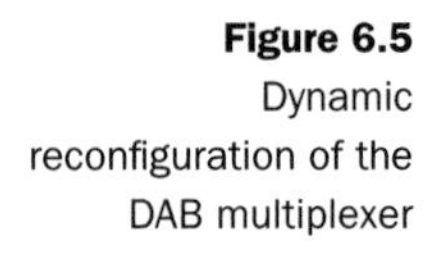

Figure 6.5 Dynamic reconfiguration of the DAB multiplexer

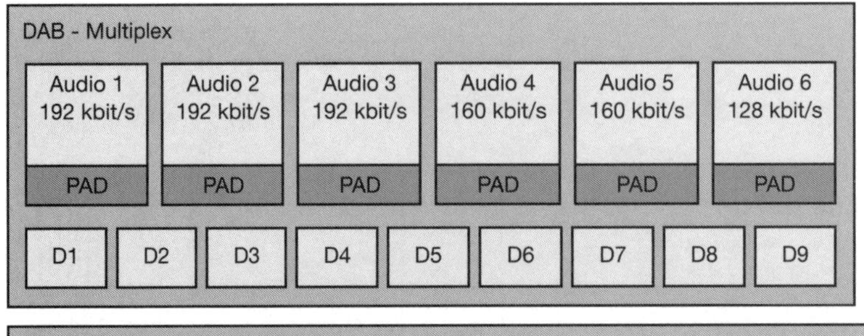

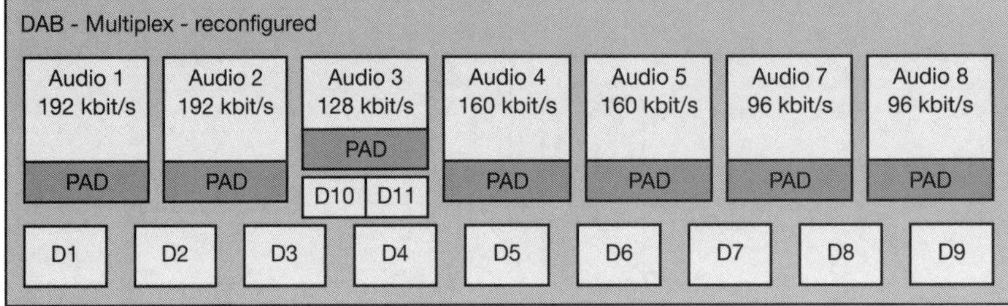

6.3.1 Multi-media object transfer protocol

A problem which technologies like DAB are facing is the broad range of different receiver capabilities. Receivers could be simple audio-only devices with single-line text displays or more advanced radios with extra color graphics displays. DAB receivers can also be adapters in multimedia PCs. However, all different types of receivers should at least be able to recognize all program-associated and program-independent data, and process some of this data.

To solve this problem, DAB defines a common standard for data transmission, the **multi-media object transfer (MOT)** protocol (ETSI, 1999a). The primary goal of MOT is the support of data formats used in other multi-media systems (e.g., on line services, Internet, CD-ROM). Example formats are multimedia and hypermedia information coding experts group (MHEG), Java, joint photographic experts group (JPEG), American standard code for information interchange (ASCII), moving pictures expert group (MPEG), hypertext markup language (HTML), hypertext transfer protocol (HTTP), bitmap (BMP), graphics interchange format (GIF).

MOT data is transferred in MOT objects consisting of a header core, a header extension, and a body (Figure 6.6).

Figure 6.6 MOT object structure

Header core	Header extension	Body

7 byte (Header core)

- **Header core:** This seven byte field contains the sizes of the header and the body, and the content type of the object. Depending on this header information, the receiver may decide if it has enough resources (memory, CPU power, display etc.) available to decode and further process the object.
- **Header extension:** The extension field of variable size contains additional handling data for the object, such as, e.g., the repetition distance to support advanced caching strategies (see section 6.2), the segmentation information, and the priority of the data. With the help of the priority information a receiver can decide which data to cache and which to replace. For example, the index HTML page may have a higher priority than an arbitrary page.
- **Body:** Arbitrary data can be transferred in the variable body as described in the header fields.

Larger MOT objects will be segmented into smaller segments. DAB can apply different interleaving and repetition schemes to objects and segments (MOT data carousel):

- **Object repetition:** DAB can repeat objects several times. If an object A consists of four segments (A_1, A_2, A_3, and A_4), a simple repetition pattern would be $A_1A_2A_3A_4A_1A_2A_3A_4A_1A_2A_3A_4$...
- **Interleaved objects:** To mitigate burst error problems, DAB can also interleave segments from different objects. Interleaving the objects A, B, and C could result in the pattern $A_1B_1C_1A_2B_2C_2A_3B_3C_3$...
- **Segment repetition:** If some segments are more important than others, DAB can repeat these segments more often (e.g. $A_1A_1A_2A_2$ $A_2A_3A_4A_4$...).
- **Header repetition:** If a receiver cannot receive the header of an MOT, it will not be able to decode the object. It can be useful to retransmit the header several times. Then, the receiver can synchronize with the data stream as soon as it receives the header and can start decoding. A pattern could be $HA_1A_2HA_3A_4HA_5A_6H$... with H being the header of the MOT object A.

Obviously, DAB can also apply all interleaving and repetition schemes at the same time.

6.4 Digital video broadcasting

The logical consequence of applying digital technology to radio broadcasting is doing the same for the traditional television system. The analog system used today has basically remained unchanged for decades. The only invention worth mentioning was the introduction of color TV for the mass market back in the 1960s. Television still uses the low resolution of 625 lines for the European PAL system or only 525 lines for the US NTSC respectively[2]. The display is interlaced with 25 or 30 frames per second respectively. So, compared with today's computer displays with resolutions of 1,280 × 1,024 and more than 75 Hz frame rate, non-interlaced, TV performance is not very impressive.

There have been many attempts to change this and to introduce digital TV with higher resolution, better sound and additional features, but no approach has yet been truly successful. One reason for this is the huge number of old systems that are installed and cannot be replaced as fast as computers (we can watch the latest movie on an old TV, but it is impossible to run new software on older computers!). Varying political and economic interests are counterproductive to a common standard for digital TV. One approach toward such a standard, which may prove useful for mobile communication, too, is presented in the following sections.

After some national failures in introducing digital TV, the so-called European Launching Group was founded in 1991 with the aim of developing a common digital television system for Europe. In 1993 these common efforts were named **digital video broadcasting (DVB)** (Reimers, 1998), (DVB, 2002). Although the name shows a certain affinity to DAB, there are some fundamental differences regarding the transmission technology, frequencies, modulation etc. The goal of DVB is to introduce digital television broadcasting using satellite transmission (DVB-S, (ETSI, 1997)), cable technology (DVB-C, (ETSI, 1998)), and also terrestrial transmission (DVB-T, (ETSI, 2001b)).

Figure 6.7 shows components that should be integrated into the DVB architecture. The center point is an integrated receiver-decoder (set-top box) connected to a high-resolution monitor. This set-top box can receive DVB signals via satellites, terrestrial local/regional senders (multi-point distribution systems, terrestrial receiver), cable, B-ISDN, ADSL, or other possible future technologies. Cable, ADSL, and B-ISDN connections also offer a return channel, i.e., a user can send data such as channel selection, authentication information, or a shopping list. Audio/video streams can be recorded, processed, and replayed using **digital versatile disk (DVD)** or multimedia PCs. Different levels of quality are envisaged: **standard definition TV (SDTV)**, **enhanced definition TV (EDTV)**, and **high definition TV (HDTV)** with a resolution of up to 1,920 × 1,080 pixels.

Similar to DAB, DVB also transmits data using flexible containers. These containers are basically MPEG-2 frames that do not restrict the type of information. DVB sends service information contained in its data stream, which specifies the content of a container. The following contents have been defined:

---

2 Only about 580 lines are visible in the PAL system, only 480 with NTSC. The horizontal resolution depends on transmission and recording quality. For VHS this results in 240 'pixels' per line, terrestrial transmission allows up to 330 and high-quality TV sets up to 500 'pixels'. However, transmission is analog and, these 'pixels' cannot directly be compared with the pixels of computer monitors.

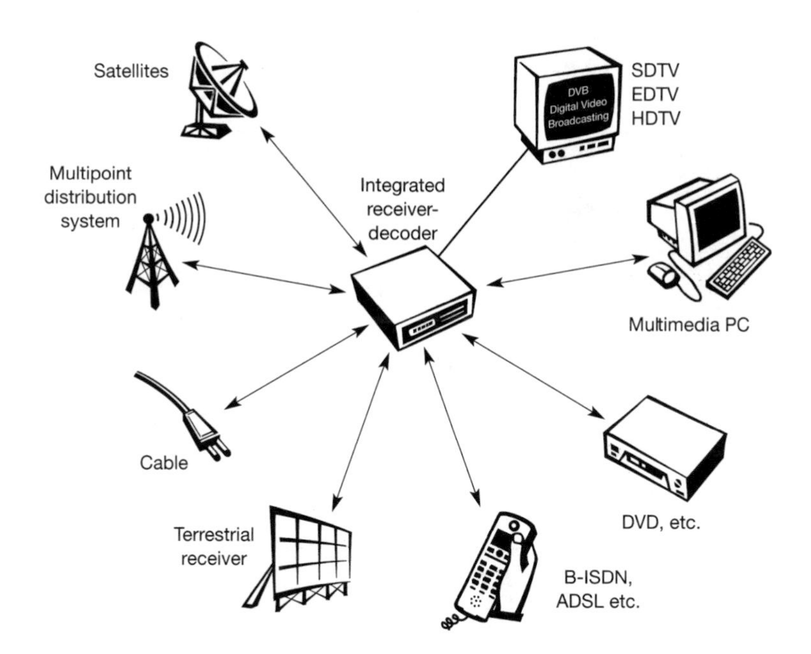

Figure 6.7
Digital video broadcasting scenario

- **Network information table (NIT):** NIT lists the services of a provider and contains additional information for set-top boxes.
- **Service description table (SDT):** SDT lists names and parameters for each service within an MPEG multiplex channel.
- **Event information table (EIT):** EIT contains status information about the current transmission and some additional information for set-top boxes.
- **Time and date table (TDT):** Finally, TDT contains update information for set-top boxes.

As shown in Figure 6.8, an MPEG-2/DVB container can store different types of data. It either contains a single channel for HDTV, multiple channels for EDTV or SDTV, or arbitrary multi-media data (data broadcasting).

6.4.1 DVB data broadcasting

As the MPEG-2 transport stream is able to carry arbitrary data within packets with a fixed length of 188 byte (184 byte payload), ETSI (1999b) and ETSI (1999c) define several profiles for data broadcasting which can be used, e.g., for high bandwidth mobile Internet services.

- **Data pipe:** simple, asynchronous end-to-end delivery of data; data is directly inserted in the payload of MPEG2 transport packets.
- **Data streaming:** streaming-oriented, asynchronous, synchronized (synchronization with other streams, e.g., audio/video possible), or synchronous (data and clock regeneration at receiver possible) end-to-end delivery of data.

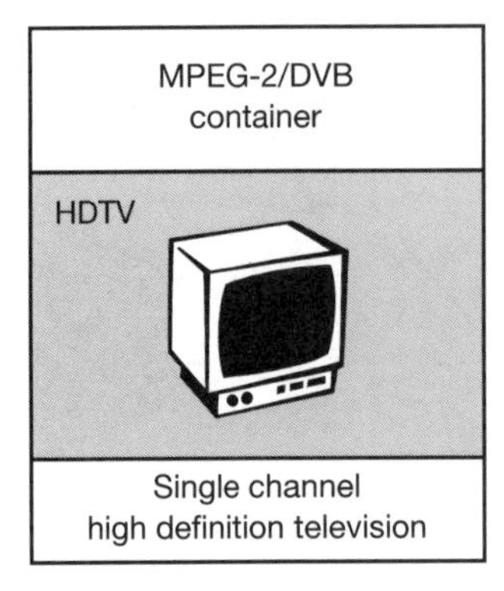

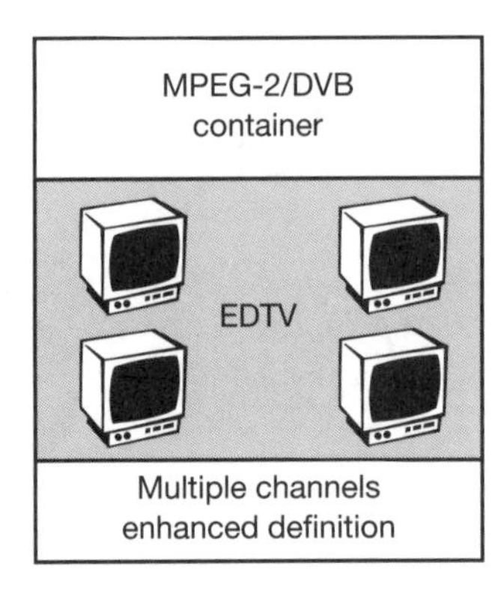

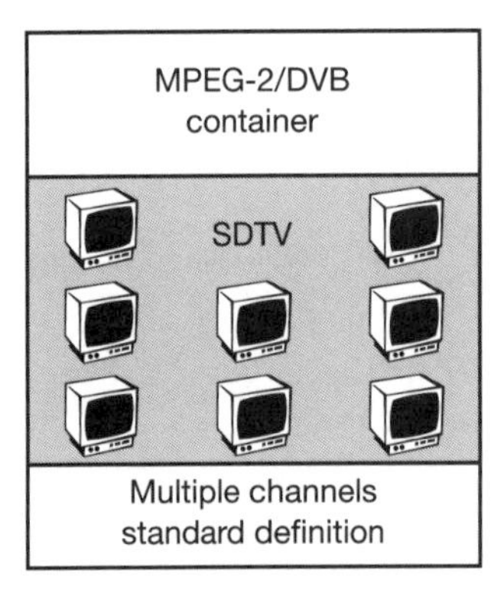

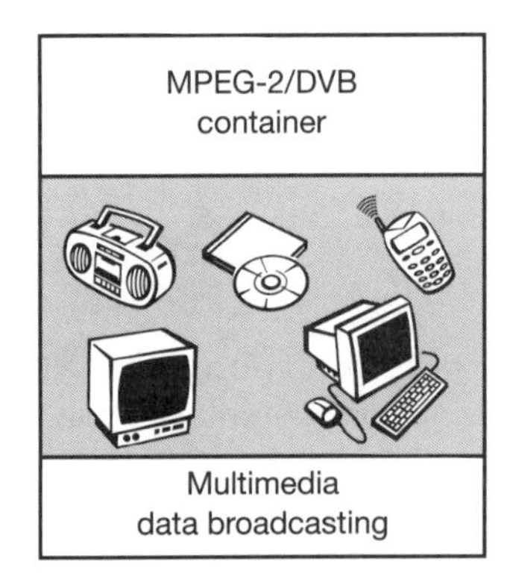

Figure 6.8
Different contents of MPEG-2/DVB containers

- **Multiprotocol encapsulation:** transport of arbitrary data network protocols on top of the MPEG-2 transport stream; optimized for IP, support for 48 bit MAC addresses, unicast, multi-cast, and broadcast.
- **Data carousels:** periodic transmission of data.
- **Object carousels:** periodic transmission of objects; platform independent, compatible with the object request broker (ORB) framework as defined by CORBA (2002).

6.4.2 DVB for high-speed Internet access

Apart from this data/multi-media broadcasting, DVB can be also used for high-bandwidth, asymmetrical Internet access. A typical scenario could be the following (see Figure 6.9): An information provider, e.g., video store, offers its data to potential customers with the help of a service provider. If a customer wants to download high-volume information, the information provider transmits this information to a satellite provider via a service provider. In fixed networks this is done using leased lines because high bandwidth and QoS guarantees are needed. The satellite provider now multiplexes this data stream together with other digital TV channels and transmits it to the customer via satellite and a satel-

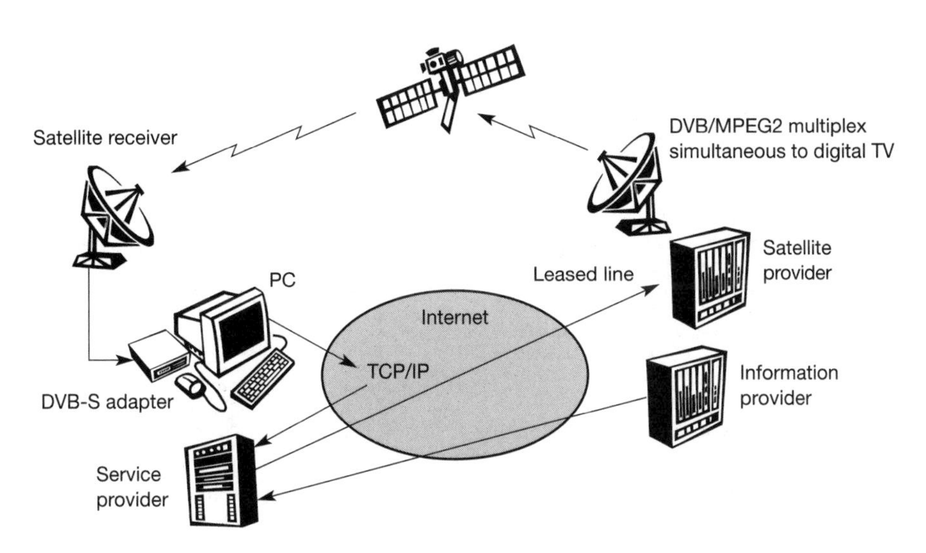

Figure 6.9
High-bandwidth Internet access using DVB

lite receiver. The customer can now receive the requested information with the help of a DVB adapter inside a multi-media PC. Typically, the information for the customer will be encrypted to ensure that only paying customers can use the information. The return channel for requests etc. can be a standard TCP/IP connection via the internet as this channel only requires a low bandwidth.

Typical data rates per user are 5–30 Mbit/s for the downlink via satellite and a return channel with 33 kbit/s using a standard modem, 64 kbit/s with ISDN, or several 100 kbit/s using DSL. One advantage of this approach is that it is transmitted along with the TV programs using free space in the transmitted data stream, so it does not require additional lines or hardware per customer. This factor is particularly important for remote areas or developing countries where high bandwidth wired access such as ADSL is not available. A clear disadvantage of the approach, however, is the shared medium 'satellite'. If a lot of users request data streams via DVB, they all have to share the satellite's bandwidth. This system cannot give hard QoS guarantees to all users without being very expensive.

6.5. Convergence of broadcasting and mobile communications

To enable the convergence of digital broadcasting systems and mobile communication systems ETSI (2000) and ETSI (1999d) define **interaction channels** through GSM for DAB and DVB, respectively. An interaction channel is not only common to DAB and DVB but covers also different fixed and mobile systems (UMTS, DECT, ISDN, PSTN etc.). 3G systems are typically characterized by very small cells, especially in densely populated areas. Although 3G systems offer higher data rates than 2G systems, their design has not fully taken into consideration the integration of

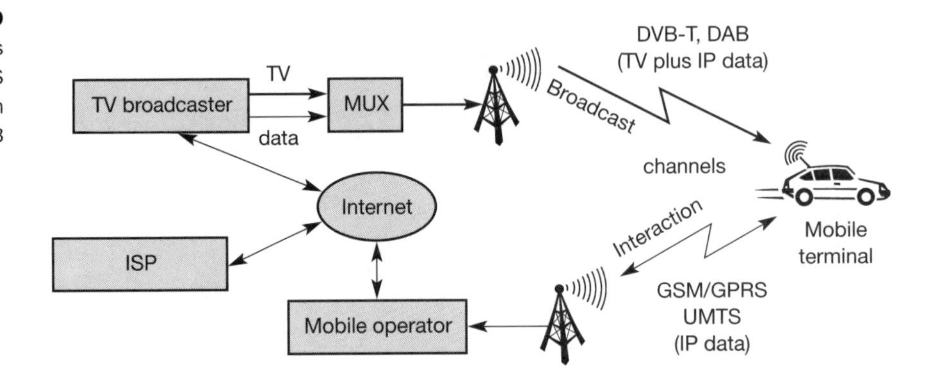

Figure 6.10
Mobile Internet services using IP over GSM/GPRS or UMTS as interaction channel for DAB or DVB

broadcast quality audio and TV services onto 3G terminals. This is true from a technical point of view (capacity per cell in bit/s) as well as from an economic point of view (very high deployment cost for full coverage, typically low return on invest for video services).

Figure 6.10 shows a scenario which benefits from the complementary characteristics of digital broadcast systems and 2.5G/3G mobile systems. High bandwidth audio and video is sent together with IP data via the broadcast channel. IP data could use multi-casting, data carousels etc. as described above. For example, IP data in a DVB-T carousel could contain the top hundred web pages of the ISP's portal. Individual pages for single users are then additionally sent via GRPS or UMTS (DRiVE, 2002).

6.6 Summary

This chapter has presented two examples of broadcast technologies that somehow stand out from the other technologies presented in this book. DAB and DVB are most likely the successors of the traditional radio and television in many countries (probably not everywhere due to varying political and economic interests). In addition to the transmission of audio or video streams, these systems allow for the broadcasting of multimedia data streams. Although both technologies only support unidirectional communication, both will be an integral part of tomorrow's mobile communication scenarios. DAB and DVB will be used to distribute mass data in a cost-effective manner and rely on other low bandwidth wireless technologies for the return channel if required. These technologies support the ongoing amalgamation of computer, communication, and entertainment industries by merging TV/radio data streams with personalized multi-media streams. We can imagine a scenario in which a movie is distributed to everyone, but for example, with individual commercials depending on the user's interests. The set-top box will merge both data streams and the user will, e.g., watch a soccer game with fully individualized billboards. Another feature, which makes DAB particularly attractive for mobile communication, is that it is the only commercial radio system suitable for high speeds and high data rates: up to 1.5 Mbit/s at 900 km/h! This makes it possible to instal, TV sets in for example, trains and other vehicles that would suffer from multi-path propagation using other technologies. Although DVB was not designed for very fast moving receivers, it has been shown in the MOTIVATE project that it still works at over 250 km/h (at reduced data rates; DVB, 2002, MOTIVATE, 2002).

As the aggregate capacity of a UMTS cell is limited (approx. 2 Mbit/s per 5 MHz bandwidth in the standard case) and shared between all active users, UMTS is preferably used for individual communication purposes. On the other hand, high bandwidth distribution of data in a point-to-multipoint fashion will be more efficient and economical on broadcast platforms such as DAB or DVB. However, as the capacity of a mobile DVB-T system is relatively small (about 10–15 Mbit/s per 8 MHz bandwidth), and cell size is large (>100 km^2) this system may not efficiently provide individual data to many users. Both 3G and broadcast platforms, can be seen as complementary, not competitive. Table 6.1 compares the main features of UMTS, DAB and DVB.

Chapter 11 will show further scenarios integrating broadcast and other 4G systems.

Table 6.1 Comparison of UMTS, DAB and DVB

	UMTS	DAB	DVB
Spectrum bands (depends on national regulations) [MHz]	2000 (terrestrial), 2500 (satellite)	1140–1504, 220–228 (UK)	130–260, 430–862 (UK)
Regulation	Telecom, licensed	Broadcast, licensed	Broadcast, licensed
Bandwidth	5 MHz	1.5 MHz	8 MHz
Effective throughput	30–300 kbit/s (per user)	1.5 Mbit/s (shared)	5–30 Mbit/s (shared)
Mobility support	Low to high	Very high	Low to high
Application	Voice, data	Audio, push Internet, images, low res. video	High res. video, audio, push Internet
Coverage	Local to wide	Wide	Wide
Deployment cost for wide coverage	Very high	Low	Low

6.7 Review exercises

1. 2G and 3G systems can both transfer data. Compare these approaches with DAB/DVB and list reasons for and against the use of DAB/DVB.
2. Which web pages would be appropriate for distribution via DAB or DVB?
3. How could location based services and broadcast systems work together?

6.8 References

Acharya, S., Franklin, M., Zdonik, S. (1995) 'Dissemination-based data delivery using broadcast disks,' *IEEE Personal Communications*, 2(6).

CORBA (2002), Common Object Request Broker Architecture, Object Management Group (OMG), http://www.corba.org/, http://www.omg.org/.

DAB (2002) World DAB Forum, http://www.worlddab.org/.

DRiVE (2002), Dynamic Radio for IP-Services in Vehicular Environments, IST-1999-12515, http://www.ist-drive.org/.

DVB (2002) DVB Project Office, http://www.dvb.org/.

ETSI (1997) *Framing structure, channel coding and modulation for 11/12 GHz satellite services*, European Telecommunications Standards Institute, EN 300 421.

ETSI (1998) *Framing structure, channel coding and modulation for cable systems*, European Telecommunications Standards Institute, EN 300 429.

ETSI (1999a) *DAB Multimedia Object Transfer (MOT) protocol*, European Telecommunications Standards Institute, EN 301 234.

ETSI (1999b) *Digital Video Broadcasting (DVB); Implementation guidelines for data broadcasting*, European Telecommunications Standards Institute, TR 101 202.

ETSI (1999c) *Digital Video Broadcasting (DVB); DVB specification for data broadcasting*, European Telecommunications Standards Institute, EN 301 192.

ETSI (1999d) *Digital Video Broadcasting (DVB); Interaction channel through the Global System for Mobile communications (GSM)*, European Telecommunications Standards Institute, EN 301 195.

ETSI (2000) *Digital Audio Broadcasting (DAB); Interaction channel through the Global System for Mobile communications (GSM), the Public Switched Telecommunications System (PSTN), Integrated Services Digital Network (ISDN) and Digital Enhanced Cordless Telecommunications (DECT)*, European Telecommunications Standards Institute, TS 101 737.

ETSI (2001a) *Digital Audio Broadcasting (DAB) to mobile, portable, and fixed receivers*, European Telecommunications Standards Institute, EN 300 401.

ETSI (2001b) *Framing structure, channel coding and modulation for digital terrestrial television*, European Telecommunications Standards Institute, EN 300 744.

ETSI (2002) *Digital Audio and Video Broadcasting*, European Telecommunications Standards Institute, http://www.etsi.org/.

MOTIVATE (2002) Mobile Television and Innovative Receivers, AC318, http://www.cordis.lu/infowin/acts/rus/projects/ac318.htm.

Reimers, U. (1998) 'Digital Video Broadcasting,' *IEEE Communications Magazine*, 36(6).

7 Wireless LAN

This chapter presents several wireless local area network (WLAN) technologies. This constitutes a fast-growing market introducing the flexibility of wireless access into office, home, or production environments. In contrast to the technologies described in chapters 4 through 6, WLANs are typically restricted in their diameter to buildings, a campus, single rooms etc. and are operated by individuals, not by large-scale network providers. The global goal of WLANs is to replace office cabling, to enable tetherless access to the internet and, to introduce a higher flexibility for ad-hoc communication in, e.g., group meetings. The following points illustrate some general advantages and disadvantages of WLANs compared to their wired counterparts.

Some **advantages** of WLANs are:

- **Flexibility:** Within radio coverage, nodes can communicate without further restriction. Radio waves can penetrate walls, senders and receivers can be placed anywhere (also non-visible, e.g., within devices, in walls etc.). Sometimes wiring is difficult if firewalls separate buildings (real firewalls made out of, e.g., bricks, not routers set up as a firewall). Penetration of a firewall is only permitted at certain points to prevent fire from spreading too fast.
- **Planning:** Only wireless ad-hoc networks allow for communication without previous planning, any wired network needs wiring plans. As long as devices follow the same standard, they can communicate. For wired networks, additional cabling with the right plugs and probably interworking units (such as switches) have to be provided.
- **Design:** Wireless networks allow for the design of small, independent devices which can for example be put into a pocket. Cables not only restrict users but also designers of small PDAs, notepads etc. Wireless senders and receivers can be hidden in historic buildings, i.e., current networking technology can be introduced without being visible.
- **Robustness:** Wireless networks can survive disasters, e.g., earthquakes or users pulling a plug. If the wireless devices survive, people can still communicate. Networks requiring a wired infrastructure will usually break down completely.

- **Cost:** After providing wireless access to the infrastructure via an access point for the first user, adding additional users to a wireless network will not increase the cost. This is, important for e.g., lecture halls, hotel lobbies or gate areas in airports where the numbers using the network may vary significantly. Using a fixed network, each seat in a lecture hall should have a plug for the network although many of them might not be used permanently. Constant plugging and unplugging will sooner or later destroy the plugs. Wireless connections do not wear out.

But WLANs also have several **disadvantages**:

- **Quality of service:** WLANs typically offer lower quality than their wired counterparts. The main reasons for this are the lower bandwidth due to limitations in radio transmission (e.g., only 1–10 Mbit/s user data rate instead of 100–1,000 Mbit/s), higher error rates due to interference (e.g., 10^{-4} instead of 10^{-12} for fiber optics), and higher delay/delay variation due to extensive error correction and detection mechanisms.
- **Proprietary solutions:** Due to slow standardization procedures, many companies have come up with proprietary solutions offering standardized functionality plus many enhanced features (typically a higher bit rate using a patented coding technology or special inter-access point protocols). However, these additional features only work in a homogeneous environment, i.e., when adapters from the same vendors are used for all wireless nodes. At least most components today adhere to the basic standards IEEE 802.11b or (newer) 802.11a (see section 7.3).
- **Restrictions:** All wireless products have to comply with national regulations. Several government and non-government institutions worldwide regulate the operation and restrict frequencies to minimize interference. Consequently, it takes a very long time to establish global solutions like, e.g., IMT-2000, which comprises many individual standards (see chapter 4). WLANs are limited to low-power senders and certain license-free frequency bands, which are not the same worldwide.
- **Safety and security:** Using radio waves for data transmission might interfere with other high-tech equipment in, e.g., hospitals. Senders and receivers are operated by laymen and, radiation has to be low. Special precautions have to be taken to prevent safety hazards. The open radio interface makes eavesdropping much easier in WLANs than, e.g., in the case of fiber optics. All standards must offer (automatic) encryption, privacy mechanisms, support for anonymity etc. Otherwise more and more wireless networks will be hacked into as is the case already (aka war driving: driving around looking for unsecured wireless networks; WarDriving, 2002).

Many different, and sometimes competing, design goals have to be taken into account for WLANs to ensure their commercial success:

- **Global operation:** WLAN products should sell in all countries so, national and international frequency regulations have to be considered. In contrast to the infrastructure of wireless WANs, LAN equipment may be carried from one country into another – the operation should still be legal in this case.
- **Low power:** Devices communicating via a WLAN are typically also wireless devices running on battery power. The LAN design should take this into account and implement special power-saving modes and power management functions. Wireless communication with devices plugged into a power outlet is only useful in some cases (e.g., no additional cabling should be necessary for the network in historic buildings or at trade shows). However, the future clearly lies in small handheld devices without any restricting wire.
- **License-free operation:** LAN operators do not want to apply for a special license to be able to use the product. The equipment must operate in a license-free band, such as the 2.4 GHz ISM band.
- **Robust transmission technology:** Compared to their wired counterparts, WLANs operate under difficult conditions. If they use radio transmission, many other electrical devices can interfere with them (vacuum cleaners, hairdryers, train engines etc.). WLAN transceivers cannot be adjusted for perfect transmission in a standard office or production environment. Antennas are typically omnidirectional, not directed. Senders and receivers may move.
- **Simplified spontaneous cooperation:** To be useful in practice, WLANs should not require complicated setup routines but should operate spontaneously after power-up. These LANs would not be useful for supporting, e.g., ad-hoc meetings.
- **Easy to use:** In contrast to huge and complex wireless WANs, wireless LANs are made for simple use. They should not require complex management, but rather work on a plug-and-play basis.
- **Protection of investment:** A lot of money has already been invested into wired LANs. The new WLANs should protect this investment by being interoperable with the existing networks. This means that simple bridging between the different LANs should be enough to interoperate, i.e., the wireless LANs should support the same data types and services that standard LANs support.
- **Safety and security:** Wireless LANs should be safe to operate, especially regarding low radiation if used, e.g., in hospitals. Users cannot keep safety distances to antennas. The equipment has to be safe for pacemakers, too. Users should not be able to read personal data during transmission, i.e., encryption mechanisms should be integrated. The networks should also take into account user privacy, i.e., it should not be possible to collect roaming profiles for tracking persons if they do not agree.

- **Transparency for applications:** Existing applications should continue to run over WLANs, the only difference being higher delay and lower bandwidth. The fact of wireless access and mobility should be hidden if it is not relevant, but the network should also support location aware applications, e.g., by providing location information.

The following sections first introduce basic transmission technologies used for WLANs, infra red and radio, then the two basic settings for WLANs: infrastructure-based and ad-hoc, are presented. The three main sections of this chapter present the IEEE standard for WLANs, IEEE 802.11, the European ETSI standard for a high-speed WLAN with QoS support, HiperLAN2, and finally, an industry approach toward wireless personal area networks (WPAN), i.e., WLANs at an even smaller range, called Bluetooth.

7.1 Infra red vs radio transmission

Today, two different basic transmission technologies can be used to set up WLANs. One technology is based on the transmission of infra red light (e.g., at 900 nm wavelength), the other one, which is much more popular, uses radio transmission in the GHz range (e.g., 2.4 GHz in the license-free ISM band). Both technologies can be used to set up ad-hoc connections for work groups, to connect, e.g., a desktop with a printer without a wire, or to support mobility within a small area.

Infra red technology uses diffuse light reflected at walls, furniture etc. or directed light if a line-of-sight (LOS) exists between sender and receiver. Senders can be simple light emitting diodes (LEDs) or laser diodes. Photodiodes act as receivers. Details about infra red technology, such as modulation, channel impairments etc. can be found in Wesel (1998) and Santamaría (1994).

- The main **advantages** of infra red technology are its simple and extremely cheap senders and receivers which are integrated into nearly all mobile devices available today. PDAs, laptops, notebooks, mobile phones etc. have an infra red data association (IrDA) interface. Version 1.0 of this industry standard implements data rates of up to 115 kbit/s, while IrDA 1.1 defines higher data rates of 1.152 and 4 Mbit/s. No licenses are needed for infra red technology and shielding is very simple. Electrical devices do not interfere with infra red transmission.
- **Disadvantages** of infra red transmission are its low bandwidth compared to other LAN technologies. Typically, IrDA devices are internally connected to a serial port limiting transfer rates to 115 kbit/s. Even 4 Mbit/s is not a particularly high data rate. However, their main disadvantage is that infra red is quite easily shielded. Infra red transmission cannot penetrate walls or other obstacles. Typically, for good transmission quality and high data rates a LOS, i.e., direct connection, is needed.

Almost all networks described in this book use **radio** waves for data transmission, e.g., GSM at 900, 1,800, and 1,900 MHz, DECT at 1,880 MHz etc.

- **Advantages** of radio transmission include the long-term experiences made with radio transmission for wide area networks (e.g., microwave links) and mobile cellular phones. Radio transmission can cover larger areas and can penetrate (thinner) walls, furniture, plants etc. Additional coverage is gained by reflection. Radio typically does not need a LOS if the frequencies are not too high. Furthermore, current radio-based products offer much higher transmission rates (e.g., 54 Mbit/s) than infra red (directed laser links, which offer data rates well above 100 Mbit/s. These are not considered here as it is very difficult to use them with mobile devices).
- Again, the main advantage is also a big **disadvantage** of radio transmission. Shielding is not so simple. Radio transmission can interfere with other senders, or electrical devices can destroy data transmitted via radio. Additionally, radio transmission is only permitted in certain frequency bands. Very limited ranges of license-free bands are available worldwide and those that are available are not the same in all countries. However, a lot of harmonization is going on due to market pressure.

Of the three WLAN technologies presented in this chapter, only one (IEEE 802.11) standardized infra red transmission in addition to radio transmission. The other two (HIPERLAN and Bluetooth) rely on radio. The main reason for this are the shielding problems of infra red. WLANs should, e.g., cover a whole floor of a building and not just the one room where LOSs exist. Future mobile devices may have to communicate while still in a pocket or a suitcase so cannot rely on infra red. The big advantage of radio transmission in everyday use is indeed the ability to penetrate certain materials and that a LOS is not required. Many users experience a lot of difficulties adjusting infra red ports of, e.g., mobile phones to the infra red port of their PDA. Using, e.g., Bluetooth is much simpler.

7.2 Infrastructure and ad-hoc networks

Many WLANs of today need an **infrastructure** network. Infrastructure networks not only provide access to other networks, but also include forwarding functions, medium access control etc. In these infrastructure-based wireless networks, communication typically takes place only between the wireless nodes and the access point (see Figure 7.1), but not directly between the wireless nodes.

The access point does not just control medium access, but also acts as a bridge to other wireless or wired networks. Figure 7.1 shows three access points with their three wireless networks and a wired network. Several wireless networks may form one logical wireless network, so the access points together with the fixed network in between can connect several wireless networks to form a larger network beyond actual radio coverage.

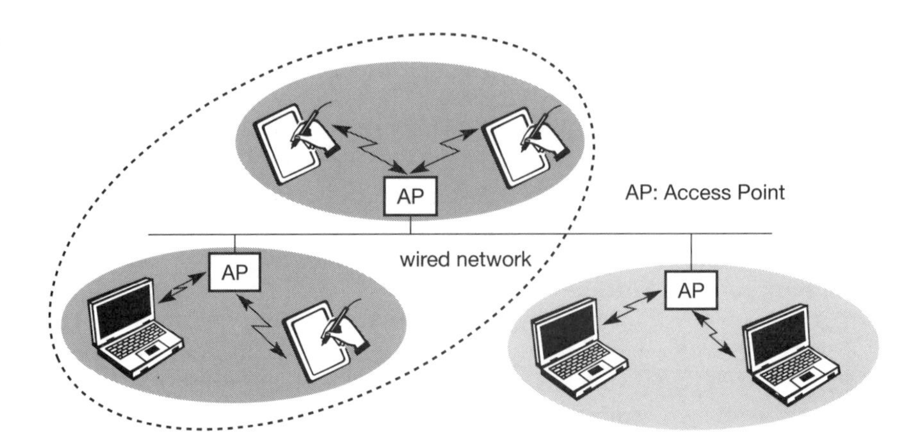

Figure 7.1 Example of three infrastructure-based wireless networks

Typically, the design of infrastructure-based wireless networks is simpler because most of the network functionality lies within the access point, whereas the wireless clients can remain quite simple. This structure is reminiscent of switched Ethernet or other star-based networks, where a central element (e.g., a switch) controls network flow. This type of network can use different access schemes with or without collision. Collisions may occur if medium access of the wireless nodes and the access point is not coordinated. However, if only the access point controls medium access, no collisions are possible. This setting may be useful for quality of service guarantees such as minimum bandwidth for certain nodes. The access point may poll the single wireless nodes to ensure the data rate.

Infrastructure-based networks lose some of the flexibility wireless networks can offer, e.g., they cannot be used for disaster relief in cases where no infrastructure is left. Typical cellular phone networks are infrastructure-based networks for a wide area (see chapter 4). Also satellite-based cellular phones have an infrastructure – the satellites (see chapter 5). Infrastructure does not necessarily imply a wired fixed network.

Ad-hoc wireless networks, however, do not need any infrastructure to work. Each node can communicate directly with other nodes, so no access point controlling medium access is necessary. Figure 7.2 shows two ad-hoc networks with three nodes each. Nodes within an ad-hoc network can only communicate if they can reach each other physically, i.e., if they are within each other's radio range or if other nodes can forward the message. Nodes from the two networks shown in Figure 7.2 cannot, therefore, communicate with each other if they are not within the same radio range.

In ad-hoc networks, the complexity of each node is higher because every node has to implement medium access mechanisms, mechanisms to handle hidden or exposed terminal problems, and perhaps priority mechanisms, to provide a certain quality of service. This type of wireless network exhibits the greatest possible flexibility as it is, for example, needed for unexpected meetings, quick replacements of infrastructure or communication scenarios far away from any infrastructure.

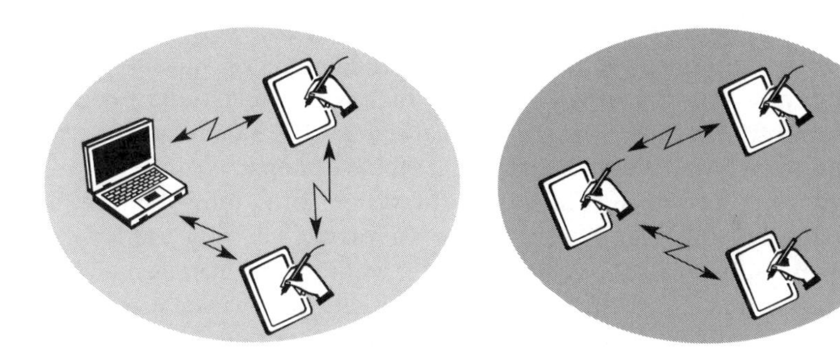

Figure 7.2
Example of two ad-hoc wireless networks

Clearly, the two basic variants of wireless networks (here especially WLANs), infrastructure-based and ad-hoc, do not always come in their pure form. There are networks that rely on access points and infrastructure for basic services (e.g., authentication of access, control of medium access for data with associated quality of service, management functions), but that also allow for direct communication between the wireless nodes.

However, ad-hoc networks might only have selected nodes with the capabilities of forwarding data. Most of the nodes have to connect to such a special node first to transmit data if the receiver is out of their range.

From the three WLANs presented, IEEE 802.11 (see section 7.3) and HiperLAN2 (see section 7.4) are typically infrastructure-based networks, which additionally support ad-hoc networking. However, many implementations only offer the basic infrastructure-based version. The third WLAN, Bluetooth (see section 7.5), is a typical wireless ad-hoc network. Bluetooth focuses precisely on spontaneous ad-hoc meetings or on the simple connection of two or more devices without requiring the setup of an infrastructure.

7.3 IEEE 802.11

The IEEE standard 802.11 (IEEE, 1999) specifies the most famous family of WLANs in which many products are available. As the standard's number indicates, this standard belongs to the group of 802.x LAN standards, e.g., 802.3 Ethernet or 802.5 Token Ring. This means that the standard specifies the physical and medium access layer adapted to the special requirements of wireless LANs, but offers the same interface as the others to higher layers to maintain interoperability.

The primary goal of the standard was the specification of a simple and robust WLAN which offers time-bounded and asynchronous services. The MAC layer should be able to operate with multiple physical layers, each of which exhibits a different medium sense and transmission characteristic. Candidates for physical layers were infra red and spread spectrum radio transmission techniques.

Additional features of the WLAN should include the support of power management to save battery power, the handling of hidden nodes, and the ability to operate worldwide. The 2.4 GHz ISM band, which is available in most countries around the world, was chosen for the original standard. Data rates envisaged for the standard were 1 Mbit/s mandatory and 2 Mbit/s optional.

The following sections will introduce the system and protocol architecture of the initial IEEE 802.11 and then discuss each layer, i.e., physical layer and medium access. After that, the complex and very important management functions of the standard are presented. Finally, this subsection presents the enhancements of the original standard for higher data rates, 802.11a (up to 54 Mbit/s at 5 GHz) and 802.11b (today the most successful with 11 Mbit/s) together with further developments for security support, harmonization, or other modulation schemes.

7.3.1 System architecture

Wireless networks can exhibit two different basic system architectures as shown in section 7.2: infrastructure-based or ad-hoc. Figure 7.3 shows the components of an infrastructure and a wireless part as specified for IEEE 802.11. Several nodes, called **stations (STA_i)**, are connected to **access points (AP)**. Stations are terminals with access mechanisms to the wireless medium and radio contact to

Figure 7.3 Architecture of an infrastructure-based IEEE 802.11

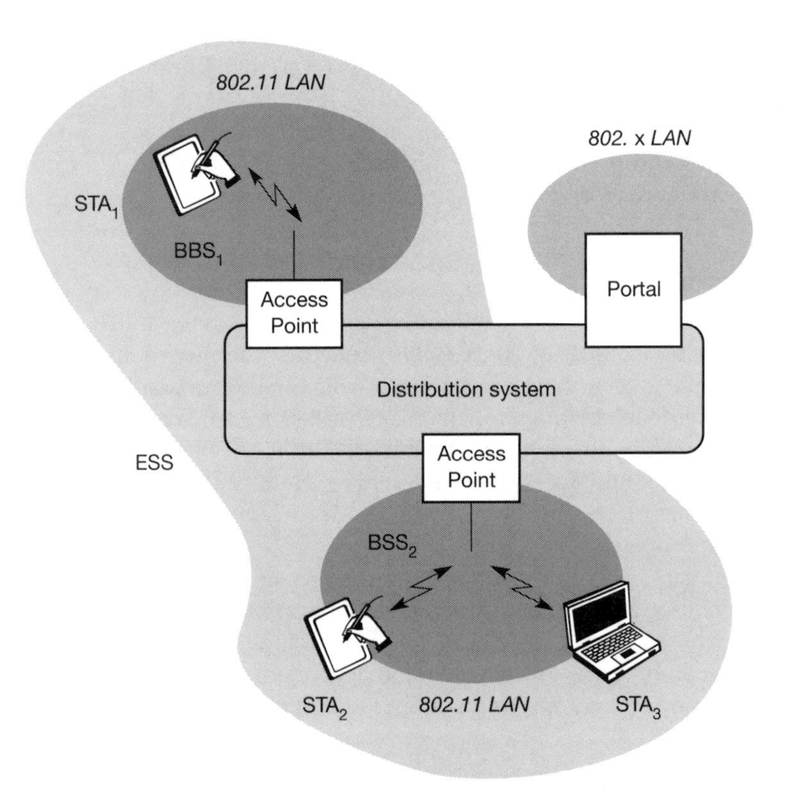

the AP. The stations and the AP which are within the same radio coverage form a **basic service set (BSS_i)**. The example shows two BSSs – BSS_1 and BSS_2 – which are connected via a **distribution system**. A distribution system connects several BSSs via the AP to form a single network and thereby extends the wireless coverage area. This network is now called an **extended service set (ESS)** and has its own identifier, the ESSID. The ESSID is the 'name' of a network and is used to separate different networks. Without knowing the ESSID (and assuming no hacking) it should not be possible to participate in the WLAN. The distribution system connects the wireless networks via the APs with a **portal**, which forms the interworking unit to other LANs.

The architecture of the distribution system is not specified further in IEEE 802.11. It could consist of bridged IEEE LANs, wireless links, or any other networks. However, **distribution system services** are defined in the standard (although, many products today cannot interoperate and needs the additional standard IEEE 802.11f to specify an inter access point protocol, see section 7.3.8).

Stations can select an AP and associate with it. The APs support roaming (i.e., changing access points), the distribution system handles data transfer between the different APs. APs provide synchronization within a BSS, support power management, and can control medium access to support time-bounded service. These and further functions are explained in the following sections.

In addition to infrastructure-based networks, IEEE 802.11 allows the building of ad-hoc networks between stations, thus forming one or more independent BSSs (IBSS) as shown in Figure 7.4. In this case, an IBSS comprises a group of stations using the same radio frequency. Stations STA_1, STA_2, and STA_3 are in $IBSS_1$, STA_4 and STA_5 in $IBSS_2$. This means for example that STA_3 can communicate

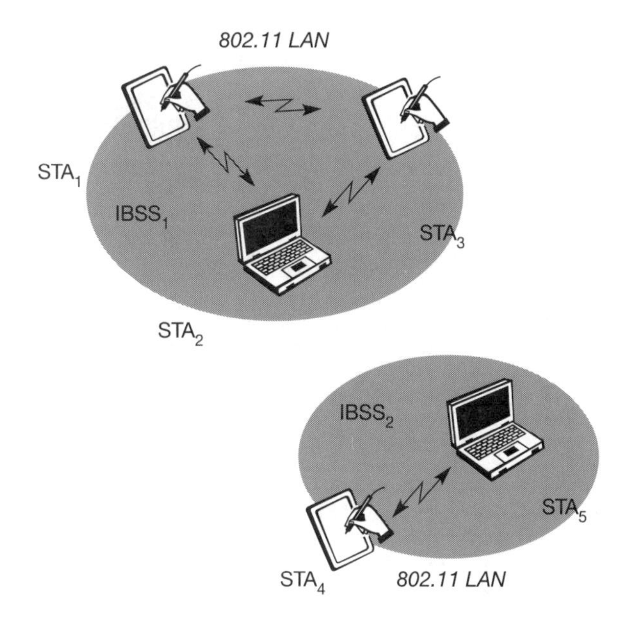

Figure 7.4
Architecture of IEEE 802.11 ad-hoc wireless LANs

directly with STA_2 but not with STA_5. Several IBSSs can either be formed via the distance between the IBSSs (see Figure 7.4) or by using different carrier frequencies (then the IBSSs could overlap physically). IEEE 802.11 does not specify any special nodes that support routing, forwarding of data or exchange of topology information as, e.g., HIPERLAN 1 (see section 7.4) or Bluetooth (see section 7.5).

7.3.2 Protocol architecture

As indicated by the standard number, IEEE 802.11 fits seamlessly into the other 802.x standards for wired LANs (see Halsall, 1996; IEEE, 1990). Figure 7.5 shows the most common scenario: an IEEE 802.11 wireless LAN connected to a switched IEEE 802.3 Ethernet via a bridge. Applications should not notice any difference apart from the lower bandwidth and perhaps higher access time from the wireless LAN. The WLAN behaves like a slow wired LAN. Consequently, the higher layers (application, TCP, IP) look the same for wireless nodes as for wired nodes. The upper part of the data link control layer, the logical link control (LLC), covers the differences of the medium access control layers needed for the different media. In many of today's networks, no explicit LLC layer is visible. Further details like Ethertype or sub-network access protocol (SNAP) and bridging technology are explained in, e.g., Perlman (1992).

The IEEE 802.11 standard only covers the physical layer **PHY** and medium access layer **MAC** like the other 802.x LANs do. The physical layer is subdivided into the **physical layer convergence protocol (PLCP)** and the **physical medium dependent** sublayer **PMD** (see Figure 7.6). The basic tasks of the MAC layer comprise medium access, fragmentation of user data, and encryption. The

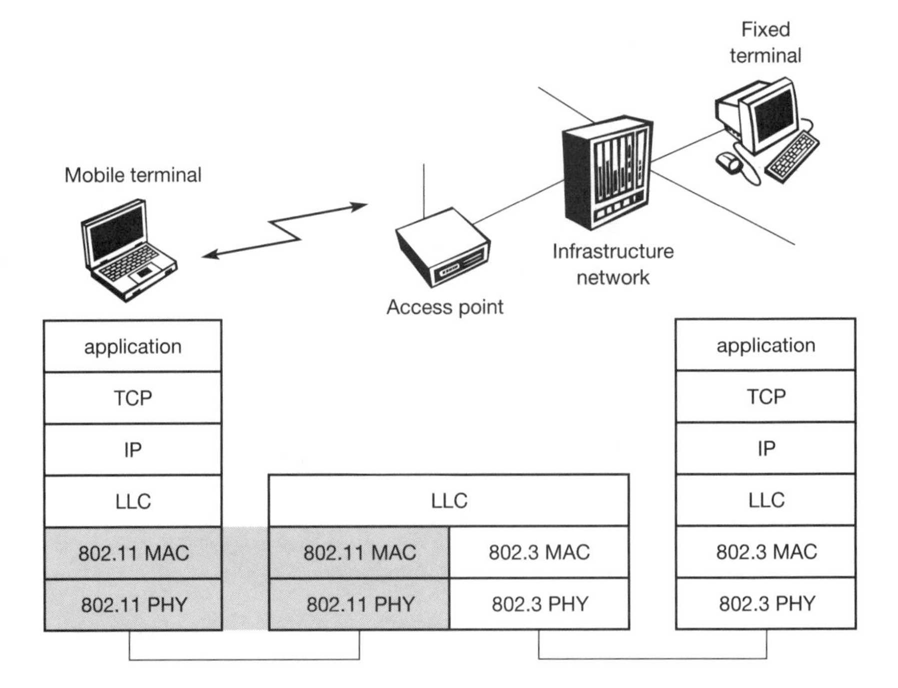

Figure 7.5
IEEE 802.11 protocol architecture and bridging

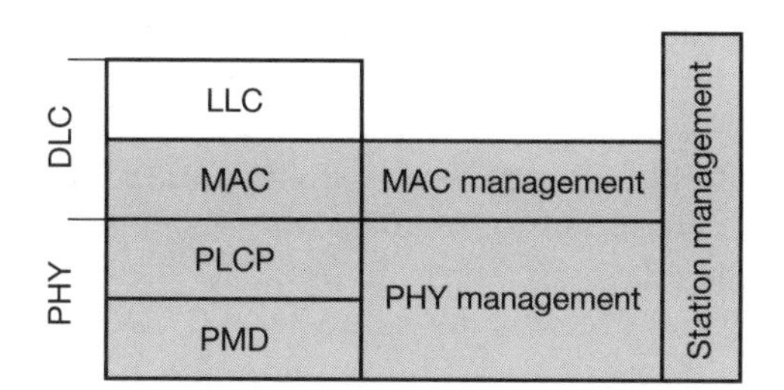

Figure 7.6
Detailed IEEE 802.11 protocol architecture and management

PLCP sublayer provides a carrier sense signal, called clear channel assessment (CCA), and provides a common PHY service access point (SAP) independent of the transmission technology. Finally, the PMD sublayer handles modulation and encoding/decoding of signals. The PHY layer (comprising PMD and PLCP) and the MAC layer will be explained in more detail in the following sections.

Apart from the protocol sublayers, the standard specifies management layers and the station management. The **MAC management** supports the association and re-association of a station to an access point and roaming between different access points. It also controls authentication mechanisms, encryption, synchronization of a station with regard to an access point, and power management to save battery power. MAC management also maintains the MAC management information base (MIB).

The main tasks of the **PHY management** include channel tuning and PHY MIB maintenance. Finally, **station management** interacts with both management layers and is responsible for additional higher layer functions (e.g., control of bridging and interaction with the distribution system in the case of an access point).

7.3.3 Physical layer

IEEE 802.11 supports three different physical layers: one layer based on infra red and two layers based on radio transmission (primarily in the ISM band at 2.4 GHz, which is available worldwide). All PHY variants include the provision of the **clear channel assessment** signal **(CCA)**. This is needed for the MAC mechanisms controlling medium access and indicates if the medium is currently idle. The transmission technology (which will be discussed later) determines exactly how this signal is obtained.

The PHY layer offers a service access point (SAP) with 1 or 2 Mbit/s transfer rate to the MAC layer (basic version of the standard). The remainder of this section presents the three versions of a PHY layer defined in the standard.

7.3.3.1 Frequency hopping spread spectrum

Frequency hopping spread spectrum (FHSS) is a spread spectrum technique which allows for the coexistence of multiple networks in the same area by separating different networks using different hopping sequences (see chapters 2 and 3). The original standard defines 79 hopping channels for North America and Europe, and 23 hopping channels for Japan (each with a bandwidth of 1 MHz in the 2.4 GHz ISM band). The selection of a particular channel is achieved by using a pseudo-random hopping pattern. National restrictions also determine further parameters, e.g., maximum transmit power is 1 W in the US, 100 mW EIRP (equivalent isotropic radiated power) in Europe and 10 mW/MHz in Japan.

The standard specifies Gaussian shaped FSK (frequency shift keying), GFSK, as modulation for the FHSS PHY. For 1 Mbit/s a 2 level GFSK is used (i.e., 1 bit is mapped to one frequency, see chapter 2), a 4 level GFSK for 2 Mbit/s (i.e., 2 bits are mapped to one frequency). While sending and receiving at 1 Mbit/s is mandatory for all devices, operation at 2 Mbit/s is optional. This facilitated the production of low-cost devices for the lower rate only and more powerful devices for both transmission rates in the early days of 802.11.

Figure 7.7 shows a frame of the physical layer used with FHSS. The frame consists of two basic parts, the PLCP part (preamble and header) and the payload part. While the PLCP part is always transmitted at 1 Mbit/s, payload, i.e. MAC data, can use 1 or 2 Mbit/s. Additionally, MAC data is scrambled using the polynomial $s(z) = z^7 + z^4 + 1$ for DC blocking and whitening of the spectrum. The fields of the frame fulfill the following functions:

- **Synchronization:** The PLCP preamble starts with 80 bit synchronization, which is a 010101... bit pattern. This pattern is used for synchronization of potential receivers and signal detection by the CCA.
- **Start frame delimiter (SFD):** The following 16 bits indicate the start of the frame and provide frame synchronization. The SFD pattern is 0000110010111101.
- **PLCP_PDU length word (PLW):** This first field of the PLCP header indicates the length of the payload in bytes including the 32 bit CRC at the end of the payload. PLW can range between 0 and 4,095.
- **PLCP signalling field (PSF):** This 4 bit field indicates the data rate of the payload following. All bits set to zero (0000) indicates the lowest data rate of 1 Mbit/s. The granularity is 500 kbit/s, thus 2 Mbit/s is indicated by 0010 and the maximum is 8.5 Mbit/s (1111). This system obviously does not accommodate today's higher data rates.
- **Header error check (HEC):** Finally, the PLCP header is protected by a 16 bit checksum with the standard ITU-T generator polynomial $G(x) = x^{16} + x^{12} + x^5 + 1$.

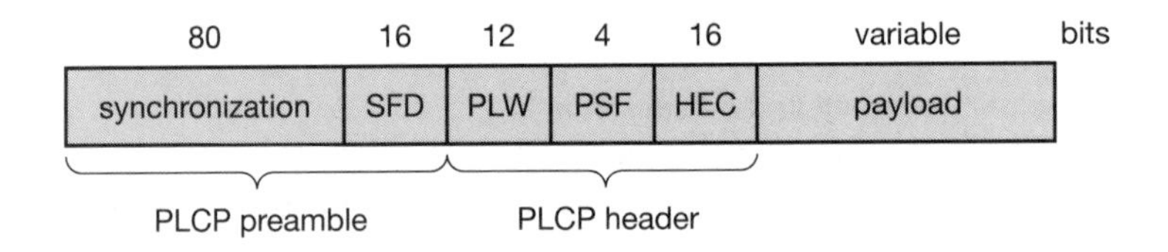

Figure 7.7 Format of an IEEE 802.11 PHY frame using FHSS

7.3.3.2 Direct sequence spread spectrum

Direct sequence spread spectrum (DSSS) is the alternative spread spectrum method separating by code and not by frequency. In the case of IEEE 802.11 DSSS, spreading is achieved using the 11-chip Barker sequence (+1, –1, +1, +1, –1, +1, +1, +1, –1, –1, –1). The key characteristics of this method are its robustness against interference and its insensitivity to multipath propagation (time delay spread). However, the implementation is more complex compared to FHSS.

IEEE 802.11 DSSS PHY also uses the 2.4 GHz ISM band and offers both 1 and 2 Mbit/s data rates. The system uses differential binary phase shift keying (DBPSK) for 1 Mbit/s transmission and differential quadrature phase shift keying (DQPSK) for 2 Mbit/s as modulation schemes. Again, the maximum transmit power is 1 W in the US, 100 mW EIRP in Europe and 10 mW/MHz in Japan. The symbol rate is 1 MHz, resulting in a chipping rate of 11 MHz. All bits transmitted by the DSSS PHY are scrambled with the polynomial $s(z) = z^7 + z^4 + 1$ for DC blocking and whitening of the spectrum. Many of today's products offering 11 Mbit/s according to 802.11b are still backward compatible to these lower data rates.

Figure 7.8 shows a frame of the physical layer using DSSS. The frame consists of two basic parts, the PLCP part (preamble and header) and the payload part. While the PLCP part is always transmitted at 1 Mbit/s, payload, i.e., MAC data, can use 1 or 2 Mbit/s. The fields of the frame have the following functions:

- **Synchronization:** The first 128 bits are not only used for synchronization, but also gain setting, energy detection (for the CCA), and frequency offset compensation. The synchronization field only consists of scrambled 1 bits.
- **Start frame delimiter (SFD):** This 16 bit field is used for synchronization at the beginning of a frame and consists of the pattern 1111001110100000.
- **Signal:** Originally, only two values have been defined for this field to indicate the data rate of the payload. The value 0x0A indicates 1 Mbit/s (and thus DBPSK), 0x14 indicates 2 Mbit/s (and thus DQPSK). Other values have been reserved for future use, i.e., higher bit rates. Coding for higher data rates is explained in sections 7.3.6 and 7.3.7.
- **Service:** This field is reserved for future use; however, 0x00 indicates an IEEE 802.11 compliant frame.
- **Length:** 16 bits are used in this case for length indication of the payload in microseconds.
- **Header error check (HEC):** Signal, service, and length fields are protected by this checksum using the ITU-T CRC-16 standard polynomial.

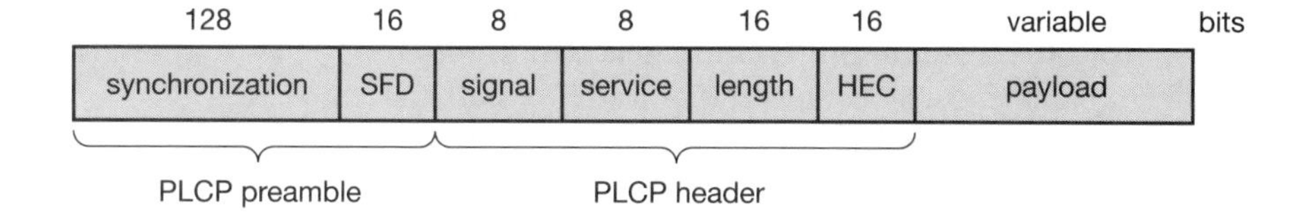

Figure 7.8 Format of an IEEE 802.11 PHY frame using DSSS

7.3.3.3 Infra red

The PHY layer, which is based on infra red (IR) transmission, uses near visible light at 850–950 nm. Infra red light is not regulated apart from safety restrictions (using lasers instead of LEDs). The standard does not require a line-of-sight between sender and receiver, but should also work with diffuse light. This allows for point-to-multipoint communication. The maximum range is about 10 m if no sunlight or heat sources interfere with the transmission. Typically, such a network will only work in buildings, e.g., classrooms, meeting rooms etc. Frequency reuse is very simple – a wall is more than enough to shield one IR based IEEE 802.11 network from another. (See also section 7.1 for a comparison between IR and radio transmission and Wesel, 1998 for more details.) Today, no products are available that offer infra red communication based on 802.11. Proprietary products offer, e.g., up to 4 Mbit/s using diffuse infra red light. Alternatively, directed infra red communication based on IrDA can be used (IrDA, 2002).

7.3.4 Medium access control layer

The MAC layer has to fulfill several tasks. First of all, it has to control medium access, but it can also offer support for roaming, authentication, and power conservation. The basic services provided by the MAC layer are the mandatory **asynchronous data service** and an optional **time-bounded service**. While 802.11 only offers the asynchronous service in ad-hoc network mode, both service types can be offered using an infrastructure-based network together with the access point coordinating medium access. The asynchronous service supports broadcast and multi-cast packets, and packet exchange is based on a 'best effort' model, i.e., no delay bounds can be given for transmission.

The following three basic access mechanisms have been defined for IEEE 802.11: the mandatory basic method based on a version of CSMA/CA, an optional method avoiding the hidden terminal problem, and finally a contention-free polling method for time-bounded service. The first two methods are also summarized as **distributed coordination function (DCF)**, the third method is called **point coordination function (PCF)**. DCF only offers asynchronous service, while PCF offers both asynchronous and time-bounded service but needs an access point to control medium access and to avoid contention. The MAC mechanisms are also called **distributed foundation wireless medium access control (DFWMAC)**.

For all access methods, several parameters for controlling the waiting time before medium access are important. Figure 7.9 shows the three different parameters that define the priorities of medium access. The values of the parameters depend on the PHY and are defined in relation to a **slot** time. Slot time is derived from the medium propagation delay, transmitter delay, and other PHY dependent parameters. Slot time is 50 μs for FHSS and 20 μs for DSSS.

The medium, as shown, can be busy or idle (which is detected by the CCA). If the medium is busy this can be due to data frames or other control frames. During a contention phase several nodes try to access the medium.

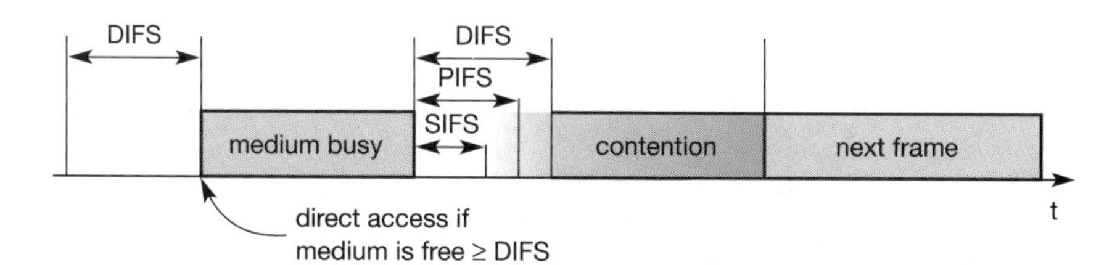

Figure 7.9
Medium access and inter-frame spacing

- **Short inter-frame spacing (SIFS):** The shortest waiting time for medium access (so the highest priority) is defined for short control messages, such as acknowledgements of data packets or polling responses. For DSSS SIFS is 10 μs and for FHSS it is 28 μs. The use of this parameter will be explained in sections 7.3.4.1 through 7.3.4.3.
- **PCF inter-frame spacing (PIFS):** A waiting time between DIFS and SIFS (and thus a medium priority) is used for a time-bounded service. An access point polling other nodes only has to wait PIFS for medium access (see section 7.3.4.3). PIFS is defined as SIFS plus one slot time.
- **DCF inter-frame spacing (DIFS):** This parameter denotes the longest waiting time and has the lowest priority for medium access. This waiting time is used for asynchronous data service within a contention period (this parameter and the basic access method are explained in section 7.3.4.1). DIFS is defined as SIFS plus two slot times.

7.3.4.1 Basic DFWMAC-DCF using CSMA/CA

The mandatory access mechanism of IEEE 802.11 is based on **carrier sense multiple access with collision avoidance** (CSMA/CA), which is a random access scheme with carrier sense and collision avoidance through random backoff. The basic CSMA/CA mechanism is shown in Figure 7.10. If the medium is idle for at least the duration of DIFS (with the help of the CCA signal of the physical layer), a node can access the medium at once. This allows for short access delay under light load. But as more and more nodes try to access the medium, additional mechanisms are needed.

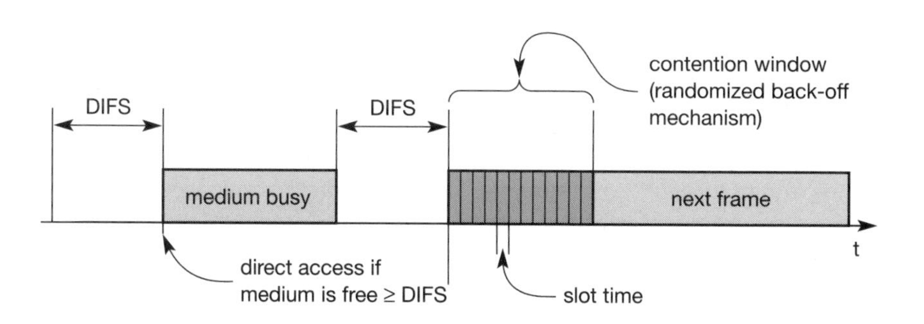

Figure 7.10
Contention window and waiting time

If the medium is busy, nodes have to wait for the duration of DIFS, entering a contention phase afterwards. Each node now chooses a **random backoff time** within a **contention window** and delays medium access for this random amount of time. The node continues to sense the medium. As soon as a node senses the channel is busy, it has lost this cycle and has to wait for the next chance, i.e., until the medium is idle again for at least DIFS. But if the randomized additional waiting time for a node is over and the medium is still idle, the node can access the medium immediately (i.e., no other node has a shorter waiting time). The additional waiting time is measured in multiples of the above-mentioned slots. This additional randomly distributed delay helps to avoid collisions – otherwise all stations would try to transmit data after waiting for the medium becoming idle again plus DIFS.

Obviously, the basic CSMA/CA mechanism is not fair. Independent of the overall time a node has already waited for transmission; each node has the same chances for transmitting data in the next cycle. To provide fairness, IEEE 802.11 adds a **backoff timer.** Again, each node selects a random waiting time within the range of the contention window. If a certain station does not get access to the medium in the first cycle, it stops its backoff timer, waits for the channel to be idle again for DIFS and starts the counter again. As soon as the counter expires, the node accesses the medium. This means that deferred stations do not choose a randomized backoff time again, but continue to count down. Stations that have waited longer have the advantage over stations that have just entered, in that they only have to wait for the remainder of their backoff timer from the previous cycle(s).

Figure 7.11 explains the basic access mechanism of IEEE 802.11 for five stations trying to send a packet at the marked points in time. Station_3 has the first request from a higher layer to send a packet (packet arrival at the MAC SAP). The station senses the medium, waits for DIFS and accesses the medium, i.e., sends the packet. Station_1, station_2, and station_5 have to wait at least until the medium is idle for DIFS again after station_3 has stopped sending. Now all three stations choose a backoff time within the contention window and start counting down their backoff timers.

Figure 7.11 shows the random backoff time of station_1 as sum of bo_e (the elapsed backoff time) and bo_r (the residual backoff time). The same is shown for station_5. Station_2 has a total backoff time of only bo_e and gets access to the medium first. No residual backoff time for station_2 is shown. The backoff timers of station_1 and station_5 stop, and the stations store their residual backoff times. While a new station has to choose its backoff time from the whole contention window, the two old stations have statistically smaller backoff values. The older values are on average lower than the new ones.

Now station_4 wants to send a packet as well, so after DIFS waiting time, three stations try to get access. It can now happen, as shown in the figure, that two stations accidentally have the same backoff time, no matter whether remaining or newly chosen. This results in a collision on the medium as shown, i.e., the trans-

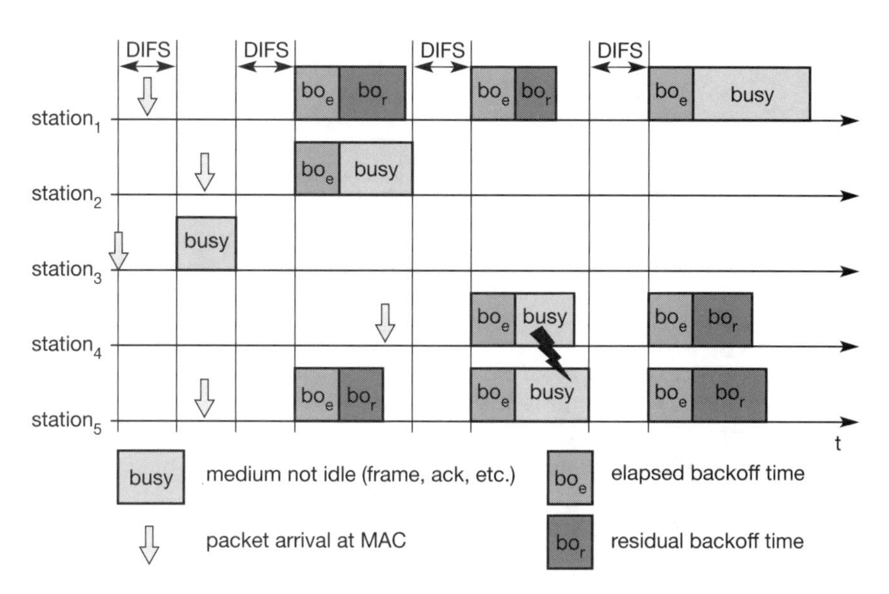

Figure 7.11 Basic DFWMAC–DCF with several competing senders

mitted frames are destroyed. Station_1 stores its residual backoff time again. In the last cycle shown station_1 finally gets access to the medium, while station_4 and station_5 have to wait. A collision triggers a retransmission with a new random selection of the backoff time. Retransmissions are not privileged.

Still, the access scheme has problems under heavy or light load. Depending on the size of the contention window (CW), the random values can either be too close together (causing too many collisions) or the values are too high (causing unnecessary delay). The system tries to adapt to the current number of stations trying to send.

The contention window starts with a size of, e.g., $CW_{min} = 7$. Each time a collision occurs, indicating a higher load on the medium, the contention window doubles up to a maximum of, e.g., $CW_{max} = 255$ (the window can take on the values 7, 15, 31, 63, 127, and 255). The larger the contention window is, the greater is the resolution power of the randomized scheme. It is less likely to choose the same random backoff time using a large CW. However, under a light load, a small CW ensures shorter access delays. This algorithm is also called **exponential backoff** and is already familiar from IEEE 802.3 CSMA/CD in a similar version.

While this process describes the complete access mechanism for broadcast frames, an additional feature is provided by the standard for unicast data transfer. Figure 7.12 shows a sender accessing the medium and sending its data. But now, the receiver answers directly with an **acknowledgement (ACK)**. The receiver accesses the medium after waiting for a duration of SIFS so no other station can access the medium in the meantime and cause a collision. The other stations have to wait for DIFS plus their backoff time. This acknowledgement ensures the correct reception (correct checksum CRC at the receiver) of a frame on the MAC layer, which is especially important in error-prone environments

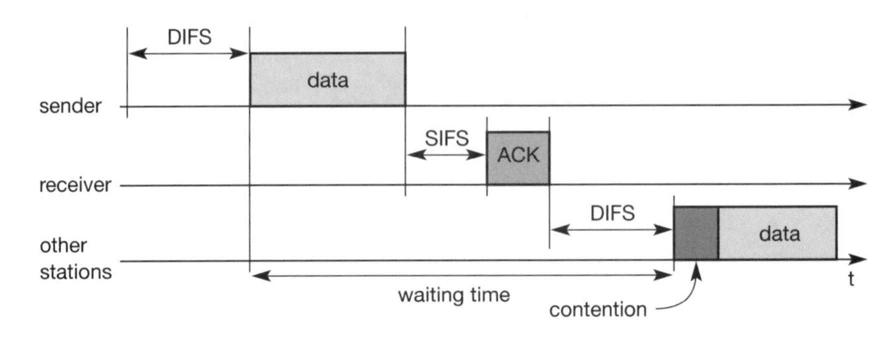

Figure 7.12
IEEE 802.11 unicast data transfer

such as wireless connections. If no ACK is returned, the sender automatically retransmits the frame. But now the sender has to wait again and compete for the access right. There are no special rules for retransmissions. The number of retransmissions is limited, and final failure is reported to the higher layer.

7.3.4.2 DFWMAC-DCF with RTS/CTS extension

Section 3.1 discussed the problem of hidden terminals, a situation that can also occur in IEEE 802.11 networks. This problem occurs if one station can receive two others, but those stations cannot receive each other. The two stations may sense the channel is idle, send a frame, and cause a collision at the receiver in the middle. To deal with this problem, the standard defines an additional mechanism using two control packets, RTS and CTS. The use of the mechanism is optional; however, every 802.11 node has to implement the functions to react properly upon reception of RTS/CTS control packets.

Figure 7.13 illustrates the use of RTS and CTS. After waiting for DIFS (plus a random backoff time if the medium was busy), the sender can issue a **request to send (RTS)** control packet. The RTS packet thus is not given any higher priority compared to other data packets. The RTS packet includes the receiver of the data transmission to come and the duration of the whole data transmission. This duration specifies the time interval necessary to transmit the whole data frame and the acknowledgement related to it. Every node receiving this RTS now has to set its **net allocation vector (NAV)** in accordance with the duration field. The NAV then specifies the earliest point at which the station can try to access the medium again.

If the receiver of the data transmission receives the RTS, it answers with a **clear to send (CTS)** message after waiting for SIFS. This CTS packet contains the duration field again and all stations receiving this packet from the receiver of the intended data transmission have to adjust their NAV. The latter set of receivers need not be the same as the first set receiving the RTS packet. Now all nodes within receiving distance around sender and receiver are informed that they have to wait more time before accessing the medium. Basically, this mechanism reserves the medium for one sender exclusively (this is why it is sometimes called a virtual reservation scheme).

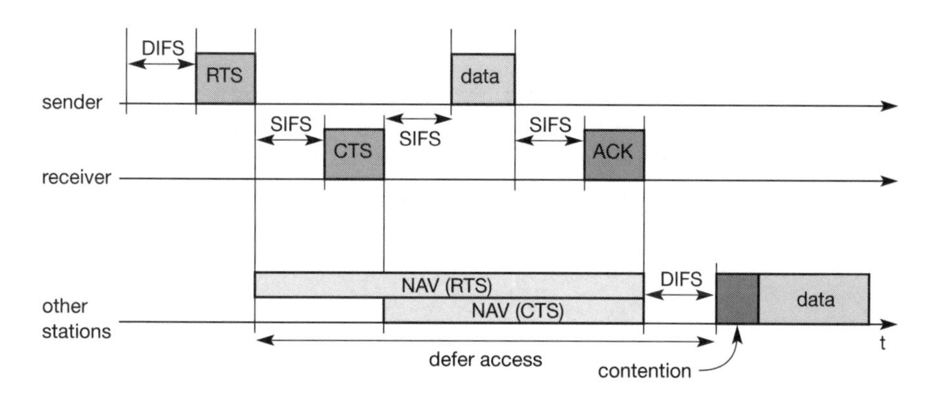

Figure 7.13 IEEE 802.11 hidden node provisions for contention-free access

Finally, the sender can send the data after SIFS. The receiver waits for SIFS after receiving the data packet and then acknowledges whether the transfer was correct. The transmission has now been completed, the NAV in each node marks the medium as free and the standard cycle can start again.

Within this scenario (i.e., using RTS and CTS to avoid the hidden terminal problem), collisions can only occur at the beginning while the RTS is sent. Two or more stations may start sending at the same time (RTS or other data packets). Using RTS/CTS can result in a non-negligible overhead causing a waste of bandwidth and higher delay. An RTS threshold can determine when to use the additional mechanism (basically at larger frame sizes) and when to disable it (short frames). Chhaya (1996) and Chhaya (1997) give an overview of the asynchronous services in 802.11 and discuss performance under different load scenarios.

Wireless LANs have bit error rates in transmission that are typically several orders of magnitude higher than, e.g., fiber optics. The probability of an erroneous frame is much higher for wireless links assuming the same frame length. One way to decrease the error probability of frames is to use shorter frames. In this case, the bit error rate is the same, but now only short frames are destroyed and, the frame error rate decreases.

However, the mechanism of fragmenting a user data packet into several smaller parts should be transparent for a user. The MAC layer should have the possibility of adjusting the transmission frame size to the current error rate on the medium. The IEEE 802.11 standard specifies a **fragmentation** mode (see Figure 7.14). Again, a sender can send an RTS control packet to reserve the medium after a waiting time of DIFS. This RTS packet now includes the duration for the transmission of the first fragment and the corresponding acknowledgement. A certain set of nodes may receive this RTS and set their NAV according to the duration field. The receiver answers with a CTS, again including the duration of the transmission up to the acknowledgement. A (possibly different) set of receivers gets this CTS message and sets the NAV.

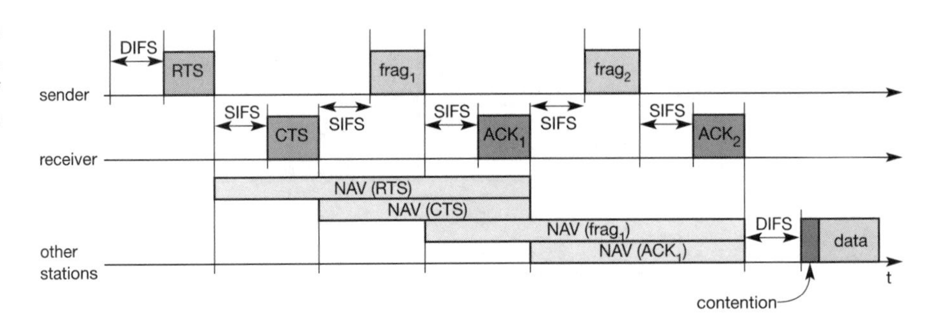

Figure 7.14 IEEE 802.11 fragmentation of user data

As shown in Figure 7.13, the sender can now send the first data frame, $frag_1$, after waiting only for SIFS. The new aspect of this fragmentation mode is that it includes another duration value in the frame $frag_1$. This duration field reserves the medium for the duration of the transmission following, comprising the second fragment and its acknowledgement. Again, several nodes may receive this reservation and adjust their NAV. If all nodes are static and transmission conditions have not changed, then the set of nodes receiving the duration field in $frag_1$ should be the same as the set that has received the initial reservation in the RTS control packet. However, due to the mobility of nodes and changes in the environment, this could also be a different set of nodes.

The receiver of $frag_1$ answers directly after SIFS with the acknowledgement packet ACK_1 including the reservation for the next transmission as shown. Again, a fourth set of nodes may receive this reservation and adjust their NAV (which again could be the same as the second set of nodes that has received the reservation in the CTS frame).

If $frag_2$ was not the last frame of this transmission, it would also include a new duration for the third consecutive transmission. (In the example shown, $frag_2$ is the last fragment of this transmission so the sender does not reserve the medium any longer.) The receiver acknowledges this second fragment, not reserving the medium again. After ACK_2, all nodes can compete for the medium again after having waited for DIFS.

7.3.4.3 DFWMAC-PCF with polling

The two access mechanisms presented so far cannot guarantee a maximum access delay or minimum transmission bandwidth. To provide a time-bounded service, the standard specifies a **point coordination function (PCF)** on top of the standard DCF mechanisms. Using PCF requires an access point that controls medium access and polls the single nodes. Ad-hoc networks cannot use this function so, provide no QoS but 'best effort' in IEEE 802.11 WLANs.

The **point co-ordinator** in the access point splits the access time into super frame periods as shown in Figure 7.15. A **super frame** comprises a **contention-free period** and a **contention period**. The contention period can be used for the two access mechanisms presented above. The figure also shows several wireless stations (all on the same line) and the stations' NAV (again on one line).

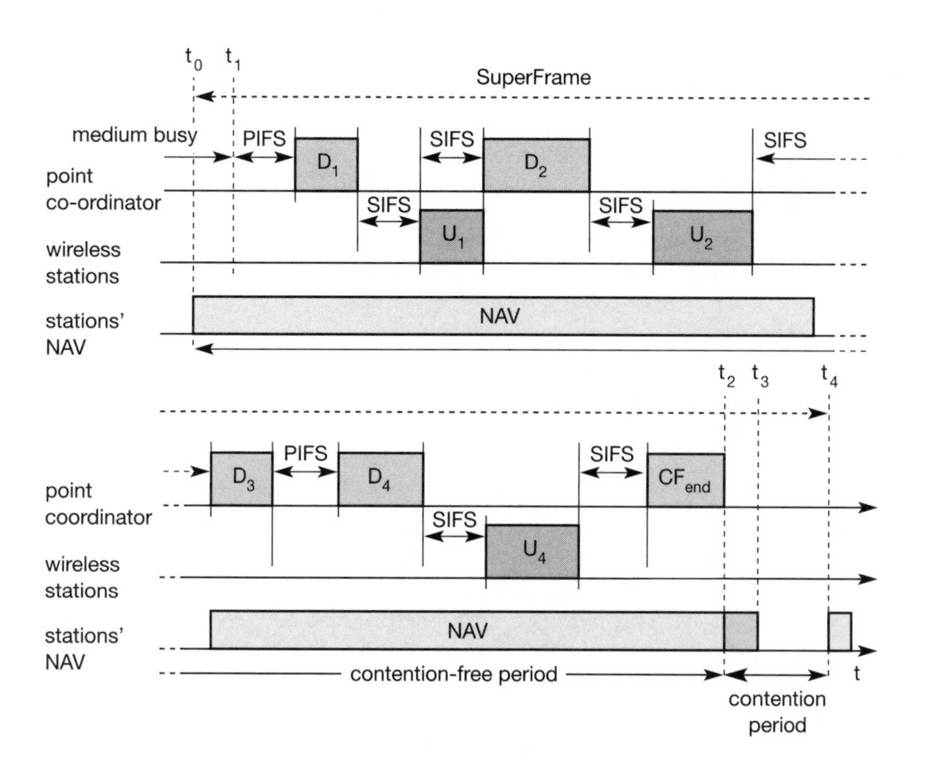

Figure 7.15
Contention-free access using polling mechanisms (PCF)

At time t_0 the contention-free period of the super frame should theoretically start, but another station is still transmitting data (i.e., the medium is busy). This means that PCF also defers to DCF, and the start of the super frame may be postponed. The only possibility of avoiding variations is not to have any contention period at all. After the medium has been idle until t_1, the point coordinator has to wait for PIFS before accessing the medium. As PIFS is smaller than DIFS, no other station can start sending earlier.

The point coordinator now sends data D_1 downstream to the first wireless station. This station can answer at once after SIFS (see Figure 7.15). After waiting for SIFS again, the point coordinator can poll the second station by sending D_2. This station may answer upstream to the coordinator with data U_2. Polling continues with the third node. This time the node has nothing to answer and the point coordinator will not receive a packet after SIFS.

After waiting for PIFS, the coordinator can resume polling the stations. Finally, the point coordinator can issue an end marker (CF_{end}), indicating that the contention period may start again. Using PCF automatically sets the NAV, preventing other stations from sending. In the example, the contention-free period planned initially would have been from t_0 to t_3. However, the point coordinator finished polling earlier, shifting the end of the contention-free period to t_2. At t_4, the cycle starts again with the next super frame.

The transmission properties of the whole wireless network are now determined by the polling behavior of the access point. If only PCF is used and polling is distributed evenly, the bandwidth is also distributed evenly among all polled nodes. This would resemble a static, centrally controlled time division multiple access (TDMA) system with time division duplex (TDD) transmission. This method comes with an overhead if nodes have nothing to send, but the access point polls them permanently. Anastasi (1998) elaborates the example of voice transmission using 48 byte packets as payload. In this case, PCF introduces an overhead of 75 byte.

7.3.4.4 MAC frames

Figure 7.16 shows the basic structure of an IEEE 802.11 MAC data frame together with the content of the frame control field. The fields in the figure refer to the following:

- **Frame control:** The first 2 bytes serve several purposes. They contain several sub-fields as explained after the MAC frame.
- **Duration/ID:** If the field value is less than 32,768, the duration field contains the value indicating the period of time in which the medium is occupied (in μs). This field is used for setting the NAV for the virtual reservation mechanism using RTS/CTS and during fragmentation. Certain values above 32,768 are reserved for identifiers.
- **Address 1 to 4:** The four address fields contain standard IEEE 802 MAC addresses (48 bit each), as they are known from other 802.x LANs. The meaning of each address depends on the DS bits in the frame control field and is explained in more detail in a separate paragraph.
- **Sequence control:** Due to the acknowledgement mechanism frames may be duplicated. Therefore a sequence number is used to filter duplicates.
- **Data:** The MAC frame may contain arbitrary data (max. 2,312 byte), which is transferred transparently from a sender to the receiver(s).
- **Checksum (CRC):** Finally, a 32 bit checksum is used to protect the frame as it is common practice in all 802.x networks.

The frame control field shown in Figure 7.16 contains the following fields:

- **Protocol version:** This 2 bit field indicates the current protocol version and is fixed to 0 by now. If major revisions to the standard make it incompatible with the current version, this value will be increased.
- **Type:** The type field determines the function of a frame: management (=00), control (=01), or data (=10). The value 11 is reserved. Each type has several subtypes as indicated in the following field.
- **Subtype:** Example subtypes for management frames are: 0000 for association request, 1000 for beacon. RTS is a control frame with subtype 1011, CTS is coded as 1100. User data is transmitted as data frame with subtype 0000. All details can be found in IEEE, 1999.

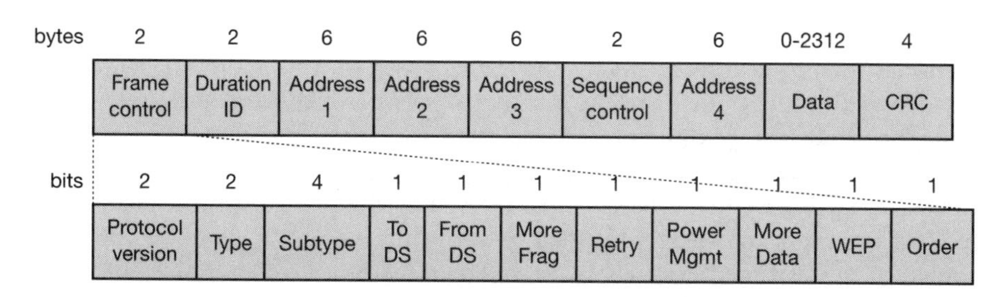

Figure 7.16 IEEE 802.11 MAC packet structure

- **To DS/From DS:** Explained in the following in more detail.
- **More fragments:** This field is set to 1 in all data or management frames that have another fragment of the current MSDU to follow.
- **Retry:** If the current frame is a retransmission of an earlier frame, this bit is set to 1. With the help of this bit it may be simpler for receivers to eliminate duplicate frames.
- **Power management:** This field indicates the mode of a station after successful transmission of a frame. Set to 1 the field indicates that the station goes into power-save mode. If the field is set to 0, the station stays active.
- **More data:** In general, this field is used to indicate a receiver that a sender has more data to send than the current frame. This can be used by an access point to indicate to a station in power-save mode that more packets are buffered. Or it can be used by a station to indicate to an access point after being polled that more polling is necessary as the station has more data ready to transmit.
- **Wired equivalent privacy (WEP):** This field indicates that the standard security mechanism of 802.11 is applied. However, due to many weaknesses found in the WEP algorithm higher layer security should be used to secure an 802.11 network (Borisov, 2001).
- **Order:** If this bit is set to 1 the received frames must be processed in strict order.

MAC frames can be transmitted between mobile stations; between mobile stations and an access point and between access points over a DS (see Figure 7.3). Two bits within the Frame Control field, '**to DS**' and '**from DS**', differentiate these cases and control the meaning of the four addresses used. Table 7.1 gives an overview of the four possible bit values of the DS bits and the associated interpretation of the four address fields.

Table 7.1 Interpretation of the MAC addresses in an 802.11 MAC frame

to DS	from DS	Address 1	Address 2	Address 3	Address 4
0	0	DA	SA	BSSID	–
0	1	DA	BSSID	SA	–
1	0	BSSID	SA	DA	–
1	1	RA	TA	DA	SA

Every station, access point or wireless node, filters on **address 1**. This address identifies the physical receiver(s) of the frame. Based on this address, a station can decide whether the frame is relevant or not. The second address, **address 2**, represents the physical transmitter of a frame. This information is important because this particular sender is also the recipient of the MAC layer acknowledgement. If a packet from a transmitter (address 2) is received by the receiver with address 1, this receiver in turn acknowledges the data packet using address 2 as receiver address as shown in the ACK packet in Figure 7.17. The remaining two addresses, **address 3** and **address 4**, are mainly necessary for the logical assignment of frames (logical sender, BSS identifier, logical receiver). If address 4 is not needed the field is omitted.

For addressing, the following four scenarios are possible:

- **Ad-hoc network:** If both DS bits are zero, the MAC frame constitutes a packet which is exchanged between two wireless nodes without a distribution system. **DA** indicates the **destination address**, **SA** the **source address** of the frame, which are identical to the physical receiver and sender addresses respectively. The third address identifies the **basic service set (BSSID)** (see Figure 7.4), the fourth address is unused.
- **Infrastructure network, from AP:** If only the 'from DS' bit is set, the frame physically originates from an access point. DA is the logical and physical receiver, the second address identifies the BSS, the third address specifies the logical sender, the source address of the MAC frame. This case is an example for a packet sent to the receiver via the access point.
- **Infrastructure network, to AP:** If a station sends a packet to another station via the access point, only the 'to DS' bit is set. Now the first address represents the physical receiver of the frame, the access point, via the BSS identifier. The second address is the logical and physical sender of the frame, while the third address indicates the logical receiver.
- **Infrastructure network, within DS:** For packets transmitted between two access points over the distribution system, both bits are set. The first **receiver address (RA)**, represents the MAC address of the receiving access point. Similarly, the second address **transmitter address (TA)**, identifies the sending access point within the distribution system. Now two more addresses are needed to identify the original destination DA of the frame and the original source of the frame SA. Without these additional addresses, some encapsulation mechanism would be necessary to transmit MAC frames over the distribution system transparently.

Figure 7.17 shows three control packets as examples for many special packets defined in the standard. The **acknowledgement packet (ACK)** is used to acknowledge the correct reception of a data frame as shown in Figure 7.12. The receiver address is directly copied from the address 2 field of the immediately previous frame. If no more fragments follow for a certain frame the duration field is set to 0. Otherwise the duration value of the previous frame (minus the time required to transmit the ACK minus SIFS) is stored in the duration field.

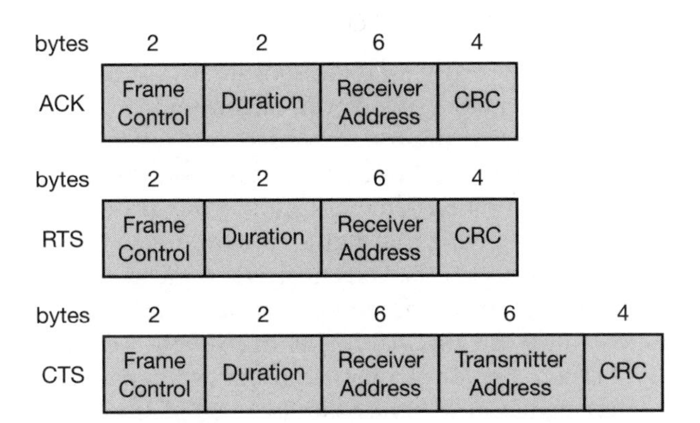

Figure 7.17
IEEE 802.11 special control packets: ACK, RTS, and CTS

For the MACA algorithm the RTS/CTS packets are needed. As Figure 7.13 shows, these packets have to reserve the medium to avoid collisions. Therefore, the **request to send (RTS)** packet contains the receiver address of the intended recipient of the following data transfer and the transmitter address of the station transmitting the RTS packet. The duration (in μs) comprises the time to send the CTS, data, and ACK plus three SIFS. The immediately following **clear to send (CTS)** frame copies the transmitter address from the RTS packet into its receiver address field. Additionally, it reads the duration field, subtracts the time to send the CTS and a SIFS and writes the result into its own duration field.

7.3.5 MAC management

MAC management plays a central role in an IEEE 802.11 station as it more or less controls all functions related to system integration, i.e., integration of a wireless station into a BSS, formation of an ESS, synchronization of stations etc. The following functional groups have been identified and will be discussed in more detail in the following sections:

- **Synchronization:** Functions to support finding a wireless LAN, synchronization of internal clocks, generation of beacon signals.
- **Power management:** Functions to control transmitter activity for power conservation, e.g., periodic sleep, buffering, without missing a frame.
- **Roaming:** Functions for joining a network (association), changing access points, scanning for access points.
- **Management information base (MIB):** All parameters representing the current state of a wireless station and an access point are stored within a MIB for internal and external access. A MIB can be accessed via standardized protocols such as the simple network management protocol (SNMP).

7.3.5.1 Synchronization

Each node of an 802.11 network maintains an internal clock. To synchronize the clocks of all nodes, IEEE 802.11 specifies a **timing synchronization function (TSF)**. As we will see in the following section, synchronized clocks are needed for power management, but also for coordination of the PCF and for synchronization of the hopping sequence in an FHSS system. Using PCF, the local timer of a node can predict the start of a super frame, i.e., the contention free and contention period. FHSS physical layers need the same hopping sequences so that all nodes can communicate within a BSS.

Within a BSS, timing is conveyed by the (quasi)periodic transmissions of a beacon frame. A **beacon** contains a timestamp and other management information used for power management and roaming (e.g., identification of the BSS). The timestamp is used by a node to adjust its local clock. The node is not required to hear every beacon to stay synchronized; however, from time to time internal clocks should be adjusted. The transmission of a beacon frame is not always periodic because the beacon frame is also deferred if the medium is busy.

Within **infrastructure-based** networks, the access point performs synchronization by transmitting the (quasi)periodic beacon signal, whereas all other wireless nodes adjust their local timer to the time stamp. This represents the simple case shown in Figure 7.18. The access point is not always able to send its beacon B periodically if the medium is busy. However, the access point always tries to schedule transmissions according to the expected beacon interval (**target beacon transmission time**), i.e., beacon intervals are not shifted if one beacon is delayed. The timestamp of a beacon always reflects the real transmit time, not the scheduled time.

For ad-hoc networks, the situation is slightly more complicated as they do not have an access point for beacon transmission. In this case, each node maintains its own synchronization timer and starts the transmission of a beacon frame after the beacon interval. Figure 7.19 shows an example where multiple stations try to send their beacon. However, the standard random backoff algorithm is also applied to the beacon frames so only one beacon wins. All other stations now adjust their internal clocks according to the received beacon and

Figure 7.18
Beacon transmission in a busy 802.11 infrastructure network

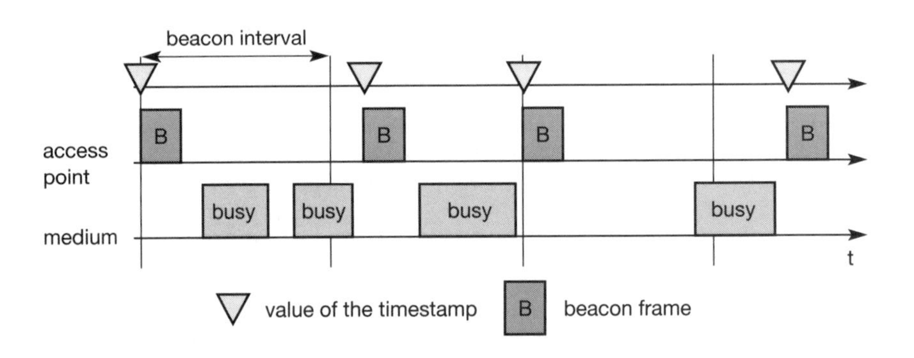

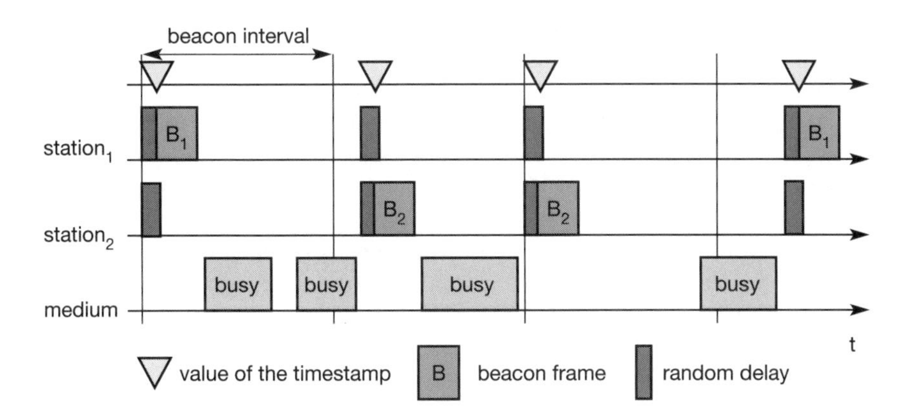

Figure 7.19
Beacon transmission in a busy 802.11 ad-hoc network

suppress their beacons for this cycle. If collision occurs, the beacon is lost. In this scenario, the beacon intervals can be shifted slightly because all clocks may vary as may the start of a beacon interval from a node's point of view. However, after successful synchronization all nodes again have the same consistent view.

7.3.5.2 Power management

Wireless devices are battery powered (unless a solar panel is used). Therefore, power-saving mechanisms are crucial for the commercial success of such devices. Standard LAN protocols assume that stations are always ready to receive data, although receivers are idle most of the time in lightly loaded networks. However, this permanent readiness of the receiving module is critical for battery life as the receiver current may be up to 100 mA (Woesner, 1998).

The basic idea of IEEE 802.11 power management is to switch off the transceiver whenever it is not needed. For the sending device this is simple to achieve as the transfer is triggered by the device itself. However, since the power management of a receiver cannot know in advance when the transceiver has to be active for a specific packet, it has to 'wake up' the transceiver periodically. Switching off the transceiver should be transparent to existing protocols and should be flexible enough to support different applications. However, throughput can be traded-off for battery life. Longer off-periods save battery life but reduce average throughput and vice versa.

The basic idea of power saving includes two states for a station: **sleep** and **awake**, and buffering of data in senders. If a sender intends to communicate with a power-saving station it has to buffer data if the station is asleep. The sleeping station on the other hand has to wake up periodically and stay awake for a certain time. During this time, all senders can announce the destinations of their buffered data frames. If a station detects that it is a destination of a buffered packet it has to stay awake until the transmission takes place. Waking up at the right moment requires the **timing synchronization function (TSF)** introduced in section 7.3.5.1. All stations have to wake up or be awake at the same time.

Power management in **infrastructure**-based networks is much simpler compared to ad-hoc networks. The access point buffers all frames destined for stations operating in power-save mode. With every beacon sent by the access point, a **traffic indication map (TIM)** is transmitted. The TIM contains a list of stations for which unicast data frames are buffered in the access point.

The TSF assures that the sleeping stations will wake up periodically and listen to the beacon and TIM. If the TIM indicates a unicast frame buffered for the station, the station stays awake for transmission. For multi-cast/broadcast transmission, stations will always stay awake. Another reason for waking up is a frame which has to be transmitted from the station to the access point. A sleeping station still has the TSF timer running.

Figure 7.20 shows an example with an access point and one station. The state of the medium is indicated. Again, the access point transmits a beacon frame each beacon interval. This interval is now the same as the TIM interval. Additionally, the access point maintains a **delivery traffic indication map (DTIM)** interval for sending broadcast/multicast frames. The DTIM interval is always a multiple of the TIM interval.

All stations (in the example, only one is shown) wake up prior to an expected TIM or DTIM. In the first case, the access point has to transmit a broadcast frame and the station stays awake to receive it. After receiving the broadcast frame, the station returns to sleeping mode. The station wakes up again just before the next TIM transmission. This time the TIM is delayed due to a busy medium so, the station stays awake. The access point has nothing to send and the station goes back to sleep.

At the next TIM interval, the access point indicates that the station is the destination for a buffered frame. The station answers with a **PS** (power saving) **poll** and stays awake to receive data. The access point then transmits the data for the station, the station acknowledges the receipt and may also send some

Figure 7.20
Power management in IEEE 802.11 infrastructure networks

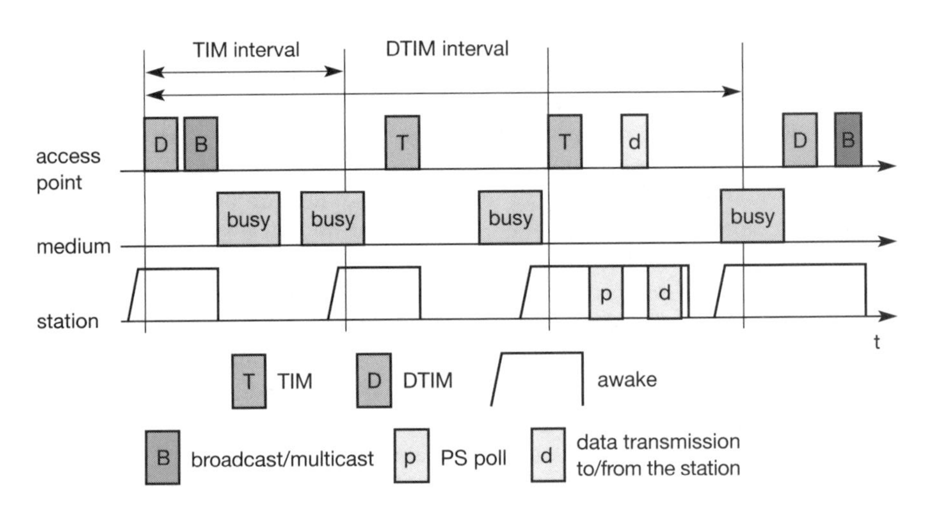

data (as shown in the example). This is acknowledged by the access point (acknowledgments are not shown in the figure). Afterwards, the station switches to sleep mode again.

Finally, the access point has more broadcast data to send at the next DTIM interval, which is again deferred by a busy medium. Depending on internal thresholds, a station may stay awake if the sleeping period would be too short. This mechanism clearly shows the trade-off between short delays in station access and saving battery power. The shorter the TIM interval, the shorter the delay, but the lower the power-saving effect.

In ad-hoc networks, power management is much more complicated than in infrastructure networks. In this case, there is no access point to buffer data in one location but each station needs the ability to buffer data if it wants to communicate with a power-saving station. All stations now announce a list of buffered frames during a period when they are all awake. Destinations are announced using **ad-hoc traffic indication map (ATIMs)** – the announcement period is called the **ATIM window**.

Figure 7.21 shows a simple ad-hoc network with two stations. Again, the beacon interval is determined by a distributed function (different stations may send the beacon). However, due to this synchronization, all stations within the ad-hoc network wake up at the same time. All stations stay awake for the ATIM interval as shown in the first two steps and go to sleep again if no frame is buffered for them. In the third step, $station_1$ has data buffered for $station_2$. This is indicated in an ATIM transmitted by $station_1$. $Station_2$ acknowledges this ATIM and stays awake for the transmission. After the ATIM window, $station_1$ can transmit the data frame, and $station_2$ acknowledges its receipt. In this case, the stations stay awake for the next beacon.

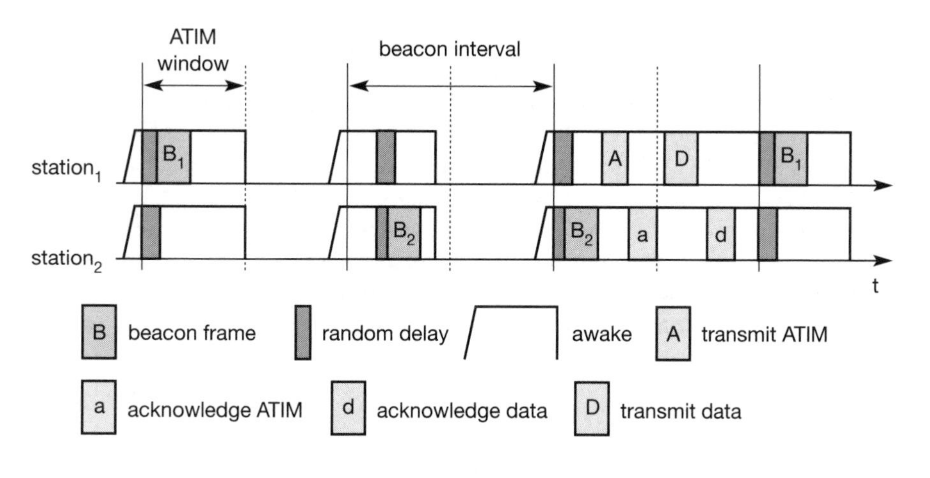

Figure 7.21
Power management in IEEE 802.11 ad-hoc networks

One problem with this approach is that of scale. If many stations within an ad-hoc network operate in power-save mode, they may also want to transmit their ATIM within the ATIM window. More ATIM transmissions take place, more collisions happen and more stations are deferred. The access delay of large networks is difficult to predict. QoS guarantees can not be given under heavy load.

7.3.5.3 Roaming

Typically, wireless networks within buildings require more than just one access point to cover all rooms. Depending on the solidity and material of the walls, one access point has a transmission range of 10–20 m if transmission is to be of decent quality. Each storey of a building needs its own access point(s) as quite often walls are thinner than floors. If a user walks around with a wireless station, the station has to move from one access point to another to provide uninterrupted service. Moving between access points is called **roaming**. The term "handover" or "handoff" as used in the context of mobile or cellular phone systems would be more appropriate as it is simply a change of the active cell. However, for WLANs roaming is more common.

The steps for roaming between access points are:

- A station decides that the current link quality to its access point AP_1 is too poor. The station then starts **scanning** for another access point.
- Scanning involves the active search for another BSS and can also be used for setting up a new BSS in case of ad-hoc networks. IEEE 802.11 specifies scanning on single or multiple channels (if available at the physical layer) and differentiates between passive scanning and active scanning. **Passive scanning** simply means listening into the medium to find other networks, i.e., receiving the beacon of another network issued by the synchronization function within an access point. **Active scanning** comprises sending a **probe** on each channel and waiting for a response. Beacon and probe responses contain the information necessary to join the new BSS.
- The station then selects the best access point for roaming based on, e.g., signal strength, and sends an **association request** to the selected access point AP_2.
- The new access point AP_2 answers with an **association response**. If the response is successful, the station has roamed to the new access point AP_2. Otherwise, the station has to continue scanning for new access points.
- The access point accepting an association request indicates the new station in its BSS to the distribution system (DS). The DS then updates its database, which contains the current location of the wireless stations. This database is needed for forwarding frames between different BSSs, i.e. between the different access points controlling the BSSs, which combine to form an ESS (see Figure 7.3). Additionally, the DS can inform the old access point AP_1 that the station is no longer within its BSS.

Unfortunately, many products implemented proprietary or incompatible versions of protocols that support roaming and inform the old access point about the change in the station's location. The standard **IEEE 802.11f (Inter Access Point Protocol, IAPP)** should provide a compatible solution for all vendors. This also includes load-balancing between access points and key generation for security algorithms based on IEEE 802.1x (IEEE, 2001).

7.3.6 802.11b

As standardization took some time, the capabilities of the physical layers also evolved. Soon after the first commercial 802.11 products came on the market some companies offered proprietary solutions with 11 Mbit/s. To avoid market segmentation, a common standard, **IEEE 802.11b** (IEEE 1999) soon followed and was added as supplement to the original standard (Higher-speed physical layer extension in the 2.4 GHz band). This standard describes a new PHY layer and is by far the most successful version of IEEE 802.11 available today. Do not get confused about the fact that 802.11b hit the market before 802.11a. The standards are named according to the order in which the respective study groups have been established.

As the name of the supplement implies, this standard only defines a new PHY layer. All the MAC schemes, management procedures etc. explained above are still used. Depending on the current interference and the distance between sender and receiver 802.11b systems offer 11, 5.5, 2, or 1 Mbit/s. Maximum user data rate is approx 6 Mbit/s. The lower data rates 1 and 2 Mbit/s use the 11-chip Barker sequence as explained in section 7.3.3.2 and DBPSK or DQPSK, respectively. The new data rates, 5.5 and 11 Mbit/s, use 8-chip **complementary code keying (CCK)** (see IEEE, 1999, or Pahlavan, 2002, for details).

The standard defines several packet formats for the physical layer. The mandatory format interoperates with the original versions of 802.11. The optional versions provide a more efficient data transfer due to shorter headers/different coding schemes and can coexist with other 802.11 versions. However, the standard states that control all frames shall be transmitted at one of the basic rates, so they will be understood by all stations in a BSS.

Figure 7.22 shows two packet formats standardized for 802.11b. The mandatory format is called **long PLCP PPDU** and is similar to the format illustrated in Figure 7.8. One difference is the rate encoded in the signal field this is encoded in multiples of 100 kbit/s. Thus, 0x0A represents 1 Mbit/s, 0x14 is used for 2 Mbit/s, 0x37 for 5.5 Mbit/s and 0x6E for 11 Mbit/s. Note that the preamble and the header are transmitted at 1 Mbit/s using DBPSK. The optional **short PLCP PPDU** format differs in several ways. The short synchronization field consists of 56 scrambled zeros instead of scrambled ones. The short start frame delimiter SFD consists of a mirrored bit pattern compared to the SFD of the long format: 0000 0101 1100 1111 is used for the short PLCP PDU instead of 1111 0011 1010 0000 for the long PLCP PPDU. Receivers that are unable to receive the short format will not detect the start of a frame (but will sense the medium

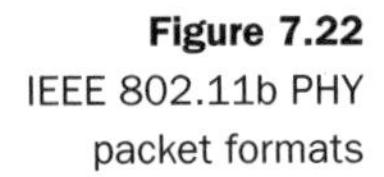

Figure 7.22 IEEE 802.11b PHY packet formats

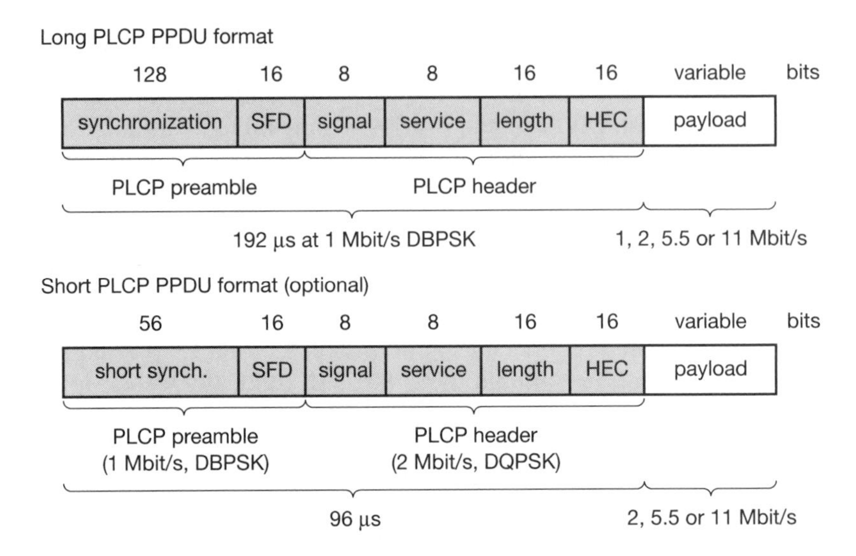

is busy). Only the preamble is transmitted at 1 Mbit/s, DBPSK. The following header is already transmitted at 2 Mbit/s, DQPSK, which is also the lowest available data rate. As Figure 7.22 shows, the length of the overhead is only half for the short frames (96 μs instead of 192 μs). This is useful for, e.g., short, but time-critical, data transmissions.

As IEEE 802.11b is the most widespread version, some more information is given for practical usage. The standards operates (like the DSSS version of 802.11) on certain frequencies in the 2.4 GHz ISM band. These depend on national regulations. Altogether 14 channels have been defined as Table 7.2 shows. For each channel the center frequency is given. Depending on national restrictions 11 (US/Canada), 13 (Europe with some exceptions) or 14 channels (Japan) can be used.

Figure 7.23 illustrates the non-overlapping usage of channels for an IEEE 802.11b installation with minimal interference in the US/Canada and Europe. The spacing between the center frequencies should be at least 25 MHz (the occupied bandwidth of the main lobe of the signal is 22 MHz). This results in the channels 1, 6, and 11 for the US/Canada or 1, 7, 13 for Europe, respectively. It may be the case that, e.g., travellers from the US cannot use the additional channels (12 and 13) in Europe as their hardware is limited to 11 channels. Some European installations use channel 13 to minimize interference. Users can install overlapping cells for WLANs using the three non-overlapping channels to provide seamless coverage. This is similar to the cell planning for mobile phone systems.

Table 7.2 Channel plan for IEEE 802.11b

Channel	Frequency [MHz]	US/Canada	Europe	Japan
1	2412	X	X	X
2	2417	X	X	X
3	2422	X	X	X
4	2427	X	X	X
5	2432	X	X	X
6	2437	X	X	X
7	2442	X	X	X
8	2447	X	X	X
9	2452	X	X	X
10	2457	X	X	X
11	2462	X	X	X
12	2467	–	X	X
13	2472	–	X	X
14	2484	–	–	X

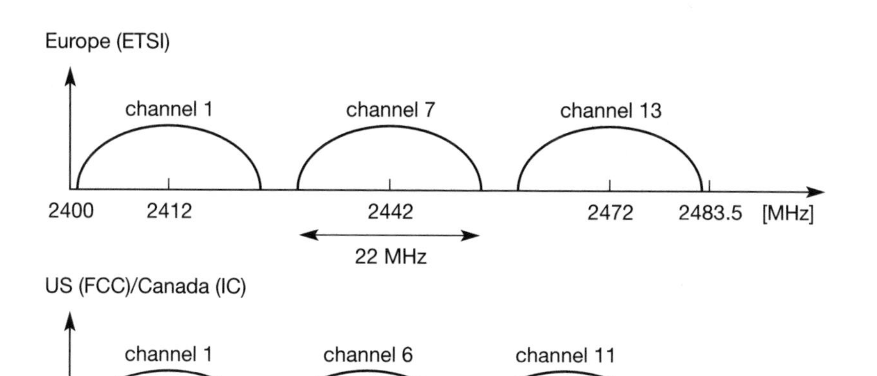

Figure 7.23 IEEE 802.11b non-overlapping channel selection

7.3.7 802.11a

Initially aimed at the US 5 GHz U-NII (Unlicensed National Information Infrastructure) bands **IEEE 802.11a** offers up to 54 Mbit/s using OFDM (IEEE, 1999). The first products were available in 2001 and can now be used (after some harmonization between IEEE and ETSI) in Europe. The FCC (US) regulations offer three different 100 MHz domains for the use of 802.11a, each with a different legal maximum power output: 5.15–5.25 GHz/50 mW, 5.25–5.35 GHz/250 mW, and 5.725–5.825 GHz/1 W. ETSI (Europe) defines different frequency bands for Europe: 5.15–5.35 GHz and 5.47–5.725 GHz and requires two additional mechanisms for operation: dynamic frequency selection (DFS) and transmit power control (TPC) which will be explained in the context of HiperLAN2 in more detail. (This is also the reason for introducing IEEE 802.11h, see section 7.3.8.) Maximum transmit power is 200 mW EIRP for the lower frequency band (indoor use) and 1 W EIRP for the higher frequency band (indoor and outdoor use). DFS and TPC are not necessary, if the transmit power stays below 50 mW EIRP and only 5.15–5.25 GHz are used. Japan allows operation in the frequency range 5.15–5.25 GHz and requires carrier sensing every 4 ms to minimize interference. Up to now, only 100 MHz are available 'worldwide' at 5.15–5.25 GHz.

The physical layer of IEEE 802.11a and the ETSI standard HiperLAN2 has been jointly developed, so both physical layers are almost identical. Most statements and explanations in the following, which are related to the transmission technology are also valid for HiperLAN2. However, HiperLAN2 differs in the MAC layer, the PHY layer packet formats, and the offered services (quality of service, real time etc.). This is discussed in more detail in section 7.4. It should be noted that most of the development for the physical layer for 802.11a was adopted from the HiperLAN2 standardization – but 802.11a products were available first and are already in widespread use.

Again, IEEE 802.11a uses the same MAC layer as all 802.11 physical layers do and, in the following, only the lowest layer is explained in some detail. To be able to offer data rates up to 54 Mbit/s IEEE 802.11a uses many different technologies. The system uses 52 subcarriers (48 data + 4 pilot) that are modulated using BPSK, QPSK, 16-QAM, or 64-QAM. To mitigate transmission errors, FEC is applied using coding rates of 1/2, 2/3, or 3/4. Table 7.3 gives an overview of the standardized combinations of modulation and coding schemes together with the resulting data rates. To offer a data rate of 12 Mbit/s, 96 bits are coded into one OFDM symbol. These 96 bits are distributed over 48 subcarriers and 2 bits are modulated per sub-carrier using QPSK (2 bits per point in the constellation diagram). Using a coding rate of 1/2 only 48 data bits can be transmitted.

Data rate [Mbit/s]	Modulation	Coding rate	Coded bits per subcarrier	Coded bits per OFDM symbol	Data bits per OFDM symbol
6	BPSK	1/2	1	48	24
9	BPSK	3/4	1	48	36
12	QPSK	1/2	2	96	48
18	QPSK	3/4	2	96	72
24	16-QAM	1/2	4	192	96
36	16-QAM	3/4	4	192	144
48	64-QAM	2/3	6	288	192
54	64-QAM	3/4	6	288	216

Table 7.3 Rate dependent parameters for IEEE 802.11a

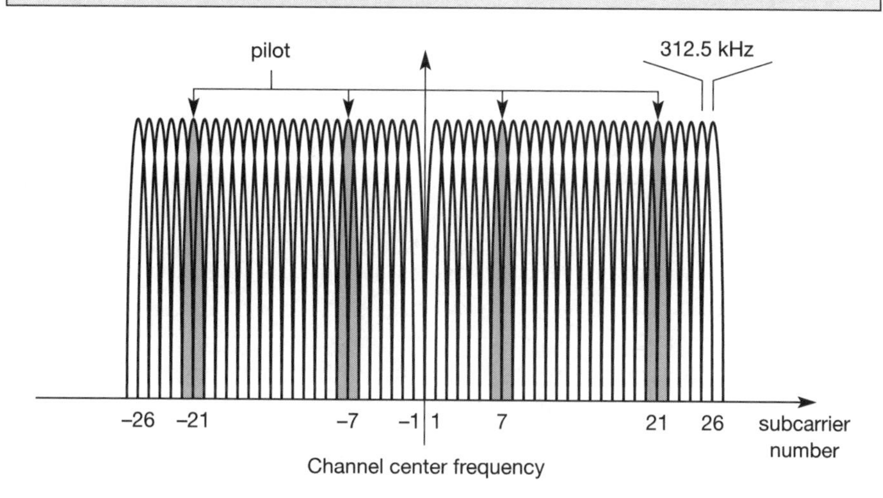

Figure 7.24 Usage of OFDM in IEEE 802.11a

Figure 7.24 shows the usage of OFDM in IEEE 802.11a. Remember, the basic idea of OFDM (or MCM in general) was the reduction of the symbol rate by distributing bits over numerous subcarriers. IEEE 802.11a uses a fixed symbol rate of 250,000 symbols per second independent of the data rate (0.8 μs guard interval for ISI mitigation plus 3.2 μs used for data results in a symbol duration of 4 μs). As Figure 7.24 shows, 52 subcarriers are equally spaced around a center frequency. (Center frequencies will be explained later). The spacing between the subcarriers is 312.5 kHz. 26 subcarriers are to the left of the center frequency and 26 are to the right. The center frequency itself is not used as subcarrier. Subcarriers with the numbers –21, –7, 7, and 21 are used for pilot signals to make the signal detection robust against frequency offsets.

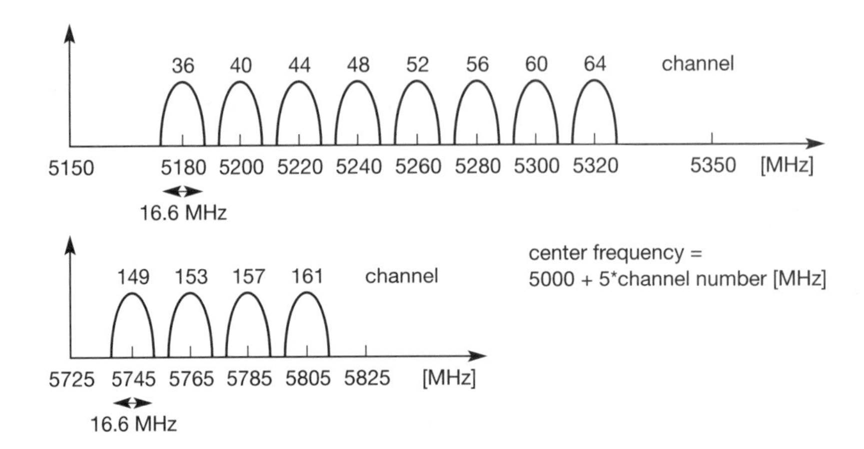

Figure 7.25 Operating channels of IEEE 802.11a in the U-NII bands

Similar to 802.11b several operating channels have been standardized to minimize interference. Figure 7.25 shows the **channel layout** for the US U-NII bands. The center frequency of a channel is 5000 + 5*channel number [MHz]. This definition provides a unique numbering of channels with 5 MHz spacing starting from 5 GHz. Depending on national regulations, different sets of channels may be used. Eight channels have been defined for the lower two bands in the U-NII (36, 40, 44, 48, 52, 56, 60, and 64); four more are available in the high band (149, 153, 157, and 161). Using these channels allows for interference-free operation of overlapping 802.11a cells. Channel spacing is 20 MHz, the occupied bandwidth of 802.11a is 16.6 MHz. How is this related to the spacing of the sub-carriers? 20 MHz/64 equals 312.5 kHz. 802.11a uses 48 carriers for data, 4 for pilot signals, and 12 carriers are sometimes called virtual subcarriers. (Set to zero, they do not contribute to the data transmission but may be used for an implementation of OFDM with the help of FFT, see IEEE, 1999, or ETSI, 2001a, for more details). Multiplying 312.5 kHz by 52 subcarriers and adding the extra space for the center frequency results in approximately 16.6 MHz occupied bandwidth per channel (details of the transmit spectral power mask neglected, see ETSI, 2001a).

Due to the nature of OFDM, the PDU on the physical layer of IEEE 802.11a looks quite different from 802.11b or the original 802.11 physical layers. Figure 7.26 shows the basic structure of an **IEEE 802.11a PPDU**.

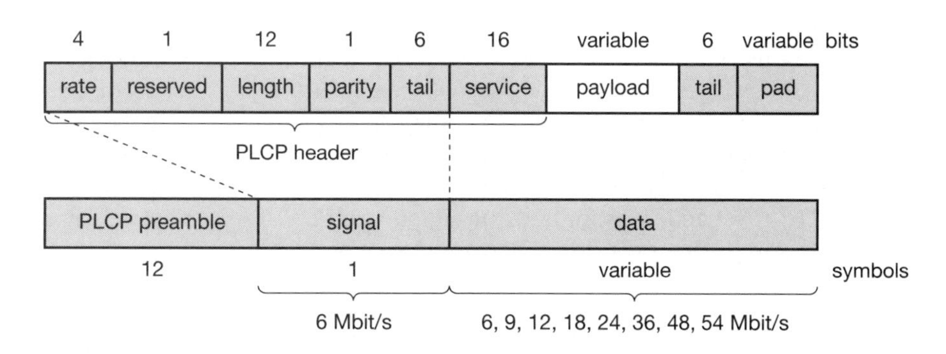

Figure 7.26
IEEE 802.11a physical layer PDU

- The **PLCP preamble** consists of 12 symbols and is used for frequency acquisition, channel estimation, and synchronization. The duration of the preamble is 16 μs.
- The following OFDM symbol, called **signal**, contains the following fields and is BPSK-modulated. The 4 bit **rate** field determines the data rate and the modulation of the rest of the packet (examples are 0x3 for 54 Mbit/s, 0x9 for 24 Mbit/s, or 0xF for 9 Mbit/s). The **length** field indicates the number of bytes in the payload field. The **parity** bit shall be an even parity for the first 16 bits of the signal field (rate, length and the reserved bit). Finally, the six **tail** bits are set to zero.
- The **data** field is sent with the rate determined in the rate field and contains a **service** field which is used to synchronize the descrambler of the receiver (the data stream is scrambled using the polynomial $x^7 + x^4 + 1$) and which contains bits for future use. The **payload** contains the MAC PDU (1-4095 byte). The **tail** bits are used to reset the encoder. Finally, the **pad** field ensures that the number of bits in the PDU maps to an integer number of OFDM symbols.

Compared to IEEE 802.11b working at 2.4 GHz IEEE 802.11a at 5 GHz offers much higher data rates. However, shading at 5 GHz is much more severe compared to 2.4 GHz and depending on the SNR, propagation conditions and the distance between sender and receiver, data rates may drop fast (e.g., 54 Mbit/s may be available only in an LOS or near LOS condition). Additionally, the MAC layer of IEEE 802.11 adds overheads. User data rates are therefore much lower than the data rates listed above. Typical user rates in Mbit/s are (transmission rates in brackets) 5.3 (6), 18 (24), 24 (36), and 32 (54). The following section presents some additional developments in the context of 802.11, which also comprise a standard for higher data rates at 2.4 GHz that can benefit from the better propagation conditions at lower frequencies.

7.3.8 Newer developments

While many products that follow the IEEE 802.11a and 802.11b standards are available, several new groups have been formed within the IEEE to discuss enhancements of the standard and new applications. As things change fast, the current status can be checked via (IEEE, 2002a). The following is only a selection of ongoing work (at the time of writing). The completed standards **IEEE 802.11c** and **802.11d** cover additions for bridging support and updates for physical layer requirements in different regulatory domains (i.e., countries).

- **802.11e (MAC enhancements):** Currently, the 802.11 standards offer no quality of service in the DCF operation mode. Some QoS guarantees can be given, only via polling using PCF. For applications such as audio, video, or media stream, distribution service classes have to be provided. For this reason, the MAC layer must be enhanced compared to the current standard.
- **802.11f (Inter-Access Point Protocol):** The current standard only describes the basic architecture of 802.11 networks and their components. The implementation of components, such as the distribution system, was deliberately not specified. Specifications of implementations should generally be avoided as they hinder improvements. However, a great flexibility in the implementation combined with a lack of detailed interface definitions and communication protocols, e.g., for management severely limits the interoperability of devices from different vendors. For example, seamless roaming between access points of different vendors is often impossible. 802.11f standardizes the necessary exchange of information between access points to support the functions of a distribution system.
- **802.11g (Data rates above 20 Mbit/s at 2.4 GHz):** Introducing new modulation schemes, forward error correction and OFDM also allows for higher data rates at 2.4 GHz. This approach should be backward compatible to 802.11b and should benefit from the better propagation characteristics at 2.4 GHz compared to 5 GHz. Currently, chips for 54 Mbit/s are available as well as first products. An alternative (or additional) proposal for 802.11g suggests the so-called packet binary convolutional coding (PBCC) to reach a data rate of 22 Mbit/s (Heegard, 2001). While the 54 Mbit/s OFDM mode is mandatory, the 22 Mbit/s PBCC mode can be used as an option. The decision between 802.11a and 802.11g is not obvious. Many 802.11a products are already available and the 5 GHz band is (currently) not as crowded as the 2.4 GHz band where not only microwave ovens, but also Bluetooth, operate (see section 7.5). Coverage is better at 2.4 GHz and fewer access points are needed, lowering the overall system cost. 802.11g access points can also communicate with 802.11b devices as the current 802.11g products show. Dual mode (or then triple mode) devices will be available covering 802.11a and b (and g). If a high traffic volume per square meter is expected (e.g., hot spots in airport terminals), the smaller cells of 802.11a access points and the higher number of available channels (to avoid interference) at 5 GHz are clear advantages.

- **802.11h (Spectrum managed 802.11a):** The 802.11a standard was primarily designed for usage in the US U-NII bands. The standardization did not consider non-US regulations such as the European requirements for power control and dynamic selection of the transmit frequency. To enable the regulatory acceptance of 5 GHz products, dynamic channel selection (DCS) and transmit power control (TPC) mechanisms (as also specified for the European HiperLAN2 standard) have been added. With this extension, 802.11a products can also be operated in Europe. These additional mechanisms try to balance the load in the 5 GHz band.
- **802.11i (Enhanced Security mechanisms):** As the original security mechanisms (WEP) proved to be too weak soon after the deployment of the first products (Borisov, 2001), this working group discusses stronger encryption and authentication mechanisms. IEEE 802.1x will play a major role in this process.

Additionally, IEEE 802.11 has several **study groups** for new and upcoming topics. The group 'Radio Resource Measurements' investigates the possibilities of 802.11 devices to provide measurements of radio resources. Solutions for even higher throughput are discussed in the 'High Throughput' study group. Both groups had their first meetings in 2002. The first study group recently became the IEEE project 802.11k 'Radio Resource Measurement Enhancements.'

7.4 HIPERLAN

In 1996, the ETSI standardized HIPERLAN 1 as a WLAN allowing for node mobility and supporting ad-hoc and infrastructure-based topologies (ETSI, 1996). (HIPERLAN stands for **high performance local area network**.) **HIPERLAN 1** was originally one out of four HIPERLANs envisaged, as ETSI decided to have different types of networks for different purposes. The key feature of all four networks is their integration of time-sensitive data transfer services. Over time, names have changed and the former HIPERLANs 2, 3, and 4 are now called HiperLAN2, HIPERACCESS, and HIPERLINK. The current focus is on HiperLAN2, a standard that comprises many elements from ETSI's **BRAN** (broadband radio access networks) and **wireless ATM** activities. Neither wireless ATM nor HIPERLAN 1 were a commercial success. However, the standardization efforts had a lot of impact on QoS supporting wireless broadband networks such as **HiperLAN2**. Before describing HiperLAN2 in more detail, the following three sections explain key features of, and the motivation behind, HIPERLAN 1, wireless ATM, and BRAN. Readers not interested in the historical background may proceed directly to section 7.4.4.

7.4.1 Historical: HIPERLAN 1

ETSI (1998b) describes HIPERLAN 1 as a wireless LAN supporting priorities and packet life time for data transfer at 23.5 Mbit/s, including forwarding mechanisms, topology discovery, user data encryption, network identification and power conservation mechanisms. HIPERLAN 1 should operate at 5.1–5.3 GHz with a range of 50 m in buildings at 1 W transmit power.

The service offered by a HIPERLAN 1 is compatible with the standard MAC services known from IEEE 802.x LANs. Addressing is based on standard 48 bit MAC addresses. A special HIPERLAN 1 identification scheme allows the concurrent operation of two or more physically overlapping HIPERLANs without mingling their communication. Confidentiality is ensured by an encryption/decryption algorithm that requires the identical keys and initialization vectors for successful decryption of a data stream encrypted by a sender.

An innovative feature of HIPERLAN 1, which many other wireless networks do not offer, is its ability to forward data packets using several relays. Relays can extend the communication on the MAC layer beyond the radio range. For power conservation, a node may set up a specific wake-up pattern. This pattern determines at what time the node is ready to receive, so that at other times, the node can turn off its receiver and save energy. These nodes are called p-savers and need so-called p-supporters that contain information about the wake-up patterns of all the p-savers they are responsible for. A p-supporter only forwards data to a p-saver at the moment the p-saver is awake. This action also requires buffering mechanisms for packets on p-supporting forwarders.

The following describes only the medium access scheme of HIPERLAN 1, a scheme that provides QoS and a powerful prioritization scheme. However, it turned out that priorities and QoS in general are not that important for standard LAN applications today. IEEE 802.11 in its standard versions does not offer priorities, the optional PCF is typically not implemented in products – yet 802.11 is very popular.

Elimination-yield non-preemptive priority multiple access (EY-NPMA) is not only a complex acronym, but also the heart of the channel access providing priorities and different access schemes. EY-NPMA divides the medium access of different competing nodes into three phases:

- **Prioritization:** Determine the highest priority of a data packet ready to be sent by competing nodes.
- **Contention:** Eliminate all but one of the contenders, if more than one sender has the highest current priority.
- **Transmission:** Finally, transmit the packet of the remaining node.

In a case where several nodes compete for the medium, all three phases are necessary (called 'channel access in **synchronized channel condition**'). If the channel is free for at least 2,000 so-called high rate bit-periods plus a dynamic extension, only the third phase, i.e. transmission, is needed (called 'channel

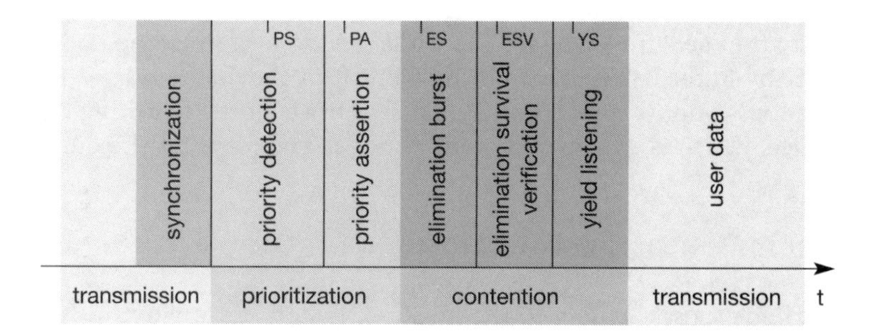

Figure 7.27 Phases of the HIPERLAN 1 EY-NPMA access scheme

access in **channel-free condition**'). The dynamic extension is randomly chosen between 0 and 3 times 200 high rate bit-periods with equal likelihood. This extension further minimizes the probability of collisions accessing a free channel if stations are synchronized on higher layers and try to access the free channel at the same time. HIPERLAN 1 also supports 'channel access in the **hidden elimination condition**' to handle the problem of hidden terminals as described in ETSI (1998b).

The contention phase is further subdivided into an **elimination phase** and a **yield phase**. The purpose of the elimination phase is to eliminate as many contending nodes as possible (but surely not all). The result of the elimination phase is a more or less constant number of remaining nodes, almost independent of the initial number of competing nodes. Finally, the yield phase completes the work of the elimination phase with the goal of only one remaining node.

Figure 7.27 gives an overview of the three main phases and some more details which will be explained in the following sections. For every node ready to send data, the access cycle starts with synchronization to the current sender. The first phase, prioritization, follows. After that, the elimination and yield part of the contention phase follow. Finally, the remaining node can transmit its data. Every phase has a certain duration which is measured in numbers of slots and is determined by the variables I_{PS}, I_{PA}, I_{ES}, I_{ESV}, and I_{YS}.

7.4.1.1 Prioritization phase

HIPERLAN 1 offers five different priorities for data packets ready to be sent. After one node has finished sending, many other nodes can compete for the right to send. The first objective of the prioritization phase is to make sure that no node with a lower priority gains access to the medium while packets with higher priority are waiting at other nodes. This mechanism always grants nodes with higher priority access to the medium, no matter how high the load on lower priorities.

In the first step of the prioritization phase, the priority detection, time is divided into five slots, slot 0 (highest priority) to slot 4 (lowest priority). Each slot has a duration of IPS = 168 high rate bit-periods. If a node has the access priority p, it has to listen into the medium for p slots (priority detection). If the node senses the medium is idle for the whole period of p slots, the node asserts the priority by immediately transmitting a burst for the duration I_{PA} = 168 high rate bit-periods (priority assertion). The burst consists of the following high rate bit sequence, which is repeated as many times as necessary for the duration of the burst:

11111010100010011100000110010110

If the node senses activity in the medium, it stops its attempt to send data in this transmission cycle and waits for the next one. The whole prioritization phase ends as soon as one node asserts the access priority with a burst. This means that the prioritization phase is not limited by a fixed length, but depends on the highest priority.

Let us assume, for example, that there are three nodes with data ready to be sent, the packets of node 1 and node 2 having the priority 2, the packet of node 3 having the priority 4. Then nodes 1, 2 and 3 listen into the medium and sense slots 0 and 1 are idle. Nodes 1 and 2 both send a burst in slot 2 as priority assertion. Node 3 stops its attempt to transmit its packet. In this example, the prioritization phase has taken three slots.

After this first phase at least one of the contending nodes will survive, the surviving nodes being all nodes with the highest priority of this cycle.

7.4.1.2 Elimination phase

Several nodes may now enter the elimination phase. Again, time is divided into slots, using the elimination slot interval I_{ES} = 212 high rate bit periods. The length of an individual elimination burst is 0 to 12 slot intervals long, the probability of bursting within a slot is 0.5. The probability $P_E(n)$ of an elimination burst to be n elimination slot intervals long is given by:

- $P_E(n) = 0.5^{n+1}$ for $0 \leq n < 12$
- $P_E(n) = 0.5^{12}$ for $n = 12$

The elimination phase now resolves contention by means of elimination bursting and elimination survival verification. Each contending node sends an elimination burst with length n as determined via the probabilities and then listens to the channel during the survival verification interval I_{ESV} = 256 high rate bit periods. The burst sent is the same as for the priority assertion. A contending node survives this elimination phase if, and only if, it senses the channel is idle during its survival verification period. Otherwise, the node is eliminated and stops its attempt to send data during this transmission cycle.

The whole elimination phase will last for the duration of the longest elimination burst among the contending nodes plus the survival verification time. One or more nodes will survive this elimination phase, and can then continue with the next phase.

7.4.1.3 Yield phase

During the yield phase, the remaining nodes only listen into the medium without sending any additional bursts. Again, time is divided into slots, this time called yield slots with a duration of I_{YS} = 168 high rate bit-periods. The length of an individual yield listening period can be 0 to 9 slots with equal likelihood. The probability $P_Y(n)$ for a yield listening period to be n slots long is 0.1 for all n, $0 \leq n \leq 9$.

Each node now listens for its yield listening period. If it senses the channel is idle during the whole period, it has survived the yield listening. Otherwise, it withdraws for the rest of the current transmission cycle. This time, the length of the yield phase is determined by the shortest yield-listening period among all the contending nodes. At least one node will survive this phase and can start to transmit data. This is what the other nodes with longer yield listening period can sense. It is important to note that at this point there can still be more than one surviving node so a collision is still possible.

7.4.1.4 Transmission phase

A node that has survived the prioritization and contention phase can now send its data, called a low bit-rate high bit-rate HIPERLAN 1 CAC protocol data unit (LBR-HBR HCPDU). This PDU can either be multicast or unicast. In case of a unicast transmission, the sender expects to receive an immediate acknowledgement from the destination, called an acknowledgement HCPDU (AK-HCPDU), which is an LBR HCPDU containing only an LBR part.

7.4.1.5 Quality of service support and other specialties

The speciality of HIPERLAN 1 is its QoS support. The quality of service offered by the MAC layer is based on three parameters (**HMQoS-parameters**). The user can set a priority for data, priority = 0 denotes a high priority, priority = 1, a low priority. The user can determine the lifetime of an MSDU to specify time-bounded delivery. The **MSDU lifetime** specifies the maximum time that can elapse between sending and receiving an MSDU. Beyond this, delivery of the MSDU becomes unnecessary. The MSDU lifetime has a range of 0–16,000 ms. The **residual MSDU lifetime** shows the remaining lifetime of a packet.

Besides data transfer, the MAC layer offers functions for looking up other HIPERLANs within radio range as well as special power conserving functions. **Power conservation** is achieved by setting up certain recurring patterns when a node can receive data instead of constantly being ready to receive. Special group-attendance patterns can be defined to enable multicasting. All nodes participating in a multicast group must be ready to receive at the same time when a sender transmits data.

HIPERLAN 1 MAC also offers user data **encryption** and **decryption** using a simple XOR-scheme together with random numbers. A key is chosen from a set of keys using a key identifier (KID) and is used together with an initialization vector IV to initialize the pseudo random number generator. This random sequence is XORed with the user data (UD) to generate the encrypted data. Decryption of the encrypted UD works the same way, using the same random number sequence. This is not a strong encryption scheme – encryption is left to higher layers.

Table 7.4 Mapping of the normalized residual lifetime to the CAC priority

NRL	MSDU priority = 0	MSDU priority = 1
NRL < 10 ms	0	1
10 ms ≤ NRL < 20 ms	1	2
20 ms ≤ NRL < 40 ms	2	3
40 ms ≤ NRL < 80 ms	3	4
80 ms ≤ NRL	4	4

It is interesting to see how the HIPERLAN 1 MAC layer selects the next PDU for transmission if several PDUs are ready and how the waiting time of a PDU before transmission is reflected in its channel access priority. The selection has to reflect the user priority (0 or 1) and the residual lifetime to guarantee a time-bounded service. The MAC layer then has to map this information onto a channel access priority used by the CAC, competing with other nodes for the transmit rights.

First of all, the MAC layer determines the **normalized residual HMPDU lifetime (NRL)**. This is the residual lifetime divided by the estimated number of hops the PDU has to travel. The computation reflects both the waiting time of a PDU in the node and the distance, and the additional waiting times in other nodes. Then the MAC layer computes the channel access priority for each PDU following the mapping shown in Table 7.4.

The final selection of the most important HMPDU (HIPERLAN 1 MAC PDU) is performed in the following order:

- HMPDUs with the highest priority are selected;
- from these, all HMPDUs with the shortest NRL are selected;
- from which finally any one without further preferences is selected from the remaining HMPDUs.

Besides transferring data from a sender to a receiver within the same radio coverage, HIPERLAN 1 offers functions to forward traffic via several other wireless nodes – a feature which is especially important in wireless ad-hoc networks without an infrastructure. This forwarding mechanism can also be used if a node can only reach an access point via other HIPERLAN 1 nodes.

7.4.2 WATM

Wireless ATM (WATM; sometimes also called wireless, mobile ATM, wmATM) does not only describe a transmission technology but tries to specify a complete communication system (Acampora, 1996), (Ayanoglu, 1996). While many aspects of the IEEE WLANs originate from the data communication community, many

WATM aspects come from the telecommunication industry (Händel, 1994). This specific situation can be compared to the case of competition and merging with regard to the concepts TCP/IP and ATM (IP-switching, MPLS). Similar to fixed networks where ATM never made it to the desktop, WATM will not make it to mobile terminals. However, many concepts found in WATM can also be found in QoS supporting WLANs such as HiperLAN2 (reference models, QoS parameters, see section 7.4.4).

7.4.2.1 Motivation for WATM

Several reasons led to the development of WATM:

- The need for seamless integration of wireless terminals into an ATM network. This is a basic requirement for supporting the same integrated services and different types of traffic streams as ATM does in fixed networks.
- ATM networks scale well from LANs to WANs – and mobility is needed in local and wide area applications. Strategies were needed to extend ATM for wireless access in local and global environments.
- For ATM to be successful, it must offer a wireless extension. Otherwise it cannot participate in the rapidly growing field of mobile communications.
- WATM could offer QoS for adequate support of multi-media data streams. Many other wireless technologies (e.g., IEEE 802.11) typically only offer best effort services or to some extent, time-bounded services. However, these services do not provide as many QoS parameters as ATM networks do.
- For telecommunication service providers, it appears natural that merging of mobile wireless communication and ATM technology leads to wireless ATM. One goal in this context is the seamless integration of mobility into B-ISDN which already uses ATM as its transfer technology.

It is clear that WATM will be much more complex than most of the other wireless systems. While, for example, IEEE 802.11 only covers local area access methods, Bluetooth only builds up piconets. Mobile IP only works on the network layer, but WATM tries to build up a comprehensive system covering physical layer, media access, routing, integration into the fixed ATM network, service integration into B-ISDN etc.

7.4.2.2 Wireless ATM working group

To develop this rather complex system, the ATM Forum formed the **Wireless ATM Working Group** in 1996, which aimed to develop a set of specifications that extends the use of ATM technology to wireless networks. These wireless networks should cover many different networking scenarios, such as private and public, local and global, mobility and wireless access (Raychaudhuri, 1996a and b).

The main goal of this working group involved ensuring the compatibility of all new proposals with existing ATM Forum standards. It should be possible to upgrade existing ATM networks, i.e., ATM switches and ATM end-systems, with certain functions to support mobility and radio access if required. Two main groups of open issues have been identified in this context: the extensions needed for the 'fixed' ATM to support mobility and all protocols and mechanisms related to the radio access.

The following more general extensions of the ATM system also need to be considered for a **mobile ATM**:

- **Location management:** Similar to other cellular networks, WATM networks must be able to locate a wireless terminal or a mobile user, i.e., to find the current access point of the terminal to the network.
- **Mobile routing:** Even if the location of a terminal is known to the system, it still has to route the traffic through the network to the access point currently responsible for the wireless terminal. Each time a user moves to a new access point, the system must reroute traffic.
- **Handover signalling:** The network must provide mechanisms which search for new access points, set up new connections between intermediate systems and signal the actual change of the access point.
- **QoS and traffic control:** In contrast to wireless networks offering only best effort traffic, and to cellular networks offering only a few different types of traffic, WATM should be able to offer many QoS parameters. To maintain these parameters, all actions such as rerouting, handover etc. have to be controlled. The network must pay attention to the incoming traffic (and check if it conforms to some traffic contract) in a similar way to today's ATM (policing).
- **Network management:** All extensions of protocols or other mechanisms also require an extension of the management functions to control the network

To ensure wireless access, the working group discussed the following topics belonging to a **radio access layer** (RAL):

- **Radio resource control:** As for any wireless network, radio frequencies, modulation schemes, antennas, channel coding etc. have to be determined.
- **Wireless media access:** Different media access schemes are possible, each with specific strengths and weaknesses for, e.g., multi-media or voice applications. Different centralized or distributed access schemes working on ATM cells can be imagined.
- **Wireless data link control:** The data link control layer might offer header compression for an ATM cell that carries almost 10 per cent overhead using a 5 byte header in a 53 byte cell. This layer can apply ARQ or FEC schemes to improve reliability.
- **Handover issues:** During handover, cells cannot only be lost but can also be out of sequence (depending on the handover mechanisms). Cells must be re-sequenced and lost cells must be retransmitted if required.

However, quite soon the ATM Forum stopped developing an own RAL but relied on other developments such as ETSI's BRAN (see section 7.4.3).

7.4.2.3 WATM services

The following paragraphs include several examples where WATM can be used from a user's perspective. These examples show that the idea behind WATM goes beyond the mere provision of wireless access or the construction of a wireless LAN. The services offered cover many aspects of today's wireless and mobile communications.

WATM systems had to be designed for transferring voice, classical data, video (from low quality to professional quality), multimedia data, short messages etc. Several service scenarios could be identified (Rauhala, 1998), (Barton, 1998), such as for example:

- **Office environments:** This includes all kinds of extensions for existing fixed networks offering a broad range of Internet/Intranet access, multi-media conferencing, online multi-media database access, and telecommuting. Using WATM technology, the office can be virtually expanded to the actual location of an employee.
- **Universities, schools, training centres:** The main foci in this scenario are distance learning, wireless and mobile access to databases, internet access, or teaching in the area of mobile multi-media computing.
- **Industry:** WATM may offer an extension of the Intranet supporting database connection, information retrieval, surveillance, but also real-time data transmission and factory management.
- **Hospitals:** Due to the quality of service offered for data transmission, WATM was thought of being the prime candidate for reliable, high-bandwidth mobile and wireless networks. Applications could include the transfer of medical images, remote access to patient records, remote monitoring of patients, remote diagnosis of patients at home or in an ambulance, as well as tele-medicine. The latter needs highly reliable networks with guaranteed quality of service to enable, e.g., remote surgery.
- **Home:** Many electronic devices at home (e.g., TV, radio equipment, CD-player, PC with internet access) could be connected using WATM technology. Here, WATM would permit various wireless connections, e.g., a PDA with TV access.
- **Networked vehicles:** All vehicles used for the transportation of people or goods will have a local network and network access in the future. Currently, vehicles such as trucks, aircraft, buses, or cars only have very limited communication capabilities (e.g., via GSM, UTMS), WATM could provide them with a high-quality access to the internet, company databases, multimedia conferencing etc. On another level, local networks among the vehicles within a certain area are of increasing importance, e.g., to prevent accidents or increase road capacity by platooning (i.e., forming a train of cars or trucks on the road with very low safety distance between single vehicles).

Mobility within an ATM network is provided by the **ATM mobility extension service (AMES)**. AMES facilitates the use of these ATM networks by different equipment and applications requiring mobility. Wireless equipment should obtain equivalent services from the network as wired terminals from a user's perspective. AMES comprises the extensions needed to support terminal portability for home and business use. Users can rearrange devices without losing access to the ATM network and retain a guaranteed service quality.

WATM should offer a personal cellular system (PCS) **access service**. PCSs like GSM, IS-95, UMTS etc. may use the mobility supporting capabilities of the fixed ATM network to route traffic to the proper base station controller. Public services for users could be a multimedia telephony service, a symmetric service offering speech and low bit rate video with medium mobility, as well as the asymmetrical service of real-time online data transfer, e.g., web browsing, e-mail and downloading of files. Private services could include a multi-media cordless telephone with higher quality compared to the public version. Special private data transfer services, e.g., carrying production data, could be deployed on a campus.

Another field of services is provided by **satellite ATM services (SATM)**. Future satellites will offer a large variety of TV, interactive video, multi-media, Internet, telephony and other services (see chapter 5). The main advantage in this context is the ubiquitous wide area coverage in remote, rural, and even urban areas. Satellites can be used directly (direct user access service), e.g., via a mobile phone or a terminal with antenna, which enables the user to access the ATM network directly. A whole network can be connected to a satellite using a mobile switch (fixed access service). For example, all computers in a school in a remote area could be connected to a switch, which connects to a satellite. Even ships can carry ATM networks and can then use the seamless integration of their onboard ATM network to a global ATM network (mobile platform service).

7.4.2.4 Generic reference model

Figure 7.28 shows a generic reference model for wireless mobile access to an ATM network. A mobile ATM (MATM) terminal uses a WATM terminal adapter to gain wireless access to a WATM RAS (Radio Access System). MATM terminals could be represented by, e.g., laptops using an ATM adapter for wired access plus software for mobility. The WATM terminal adapter enables wireless access, i.e., it includes the transceiver etc., but it does not support mobility. The RAS with the radio transceivers is connected to a mobility enhanced ATM switch (EMAS-E), which in turn connects to the ATM network with mobility aware switches (EMAS-N) and other standard ATM switches. Finally, a wired, non-mobility aware ATM end system may be the communication partner in this example.

The radio segment spans from the terminal and the terminal adapter to the access point, whereas the fixed network segment spans from the access point to the fixed end system. The fixed mobility support network, comprising all mobility aware switches EMAS-E and EMAS-N, can be distinguished from the standard ATM network with its non-mobility aware switches and end systems.

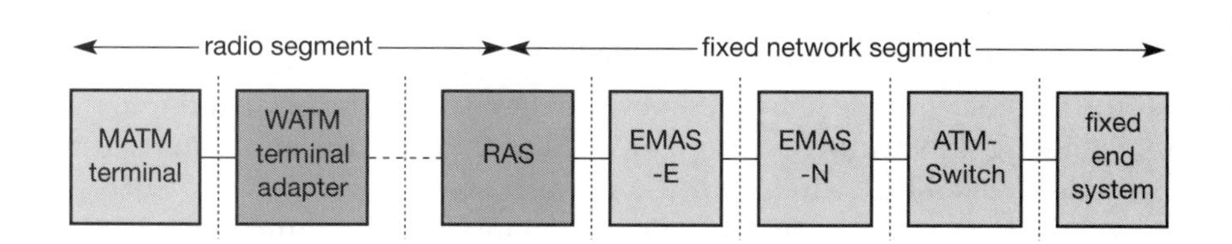

Figure 7.28
Example of a generic WATM reference model

7.4.2.5 Handover

One of the most important topics in a WATM environment is handover. Connectionless, best-effort protocols supporting handover, such as mobile IP on layer 3 and IEEE 802.11 with IAPP on layer 2, do not have to take too much care about handover quality. These protocols do not guarantee certain traffic parameters as WATM does. The main problem for WATM during the handover is rerouting all connections and maintaining connection quality. While in connectionless, best-effort environments, handover mainly involves rerouting of a packet stream without reliable transport, an end-system in WATM networks could maintain many connections, each with a different quality of service requirements (e.g., limited delay, bounded jitter, minimum bandwidth etc.). Handover not only involves rerouting of connections, it also involves reserving resources in switches, testing of availability of radio bandwidth, tracking of terminals to perform look-ahead reservations etc.

Many different requirements have been set up for handover. The following list presents some of the requirements according to Toh (1997), Bhat (1998):

- **Handover of multiple connections:** As ATM is a connection-oriented technology where end-systems can support many connections at the same time, handover in WATM must support more than only one connection. This results in the rerouting of every connection after handover. However, resource availability may not allow rerouting of all connections or forces QoS degradation. The terminal may then decide to accept a lower quality or to drop single connections.
- **Handover of point-to-multi-point connections:** Seamless support of point-to-multi-point connections is one of the major advantages of the ATM technology. WATM handover should also support these types of connection. However, due to the complexity of the scheme, some restrictions might be necessary.
- **QoS support:** Handover should aim to preserve the QoS of all connections during handover. However, due to limited resources, this is not always possible. Functions for QoS re-negotiation and dropping of connections on a priority basis may be required. Candidate access points should advertise their resources to the terminal, and this information could then be used by a handover algorithm to optimize handover and to balance the load between different access points.

- **Data integrity and security:** WATM handover should minimize cell loss and avoid all cell duplication or re-ordering. Security associations between the terminal and the network should not be compromised by handover.
- **Signaling and routing support:** WATM must provide the means to identify mobility-enabled switches in the network, to determine radio adjacent switches by another switch, and to reroute partial connections in the handover domain.
- **Performance and complexity:** The fact that WATM systems are complex by nature is mainly due to their support of connections with QoS. The simplicity of the handover functionality should be the central goal of the handover design. Modifications to the mobility-enabled switches should be extremely limited, but the functions required could have rather stringent processing time requirements. Due to performance reasons, ATM switches are very much hardware based and it is more difficult to integrate updates and new features. The handover code needed for the terminals should be rather simple due to the fact that increasing code size also requires more processing power, i.e., more battery power, which is typically a serious limitation in the design of mobile terminals.

7.4.2.6 Location management

As for all networks supporting mobility, special functions are required for looking up the current position of a mobile terminal, for providing the moving terminal with a permanent address, and for ensuring security features such as privacy, authentication, or authorization. These and more functions are grouped under the term **location management**.

Several requirements for location management have been identified (Bhat, 1998):

- **Transparency of mobility:** A user should not notice the location management function under normal operation. Any change of location should be performed without user activity. This puts certain constraints on the permissible time delay of the functions associated with location management. Transparent roaming between different domains (private/private, private/public, public/public) should be possible. This may include roaming between networks based on different technologies using, for example, a dual mode terminal.
- **Security:** To provide a security level high enough to be accepted for mission-critical use (business, emergency etc.), a WATM system requires special features. All location and user information collected for location management and accounting should be protected against unauthorized disclosure. This protection is particularly important for roaming profiles that allow the precise tracking of single terminals. As the air interface is very simple to access, special access restrictions must be implemented to, e.g., keep public

users out of private WATM networks. Users should also be able to determine the network their terminal is allowed to access. Essential security features include authentication of users and terminals, but also of access points. Encryption is also necessary, at least between terminal and access point, but preferably end-to-end.

- **Efficiency and scalability:** Imagine WATM networks with millions of users like today's mobile phone networks. Every function and system involved in location management must be scalable and efficient. This includes distributed servers for location storage, accounting and authentication. The performance of all operations should be practically independent of network size, number of current connections and network load. The clustering of switches and hierarchies of domains should be possible to increase the overall performance of the system by dividing the load. In contrast to many existing cellular networks, WATM should work with a more efficient, integrated signaling scheme. All signaling required for location management should therefore be incorporated into existing signaling mechanisms, e.g., by adding new information elements to existing messages. This allows for the utilization of the existing signaling mechanisms in the fixed ATM network which are efficient.
- **Identification**: Location management must provide the means to identify all entities of the network. Radio cells, WATM networks, terminals, and switches need unique identifiers and mechanisms to exchange identity information. This requirement also includes information for a terminal concerning its current location (home network or foreign network) and its current point of attachment. In addition to the permanent **ATM end system address (AESA)**, a terminal also needs a routable temporary AESA as soon as it is outside its home network. This temporary AESA must be forwarded to the terminal's home location.
- **Inter-working and standards:** All location management functions must cooperate with existing ATM functions from the fixed network, especially routing. Location management in WATM has to be harmonized with other location management schemes, such as location management in GSM and UMTS networks, the internet using Mobile IP, or Intranets with special features. This harmonization could, for instance, lead to a two-level location management if Mobile IP is used on top of WATM. All protocols used in WATM for database updates, registration etc. have to be standardized to permit mobility across provider network boundaries. However, inside an administrative domain, proprietary enhancements and optimizations could be applied.

7.4.2.7 Mobile quality of service

Quality of service (QoS) guarantees are one of the main advantages envisaged for WATM networks compared to, e.g., mobile IP working over packet radio networks. While the internet protocol IP does not guarantee QoS, ATM networks do (at the cost of higher complexity). WATM networks should provide mobile QoS (M-QoS). M-QoS is composed of three different parts:

- **Wired QoS:** The infrastructure network needed for WATM has the same QoS properties as any wired ATM network. Typical traditional QoS parameters are link delay, cell delay variation, bandwidth, cell error rate etc.
- **Wireless QoS:** The QoS properties of the wireless part of a WATM network differ from those of the wired part. Again, link delay and error rate can be specified, but now error rate is typically some order of magnitude that is higher than, e.g., fiber optics. Channel reservation and multiplexing mechanisms at the air interface strongly influence cell delay variation.
- **Handover QoS:** A new set of QoS parameters are introduced by handover. For example, handover blocking due to limited resources at target access points, cell loss during handover, or the speed of the whole handover procedure represent critical factors for QoS.

The WATM system has to map the QoS specified by an application onto these sets of QoS parameters at connection setup and has to check whether the QoS requested can be satisfied. However, applications will not specify single parameters in detail, but end-to-end requirements, such as delay or bandwidth. The WATM system must now map, e.g., end-to-end delay onto the cell delays on each segment, wired and wireless. To handle the complexity of such a system, WATM networks will initially only offer a set of different service classes to applications.

Additionally, applications must be adaptive to some degree to survive the effects of mobility, such as higher cell loss, delay variations etc. Applications could, for example, negotiate windows of QoS parameters where they can adapt without breaking the connection.

A crucial point in maintaining QoS over time is QoS support in hand-over protocols. These protocols can support two different types of QoS during handover:

- **Hard handover QoS:** While the QoS with the current RAS may be guaranteed due to the current availability of resources, no QoS guarantees are given after the handover. This is comparable to the traditional approach for, e.g., GSM networks with voice connections. If a terminal can set up a connection, the connection's quality is guaranteed. If there are not enough resources after handover (too many users are already in the target cell), the system cuts off the connection. This is the only possible solution if the applications and terminals cannot adapt to the new situation.
- **Soft handover QoS:** Even for the current wireless segment, only statistical QoS guarantees can be given, and the applications also have to adapt after the handover. This assumes adaptive applications and at least allows for some remaining QoS guarantees during, e.g., periods of congestion or strong interference.

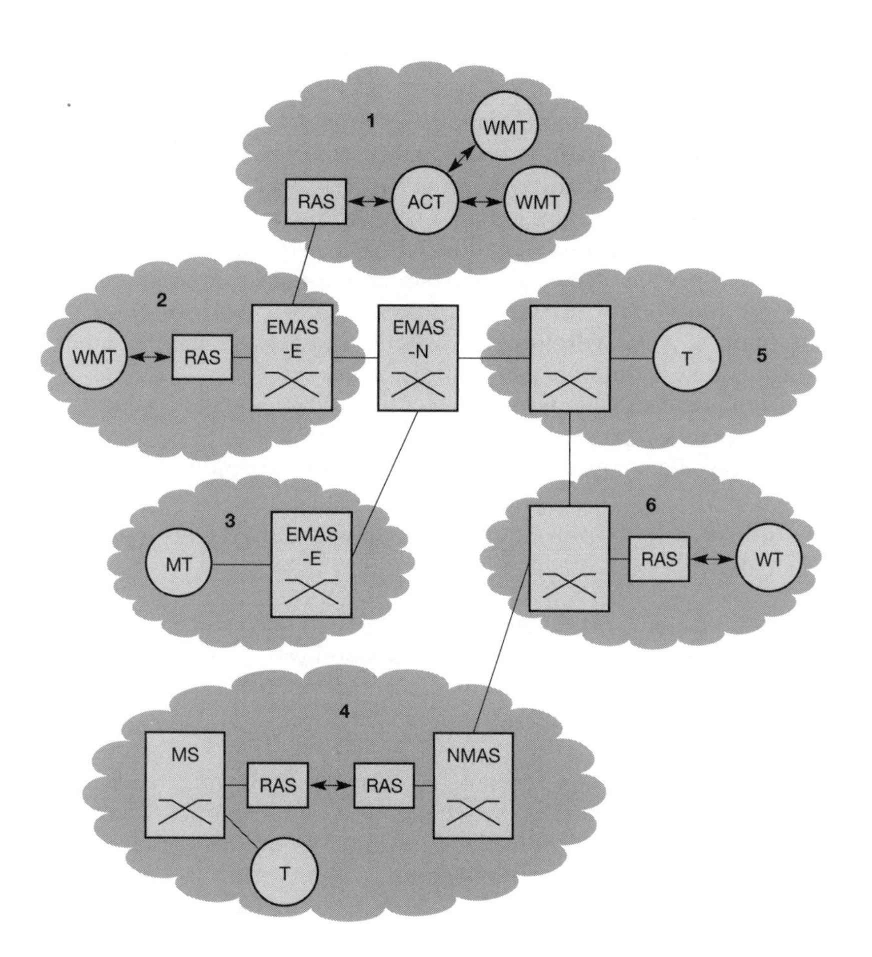

Figure 7.29
WATM reference model with several access scenarios

7.4.2.8 Access scenarios

Figure 7.29 shows possible access scenarios for WATM and illustrates what was planned during the specification of WATM. While this section has focused on the wireless access of mobile ATM terminals, several other configurations are possible (Bhat, 1998). As additional entities, Figure 7.29 shows the following components:

- **T (terminal):** A standard ATM terminal offering ATM services defined for fixed ATM networks.
- **MT (mobile terminal):** A standard ATM terminal with the additional capability of reconnecting after access point change. The terminal can be moved between different access points within a certain domain.
- **WT (wireless terminal):** This terminal is accessed via a wireless link, but the terminal itself is fixed, i.e., the terminal keeps its access point to the network.
- **WMT (wireless mobile terminal):** The combination of a wireless and a mobile terminal results in the WMT. This is exactly the type of terminal

presented throughout this WATM section, as it has the ability to change its access point and uses radio access.

- **RAS (radio access system):** Point of access to a network via a radio link as explained in this chapter.
- **EMAS (end-user mobility supporting ATM switch, -E: edge, -N: network):** Switches with the support of end-user mobility.
- **NMAS (network mobility-supporting ATM switch):** A whole network can be mobile not just terminals. Certain additional functions are needed to support this mobility from the fixed network.
- **MS (mobile ATM switch):** ATM switches can also be mobile and can use wireless access to another part of the ATM network.
- **ACT (ad-hoc controller terminal):** For the configuration of ad-hoc networks, special terminal types might be required within the wireless network. These terminals could, for example, control wireless access without an RAS.

Based on these entities, we can define several scenarios which should be supported by WATM if fully specified.

- **Wireless ad-hoc ATM network (scenario 1):** WMTs can communicate with each other without a fixed network. Communication can be set up without any infrastructure. Access control can be accomplished via the ACT. If the ad-hoc network needs a connection to a fixed network, this can be provided by means of an RAS.
- **Wireless mobile ATM terminals (scenario 2):** The configuration discussed throughout this chapter is the wireless and mobile terminal accessing the fixed network via an RAS. In this configuration, a WMT cannot communicate without the support provided by entities within the fixed network, such as an EMAS-E.
- **Mobile ATM terminals (scenario 3):** This configuration supports device portability and allows for simple network reconfiguration. Users can change the access points of their ATM equipment over time without the need for reconfiguration by hand. Again, this scenario needs support through entities in the fixed network (e.g., EMAS-E).
- **Mobile ATM switches (scenario 4):** An even more complex configuration comprises mobile switches using wireless access to other fixed ATM networks. Now entities supporting switch mobility are needed within the fixed network (NMAS). There are many applications for this scenario, e.g., networks in aircraft, trains, or ships. Within the mobile network either fixed, mobile, wireless, or mobile and wireless terminals can be used. This is the most complex configuration ever envisaged within an ATM environment.
- **Fixed ATM terminals (scenario 5):** This configuration is the standard case. Terminals and switches do not include capabilities for mobility or wireless access. This is also the reference configuration for applications which work on top of an ATM network. Convergence layers have to hide the special characteristics of mobility and wireless access because no special applications should be required for the scenarios presented here.

- **Fixed wireless ATM terminals (scenario 6):** To provide simple access to ATM networks without wiring, a fixed wireless link is the ideal solution. Many alternative carriers are using or planning to use this way of accessing customers as they do not own the wired infrastructure. This scenario does not require any changes or enhancements in the fixed network.

The main difference between WATM and other approaches is the integration of a whole system into the specification. WATM specifies radio access, mobility management, handover schemes, mobile QoS, security etc. The main complexity of WATM lies within the functions and protocols needed for handover, due to its desired ability to maintain QoS parameters for connections during handover, and the connection-oriented paradigm of ATM. Consequencently there is a need for resource reservation, checking for available resources at access points, and rerouting of connections.

As WATM was planned as an integrated approach, issues like location management, security, and efficiency of the whole system had to be considered. To minimize overheads, WATM tried to harmonize the functions required with those available in fixed ATM. Overall, the approach was already too ambitious to be realized as a stand-alone network. All configurations should have been able to interact with existing cellular systems and Internet technology. Chakraborty (1998) discusses many problems that already arise when interworking with other narrowband networks like GSM, DECT, UMTS (see chapter 4) and standard TCP/IP networks.

7.4.3 BRAN

The broadband radio access networks (BRAN), which have been standardized by the European Telecommunications Standards Institute (ETSI), could have been an RAL for WATM (ETSI, 2002b).

The main motivation behind BRAN is the deregulation and privatization of the telecommunication sector in Europe. Many new providers experience problems getting access to customers because the telephone infrastructure belongs to a few big companies. One possible technology to provide network access for customers is radio. The advantages of radio access are high flexibility and quick installation. Different types of traffic are supported, one can multiplex traffic for higher efficiency, and the connection can be asymmetrical (as, e.g., in the typical www scenario where many customers pull a lot of data from servers but only put very small amounts of data onto them). Radio access allows for economical growth of access bandwidth. If more bandwidth is needed, additional transceiver systems can be installed easily. For wired transmission this would involve the installation of additional wires. The primary market for BRAN includes private customers and small to medium-sized companies with Internet applications, multi-media conferencing, and virtual private networks. The BRAN standard and IEEE 802.16 (Broadband wireless access, IEEE, 2002b) have similar goals.

BRAN standardization has a rather large scope including indoor and campus mobility, transfer rates of 25–155 Mbit/s, and a transmission range of 50 m–5 km. Standardization efforts are coordinated with the ATM Forum, the IETF, other groups from ETSI, the IEEE etc. BRAN has specified four different network types (ETSI, 1998a):

- **HIPERLAN 1:** This high-speed WLAN supports mobility at data rates above 20 Mbit/s. Range is 50 m, connections are multi-point-to-multi-point using ad-hoc or infrastructure networks (see section 7.4.1 and ETSI, 1998b).
- **HIPERLAN/2:** This technology can be used for wireless access to ATM or IP networks and supports up to 25 Mbit/s user data rate in a point-to-multi-point configuration. Transmission range is 50 m with support of slow (< 10 m/s) mobility (ETSI, 1997). This standard has been modified over time and is presented in section 7.4.4 as a high performance WLAN with QoS support.
- **HIPERACCESS:** This technology could be used to cover the 'last mile' to a customer via a fixed radio link, so could be an alternative to cable modems or xDSL technologies (ETSI, 1998c). Transmission range is up to 5 km, data rates of up to 25 Mbit/s are supported. However, many proprietary products already offer 155 Mbit/s and more, plus QoS.
- **HIPERLINK:** To connect different HIPERLAN access points or HIPERACCESS nodes with a high-speed link, HIPERLINK technology can be chosen. HIPERLINK provides a fixed point-to-point connection with up to 155 Mbit/s. Currently, there are no plans regarding this standard.

Common characteristics of HIPERLAN/2, HIPERACCESS, and HIPERLINK include their support of the ATM service classes CBR, VBR-rt, VBR-nrt, UBR, and ABR. It is clear that only HiperLAN2 can be a candidate for the RAL of WATM. This technology fulfills the requirements of ATM QoS support, mobility, wireless access, and high bandwidth.

As an access network, BRAN technology is independent from the protocols of the fixed network. BRAN can be used for ATM and TCP/IP networks as illustrated in Figure 7.30 and explained in more detail in section 7.4.4.2. Based on possibly different physical layers, the DLC layer of BRAN offers a common interface to higher

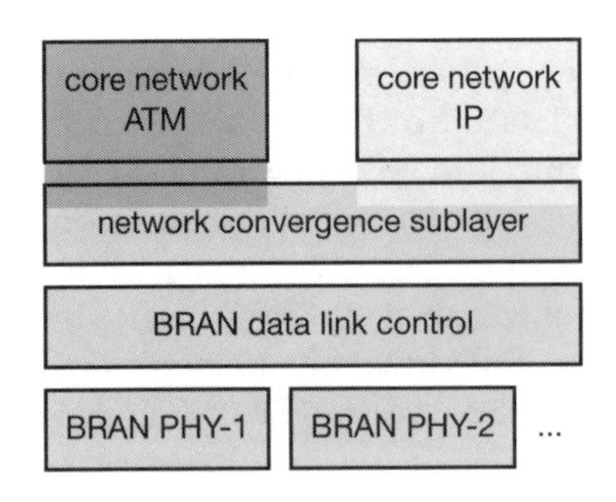

Figure 7.30 Layered model of BRAN wireless access networks

layers. To cover special characteristics of wireless links and to adapt directly to different higher layer network technologies, BRAN provides a network convergence sublayer. This is the layer which can be used by a wireless ATM network, Ethernet, Firewire, or an IP network. In the case of BRAN as the RAL for WATM, the core ATM network would use services of the BRAN network convergence sublayer.

7.4.4 HiperLAN2

While HIPERLAN 1 did not succeed HiperLAN2 might have a better chance. (This is also written as HIPERLAN/2, HiperLAN/2, H/2; official name: HIPERLAN Type 2.) Standardized by ETSI (2000a) this wireless network works at 5 GHz (Europe: 5.15–5.35 GHz and 5.47–5.725 GHz license exempt bands; US: license free U-NII bands, see section 7.3.7) and offers data rates of up to 54 Mbit/s including QoS support and enhanced security features. In comparison with basic IEEE 802.11 LANs, HiperLAN2 offers more features in the mandatory parts of the standard (HiperLAN2, 2002). A comparison is given in section 7.6.

- **High-throughput transmission:** Using OFDM in the physical layer and a dynamic TDMA/TDD-based MAC protocol, HiperLAN2 not only offers up to 54 Mbit/s at the physical layer but also about 35 Mbit/s at the network layer. The overheads introduced by the layers (medium access, packet headers etc.) remains almost constant over a wide rage of user packet sizes and data rates. HiperLAN2 uses MAC frames with a constant length of 2 ms.
- **Connection-oriented:** Prior to data transmission HiperLAN2 networks establish logical connections between a sender and a receiver (e.g., mobile device and access point). Connection set-up is used to negotiate QoS parameters. All connections are time-division-multiplexed over the air interface (TDMA with TDD for separation of up/downlink). Bidirectional point-to-point as well as unidirectional point-to-multipoint connections are offered. Additionally, a broadcast channel is available to reach all mobile devices in the transmission range of an access point.
- **Quality of service support:** With the help of connections, support of QoS is much simpler. Each connection has its own set of QoS parameters (bandwidth, delay, jitter, bit error rate etc.). A more simplistic scheme using priorities only is available.
- **Dynamic frequency selection:** HiperLAN2 does not require frequency planning of cellular networks or standard IEEE 802.11 networks. All access points have built-in support which automatically selects an appropriate frequency within their coverage area. All APs listen to neighboring APs as well as to other radio sources in the environment. The best frequency is chosen depending on the current interference level and usage of radio channels.
- **Security support:** Authentication as well as encryption are supported by HiperLAN2. Both, mobile terminal and access point can authenticate each other. This ensures authorized access to the network as well as a valid network operator. However, additional functions (directory services, key

exchange schemes etc.) are needed to support authentication. All user traffic can be encrypted using DES, Triple-DES, or AES to protect against eavesdropping or man-in-the-middle attacks.

- **Mobility support:** Mobile terminals can move around while transmission always takes place between the terminal and the access point with the best radio signal. Handover between access points is performed automatically. If enough resources are available, all connections including their QoS parameters will be supported by a new access point after handover. However, some data packets may be lost during handover.
- **Application and network independence:** HiperLAN2 was not designed with a certain group of applications or networks in mind. Access points can connect to LANs running ethernet as well as IEEE 1394 (Firewire) systems used to connect home audio/video devices. Interoperation with 3G networks is also supported, so not only best effort data is supported but also the wireless connection of, e.g., a digital camera with a TV set for live streaming of video data.
- **Power save:** Mobile terminals can negotiate certain wake-up patterns to save power. Depending on the sleep periods either short latency requirements or low power requirements can be supported.

The following sections show the reference model of HiperLAN2 and illustrate some more features.

7.4.4.1 Reference model and configurations

Figure 7.31 shows the standard architecture of an infrastructure-based HiperLAN2 network. In the example, two **access points** (AP) are attached to a core network. Core networks might be Ethernet LANs, Firewire (IEEE 1394) connections

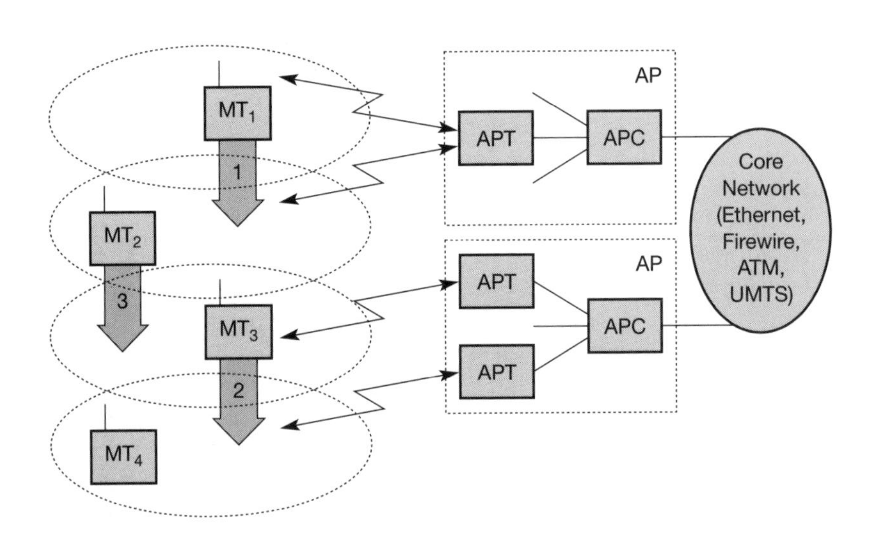

Figure 7.31 HiperLAN2 basic structure and handover scenarios

between audio and video equipment, ATM networks, UMTS 3G cellular phone networks etc. Each AP consists of an **access point controller** (APC) and one or more **access point transceivers** (APT). An APT can comprise one or more sectors (shown as cell here). Finally, four **mobile terminals** (MT) are also shown. MTs can move around in the cell area as shown. The system automatically assigns the APT/AP with the best transmission quality. No frequency planning is necessary as the APs automatically select the appropriate frequency via **dynamic frequency selection** (DFS, compare with IEEE 802.11h, section 7.3.8).

Three handover situations may occur:

- **Sector handover** (Inter sector): If sector antennas are used for an AP, which is optional in the standard, the AP shall support sector handover. This type of handover is handled inside the DLC layer so is not visible outside the AP (as long as enough resources are available in the new sector).
- **Radio handover** (Inter-APT/Intra-AP): As this handover type, too, is handled within the AP, no external interaction is needed. In the example of Figure 7.31 the terminal MT_3, moves from one APT to another of the same AP. All context data for the connections are already in the AP (encryption keys, authentication, and connection parameters) and does not have to be renegotiated.
- **Network handover** (Inter-AP/Intra-network): This is the most complex situation: MT_2 moves from one AP to another. In this case, the core network and higher layers are also involved. This handover might be supported by the core network (similar to the IAPP, IEEE 802.11f). Otherwise, the MT must provide the required information similar to the situation during a new association.

HiperLAN2 networks can operate in two different modes (which may be used simultaneously in the same network).

- **Centralized mode** (CM): This infrastructure-based mode is shown again in a more abstract way in Figure 7.32 (left side). All APs are connected to a core network and MTs are associated with APs. Even if two MTs share the same cell, all data is transferred via the AP. In this mandatory mode the AP takes complete control of everything.
- **Direct mode** (DM): The optional ad-hoc mode of HiperLAN2 is illustrated on the right side of Figure 7.32. Data is directly exchanged between MTs if they can receive each other, but the network still has to be controlled. This can be done via an AP that contains a central controller (CC) anyway or via an MT that contains the CC functionality. There is no real difference between an AP and a CC besides the fact that APs are always connected to an infrastructure but here only the CC functionality is needed. This is why the standard coined two different names. IEEE 802.11, too, offers an ad-hoc mode, but not the CC functionality for QoS support.

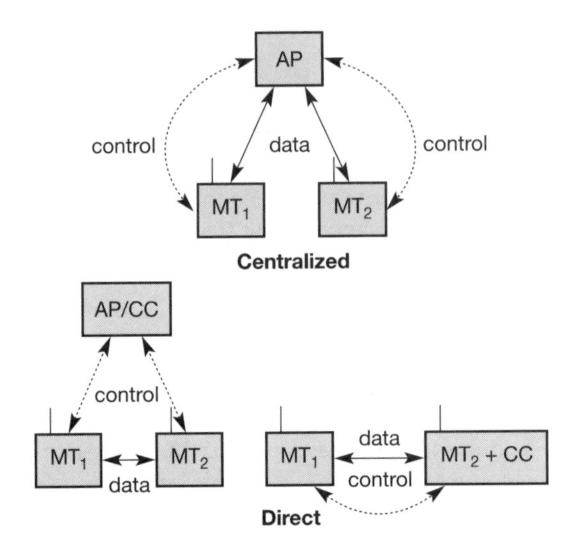

Figure 7.32
HiperLAN2 centralized vs direct mode

Figure 7.33 shows the HiperLAN2 protocol stack as used in access points. Protocol stacks in mobile terminals differ with respect to the number of MAC and RLC instances (only one of each). The lowest layer, the **physical layer**, handles as usual all functions related to modulation, forward error correction, signal detection, synchronization etc. Section 7.4.4.2 describes the physical layer in more detail. The **data link control** (DLC) layer contains the MAC functions, the RLC sublayer and error control functions. If an AP comprises several APTs then each APT requires an own MAC instance. The **MAC** of an AP assigns each MT a certain capacity to guarantee connection quality depending on available resources. Above the MAC DLC is divided into a control and a user part. This separation is common in classical connection-oriented systems such as cellular phones or PSTN. The user part contains **error control** mechanisms. HiperLAN2 offers reliable data transmission using acknowledgements and retransmissions. For broadcast transmissions a repetition mode can be used that provides increased reliability by repeating data packets. Additionally, unacknowledged data transmission is available. The **radio link control** (RLC) sublayer comprises most control functions in the DLC layer (the CC part of an AP). The **association control function** (ACF) controls association and authentication of new MTs as well as synchronization of the radio cell via beacons. The **DLC user connection control** (DCC or DUCC) service controls connection setup, modification, and release. Finally, the **radio resource control** (RRC) handles handover between APs and within an AP. These functions control the dynamic frequency selection and power save mechanisms of the MTs.

On top of the DLC layer there is the **convergence layer**. This highest layer of HiperLAN2 standardization may comprise segmentation and reassembly functions and adaptations to fixed LANs, 3G networks etc. The following sections give some more insight into the 3 HiperLAN2 layers.

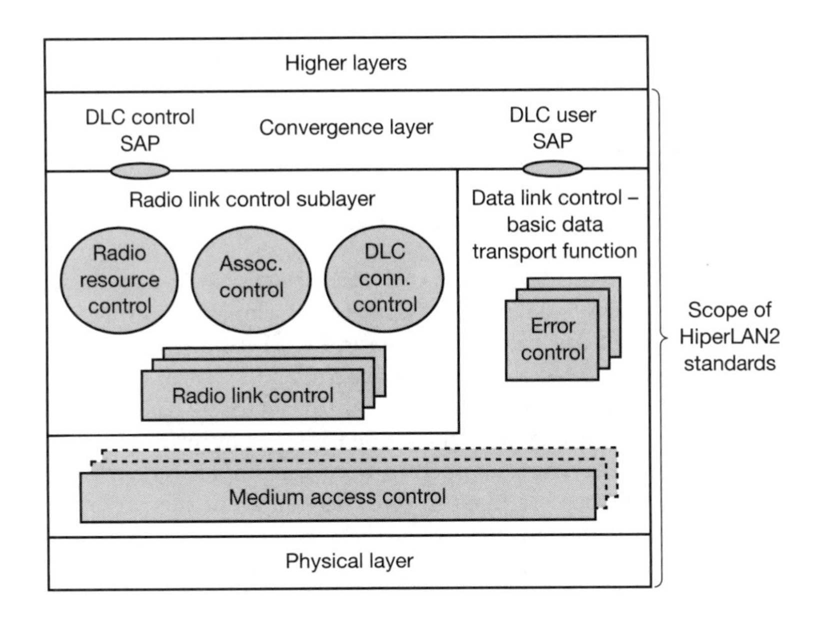

Figure 7.33 HiperLAN2 protocol stack

7.4.4.2 Physical layer

Many functions and features of HiperLAN2's physical layer (ETSI, 2001a) served as an example for IEEE 802.11a as described in section 7.3.7. It is not surprising that both standards offer similar data rates and use identical modulation schemes. Table 7.5 gives an overview of the data rates offered by HiperLAN2 together with other parameters such as coding (compare this with Table 7.3).

Figure 7.34 illustrates the reference configuration of the transmission chain of a HiperLAN2 device. After selecting one of the above transmission modes, the DLC layer passes a PSDU to the physical layer (PSDUs are called DLC PDU trains

Table 7.5 Rate dependent parameters for HiperLAN2

Data rate [Mbit/s]	Modulation	Coding rate	Coded bits per sub-carrier	Coded bits per OFDM symbol	Data bits per OFDM symbol
6	BPSK	1/2	1	48	24
9	BPSK	3/4	1	48	36
12	QPSK	1/2	2	96	48
18	QPSK	3/4	2	96	72
27	16-QAM	9/16	4	192	108
36	16-QAM	3/4	4	192	144
54	64-QAM	3/4	6	288	216

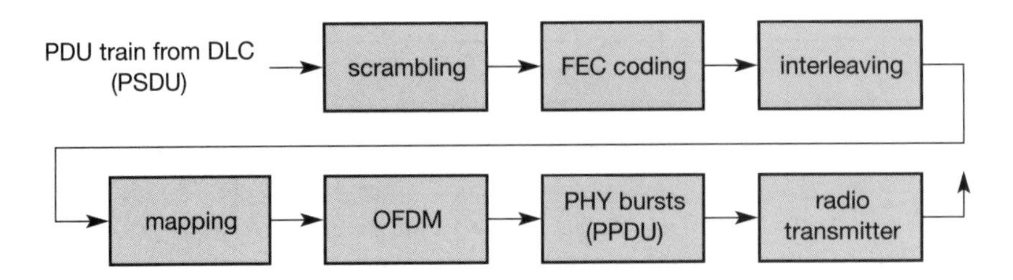

Figure 7.34 HiperLAN2 physical layer reference configuration

in the HiperLAN2 context). The first step then is **scrambling** of all data bits with the generator polynomial $x^7 + x^4 + 1$ for DC blocking and whitening of the spectrum. The result of this first step are **scrambled bits**. The next step applies **FEC coding** for error protection. Coding depends on the type of data (broadcast, uplink, downlink etc.) and the usage of sector or omni-directional antennas. The result of this step is an **encoded bit**. For mitigation of frequency selective fading **interleaving** is applied in the third step. Interleaving ensures that adjacent encoded bits are mapped onto non-adjacent subcarriers (48 subcarriers are used for data transmission). Adjacent bits are mapped alternately onto less and more significant bits of the constellation. The result is an **interleaved bit**.

The following **mapping** process first divides the bit sequence in groups of 1, 2, 4, or 6 bits depending on the modulation scheme (BPSK, QPSK, 16-QAM, or 64-QAM). These groups are mapped onto the appropriate modulation symbol according to the constellation diagrams standardized in (ETSI, 2001a). The results of this mapping are **subcarrier modulation symbols**. The **OFDM** modulation step converts these symbols into a baseband signal with the help of the inverse FFT. (For the usage of the subcarriers compare with Figure 7.24 and its description in section 7.3.7). The symbol interval is 4 μs with 3.2 μs useful part and 0.8 μs guard time. Pilot sub-carriers (sub-carriers –21, –7, 7, 21) are added. The last step before radio transmission is the creation of **PHY bursts** (PPDUs in ISO/OSI terminology). Each burst consists of a preamble and a payload. Five different PHY bursts have been defined: broadcast, downlink, uplink with short preamble, uplink with long preamble, and direct link (optional). The bursts differ in their preambles.

The final **radio transmission** shifts the baseband signal to a carrier frequency depending on the channel number and the formula already used for 802.11a: carrier_number = (carrier_frequency – 5000 MHz)/5 MHz. All nominal carrier frequencies are spaced 20 MHz apart, resulting in a frequency allocation table for Europe as illustrated in Figure 7.35.

Maximum transmit power is 200 mW EIRP for the lower frequency band (indoor use) and 1 W EIRP for the higher frequency band (indoor and outdoor use). DFS and TPC are not necessary, if the transmit power stays below 50 mW EIRP and only 5.15–5.25 GHz are used (be aware that national differences exist even within Europe and regulation may change over time).

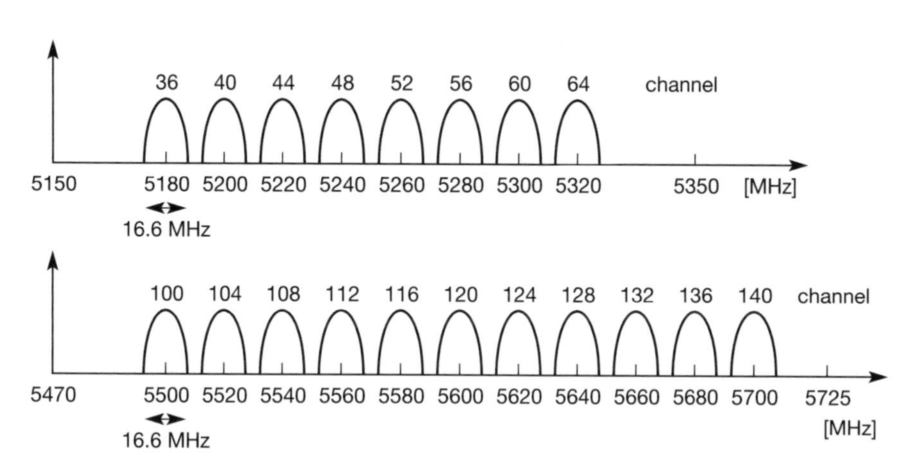

Figure 7.35 Operating channels of HiperLAN2 in Europe

7.4.4.3 Data link control layer

As described above, the DLC layer is divided into MAC, control and data part (which would fit into the LLC sublayer according to ISO/OSI). ETSI (2001b) standardizes the basic data transport functions, i.e., user part with error control and MAC, while ETSI (2002a) defines RLC functionality.

The medium access control creates frames of 2 ms duration as shown in Figure 7.36. With a constant symbol length of four μs this results in 500 OFDM symbols. Each MAC frame is further sub-divided into four phases with variable boundaries:

- **Broadcast phase:** The AP of a cell broadcasts the content of the current frame plus information about the cell (identification, status, resources).
- **Downlink phase:** Transmission of user data from an AP to the MTs.
- **Uplink phase:** Transmission of user data from MTs to an AP.
- **Random access phase:** Capacity requests from already registered MTs and access requests from non-registered MTs (slotted Aloha).

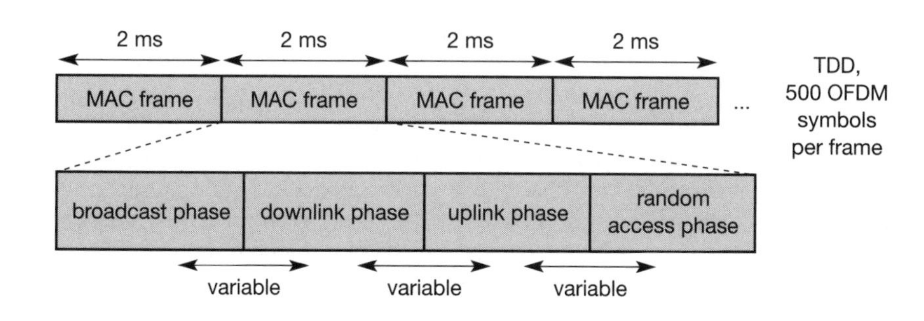

Figure 7.36 Basic structure of HiperLAN2 MAC frames

An optional **direct link phase** can be inserted between the downlink and the uplink phase. The access to the common physical medium is always controlled by the CC (typically in an AP).

HiperLAN2 defines six different so-called transport channels for data transfer in the above listed phases. These transport channels describe the basic message format within a MAC frame.

- **Broadcast channel (BCH):** This channel conveys basic information for the radio cell to all MTs. This comprises the identification and current transmission power of the AP. Furthermore, the channel contains pointers to the FCH and RCH which allows for a flexible structure of the MAC frame. The length is 15 bytes.
- **Frame channel (FCH):** This channel contains a directory of the downlink and uplink phases (LCHs, SCHs, and empty parts). This also comprises the PHY mode used. The length is a multiple of 27 bytes.
- **Access feedback channel (ACH):** This channel gives feedback to MTs regarding the random access during the RCH of the previous frame. As the access during the RCHs is based on slotted Aloha, collision at the AP may occur. The ACH signals back which slot was successfully transmitted. The length is 9 bytes.
- **Long transport channel (LCH):** This channel transports user and control data for downlinks and uplinks. The length is 54 bytes.
- **Short transport channel (SCH):** This channel transports control data for downlinks and uplinks. The length is 9 bytes.
- **Random channel (RCH):** This channel is needed to give an MT the opportunity to send information to the AP/CC even without a granted SCH. Access is via slotted Aloha so, collisions may occur. Collision resolution is performed with the help of an exponential back-off scheme (ETSI, 2001b). The length is 9 bytes. A maximum number of 31 RCHs is currently supported.

BCH, FCH and ACH are used in the broadcast phase only and use BPSK with code rate 1/2. LCH and SCH can be used in the downlink, uplink or (optional) direct link phase. RCH is used in the uplink only for random access (BPSK, code rate 1/2). HiperLAN2 defines further how many of the channels are used within a MAC frame. This configuration may change from MAC frame to MAC frame depending on the connection QoS, resource requests, number of MTs etc. Figure 7.37 shows valid combinations of channels/transfer phases within MAC frames. It is required that the transport channels BCH, FCH and ACH are present plus at least one RCH. While the duration of the BCH is fixed (15 byte), the duration of the others may vary (either due to a variable size of the channel or due to the multiple use of channels). However, the order BCH-FCH-ACH-DL phase-UL phase-RCH must be kept from an MT's point of view (centralized mode). For the direct mode the DiL phase is inserted between the DL and UL phases.

Data between entities of the DLC layer are transferred over so-called **logical channels** (just another name for any distinct data path). The type of a logical

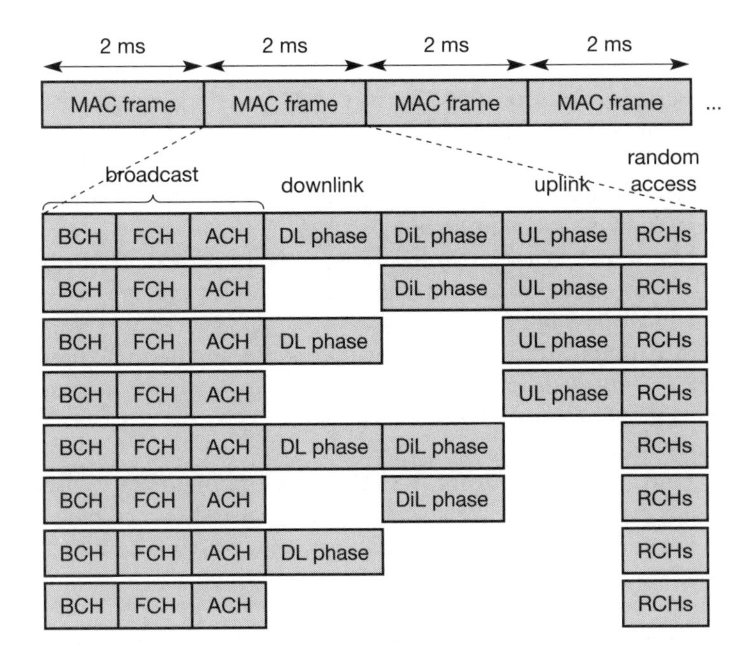

Figure 7.37
Valid configurations of MAC frames

channel is defined by the type of information it carries and the interpretation of the values in the corresponding messages. This is a well-known concept from, e.g., cellular phone systems like GSM (see chapter 4). The following logical channels are defined in HiperLAN2 (logical channels use 4 letter acronyms):

- **Broadcast control channel (BCCH):** This channel on the downlink conveys a constant amount of broadcast information concerning the whole radio cell. Examples are the seed for the scrambler, network/access point/sector identifiers, AP transmission power, expected AP reception power, pointers to the FCH/RCH, number of RCHs (1 to 31), load indicator, number of sectors etc.
- **Frame control channel (FCCH):** The FCCH describes the structure of the remaining parts of the MAC frame. This comprises resource grants for SCHs and LCHs belonging to certain MTs. Resource grants contain the MAC address the grant belongs to, the number of LCHs and SCHs, their PHY modes etc. This scheme allows for a precise reservation of the medium with associated QoS properties.
- **Random access feedback channel (RFCH):** This channel informs MTs that have used an RCH in the previous frame about the success of their access attempt.
- **RLC broadcast channel (RBCH):** This channel transfers information regarding RLC control information, MAC IDs during an association phase, information from the convergence layer, or seeds for the encryption function only if necessary.

- **Dedicated control channel (DCCH):** This channel carries RLC messages related to a certain MT and is established during the association of an MT.
- **User broadcast channel (UBCH):** A UBCH transfers broadcast messages from the convergence layer. Transmission is performed in the unacknowledged or repetition mode.
- **User multi-cast channel (UMCH):** This channel performs unacknowledged transmission of data to a group of MTs.
- **User data channel (UDCH):** Point-to-point data between an AP and an MT (CM) or between two MTs (DM) use this channel. Error protection via an ARQ scheme is possible.
- **Link control channel (LCCH):** This bi-directional channel conveys ARQ feedback and discards messages between the error control functions of an AP and an MT (CM) or between two MTs (DM). A LCCH is typically assigned to a UDCH.
- **Association control channel (ASCH):** This channel is only used in the uplink and for currently non-associated MTs (related to a certain AP). This is the case for a new association request (new MT in the network) or a handover request on behalf of the RLC.

The reader may have noticed that some transport channels transfer exactly the information of one logical channel as their descriptions were identical. This is indeed the case for some channels (BCCH-BCH, FCCH-FCH, RFCH-ACH) as the scheme of mapping logical and transport channels shows (see Figure 7.38). This figure also shows in which mode which channel can be used (uplink and downlink in the centralized mode, direct link in the direct mode).

Figure 7.39 gives an example for mapping the logical channel UDCH to the transport channel LCH. The payload of the LCH is used for a sequence number plus the payload of the UDCH.

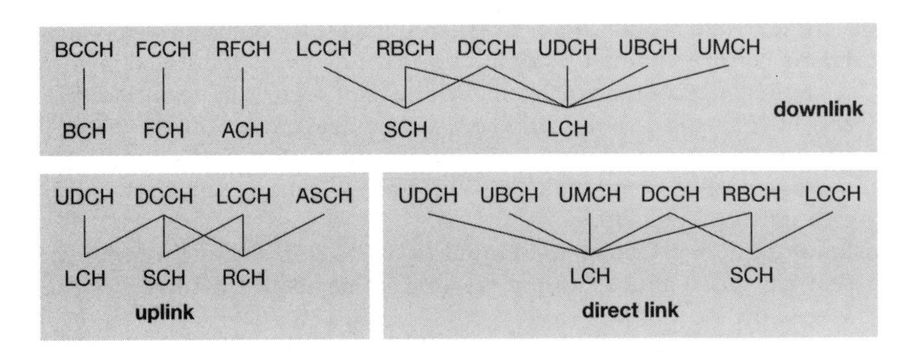

Figure 7.38
Mapping of logical and transport channels

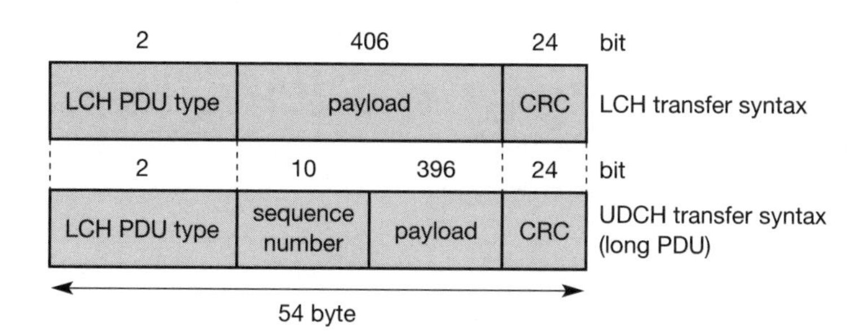

Figure 7.39
HiperLAN2 LCH and UDCH transfer syntax

The **radio link control** sublayer in connection oriented systems offering QoS like HiperLAN2 is quite complex and comprises many protocols, functions, and messages. ETSI (2002a) defines three main services for the RLC sublayer:

- **Association control function (ACF):** ACF contains all procedures for association, authentication, and encryption. An MT starts the association process. The first step is the synchronization with a beacon signal transmitted in each BCCH of a MAC frame. The network ID may be obtained via the RBCH. The next step is the MAC ID assignment. This unique ID is used to address the MT. From this point on, all RLC control messages are transmitted via a DCCH. During the following link capability negotiation, lists of supported convergence layers, authentication and encryption procedures are exchanged. Depending on these parameters the following steps may take place: encryption start-up, authentication, obtaining the ID of the MT. If all necessary steps are successful the MT is associated with the AP. Disassociation may take place at any time, either explicitly (MT or AP initiated) or implicitly (loss of the radio connection). The AP may send MT-alive messages to check if an MT is still available.
- **Radio resource control (RRC):** An important function of the RRC is handover support as already shown in Figure 7.31. Each associated MT continuously measures the link quality. To find handover candidates the MT additionally checks other frequencies. If only one transceiver is available the MT announces to the AP that it is temporarily unavailable (MT absence). Based on radio quality measurements, an AP can change the carrier frequency dynamically (DFS). The RLC offers procedures to inform all MTs. To minimize interference with other radio sources operating at the same frequency (HiperLAN2s or other WLANs) transmission power control (TPC) must be applied by the RRC. An MT can save power by negotiating with an AP a sleeping period of *n* MAC frames. After these *n* frames the MT may wake up because data is ready to be sent, or the AP signals data to be received. If the MT misses the wakeup message from the AP it starts the MT alive procedure. If no data has to be transmitted the MT can again fall asleep for *n* frames.

- **DLC user connection control (DCC or DUCC):** This service is used for setting up, releasing, or modifying unicast connections. Multi-cast and broadcast connections are implicitly set-up by a group/broadcast join during the association procedure.

7.4.4.4 Convergence layer

As the physical layer and the data link layer are independent of specific core network protocols, a special **convergence layer (CL)** is needed to adapt to the specific features of these network protocols. HiperLAN2 supports two different types of CLs: cell-based and packet-based. The **cell-based** CL (ETSI, 2000b) expects data packets of fixed size (cells, e.g., ATM cells), while the **packet-based** CL (ETSI, 2000d) handles packets that are variable in size (e.g., Ethernet or Firewire frames). For the packet-based CL additional functionality is necessary for segmentation and reassembling of packets that do not fit into the DLC payload of HiperLAN2 (49.5 byte). Three examples of convergence layers follow:

- **Ethernet:** This sublayer supports the transparent transport of Ethernet frames over a HiperLAN2 wireless network (ETSI, 2001d). This includes the mapping of Ethernet multicast and broadcast messages onto HiperLAN2 multicast and broadcast messages. A collision domain can also be emulated. This sublayer also supports priorities according to IEEE 802.1p. The standard supports the traffic classes best effort, background, excellent effort, controlled load, video, voice, and network control. The sublayer does not transmit the Ethernet preamble, start of frame delimiter, and frame check sequence. These fields of an Ethernet frame are not necessary during transmission and will be appended in the receiver's Ethernet sublayer.
- **IEEE 1394 (Firewire):** As a high-speed real-time bus for connecting, e.g., audio and video devices, timing and synchronization is of special importance for IEEE 1394. ETSI (2001e) supports synchronization of timers via the air and treats isochronous data streams with special regard to jitter.
- **ATM:** The cell-based CL is used for this type of network (ETSI, 2000c). As the payload of an ATM cell is only 48 byte, which fits into the 49.5 byte of a DLC-PDU, segmentation and reassembly is not necessary. In this case, the sublayer only has to control connection identifiers and MAC IDs.

Many people doubt that HiperLAN2 will ever be a commercial success. Wireless networks following IEEE 802.11 are already in widespread use. The story of wireless LANs could follow the race between Ethernet and Token Ring in the early days of LANs. In this case, the much simpler Ethernet succeeded although it could not offer any quality of service compared to Token Ring. However, plans for interworking between HiperLAN2 and 3G cellular systems have already been made (ETSI, 2001c). HiperLAN2 is to be used to provide high-speed access to the internet with QoS guarantees. One difference from other

WLAN solutions, besides QoS, is the interworking of HiperLAN2 security and accounting mechanisms with the mechanisms of, e.g., UMTS. A more detailed comparison of the IEEE 802.11a WLAN approach and HiperLAN2 is given at the end of this chapter.

7.5 Bluetooth

Compared to the WLAN technologies presented in sections 7.3 and 7.4, the Bluetooth technology discussed here aims at so-called **ad-hoc piconets**, which are local area networks with a very limited coverage and without the need for an infrastructure. This is a different type of network is needed to connect different small devices in close proximity (about 10 m) without expensive wiring or the need for a wireless infrastructure (Bisdikian, 1998). The envisaged gross data rate is 1 Mbit/s, asynchronous (data) and synchronous (voice) services should be available. The necessary transceiver components should be cheap – the goal is about €5 per device. (In 2002, separate adapters are still at €50, however, the additional cost of the devices integrated in, e.g., PDAs, almost reached the target.) Many of today's devices offer an infra red data association (IrDA) interface with transmission rates of, e.g., 115 kbit/s or 4 Mbit/s. There are various problems with IrDA: its very limited range (typically 2 m for built-in interfaces), the need for a line-of-sight between the interfaces, and, it is usually limited to two participants, i.e., only point-to-point connections are supported. IrDA has no internet working functions, has no media access, or any other enhanced communication mechanisms. The big advantage of IrDA is its low cost, and it can be found in almost any mobile device (laptops, PDAs, mobile phones).

The **history** of Bluetooth starts in the tenth century, when Harald Gormsen, King of Denmark (son of Gorm), erected a rune stone in Jelling, Denmark, in memory of his parents. The stone has three sides with elaborate carvings. One side shows a picture of Christ, as Harald did not only unite Norway and Denmark, but also brought Christianity to Scandinavia. Harald had the common epithet of 'Blåtand', meaning that he had a rather dark complexion (not a blue tooth).

It took a thousand years before the Swedish IT-company Ericsson initiated some studies in 1994 around a so-called multi-communicator link (Haartsen, 1998). The project was renamed (because a friend of the designers liked the Vikings) and Bluetooth was born. In spring 1998 five companies (Ericsson, Intel, IBM, Nokia, Toshiba) founded the Bluetooth consortium with the goal of developing a single-chip, low-cost, radio-based wireless network technology. Many other companies and research institutions joined the special interest group around Bluetooth (2002), whose goal was the development of mobile phones, laptops, notebooks, headsets etc. including Bluetooth technology, by the end of 1999. In 1999, Ericsson erected a rune stone in Lund, Sweden, in memory of Harald Gormsen, called Blåtand, who gave his epithet for this new wireless communication technology. This new carving shows a man holding a laptop and a cellular phone, a picture which is quite often cited (of course there are no such things visible on the original stone, that's just a nice story!)

In 2001, the first products hit the mass market, and many mobile phones, laptops, PDAs, video cameras etc. are equipped with Bluetooth technology today.

At the same time the Bluetooth development started, a study group within IEEE 802.11 discussed **wireless personal area networks (WPAN)** under the following five criteria:

- **Market potential:** How many applications, devices, vendors, customers are available for a certain technology?
- **Compatibility:** Compatibility with IEEE 802.
- **Distinct identity:** Originally, the study group did not want to establish a second 802.11 standard. However, topics such as, low cost, low power, or small form factor are not addressed in the 802.11 standard.
- **Technical feasibility:** Prototypes are necessary for further discussion, so the study group would not rely on paper work.
- **Economic feasibility:** Everything developed within this group should be cheaper than other solutions and allow for high-volume production.

Obviously, Bluetooth fulfills these criteria so the WPAN group cooperated with the Bluetooth consortium. IEEE founded its own group for WPANs, IEEE 802.15, in March 1999. This group should develop standards for wireless communications within a **personal operating space** (POS, IEEE, 2002c). A POS has been defined as a radius of 10 m around a person in which the person or devices of this person communicate with other devices. Section 7.5.10 gives an overview of 802.15 activities and their relation to Bluetooth.

7.5.1 User scenarios

Many different user scenarios can be imagined for wireless piconets or WPANs:

- **Connection of peripheral devices:** Today, most devices are connected to a desktop computer via wires (e.g., keyboard, mouse, joystick, headset, speakers). This type of connection has several disadvantages: each device has its own type of cable, different plugs are needed, wires block office space. In a wireless network, no wires are needed for data transmission. However, batteries now have to replace the power supply, as the wires not only transfer data but also supply the peripheral devices with power.
- **Support of ad-hoc networking:** Imagine several people coming together, discussing issues, exchanging data (schedules, sales figures etc.). For instance, students might join a lecture, with the teacher distributing data to their personal digital assistants (PDAs). Wireless networks can support this type of interaction; small devices might not have WLAN adapters following the IEEE 802.11 standard, but cheaper Bluetooth chips built in.

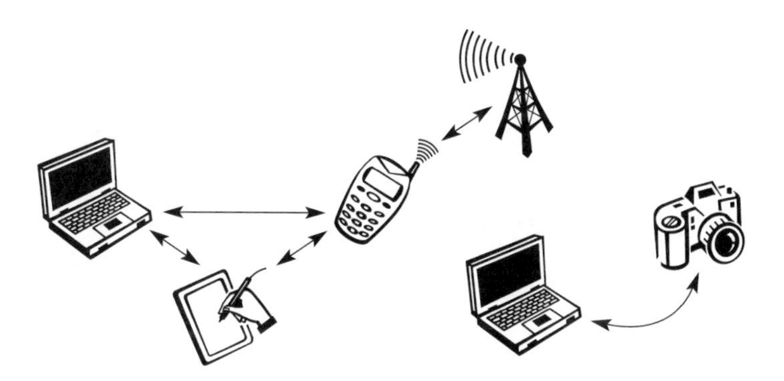

Figure 7.40
Example configurations with a Bluetooth-based piconet

- **Bridging of networks:** Using wireless piconets, a mobile phone can be connected to a PDA or laptop in a simple way. Mobile phones will not have full WLAN adapters built in, but could have a Bluetooth chip. The mobile phone can then act as a bridge between the local piconet and, e.g., the global GSM network (see Figure 7.40). For instance, on arrival at an airport, a person's mobile phone could receive e-mail via GSM and forward it to the laptop which is still in a suitcase. Via a piconet, a fileserver could update local information stored on a laptop or PDA while the person is walking into the office.

When comparing Bluetooth with other WLAN technology we have to keep in mind that one of its goals was to provide local wireless access at very low cost. From a technical point of view, WLAN technologies like those above could also be used, however, WLAN adapters, e.g., for IEEE 802.11, have been designed for higher bandwidth and larger range and are more expensive and consume a lot more power.

7.5.2 Architecture

Like IEEE 802.11b, Bluetooth operates in the 2.4 GHz ISM band. However, MAC, physical layer and the offered services are completely different. After presenting the overall architecture of Bluetooth and its specialty, the piconets, the following sections explain all protocol layers and components in more detail.

7.5.2.1 Networking

To understand the networking of Bluetooth devices a quick introduction to its key features is necessary. Bluetooth operates on 79 channels in the 2.4 GHz band with 1 MHz carrier spacing. Each device performs frequency hopping with 1,600 hops/s in a pseudo random fashion. Bluetooth applies FHSS for interference mitigation (and FH-CDMA for separation of networks). More about Bluetooth's radio layer in section 7.5.3.

A very important term in the context of Bluetooth is a **piconet**. A piconet is a collection of Bluetooth devices which are synchronized to the same hopping sequence. Figure 7.41 shows a collection of devices with different roles. One device in the piconet can act as **master** (M), all other devices connected to the

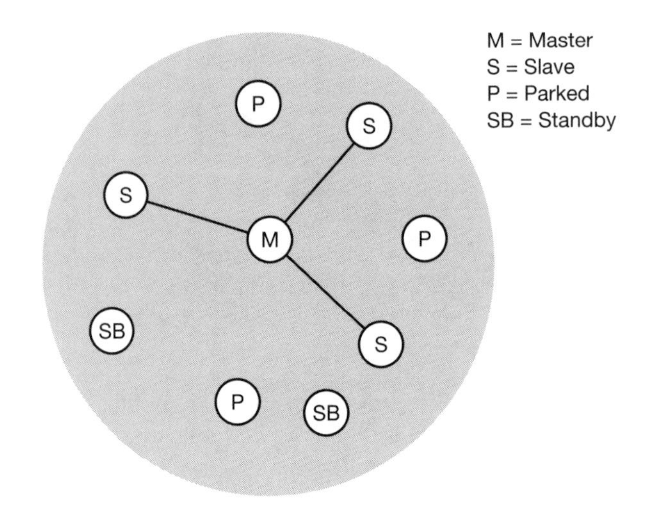

Figure 7.41
Simple Bluetooth piconet

master must act as **slaves** (S). The master determines the hopping pattern in the piconet and the slaves have to synchronize to this pattern. Each piconet has a unique hopping pattern. If a device wants to participate it has to synchronize to this. Two additional types of devices are shown: parked devices (P) can not actively participate in the piconet (i.e., they do not have a connection), but are known and can be reactivated within some milliseconds (see section 7.5.5). Devices in stand-by (SB) do not participate in the piconet. Each piconet has exactly one master and up to seven simultaneous slaves. More than 200 devices can be parked. The reason for the upper limit of eight active devices, is the 3-bit address used in Bluetooth. If a parked device wants to communicate and there are already seven active slaves, one slave has to switch to park mode to allow the parked device to switch to active mode.

Figure 7.42 gives an overview of the formation of a piconet. As all active devices have to use the same hopping sequence they must be synchronized. The first step involves a master sending its clock and device ID. All Bluetooth devices have the same networking capabilities, i.e., they can be master or slave. There is no distinction between terminals and base stations, any two or more devices can form a piconet. The unit establishing the piconet automatically becomes the master, all other devices will be slaves. The hopping pattern is determined by the device ID, a 48-bit worldwide unique identifier. The phase in the hopping pattern is determined by the master's clock. After adjusting the internal clock according to the master a device may participate in the piconet. All active devices are assigned a 3-bit **active member address** (AMA). All parked devices use an 8-bit **parked member address** (PMA). Devices in stand-by do not need an address.

All users within one piconet have the same hopping sequence and share the same 1 MHz channel. As more users join the piconet, the throughput per user drops quickly (a single piconet offers less than 1 Mbit/s gross data rate). (Only having one piconet available within the 80 MHz in total is not very efficient.) This

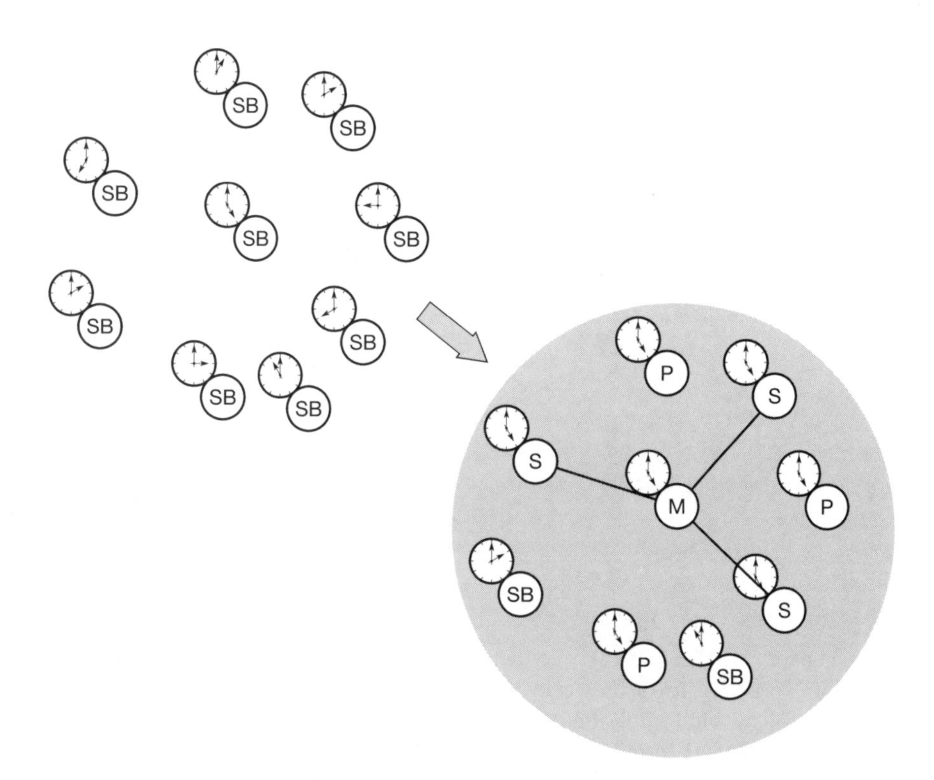

Figure 7.42
Forming a Bluetooth piconet

led to the idea of forming groups of piconets called **scatternet** (see Figure 7.43). Only those units that really must exchange data share the same piconet, so that many piconets with overlapping coverage can exist simultaneously.

In the example, the scatternet consists of two piconets, in which one device participates in two different piconets. Both piconets use a different hopping sequence, always determined by the master of the piconet. Bluetooth applies **FH-CDMA** for separation of piconets. In an average sense, all piconets can share the total of 80 MHz bandwidth available. Adding more piconets leads to a graceful performance degradation of a single piconet because more and more collisions may occur. A collision occurs if two or more piconets use the same carrier frequency at the same time. This will probably happen as the hopping sequences are not coordinated.

If a device wants to participate in more than one piconet, it has to synchronize to the hopping sequence of the piconet it wants to take part in. If a device acts as slave in one piconet, it simply starts to synchronize with the hopping sequence of the piconet it wants to join. After synchronization, it acts as a slave in this piconet and no longer participates in its former piconet. To enable synchronization, a slave has to know the identity of the master that determines the hopping sequence of a piconet. Before leaving one piconet, a slave informs the current master that it will be unavailable for a certain amount of time. The remaining devices in the piconet continue to communicate as usual.

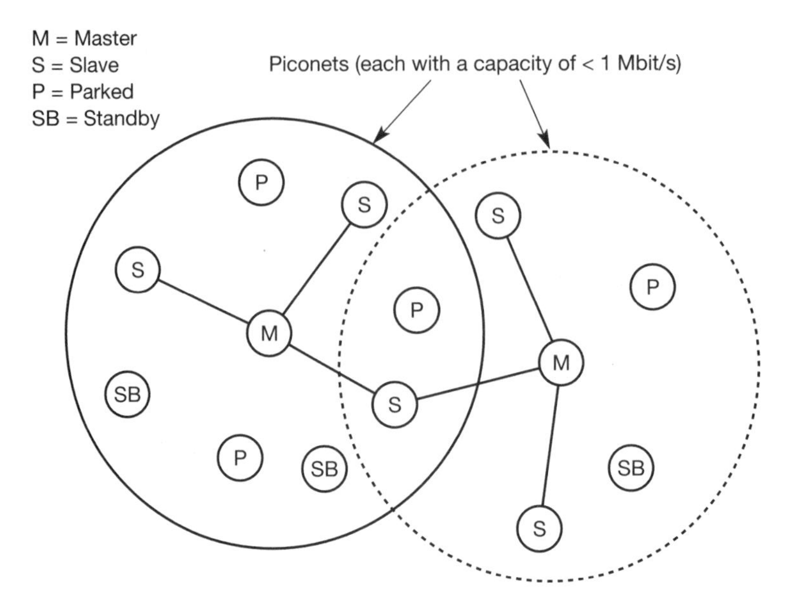

Figure 7.43 Bluetooth scatternet

A master can also leave its piconet and act as a slave in another piconet. It is clearly not possible for a master of one piconet to act as the master of another piconet as this would lead to identical behavior (both would have the same hopping sequence, which is determined by the master per definition). As soon as a master leaves a piconet, all traffic within this piconet is suspended until the master returns.

Communication between different piconets takes place by devices jumping back and forth between theses nets. If this is done periodically, for instance, isochronous data streams can be forwarded from one piconet to another. However, scatternets are not yet supported by all devices.

7.5.2.2 Protocol stack

As Figure 7.44 shows, the Bluetooth specification already comprises many protocols and components. Starting as a simple idea, it now covers over 2,000 pages dealing with not only the Bluetooth protocols but many adaptation functions and enhancements. The Bluetooth protocol stack can be divided into a **core specification** (Bluetooth, 2001a), which describes the protocols from physical layer to the data link control together with management functions, and **profile specifications** (Bluetooth, 2001b). The latter describes many protocols and functions needed to adapt the wireless Bluetooth technology to legacy and new applications (see section 7.5.9).

The **core protocols** of Bluetooth comprise the following elements:

- **Radio:** Specification of the air interface, i.e., frequencies, modulation, and transmit power (see section 7.5.3).
- **Baseband:** Description of basic connection establishment, packet formats, timing, and basic QoS parameters (see section 7.5.4).

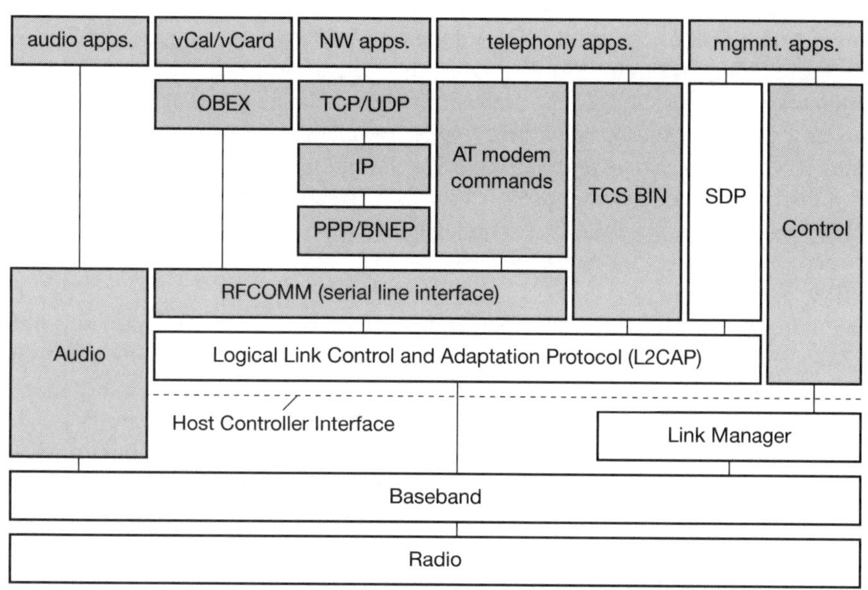

Figure 7.44
Bluetooth protocol stack

- **Link manager protocol:** Link set-up and management between devices including security functions and parameter negotiation (see section 7.5.5).
- **Logical link control and adaptation protocol (L2CAP):** Adaptation of higher layers to the baseband (connectionless and connection-oriented services, see section 7.5.6).
- **Service discovery protocol:** Device discovery in close proximity plus querying of service characteristics (see section 7.5.8).

On top of L2CAP is the **cable replacement protocol** RFCOMM that emulates a serial line interface following the EIA-232 (formerly RS-232) standards. This allows for a simple replacement of serial line cables and enables many legacy applications and protocols to run over Bluetooth. RFCOMM supports multiple serial ports over a single physical channel. The **telephony control protocol specification – binary** (TCS BIN) describes a bit-oriented protocol that defines call control signaling for the establishment of voice and data calls between Bluetooth devices. It also describes mobility and group management functions.

The **host controller interface** (HCI) between the baseband and L2CAP provides a command interface to the baseband controller and link manager, and access to the hardware status and control registers. The HCI can be seen as the hardware/software boundary.

Many **protocols** have been **adopted** in the Bluetooth standard. Classical Internet applications can still use the standard TCP/IP stack running over PPP or use the more efficient Bluetooth network encapsulation protocol (BNEP). Telephony applications can use the AT modem commands as if they were using a standard modem. Calendar and business card objects (vCalendar/vCard) can be exchanged using the object exchange protocol (OBEX) as common with IrDA interfaces.

A real difference to other protocol stacks is the support of **audio**. Audio applications may directly use the baseband layer after encoding the audio signals.

7.5.3 Radio layer

The radio specification is a rather short document (less than ten pages) and only defines the carrier frequencies and output power. Several limitations had to be taken into account when Bluetooth's radio layer was designed. Bluetooth devices will be integrated into typical mobile devices and rely on battery power. This requires small, low power chips which can be built into handheld devices. Worldwide operation also requires a frequency which is available worldwide. The combined use for data and voice transmission has to be reflected in the design, i.e., Bluetooth has to support multi-media data.

Bluetooth uses the license-free frequency band at 2.4 GHz allowing for worldwide operation with some minor adaptations to national restrictions. A frequency-hopping/time-division duplex scheme is used for transmission, with a fast hopping rate of 1,600 hops per second. The time between two hops is called a slot, which is an interval of 625 μs. Each slot uses a different frequency. Bluetooth uses 79 hop carriers equally spaced with 1 MHz. After worldwide harmonization, Bluetooth devices can be used (almost) anywhere.

Bluetooth transceivers use Gaussian FSK for modulation and are available in three classes:

- **Power class 1:** Maximum power is 100 mW and minimum is 1 mW (typ. 100 m range without obstacles). Power control is mandatory.
- **Power class 2:** Maximum power is 2.5 mW, nominal power is 1 mW, and minimum power is 0.25 mW (typ. 10 m range without obstacles). Power control is optional.
- **Power class 3:** Maximum power is 1 mW.

7.5.4 Baseband layer

The functions of the baseband layer are quite complex as it not only performs frequency hopping for interference mitigation and medium access, but also defines physical links and many packet formats. Figure 7.45 shows several examples of frequency selection during data transmission. Remember that each device participating in a certain piconet hops at the same time to the same carrier frequency (f_i in Figure 7.45). If, for example, the master sends data at f_k, then a slave may answer at f_{k+1}. This scenario shows another feature of Bluetooth. **TDD** is used for separation of the transmission directions. The upper part of Figure 7.45 shows so-called **1-slot**

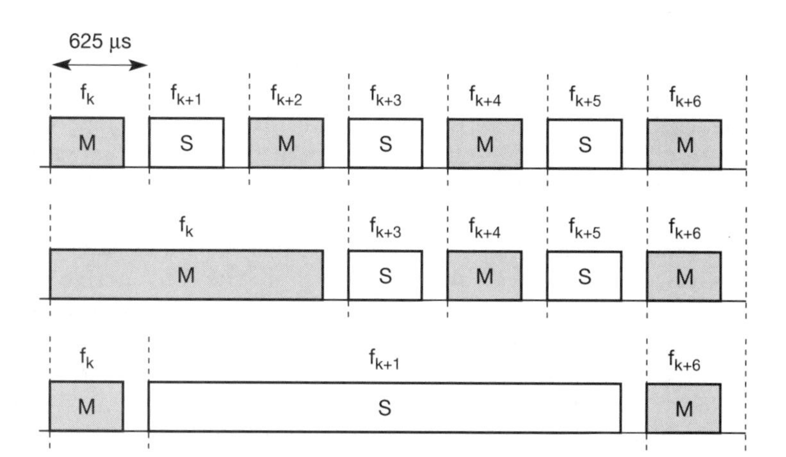

Figure 7.45 Frequency selection during data transmission (1, 3, 5 slot packets)

packets as the data transmission uses one 625 μs slot. Within each slot the master or one out of seven slaves may transmit data in an alternating fashion. The control of medium access will be described later. Bluetooth also defines **3-slot** and **5-slot** packets for higher data rates (multi-slot packets). If a master or a slave sends a packet covering three or five slots, the radio transmitter remains on the same frequency. No frequency hopping is performed within packets. After transmitting the packet, the radio returns to the frequency required for its hopping sequence. The reason for this is quite simple: not every slave might receive a transmission (hidden terminal problem) and it can not react on a multi-slot transmission. Those slaves not involved in the transmission will continue with the hopping sequence. This behavior is important so that all devices can remain synchronized, because the piconet is uniquely defined by having the same hopping sequence with the same phase. Shifting the phase in one device would destroy the piconet.

Figure 7.46 shows the components of a Bluetooth packet at baseband layer. The packet typically consists of the following three fields:

- **Access code:** This first field of a packet is needed for timing synchronization and piconet identification (channel access code, CAC). It may represent special codes during paging (device access code, DAC) and inquiry (inquiry access code, IAC, see section 7.5.5). The access code consists of a 4 bit **preamble**, a **synchronization** field, and a **trailer** (if a packet header follows). The 64-bit synchronization field is derived from the lower 24 bit of

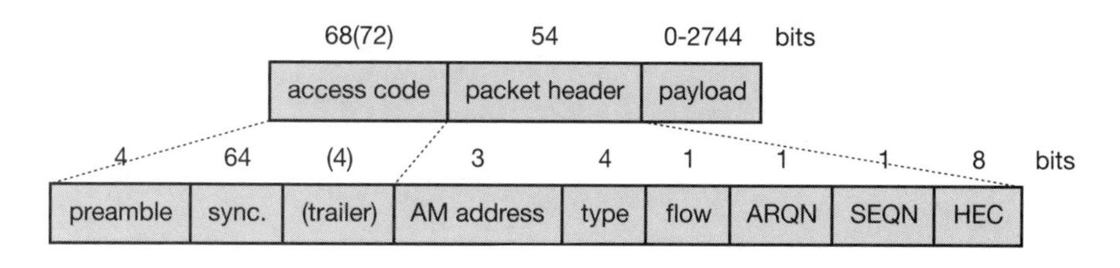

Figure 7.46 Baseband packet format

an address (lower address part, LAP). If the access code is used for channel access (i.e., data transmission between a master and a slave or vice versa), the LAP is derived from the master's globally unique 48-bit address. In case of paging (DAC) the LAP of the paged device is used. If a Bluetooth device wants to discover other (arbitrary) devices in transmission range (general inquiry procedure) it uses a special reserved LAP. Special LAPs can be defined for inquiries of dedicated groups of devices.

- **Packet header:** This field contains typical layer 2 features: address, packet type, flow and error control, and checksum. The 3-bit **active member address** represents the active address of a slave. Active addresses are temporarily assigned to a slave in a piconet. If a master sends data to a slave the address is interpreted as receiver address. If a slave sends data to the master the address represents the sender address. As only a master may communicate with a slave this scheme works well. Seven addresses may be used this way. The zero value is reserved for a broadcast from the master to all slaves. The 4-bit **type** field determines the type of the packet. Examples for packet types are given in Table 7.6. Packets may carry control, synchronous, or asynchronous data. A simple flow control mechanism for asynchronous traffic uses the 1-bit **flow** field. If a packet is received with flow=0 asynchronous data, transmission must stop. As soon as a packet with flow=1 is received, transmission may resume. If an acknowledgement of packets is required, Bluetooth sends this in the slot following the data (using its time

Table 7.6 Bluetooth baseband data rules

Type	Payload header [byte]	User payload [byte]	FEC	CRC	Symmetric max. rate [kbit/s]	Asymmetric forward	Max. rate [kbit/s] reverse
DM1	1	0–17	2/3	yes	108.8	108.8	108.8
DH1	1	0–27	no	yes	172.8	172.8	172.8
DM3	2	0–121	2/3	yes	258.1	387.2	54.4
DH3	2	0–183	no	yes	390.4	585.6	86.4
DM5	2	0–224	2/3	yes	286.7	477.8	36.3
DH5	2	0–339	no	yes	433.9	723.2	57.6
AUX1	1	0–29	no	no	185.6	185.6	185.6
HV1	na	10	1/3	no	64.0	na	na
HV2	na	20	2/3	no	64.0	na	na
HV3	na	30	no	no	64.0	na	na
DV	1 D	10+ (0–9) D	2/3 D	yes D	64.0+ 57.6 D	na	na

division duplex scheme). A simple alternating bit protocol with a single bit sequence number **SEQN** and acknowledgement number **ARQN** can be used. An 8-bit **header error check** (HEC) is used to protect the packet header. The packet header is also protected by a one-third rate forward error correction (FEC) code because it contains valuable link information and should survive bit errors. Therefore, the 18-bit header requires 54 bits in the packet.

- **Payload:** Up to 343 bytes payload can be transferred. The structure of the payload field depends on the type of link and is explained in the following sections.

7.5.4.1 Physical links

Bluetooth offers two different types of links, a synchronous connection-oriented link and an asynchronous connectionless link:

- **Synchronous connection-oriented link (SCO):** Classical telephone (voice) connections require symmetrical, circuit-switched, point-to-point connections. For this type of link, the master reserves two consecutive slots (forward and return slots) at fixed intervals. A master can support up to three simultaneous SCO links to the same slave or to different slaves. A slave supports up to two links from different masters or up to three links from the same master. Using an SCO link, three different types of single-slot packets can be used (Figure 7.47). Each SCO link carries voice at 64 kbit/s, and no **forward error correction** (FEC), 2/3 FEC, or 1/3 FEC can be selected. The 1/3 FEC is as strong as the FEC for the packet header and triples the amount of data. Depending on the error rate of the channel, different FEC schemes can be applied. FEC always causes an overhead, but avoids retransmission of data with a higher probability. However, voice data over an SCO is never retransmitted. Instead, a very robust voice-encoding scheme, **continuous variable slope delta (CVSD)**, is applied (Haartsen, 1998).

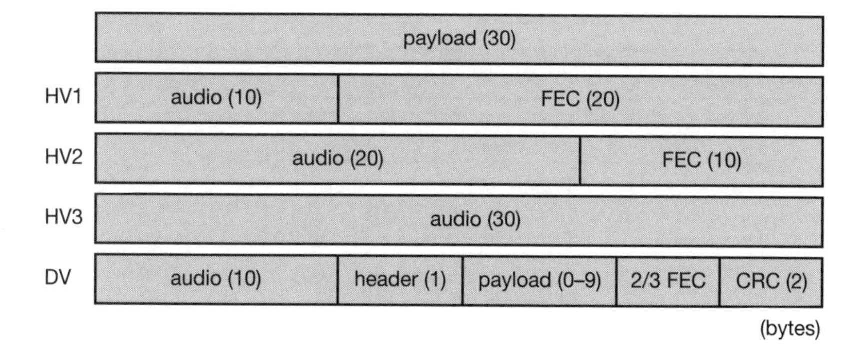

Figure 7.47
SCO payload types

- **Asynchronous connectionless link (ACL):** Typical data applications require symmetrical or asymmetrical (e.g., web traffic), packet-switched, point-to-multipoint transfer scenarios (including broadcast). Here the master uses a polling scheme. A slave may only answer if it has been addressed in the preceding slot. Only one ACL link can exist between a master and a slave. For ACLs carrying data, 1-slot, 3-slot or 5-slot packets can be used (Figure 7.48). Additionally, data can be protected using a 2/3 FEC scheme. This FEC protection helps in noisy environments with a high link error rate. However, the overhead introduced by FEC might be too high. Bluetooth therefore offers a fast automatic repeat request (ARQ) scheme for reliable transmission. The **payload header** (1 byte for 1-slot packets, 2 bytes for multi-slot packets) contains an identifier for a logical channel between L2CAP entities, a flow field for flow control at L2CAP level, and a length field indicating the number of bytes of data in the payload, excluding payload header and CRC. Payload is always CRC protected except for the AUX1 packet.

Table 7.6 lists Bluetooth's ACL and SCO packets. Additionally, control packets are available for polling slaves, hopping synchronization, or acknowledgement. The ACL types DM1 (data medium rate) and DH1 (data high rate) use a single slot and a one byte header. DM3 and DH3 use three slots, DM5 and DH5 use five

Figure 7.48
ACL payload types

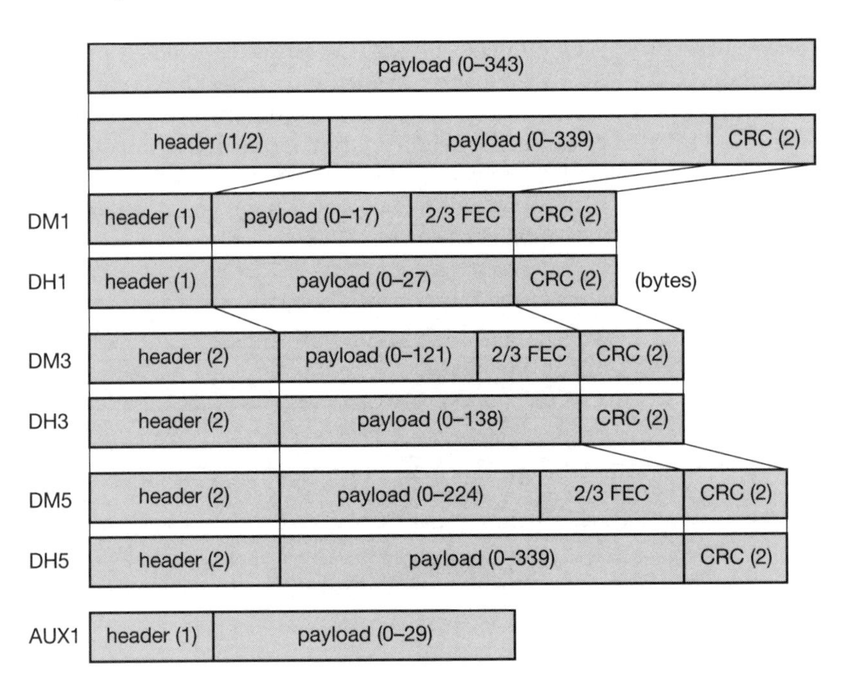

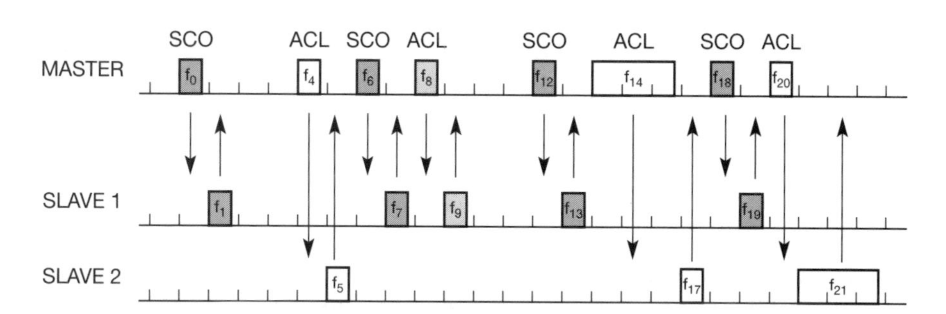

Figure 7.49 Example data transmission

slots. Medium rates are always FEC protected, the high rates rely on CRC only for error detection. The highest available data rates for Bluetooth devices are 433.9 kbit/s (symmetric) or 723.3/57.6 kbit/s (asymmetric). High quality voice (HV) packets always use a single slot but differ with respect to the amount of redundancy for FEC. DV (data and voice) is a combined packet where CRC, FEC, and payload header are valid for the data part only.

Figure 7.49 shows an example transmission between a master and two slaves. The master always uses the even frequency slots, the odd slots are for the slaves. In this example every sixth slot is used for an SCO link between the master and slave 1. The ACL links use single or multiple slots providing asymmetric bandwidth for connectionless packet transmission. This example again shows the hopping sequence which is independent of the transmission of packets.

The robustness of Bluetooth data transmissions is based on several technologies. FH-CDMA separates different piconets within a scatternet. FHSS mitigates interference from other devices operating in the 2.4 GHz ISM band. Additionally, FEC can be used to correct transmission errors. Bluetooth's 1/3 FEC simply sends three copies of each bit. The receiver then performs a majority decision: each received triple of bits is mapped into whichever bit is in majority. This simple scheme can correct all single bit errors in these triples. The 2/3 FEC encoding detects all double errors and can correct all single bit errors in a codeword.

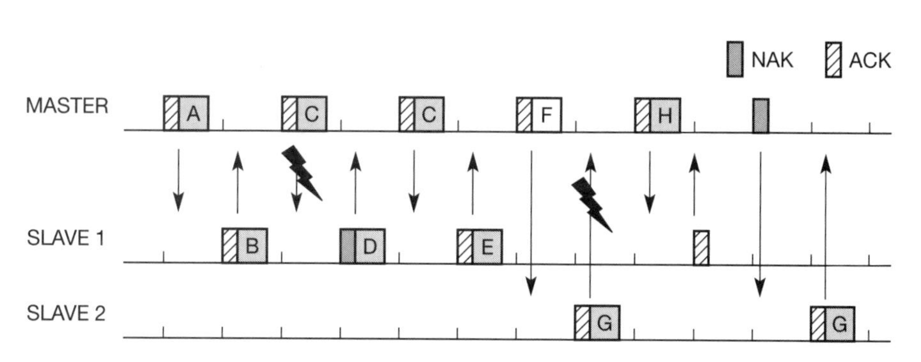

Figure 7.50 Error recovery

ACL links can additionally be protected using an ARQ scheme and a checksum. Each packet can be acknowledged in the slot following the packet. If a packet is lost, a sender can retransmit it immediately in the next slot after the negative acknowledgement, so it is called a fast ARQ scheme. This scheme hardly exhibits any overheads in environments with low error rates, as only packets which are lost or destroyed have to be retransmitted. Retransmission is triggered by a negative acknowledgement or a time-out.

7.5.5 Link manager protocol

The link manager protocol (LMP) manages various aspects of the radio link between a master and a slave and the current parameter setting of the devices. LMP enhances baseband functionality, but higher layers can still directly access the baseband. The following groups of functions are covered by the LMP:

- **Authentication, pairing, and encryption:** Although basic authentication is handled in the baseband, LMP has to control the exchange of random numbers and signed responses. The pairing service is needed to establish an initial trust relationship between two devices that have never communicated before. The result of pairing is a link key. This may be changed, accepted or rejected. LMP is not directly involved in the encryption process, but sets the encryption mode (no encryption, point-to-point, or broadcast), key size, and random speed. Section 7.5.7 gives an overview of Bluetooth's security mechanisms.
- **Synchronization:** Precise synchronization is of major importance within a Bluetooth network. The clock offset is updated each time a packet is received from the master. Additionally, special synchronization packets can be received. Devices can also exchange timing information related to the time differences (slot boundaries) between two adjacent piconets.
- **Capability negotiation:** Not only the version of the LMP can be exchanged but also information about the supported features. Not all Bluetooth devices will support all features that are described in the standard, so devices have to agree the usage of, e.g., multi-slot packets, encryption, SCO links, voice encoding, park/sniff/hold mode (explained below), HV2/HV3 packets etc.
- **Quality of service negotiation:** Different parameters control the QoS of a Bluetooth device at these lower layers. The poll interval, i.e., the maximum time between transmissions from a master to a particular slave, controls the latency and transfer capacity. Depending on the quality of the channel, DM or DH packets may be used (i.e., 2/3 FEC protection or no protection). The number of repetitions for broadcast packets can be controlled. A master can also limit the number of slots available for slaves' answers to increase its own bandwidth.
- **Power control:** A Bluetooth device can measure the received signal strength. Depending on this signal level the device can direct the sender of the measured signal to increase or decrease its transmit power.

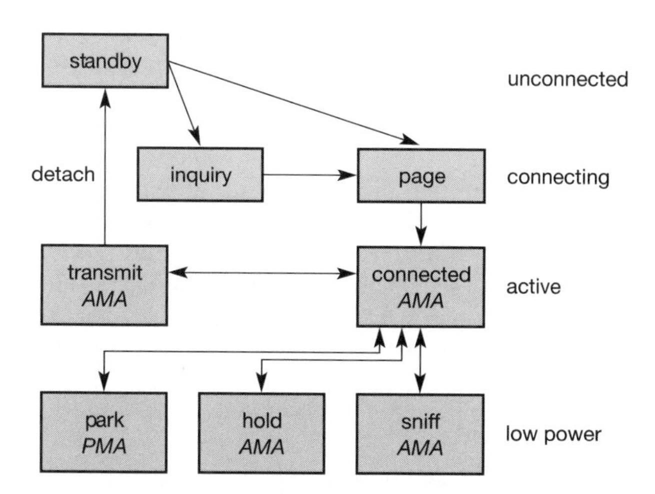

Figure 7.51
Major baseband states of a Bluetooth device

- **Link supervision:** LMP has to control the activity of a link, it may set up new SCO links, or it may declare the failure of a link.
- **State and transmission mode change:** Devices might switch the master/slave role, detach themselves from a connection, or change the operating mode. The available modes will be explained together with Figure 7.51.

With transmission power of up to 100 mW, Bluetooth devices can have a range of up to 100 m. Having this power and relying on batteries, a Bluetooth device cannot be in an active transmit mode all the time. Bluetooth defines several low-power states for a device. Figure 7.51 shows the major states of a Bluetooth device and typical transitions.

Every device, which is currently not participating in a piconet (and not switched off), is in **standby** mode. This is a low-power mode where only the native clock is running. The next step towards the **inquiry** mode can happen in two different ways. Either a device wants to establish a piconet or a device just wants to listen to see if something is going on.

- A device wants to establish a piconet: A user of the device wants to scan for other devices in the radio range. The device starts the inquiry procedure by sending an inquiry access code (IAC) that is common to all Bluetooth devices. The IAC is broadcast over 32 so-called wake-up carriers in turn.
- Devices in standby that listen periodically: Devices in standby may enter the inquiry mode periodically to search for IAC messages on the wake-up carriers. As soon as a device detects an inquiry it returns a packet containing its device address and timing information required by the master to initiate a connection. From that moment on, the device acts as slave.

If the inquiry was successful, a device enters the page mode. The inquiry phase is not coordinated; inquiry messages and answers to these messages may collide, so it may take a while before the inquiry is successful. After a while (typically seconds but sometimes up to a minute) a Bluetooth device sees all the devices in its radio range.

During the **page** state two different roles are defined. After finding all required devices the master is able to set up connections to each device, i.e., setting up a piconet. Depending on the device addresses received the master calculates special hopping sequences to contact each device individually. The slaves answer and synchronize with the master's clock, i.e., start with the hopping sequence defined by the master. The master may continue to page more devices that will be added to the piconet. As soon as a device synchronizes to the hopping pattern of the piconet it also enters the connection state.

The connection state comprises the active state and the low power states park, sniff, and hold. In the **active** state the slave participates in the piconet by listening, transmitting, and receiving. ACL and SCO links can be used. A master periodically synchronizes with all slaves. All devices being active must have the 3-bit **active member address** (AMA). Within the active state devices either transmit data or are simply connected. A device can enter standby again, via a detach procedure

To save battery power, a Bluetooth device can go into one of three low power states:

- **Sniff state:** The sniff state has the highest power consumption of the low power states. Here, the device listens to the piconet at a reduced rate (not on every other slot as is the case in the active state). The interval for listening into the medium can be programed and is application dependent. The master designates a reduced number of slots for transmission to slaves in sniff state. However, the device keeps its AMA.
- **Hold state:** The device does not release its AMA but stops ACL transmission. A slave may still exchange SCO packets. If there is no activity in the piconet, the slave may either reduce power consumption or participate in another piconet.
- **Park state:** In this state the device has the lowest duty cycle and the lowest power consumption. The device releases its AMA and receives a parked member address (PMA). The device is still a member of the piconet, but gives room for another device to become active (AMA is only 3 bit, PMA 8 bit). Parked devices are still FH synchronized and wake up at certain beacon intervals for re-synchronization. All PDUs sent to parked slaves are broadcast.

Table 7.7 Example power consumption (CSR, 2002)

Operating mode	Average current [mA]
SCO, HV1	53
SCO, HV3, 1 s interval sniff mode	26
ACL, 723.2 kbit/s	53
ACL, 115.2 kbit/s	15.5
ACL, 38.4 kbit/s, 40 ms interval sniff mode	4
ACL, 38.4 kbit/s, 1.28 s interval sniff mode	0.5
Park mode, 1.28 s beacon interval	0.6
Standby (no RF activity)	0.047

The effect of the low power states is shown in Table 7.7. This table shows the typical average power consumption of a Bluetooth device (BlueCore2, CSR, 2002). It is obvious that higher data rates also require more transmission power. The intervals in sniff mode also influence power consumption. Typical IEEE 802.11b products have an average current in the order of 200 mA while receiving, 300 mA while sending, and 20 mA in standby.

7.5.6 L2CAP

The **logical link control and adaptation protocol (L2CAP)** is a data link control protocol on top of the baseband layer offering logical channels between Bluetooth devices with QoS properties. L2CAP is available for ACLs only. Audio applications using SCOs have to use the baseband layer directly (see Figure 7.44). L2CAP provides three different types of logical channels that are transported via the ACL between master and slave:

- **Connectionless:** These unidirectional channels are typically used for broadcasts from a master to its slave(s).
- **Connection-oriented:** Each channel of this type is bi-directional and supports QoS flow specifications for each direction. These flow specs follow RFC 1363 (Partridge, 1992) and define average/peak data rate, maximum burst size, latency, and jitter.
- **Signaling:** This third type of logical channel is used to exchanging signaling messages between L2CAP entities.

Each channel can be identified by its **channel identifier (CID)**. Signaling channels always use a CID value of 1, a CID value of 2 is reserved for connectionless channels. For connection-oriented channels a unique CID (>= 64) is dynamically assigned at each end of the channel to identify the connection

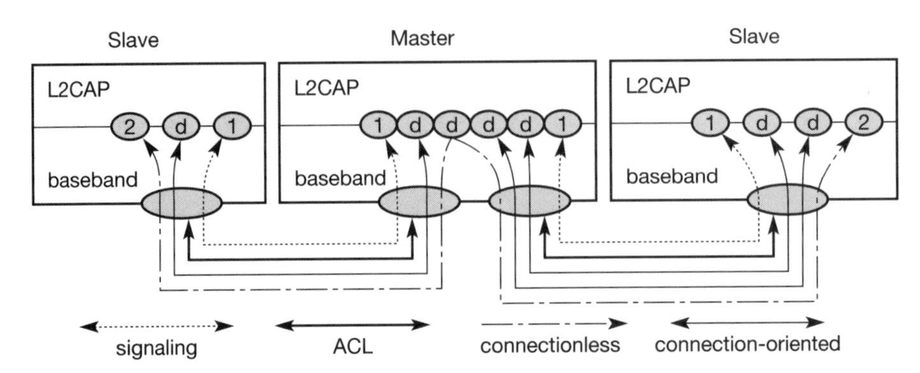

Figure 7.52 Logical channels between devices

(CIDs 3 to 63 are reserved). Figure 7.52 gives an example for logical channels using the ACL link between master and slave. The master has a bi-directional signaling channel to each slave. The CID at each end is 1. Additionally, the master maintains a connectionless, unidirectional channel to both slaves. The CID at the slaves is 2, while the CID at the beginning of the connectionless channel is dynamically assigned. L2CAP provides mechanisms to add slaves to, and remove slaves from, such a multicast group. The master has one connection oriented channel to the left slave and two to the right slave. All CIDs for these channels are dynamically assigned (between 64 and 65535).

Figure 7.53 shows the three packet types belonging to the three logical channel types. The **length** field indicates the length of the payload (plus PSM for connectionless PDUs). The **CID** has the multiplexing/demultiplexing function as explained above. For connectionless PDUs a **protocol/service multiplexor (PSM)** field is needed to identify the higher layer recipient for the payload. For connection-oriented PDUs the CID already fulfills this function. Several PSM values have been defined, e.g., 1 (SDP), 3 (RFCOMM), 5 (TCS-BIN). Values above 4096 can be assigned dynamically. The payload of the signaling PDU contains one or more **commands**. Each command has its own **code** (e.g., for command reject, connection request, disconnection response etc.) and an **ID** that matches a request with its reply. The **length** field indicates the length of the **data** field for this command.

Besides protocol multiplexing, flow specification, and group management, the L2CAP layer also provides segmentation and reassembly functions. Depending on the baseband capabilities, large packets have to be chopped into smaller segments. DH5 links, for example, can carry a maximum of 339 bytes while the L2CAP layer accepts up to 64 kbyte.

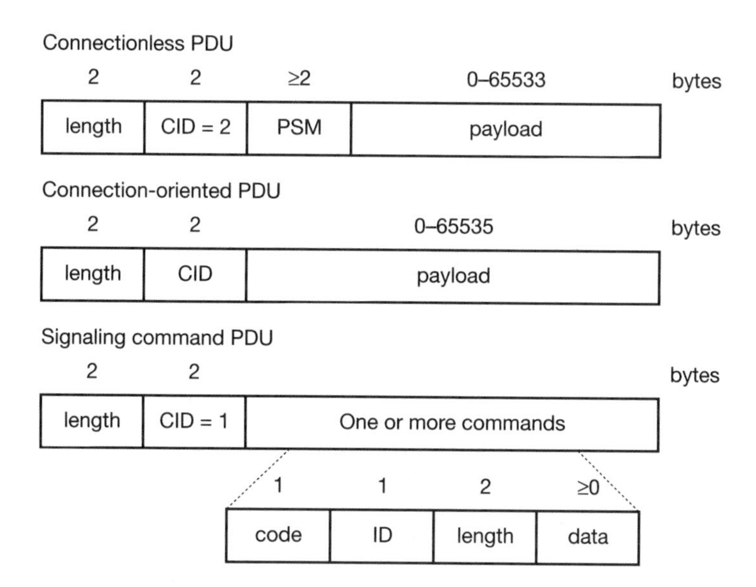

Figure 7.53 L2CAP packet formats

7.5.7 Security

A radio interface is by nature easy to access. Bluetooth devices can transmit private data, e.g., schedules between a PDA and a mobile phone. A user clearly does not want another person to eavesdrop the data transfer. Just imagine a scenario where two Bluetooth enabled PDAs in suitcases 'meet' on the conveyor belt of an airport exchanging personal information! Bluetooth offers mechanisms for authentication and encryption on the MAC layer, which must be implemented in the same way within each device.

The main security features offered by Bluetooth include a challenge-response routine for authentication, a stream cipher for encryption, and a session key generation. Each connection may require a one-way, two-way, or no authentication using the challenge-response routine. All these schemes have to be implemented in silicon, and higher layers should offer stronger encryption if needed. The security features included in Bluetooth only help to set up a local domain of trust between devices.

The security algorithms use the public identity of a device, a secret private user key, and an internally generated random key as input parameters. For each transaction, a new random number is generated on the Bluetooth chip. Key management is left to higher layer software.

Figure 7.54 shows several steps in the security architecture of Bluetooth. The illustration is simplified and the interested reader is referred to Bluetooth (2001a) for further details. The first step, called **pairing**, is necessary if two Bluetooth devices have never met before. To set up trust between the two devices a user can enter a secret PIN into both devices. This PIN can have a length of up to 16 byte. Unfortunately, most devices limit the length to four digits or, even worse, program

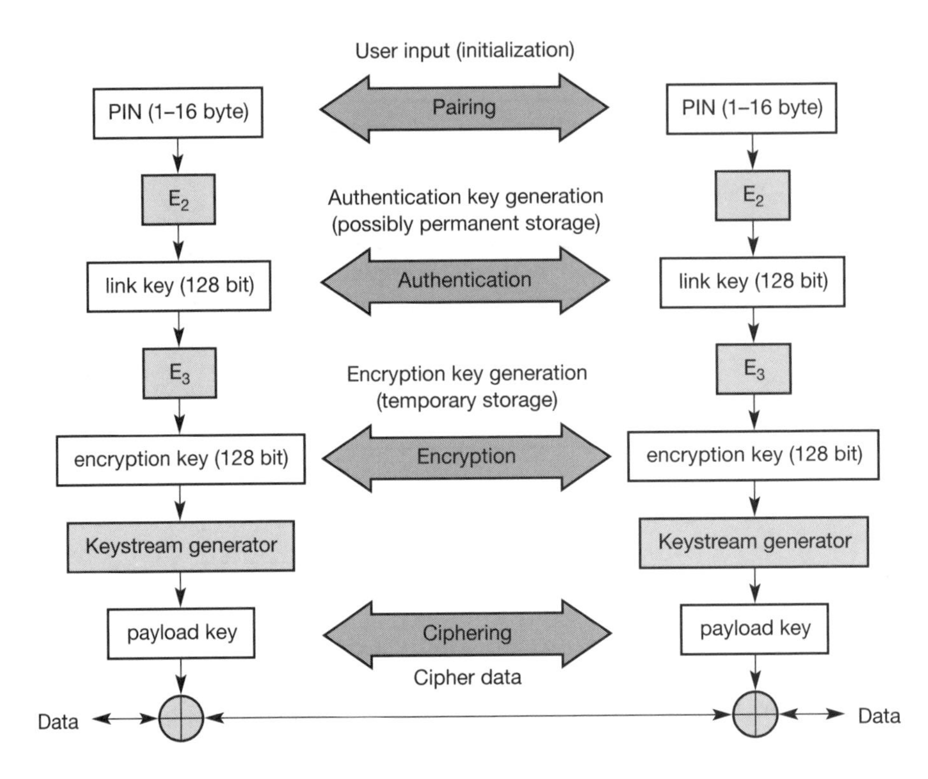

Figure 7.54 Bluetooth security components and protocols

the devices with the fixed PIN '0000' rendering the whole security concept of Bluetooth questionable at least. Based on the PIN, the device address, and random numbers, several keys can be computed which can be used as link key for **authentication**. Link keys are typically stored in a persistent storage. The authentication is a challenge-response process based on the link key, a random number generated by a verifier (the device that requests authentication), and the device address of the claimant (the device that is authenticated).

Based on the link key, values generated during the authentication, and again a random number an encryption key is generated during the **encryption** stage of the security architecture. This key has a maximum size of 128 bits and can be individually generated for each transmission. Based on the encryption key, the device address and the current clock a payload key is generated for ciphering user data. The payload key is a stream of pseudo-random bits. The **ciphering** process is a simple XOR of the user data and the payload key.

Compared to WEP in 802.11, Bluetooth offers a lot more security. However, Bluetooth, too, has some weaknesses when it comes to real implementations. The PINs are quite often fixed. Some of the keys are permanently stored on the devices and the quality of the random number generators has not been specified. If Bluetooth devices are switched on they can be detected unless they operate in the non-discoverable mode (no answers to inquiry requests). Either a user can use all services as intended by the Bluetooth system, or the devices are hidden to protect privacy. Either roaming profiles can be established, or devices are hidden and, thus many services will not work. If a lot of people carry Bluetooth devices (mobile phones, PDAs etc.) this could give, e.g., department stores, a lot of information regarding consumer behavior.

7.5.8 SDP

Bluetooth devices should work together with other devices in unknown environments in an ad-hoc fashion. It is essential to know what devices, or more specifically what services, are available in radio proximity. To find new services, Bluetooth defined the **service discovery protocol (SDP)**. SDP defines only the discovery of services, not their usage. Discovered services can be cached and gradual discovery is possible. Devices that want to offer a service have to instal an SDP server. For all other devices an SDP client is sufficient.

All the information an SDP server has about a service is contained in a **service record**. This consists of a list of service attributes and is identified by a 32-bit service record handle. SDP does not inform clients of any added or removed services. There is no service access control or service brokerage. A **service attribute** consists of an attribute ID and an attribute value. The 16-bit **attribute ID** distinguishes each service attribute from other service attributes within a service record. The attribute ID also identifies the semantics of the associated attribute value. The **attribute value** can be an integer, a UUID (universally unique identifier), a string, a Boolean, a URL (uniform resource locator) etc. Table 7.8 gives some example attributes. The service handle as well as the ID list must be present. The ID list contains the UUIDs of the service classes.in increasing generality (from the specific color postscript printer to

Attribute name	Attribute ID	Attribute value type	Example
ServiceRecordHandle	0000	32-bit unsigned integer	1f3e4723
ServiceClassIDList	0001	Data element sequence (UUIDs)	ColorPostscriptPrinterService ClassID, PostscriptPrinterService ClassID, PrinterServiceClassID
ProtocolDescriptorList	0004	Data element sequence	((L2CAP, PSM=RFCOMM), (RFCOMM, CN=2), (PPP), (IP), (TCP), (IPP))
DocumentationURL	000A	URL	www.xy.zz/print/srvs.html
IconURL	000C	URL	www.xy.zz/print/ico.png
ServiceName	0100	String	Color Printer

Table 7.8 Example attributes for an SDP service record

printers in general). The protocol descriptor list comprises the protocols needed to access this service. Additionally, the URLs for service documentation, an icon for the service and a service name which can be displayed together with the icon are stored in the example service record.

7.5.9 Profiles

Although Bluetooth started as a very simple architecture for spontaneous ad-hoc communication, many different protocols, components, extensions, and mechanisms have been developed over the last years. Application designers and vendors can implement similar, or even identical, services in many different ways using different components and protocols from the Bluetooth core standard. To provide compatibility among the devices offering the same services, Bluetooth specified many profiles in addition to the core protocols. Without the profiles too many parameters in Bluetooth would make interoperation between devices from different manufacturers almost impossible.

Profiles represent default solutions for a certain usage model. They use a selection of protocols and parameter set to form a basis for interoperability. Protocols can be seen as horizontal layers while profiles are vertical slices (as illustrated in Figure 7.55). The following **basic profiles** have been specified: generic access, service discovery, cordless telephony, intercom, serial port, headset, dial-up networking, fax, LAN access, generic object exchange, object push, file transfer, and synchronization. **Additional profiles** are: advanced audio distribution, PAN, audio video remote control, basic printing, basic imaging, extended service discovery, generic audio video distribution, hands-free, and hardcopy cable replacement. Each profile selects a set of protocols. For example, the serial port profile needs RFCOMM, SDP, LMP, L2CAP. Baseband and radio are always required. The profile further defines all interoperability requirements, such as RS232 control signals for RFCOMM or configuration options for L2CAP (QoS, max. transmission unit).

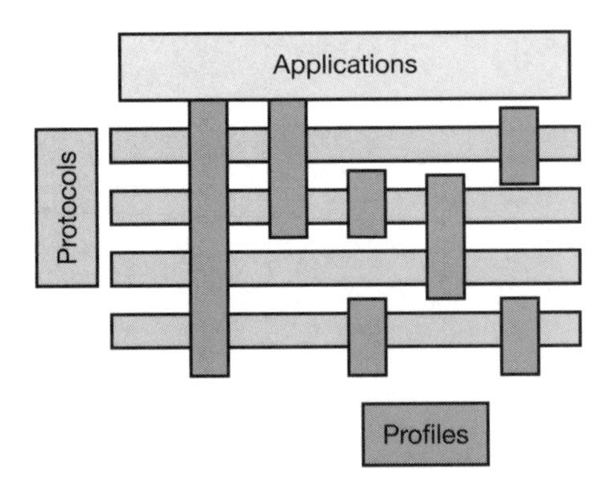

Figure 7.55
Bluetooth profiles

7.5.10 IEEE 802.15

In 1999 the IEEE established a working group for wireless personal area networks (WPAN) with similar goals to Bluetooth. The working group was divided into several subgroups focusing on different aspects of WPANs (IEEE, 2002c). The following gives a quick overview and presents the standard for low-rate WPANs, 802.15.4, in some more detail:

- **IEEE 802.15.1:** This group standardizes the lower layers of **Bluetooth** together with the Bluetooth consortium. IEEE LANs focus only on the physical and data link layer, while the Bluetooth standard also comprises higher layers, application profiles, service description etc. as explained above.
- **IEEE 802.15.2:** The **coexistence** of wireless personal area networks (WPAN) and wireless local area networks (WLAN) is the focus of this group. One task is to quantify mutual interference and to develop algorithms and protocols for coexistence. Without additional mechanisms, Bluetooth/802.15.1 may act like a rogue member of an IEEE 802.11 network. Bluetooth is not aware of gaps, inter-frame spacing, frame structures etc. Figure 7.56 illustrates the problem. As explained in section 7.3, WLANs following the IEEE 802.11b standard may use three non-overlapping channels that are chosen during installation of the access points. Bluetooth/802.15.1 networks use a frequency hopping pattern to separate different piconets – 79 channels can be used. Without additional mechanisms, the hopping pattern of Bluetooth is independent of 802.11b's channel selection. Both systems work in the 2.4 GHz ISM band and might interfere with each other. Figure 7.56 shows two hopping sequences of two piconets interfering with several data packets, acknowledgements, and inter-frame spacings of 802.11b. The real effects of the interference range from 'almost no effect' to 'complete breakdown of the WLAN'. Publications on this issue differ depending on the test scenario, traffic load, signal power, propagation conditions etc. (Lansford, 2001). However, it seems that Bluetooth with its FHSS scheme is more robust than 802.11b with CSMA/CA (Pahlavan, 2002). To overcome the interference problems between 802.11b and 802.15.1, however severe they might be, the 802.15.2 working group proposes **adaptive frequency hopping**. This

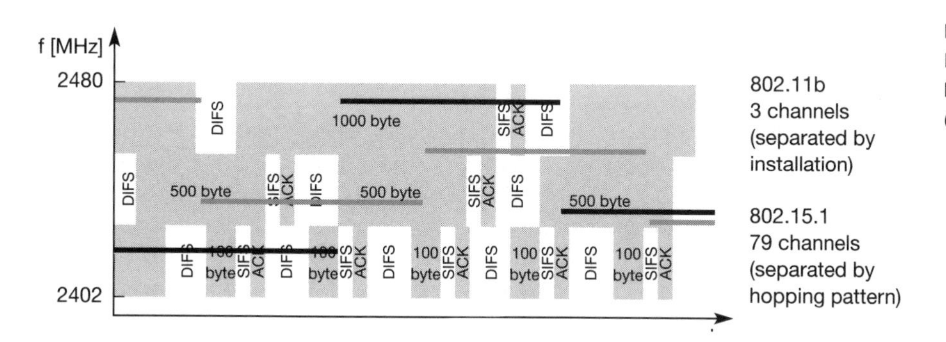

Figure 7.56
Possible interference between 802.15.1 (Bluetooth) and 802.11b

coexistence mechanism is non-collaborative in the sense that Bluetooth devices do not have to interact with the WLAN. However, the WPAN devices can check for the occupied channels and exclude them from their list of channels used for hopping. This mechanism avoids hopping into a channel occupied by 802.11b, but still offers enough channels for FHSS. The lower number of FHSS channels increases the interference among the WPANs due to a higher probability of collisions. However, if not too many piconets overlap this effect will be negligible. This type of interference in the crowded 2.4 GHz band is a strong argument for 5 GHz WLANs.

- **IEEE 802.15.3:** A **high-rate** study group looks for a standard providing data rates of 20 Mbit/s or greater while still working with low-power at low-cost. The standard should support isochronous data delivery, ad-hoc peer-to-peer networking, security features, and should meet the demanding requirements of portable consumer imaging and multi-media applications.
- **IEEE 802.15.4:** The fourth working group goes in the opposite direction for data rates. This group standardizes **low-rate wireless personal area networks (LR-WPAN)**, which are explained in the following section in more detail. The ZigBee consortium tries to standardize the higher layers of 802.15.4 similar to the activities of the Bluetooth consortium for 802.15.1 (ZigBee, 2002).

7.5.10.1 IEEE 802.15.4 – Low-rate WPANs

The reason for having low data rates is the focus of the working group on extremely low power consumption enabling multi-year battery life (Callaway, 2002). Compared to 802.11 or Bluetooth, the new system should have a much lower complexity making it suitable for low-cost wireless communication (remember that Bluetooth started with similar goals with respect to the idea of cable replacement). Example **applications** include industrial control and monitoring, smart badges, interconnection of environmental sensors, interconnection of peripherals (also an envisaged application area for Bluetooth!), remote controls etc. The new standard should offer data rates between 20 and 250 kbit/s as maximum and latencies down to 15 ms. This is enough for many home automation and consumer electronics applications.

IEEE 802.15.4 offers two different PHY options using DSSS. The **868/915 MHz PHY** operates in Europe at 868.0–868.6 MHz and in the US at 902–928 MHz. At 868 MHz one channel is available offering a data rate of 20 kbit/s. At 915 MHz 10 channels with 40 kbit/s per channel are available (in Europe GSM uses these frequencies). The advantages of the lower frequencies are better propagation conditions. However, there is also interference in these bands as many analog transmission systems use them. The **2.4 GHz PHY** operates at 2.4–2.4835 GHz and offers 16 channels with 250 kbit/s per channel. This PHY offers worldwide operation but suffers from interference in the 2.4 GHz ISM band and higher propagation loss. Typical devices with 1 mW output power are expected to cover a 10–20 m range. All PHY PDUs start with a 32 bit preamble for synchronization. After a start-of-packet delimiter, the PHY header indicates the length of the payload (maximum 127 bytes).

Compared to Bluetooth the **MAC layer** of 802.15.4 is much simpler. For example, no synchronous voice links are supported. MAC frames start with a 2-byte frame control field, which specifies how the rest of the frame looks and what it contains. The following 1-byte sequence number is needed to match acknowledgements with a previous data transmission. The variable address field (0–20 bytes) may contain source and/or destination addresses in various formats. The payload is variable in length; however, the whole MAC frame may not exceed 127 bytes in length. A 16-bit FCS protects the frame. Four different MAC frames have been defined: beacon, data, acknowledgement, and MAC command.

Optionally, this LR-WPAN offers a **superframe mode**. In this mode, a PAN coordinator transmits beacons in predetermined intervals (15 ms–245 s). With the help of beacons, the medium access scheme can have a period when contention is possible and a period which is contention free. Furthermore, with beacons a slotted **CSMA/CA** is available. Without beacons standard CSMA/CA is used for medium access. Acknowledgement frames confirming a previous transmission do not use the CSMA mechanism. These frames are sent immediately following the previous packet.

IEEE 802.15.4 specifies three levels of **security**: no security, access control lists, and symmetric encryption using AES-128. Key distribution is not specified further. Security is a must for home automation or industry control applications. Up to now, the success of this standard is unclear as it is squeezed between Bluetooth, which also aims at cable replacement, and enhanced RFIDs/RF controllers. These will be explained in the summary in more detail.

7.6 Summary

This chapter has introduced three different technologies designed for WLANs (or WPAN in the case of Bluetooth, but there is no real border between WLAN and WPAN). The basic goals of all three LAN types are the provision of much higher flexibility for nodes within a network. All WLANs suffer from limitations of the air interface and higher complexity compared to their wired counterparts, but allow for a new degree of freedom for their users within rooms, buildings, or production halls. WLANs are already in widespread use in, e.g., warehouses, classrooms, meeting rooms and hospitals.

However, the three technologies also differ in some respects. Whereas in the beginning of WLANs several proprietary products existed, nowadays they typically offer support for IEEE 802.11b (with .11a and .11g upcoming). Although the IEEE 802.11b standard is much simpler compared to others, it still leaves room for different implementations. The wireless Ethernet compatibility alliance (WECA, 2002) certifies interoperability of 802.11 products (WiFi, wireless fidelity). Today, millions of wireless adapters follow this standard and many laptops come with 802.11b WLAN adapters already integrated. One reason for this is that today the 2.4 GHz band is available worldwide (with some minor differences). The big restrictions of the past in some countries are gone.

For HiperLAN2, the history is different. Here, a standardization body (ETSI) developed a completely new standard, but no products are available yet. HiperLAN2 comprises many interesting features, particularly on the MAC layer, compared to 802.11a. Main features are QoS support, integrated security, and convergence sub-layers to different networks, e.g., Firewire, which is used for audio/video connection. However, IEEE 802.11a is available and installed in many places (using the PHY features of HiperLAN2). Up to now it is not clear if HiperLAN2 will be a success or follow HIPERLAN 1 which never made it to the market although its technical parameters were superior compared to IEEE 802.11. Anastasi (1998) gives a good overview of the capabilities of the two MAC schemes used in IEEE 802.11 and HIPERLAN 1 respectively, and investigates whether those access schemes can be used for QoS provision as it was thought of for, e.g., wireless ATM.

For Bluetooth, the situation is completely different. Here several companies founded a consortium and set up a de facto industry standard (version 1.1 in 2001). Then IEEE, as a standardization body followed with the IEEE 802.15.1 standard in 2002. Bluetooth is already available in many products (PDAs, video cameras, digital still cameras, laptops etc.) and, is clearly the most widespread WPAN technology today. The primary goal of Bluetooth was not a complex standard covering many aspects of wireless networking, but a quick and very cheap solution enabling ad-hoc personal communication within a short range in the license-free 2.4 GHz band. Today the standard covers several thousand pages and defines many usage scenarios, services definition, protocols etc. Most devices implement a basic set of functionality as the complexity of all features (e.g., support for several scatternets, jumping back and forth between piconets) is too much for embedded devices with a small footprint.

Table 7.9 gives a (simplified) comparison of IEEE 802.11b, .11a, HiperLAN2, and Bluetooth. The main differences between the 802.11a/b standards and the other two are the scope of the standardization and the initial content. Not only do both IEEE standards share the same MAC layer, they also describe the raw data transfer without the elaborate security or authentication mechanisms or adaptation layers/profiles necessary for interoperation with other networks or applications. Both standards assume an Ethernet backbone and, typically, best effort IP running on top of the MAC layer. This is perfect for most office applications, indeed for most of today's Internet applications. Special features like security or frequency selection are add-ons to the standards.

HiperLAN2 and Bluetooth want to cover almost all aspects related to wireless communication: physical layer, medium access, many different services, adaptation layers to different backbones (HiperLAN2) or profiles for different applications (Bluetooth).

This chapter left out some standards and approaches that could be mentioned in the context of WLANs. One example is **HomeRF**. This is another WLAN standard operating at 2.4 GHz. HomeRF uses a FHSS scheme with 50 hops per second. Transmission rates have been standardized up to 10 Mbit/s; higher rates are planned. The MAC layer of HomeRF combines 802.11 and DECT functionality: a TDMA/CSMA frame offers TDMA for isochronous (voice)

Table 7.9 Comparison of wireless networks presented in chapter 7

Criterion	IEEE 802.11b	IEEE 802.11a	HiperLAN2	Bluetooth
Frequency	2.4 GHz	5 GHz	5 GHz	2.4 GHz
Max. trans. rate	11 Mbit/s	54 Mbit/s	54 Mbit/s	< 1 Mbit/s
User throughput	6 Mbit/s	34 Mbit/s	34 Mbit/s	< 1 Mbit/s
Medium access	CSMA/CA	CSMA/CA	AP centralized	Master centralized
Frequency management	None	802.11h	DFS	FHSS
Authentication	None/802.1x	None/802.1x	X.509	Yes
Encryption	WEP, 802.11i	WEP, 802.11i	DES, 3DES	Yes
QoS support	Optional (PCF)	Optional (PCF)	ATM, 802.1p, RSVP	Flow spec, isochronous
Connectivity	Connectionless	Connectionless	Connection-oriented	Connectionless + connection-oriented
Available channels	3	12 (US)	19 (EU)	Soft – increasing interference
Typ. transmit power	100 mW	0.05/0.25/1W, TPC with 802.11h	0.2/1W, TPC	1/2.5/100 mW
Error control	ARQ	ARQ, FEC (PHY)	ARQ, FEC (PHY)	ARQ, FEC (MAC)

transmission and CSMA for asynchronous (data) transmission. Different QoS schemes are supported. Compared to 802.11 stronger security features are integrated from the beginning. Host/client and peer/peer networking are possible. However, due to the tremendous success of IEEE 802.11b, HomeRF could not succeed and development was stopped in 2002 (although HomeRF products were available).

Some technologies might influence WLANs/WPANs in the future:

- **Wireless sensor networks:** The technology required for sensor networks is located somewhere between 802.15.1 or .4 technology and the RFIDs (presented in the following paragraph). Sensor networks consist of many (thousands or more) nodes that are densely deployed, prone to failures, have very limited computing capabilities, and change their topology frequently. Sensor networks can be seen as an extreme form of ad-hoc networking with very low-power devices. Applications comprise those of 802.15.4

(environmental sensoring) but due to the number of devices whole areas could be 'covered' with sensors, computer, and networking power. Main research topics are: routing of data within the sensor network, management of the nodes, fault tolerance/reliability, low-power design, and medium access control. Akyildiz (2002) gives an excellent survey of this topic.

- **Radio frequency identification (RFID):** RF controllers have been well known for many years. They offer transmission rates of up to 115 kbit/s (wireless extension of a serial interface) and operate on many different ISM bands (depending on national regulations, e.g., 27, 315, 418, 426, 433, 868, 915 MHz). Applications include garage door openers, wireless mice/keyboards, car locks etc. RF controllers, typically, do not have a MAC layer, but simply act as modem. Collisions have to be detected on higher layers. The first RFIDs emerged during the 1980s (RFID, 2002). In the beginning, these very cheap tags were used for asset tracking only. As soon as a product with an RFID tag passed a reader, the product was registered. Today, RFIDs are available in dozens of different styles with very different properties. RFIDs can respond to a radio signal and transmit their tag. They can store additional data, employ collision avoidance schemes, and comprise smart-card capabilities with simple processing power. While RFIDs are not communication devices, the borders are blurring as more and more computing power is available on small embedded systems and the communication industry is looking for low power systems, such as 802.15.4.
- **Ultra wideband technology (UWB):** This technology goes one step further related to spread spectrum used in WLANs as it transmits digital data over a wide spectrum of frequency bands with very low power (UWB, 2002). Typically, the occupied spectrum is at least 25 per cent of the center frequency (e.g., 500 MHz for a 2 GHz system). Instead of sending a sine wave, UWB broadcasts a very short digital pulse (less than 1 ns) that is timed very precisely. Sender and receiver must be synchronized with very high accuracy. If the sender exactly knows when a pulse should arrive, multi-path propagation is no longer an issue (e.g., only the strongest signal will be detected within a very short time-slot). Besides radar applications (where UWB comes from), it can be used for LANs transmitting very high data rates over short distances.

Each standard presented in this chapter has its pros and cons. If the focus is on battery life, then Bluetooth is the choice as the power consumption of .11b, and particularly .11a and HiperLAN2 is too high. If isochronous traffic, QoS, and high data rates have to be supported, then HiperLAN2 is the choice. If the solution should be simple and fit into an office environment, then .11b/g or .11a are possible solutions. If interference is a topic, then .11a is better than .11b. If large cells are required, then .11b is better then .11a due to the lower propagation loss (.11g is even better). IEEE starts even more working groups, e.g., 802.20, the 'Mobile Broadband Wireless Access (MBWA)' group. This group will continue the work that was previously conducted by the 802.16 MBWA study group for cellular wireless data services supporting full vehicular mobility.

The typical mobile device of tomorrow will comprise several technologies with the ability of connecting to different networks, e.g., a GSM follow-on (such as UMTS) for wide area communication, possibly a satellite antenna, and different WLAN adapters (e.g., IEEE 802.11b/g and Bluetooth). Depending on cost, application, and location, the device will automatically choose the optimal communication device and network. Roaming between those different networks is still difficult (in particular, ensuring a certain quality of service and security), but first solutions in this direction are currently presented and integrated 802.11b/GSM adapters are available where the WLAN uses GSM functionality for providing secure access.

7.7 Review exercises

1. How is mobility restricted using WLANs? What additional elements are needed for roaming between networks, how and where can WLANs support roaming? In your answer, think of the capabilities of layer 2 where WLANs reside.
2. What are the basic differences between wireless WANs and WLANs, and what are the common features? Consider mode of operation, administration, frequencies, capabilities of nodes, services, national/international regulations.
3. With a focus on security, what are the problems of WLANs? What level of security can WLANs provide, what is needed additionally and how far do the standards go?
4. Compare IEEE 802.11, HiperLAN2, and Bluetooth with regard to their ad-hoc capabilities. Where is the focus of these technologies?
5. If Bluetooth is a commercial success, what are remaining reasons for the use of infra red transmission for WLANs?
6. Why is the PHY layer in IEEE 802.11 subdivided? What about HiperLAN2 and Bluetooth?
7. Compare the power saving mechanisms in all three LANs introduced in this chapter. What are the negative effects of the power saving mechanisms, what are the trade-offs between power consumption and transmission QoS?
8. Compare the QoS offered in all three LANs in ad-hoc mode. What advantages does an additional infrastructure offer? How is QoS provided in Bluetooth? Can one of the LAN technologies offer hard QoS (i.e., not only statistical guarantees regarding a QoS parameter)?
9. How do IEEE 802.11, HiperLAN2 and Bluetooth, respectively, solve the hidden terminal problem?
10. How are fairness problems regarding channel access solved in IEEE 802.11, HiperLAN2, and Bluetooth respectively? How is the waiting time of a packet ready to transmit reflected?

11 What different solutions do all three networks offer regarding an increased reliability of data transfer?

12 In what situations can collisions occur in all three networks? Distinguish between collisions on PHY and MAC layer. How do the three wireless networks try to solve the collisions or minimise the probability of collisions?

13 Compare the overhead introduced by the three medium access schemes and the resulting performance at zero load, light load, high load of the medium. How does the number of collisions increase with the number of stations trying to access the medium, and how do the three networks try to solve the problems? What is the overall scalability of the schemes in number of nodes?

14 How is roaming on layer 2 achieved, and how are changes in topology reflected? What are the differences between infrastructure based and ad-hoc networks regarding roaming?

15 What are advantages and problems of forwarding mechanisms in Bluetooth networks regarding security, power saving, and network stability?

16 Name reasons for the development of wireless ATM. What is one of the main differences to Internet technologies from this point of view? Why did WATM not succeed as stand-alone technology, what parts of WATM succeeded?

7.8 References

Acampora, A. (1996) 'Wireless ATM: a perspective on issues and prospects,' *IEEE Personal Communications*, 3(4).

Akyildiz, I., Su, W., Sankarasubramaniam, Y., Cayirci, E. (2002) 'Wireless sensor networks: a survey,' *Computer Networks*, Elsevier Science, 38(2002).

Anastasi, G., Lenzini, L., Mingozzi, E., Hettich, A., Krämling, A. (1998) 'MAC protocols for wideband wireless local access: evolution toward wireless ATM,' *IEEE Personal Communications*, 5(5).

Ayanoglu, E., Eng, K.Y., Karol, M.J.(1996) 'Wireless ATM: limits, challenges, and proposals,' *IEEE Personal Communications*, 3(4).

Barton, M., Paine, R., Chow, A. (1998) *Description of Wireless ATM Service Scenarios*, ATM Forum Contribution, July.

Bhat, R.R. (1998) *Wireless ATM Requirements Specification*, ATM Forum, RTD-WATM-01.02.

Bisdikian, C., Bhagwat, P., Gaucher, B.P., Janniello, F.J., Naghshineh, M., Pandoh, P., Korpeoglu, I. (1998) 'WiSAP – A wireless personal access network for handheld computing devices,' *IEEE Personal Communications*, 5(6).

Bluetooth (2001a) *Specification of the Bluetooth System*, volume 1, Core, Bluetooth SIG, version 1.1.

Bluetooth (2001b) *Specification of the Bluetooth System*, volume 2, Profiles, Bluetooth SIG, version 1.1.

Bluetooth (2002) Bluetooth Special Interest Group, http://www.bluetooth.com/.

Borisov, N., Goldberg, I., Wagner, D. (2001) *Intercepting Mobile Communications: The Insecurity of 802.11*, seventh Annual International Conference on Mobile Computing and Networking, ACM SIGMOBILE, Rome, Italy.

Callaway, E., Gorday, P., Hester, L., Gutierrez, J., Naeve, M., Heile, B., Bahl, V. (2002) 'Home Networking with IEEE 802.15.4: A Developing Standard for Low-Rate Wireless Personal Area Networks,' *IEEE Communications Magazine*, 40(8).

Chakraborty, S.S. (1998) 'The interworking approach for narrowband access to ATM transport-based multiservice mobile networks,' *IEEE Personal Communications*, 5(4).

Chhaya, H.S., Gupta, S. (1996) 'Performance of asynchronous data transfer methods of IEEE 802.11 MAC protocol,' *IEEE Personal Communications*, 3(5).

Chhaya, H.S., Gupta, S. (1997) 'Performance modeling of asynchronous data transfer methods of IEEE 802.11 MAC protocol,' *Wireless Networks* 3(1997), Baltzer Science Publishers.

CSR (2002), Cambridge Silicon Radio, http://www.csr.com/.

ETSI (1996) *Radio Equipment and Systems (RES), High Performance Radio Local Area Network (HIPERLAN) Type 1, Functional specification*, European Telecommunication Standard, ETS 300 652, European Telecommunications Standards Institute.

ETSI (1997) *Radio Equipment and Systems (RES), High Performance Radio Local Area Networks (HIPERLAN), Requirements and architectures for wireless ATM access and interconnection*, TR 101 031 v1.1.1, European Telecommunication Standards Institute.

ETSI (1998a) *Broadband Radio Access Networks (BRAN): Inventory of broadband radio technologies and techniques*, TR 101 173 v1.1.1, European Telecommunication Standards Institute.

ETSI (1998b) *Broadband Radio Access Networks (BRAN); High Performance Radio Local Area Network (HIPERLAN) Type 1; Functional specification*, EN 300 652 v1.2.1, European Telecommunications Standards Institute.

ETSI (1998c) *Broadband Radio Access Networks (BRAN): Requirements and architectures for broadband fixed radio access networks* (HIPERACCESS), TR 101 177 v1.1.1, European Telecommunication Standards Institute.

ETSI (2000a) *Broadband Radio Access Networks (BRAN); High Performance Radio Local Area Network (HIPERLAN) Type 2; System Overview*, TR 101 683 v1.1.1, European Telecommunications Standards Institute.

ETSI (2000b) *Broadband Radio Access Networks (BRAN); High Performance Radio Local Area Network (HIPERLAN) Type 2; Cell based Convergence layer; Part 1: Common Part*, TS 101 763-1 v1.1.1, European Telecommunications Standards Institute.

ETSI (2000c) *Broadband Radio Access Networks (BRAN); High Performance Radio Local Area Network (HIPERLAN) Type 2; Cell based Convergence layer; Part 2: UNI Service Specific Convergence Sublayer (SSCS)*, TS 101 763-2 v1.1.1, European Telecommunications Standards Institute.

ETSI (2000d) *Broadband Radio Access Networks (BRAN); High Performance Radio Local Area Network (HIPERLAN) Type 2; Packet based Convergence layer; Part 1: Common Part*, TS 101 493-1 v1.1.1, European Telecommunications Standards Institute.

ETSI (2001a) *Broadband Radio Access Networks (BRAN); High Performance Radio Local Area Network (HIPERLAN) Type 2; Physical (PHY) layer*, TS 101 475 v1.3.1, European Telecommunications Standards Institute.

ETSI (2001b) *Broadband Radio Access Networks (BRAN); High Performance Radio Local Area Network (HIPERLAN) Type 2; Data Link Control (DLC) layer; Part 1: Basic Data Transport Functions*, TS 101 761-1 v1.3.1, European Telecommunications Standards Institute.

ETSI (2001c) *Broadband Radio Access Networks (BRAN); High Performance Radio Local Area Network (HIPERLAN) Type 2; Requirements and Architecture for Interworking between HIPERLAN/2 and 3rd Generation Cellular systems*, TR 101 957 v1.1.1, European Telecommunications Standards Institute.

ETSI (2001d) *Broadband Radio Access Networks (BRAN); High Performance Radio Local Area Network (HIPERLAN) Type 2; Packet based Convergence layer; Part 2: Ethernet Service Specific Convergence Sublayer (SSCS)*, TS 101 493-2 v1.2.1, European Telecommunications Standards Institute.

ETSI (2001e) *Broadband Radio Access Networks (BRAN); High Performance Radio Local Area Network (HIPERLAN) Type 2; Packet based Convergence layer; Part 3: IEEE 1394 Service Specific Convergence Sublayer (SSCS)*, TS 101 493-3 v1.2.1, European Telecommunications Standards Institute.

ETSI (2002a) *Broadband Radio Access Networks (BRAN); High Performance Radio Local Area Network (HIPERLAN) Type 2; Data Link Control (DLC) layer; Part 2: Radio Link Control (RLC) sublayer*, TS 101 761-2 v1.3.1, European Telecommunications Standards Institute.

ETSI (2002b) European Telecommunications Standards Institute, http://www.etsi.org/.

Haartsen, J. (1998) 'Bluetooth – the universal radio interface for ad-hoc, wireless connectivity', *Ericsson Review* No. 3, http://www.ericsson.com/.

Händel, R., Huber, N., Schröder, S. (1994) *ATM Networks: concepts, protocols, applications*. Addison-Wesley.

Halsall, F. (1996) *Data communications, computer networks and open systems*. Addison-Wesley.

Heegard, C., Coffey, J., Gummadi, S., Murphy, P., Provencio, R., Rossin, E., Schrum, S., Shoemake, M. (2001) 'High-Performance Wireless Ethernet,' *IEEE Communications Magazine*, 39(11).

HiperLAN2 (2002) HiperLAN2 Global Forum, http://www.hiperlan2.com/.

IEEE (1990) *Local Area Network and Metropolitan Area Network – Overview and Architecture*, The Institute of Electrical and Electronics Engineers, IEEE 802.1a.

IEEE (1999) *Wireless LAN Medium Access Control (MAC) and Physical Layer (PHY) specifications*, The Institute of Electrical and Electronics Engineers, IEEE 802.11.

IEEE (2001) *Port-based Network Access Control*, The Institute of Electrical and Electronics Engineers, IEEE 802.1x.

IEEE (2002a) IEEE P802.11, The Working Group for Wireless LANs, The Institute of Electrical and Electronics Engineers, http://www.ieee802.org/11/.

IEEE (2002b) IEEE P802.16, The Working Group on Broadband Wireless Access Standards, The Institute of Electrical and Electronics Engineers, http://www.ieee802.org/16/.

IEEE (2002c) IEEE P802.15, The Working Group for WPAN, The Institute of Electrical and Electronics Engineers, http://www.ieee802.org/15/.

IrDA (2002) Infrared Data Association, http://www.irda.org/.

Lansford, J., Stephens, A., Nevo, R. (2001) 'Wi-Fi (802.11b) and Bluetooth: Enabling Coexistence,' *IEEE Network*, 15(5).

Pahlavan, K., Krishnamurthy, P. (2002) *Principles of Wireless Networks*. Prentice Hall.

Perlman, R. (1992) *Interconnections: bridges and routers*. Addison Wesley Longman.

Partridge, C. (1992) *A proposed flow specification*, RFC 1363.

Rauhala, K. (1998) *Baseline Text for Wireless ATM specifications*, ATM Forum, BTD-WATM-01.07.

Raychaudhuri, D., Dellaverson, L. (1996a) *Charter, scope and work plan for proposed wireless ATM working group*, ATM Forum Document, 96-0530.

Raychaudhuri, D. (1996b) 'Wireless ATM networks: architecture, system design and prototyping,' *IEEE Personal Communications*, 3(4).

RFID (2002) Radio Frequency Identification, http://www.rfid.org/.

Santamaría, A., López-Hernández, F. (1994) *Wireless LAN systems*, Artech House.

Toh, C.-K. (1997) *Wireless ATM and ad hoc networks*. Kluwer Academic Publishers.

UWB (2002) Ultra-Wideband working group, http://www.uwb.org/.

WarDriving (2002) http://www.wardriving.com/.

WECA (2002) *Wireless Ethernet Compatibility Alliance*. http://www.weca.net/.

Wesel, E.K. (1998) Wireless multimedia communications. Addison-Wesley.

Woesner, H., Ebert, J.P., Schläger, M., Wolisz, A. (1998) 'Power-saving mechanisms in emerging standards for wireless LANs,' *IEEE Personal Communications*, 5(3).

ZigBee (2002) Zigbee Alliance, http://www.zigbee.org/.

8 Mobile network layer

This chapter introduces protocols and mechanisms developed for the network layer to support mobility. The most prominent example is Mobile IP, discussed in the first section, which adds mobility support to the internet network layer protocol IP. While systems like GSM have been designed with mobility in mind, the internet started at a time when no one had thought of mobile computers. Today's internet lacks any mechanisms to support users traveling around the world. IP is the common base for thousands of applications and runs over dozens of different networks. This is the reason for supporting mobility at the IP layer; mobile phone systems, for example, cannot offer this type of mobility for heterogeneous networks. To merge the world of mobile phones with the internet and to support mobility in the small more efficiently, so-called micro mobility protocols have been developed.

Another kind of mobility, portability of equipment, is supported by the dynamic host configuration protocol (DHCP) presented in section 8.2. In former times, computers did not often change their location. Today, due to laptops or notebooks, students show up at a university with their computers, and want to plug them in or use wireless access. A network administrator does not want to configure dozens of computers every day or hand out lists of valid IP addresses, DNS servers, subnet prefixes, default routers etc. DHCP sets in at this point to support automatic configuration of computers.

The chapter concludes with a look at ad-hoc networks in combination with the network layer. This is a fast-growing field of research with standards that are unclear as yet. How can routing be done in a dynamic network with permanent changes in connectivity? What if there are no dedicated routers or databases telling us where a node currently is? The last section deals with some approaches offering routing by extending standard algorithms known from the internet. Knowledge of the current situation of the physical medium or of the current location can be utilized.

8.1 Mobile IP

The following gives an overall view of Mobile IP, and the extensions needed for the internet to support the mobility of hosts. A good reference for the original standard (RFC 2002, Perkins, 1996a) is Perkins (1997) and Solomon (1998) which describe the development of mobile IP, all packet formats, mechanisms, discussions of the protocol and alternatives etc. in detail. The new version of Mobile IP does not involve major changes in the basic architecture but corrects some minor problems (RFC 3344, Perkins, 2002). The following material requires some familiarity with Internet protocols, especially IP. A very good overview which includes detailed descriptions of classical Internet protocols is given in Stevens (1994). Many new approaches related to Internet protocols, applications, and architectures can be found in Kurose (2003).

8.1.1 Goals, assumptions and requirements

As shown in chapter 1, mobile computing is clearly the paradigm of the future. The internet is the network for global data communication with hundreds of millions of users. So why not simply use a mobile computer in the internet?

The reason is quite simple: you will not receive a single packet as soon as you leave your home network, i.e., the network your computer is configured for, and reconnect your computer (wireless or wired) at another place (if no additional mechanisms are available). The reason for this is quite simple if you consider routing mechanisms on the internet. A host sends an IP packet with the header containing a destination address with other fields. The destination address not only determines the receiver of the packet, but also the physical subnet of the receiver. For example, the destination address 129.13.42.99 shows that the receiver must be connected to the physical subnet with the network prefix 129.13.42 (unless CIDR is used, RFC 1519, Fuller, 1993). Routers in the internet now look at the destination addresses of incoming packets and forward them according to internal look-up tables. To avoid an explosion of routing tables, only prefixes are stored and further optimizations are applied. A router would otherwise have to store the addresses of all computers in the internet, which is obviously not feasible. As long as the receiver can be reached within its physical subnet, it gets the packets; as soon as it moves outside the subnet, a packet will not reach it. A host needs a so-called **topologically correct address.**

8.1.1.1 Quick 'solutions'

One might think that a quick solution to this problem would be to assign to the computer a new, topologically correct IP address. This is what many users do with the help of DHCP (see section 8.2). So moving to a new location would mean assigning a new IP address. The problem is that nobody knows about this new address. It is almost impossible to find a (mobile) host on the internet which has just changed its address.

One could argue that with the help of dynamic DNS (DDNS, RFC 2136, Vixie, 1997) an update of the mapping logical name – IP address is possible. This is what many computer users do if they have a dynamic IP address and still want to be permanently reachable using the same logical computer name. It is important to note that these considerations, indeed most of mobile IP's motivation, are important if a user wants to offer services from a mobile node, i.e., the node should act as server. Typically, the IP address is of no special interest for service usage: in this case DHCP is sufficient. Another motivation for permanent IP addresses is emergency communication with permanent and quick reachability via the same IP address.

So what about dynamically adapting the IP address with regard to the current location? The problem is that the domain name system (DNS) needs some time before it updates the internal tables necessary to map a logical name to an IP address. This approach does not work if the mobile node moves quite often. The internet and DNS have not been built for frequent updates. Just imagine millions of nodes moving at the same time. DNS could never present a consistent view of names and addresses, as it uses caching to improve scalability. It is simply too expensive to update quickly.

There is a severe problem with higher layer protocols like TCP which rely on IP addresses. Changing the IP address while still having a TCP connection open means breaking the connection. A TCP connection is identified by the tuple (source IP address, source port, destination IP address, destination port), also known as a **socket pair** (a socket consists of address and port). Therefore, a TCP connection cannot survive any address change. Breaking TCP connections is not an option, using even simple programs like telnet would be impossible. The mobile node would also have to notify all communication partners about the new address.

Another approach is the creation of specific routes to the mobile node. Routers always choose the best-fitting prefix for the routing decision. If a router now has an entry for a prefix 129.13.42 and an address 129.13.42.99, it would choose the port associated with the latter for forwarding, if a packet with the destination address 129.13.42.99 comes in. While it is theoretically possible to change routing tables all over the world to create specific routes to a mobile node, this does not scale at all with the number of nodes in the internet. Routers are built for extremely fast forwarding, but not for fast updates of routing tables. While the first is done with special hardware support, the latter is typically a piece of software which cannot handle the burden of frequent updates. Routers are the 'brains' of the internet, holding the whole net together. No service provider or system administrator would allow changes to the routing tables, probably sacrificing stability, just to provide mobility for individual users.

8.1.1.2 Requirements

Since the quick 'solutions' obviously did not work, a more general architecture had to be designed. Many field trials and proprietary systems finally led to mobile IP as a standard to enable mobility in the internet. Several requirements accompanied the development of the standard:

- **Compatibility:** The installed base of Internet computers, i.e., computers running TCP/IP and connected to the internet, is huge. A new standard cannot introduce changes for applications or network protocols already in use. People still want to use their favorite browser for www and do not want to change applications just for mobility, the same holds for operating systems. Mobile IP has to be integrated into existing operating systems or at least work with them (today it is available for many platforms). Routers within the internet should not necessarily require other software. While it is possible to enhance the capabilities of some routers to support mobility, it is almost impossible to change all of them. Mobile IP has to remain compatible with all lower layers used for the standard, non-mobile, IP. Mobile IP must not require special media or MAC/LLC protocols, so it must use the same interfaces and mechanisms to access the lower layers as IP does. Finally, end-systems enhanced with a mobile IP implementation should still be able to communicate with fixed systems without mobile IP. Mobile IP has to ensure that users can still access all the other servers and systems in the internet. But that implies using the same address format and routing mechanisms.
- **Transparency:** Mobility should remain 'invisible' for many higher layer protocols and applications. Besides maybe noticing a lower bandwidth and some interruption in service, higher layers should continue to work even if the mobile computer has changed its point of attachment to the network. For TCP this means that the computer must keep its IP address as explained above. If the interruption of the connectivity does not take too long, TCP connections survive the change of the attachment point. Problems related to the performance of TCP are discussed in chapter 9. Clearly, many of today's applications have not been designed for use in mobile environments, so the only effects of mobility should be a higher delay and lower bandwidth. However, there are some applications for which it is better to be 'mobility aware'. Examples are cost-based routing or video compression. Knowing that it is currently possible to use different networks, the software could choose the cheapest one. Or if a video application knows that only a low bandwidth connection is currently available, it could use a different compression scheme. Additional mechanisms are necessary to inform these applications about mobility (Brewer, 1998).
- **Scalability and efficiency:** Introducing a new mechanism to the internet must not jeopardize its efficiency. Enhancing IP for mobility must not generate too many new messages flooding the whole network. Special care has to be taken considering the lower bandwidth of wireless links. Many mobile systems will have a wireless link to an attachment point, so only some additional packets should be necessary between a mobile system and a node in the network. Looking at the number of computers connected to the internet and at the growth rates of mobile communication, it is clear that myriad devices will participate in the internet as mobile components. Just

think of cars, trucks, mobile phones, every seat in every plane around the world etc. – many of them will have some IP implementation inside and move between different networks and require mobile IP. It is crucial for a mobile IP to be scalable over a large number of participants in the whole internet, worldwide.

- **Security:** Mobility poses many security problems. The minimum requirement is that of all the messages related to the management of Mobile IP are authenticated. The IP layer must be sure that if it forwards a packet to a mobile host that this host receives the packet. The IP layer can only guarantee that the IP address of the receiver is correct. There are no ways of preventing fake IP addresses or other attacks. According to Internet philosophy, this is left to higher layers (keep the core of the internet simple, push more complex services to the edge).

The goal of a mobile IP can be summarized as: 'supporting end-system mobility while maintaining scalability, efficiency, and compatibility in all respects with existing applications and Internet protocols'.

8.1.2 Entities and terminology

The following defines several entities and terms needed to understand mobile IP as defined in RFC 3344 (Perkins, 2002; was: RFC 2002, Perkins, 1996a). Figure 8.1 illustrates an example scenario.

- **Mobile node (MN):** A mobile node is an end-system or router that can change its point of attachment to the internet using mobile IP. The MN keeps its IP address and can continuously communicate with any other system in the internet as long as link-layer connectivity is given. Mobile nodes are not necessarily small devices such as laptops with antennas or mobile phones; a router onboard an aircraft can be a powerful mobile node.

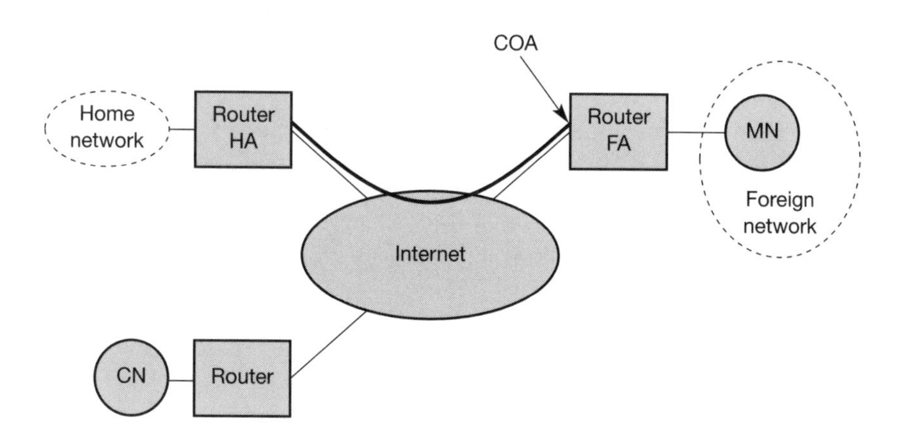

Figure 8.1 Mobile IP example network

- **Correspondent node (CN):** At least one partner is needed for communication. In the following the CN represents this partner for the MN. The CN can be a fixed or mobile node.
- **Home network:** The home network is the subnet the MN belongs to with respect to its IP address. No mobile IP support is needed within the home network.
- **Foreign network:** The foreign network is the current subnet the MN visits and which is not the home network.
- **Foreign agent (FA):** The FA can provide several services to the MN during its visit to the foreign network. The FA can have the COA (defined below), acting as tunnel endpoint and forwarding packets to the MN. The FA can be the default router for the MN. FAs can also provide security services because they belong to the foreign network as opposed to the MN which is only visiting. For mobile IP functioning, FAs are not necessarily needed. Typically, an FA is implemented on a router for the subnet the MN attaches to.
- **Care-of address (COA):** The COA defines the current location of the MN from an IP point of view. All IP packets sent to the MN are delivered to the COA, not directly to the IP address of the MN. Packet delivery toward the MN is done using a tunnel, as explained later. To be more precise, the COA marks the tunnel endpoint, i.e., the address where packets exit the tunnel. There are two different possibilities for the location of the COA:
  - **Foreign agent COA:** The COA could be located at the FA, i.e., the COA is an IP address of the FA. The FA is the tunnel end-point and forwards packets to the MN. Many MN using the FA can share this COA as common COA.
  - **Co-located COA:** The COA is co-located if the MN temporarily acquired an additional IP address which acts as COA. This address is now topologically correct, and the tunnel endpoint is at the MN. Co-located addresses can be acquired using services such as DHCP (see section 8.2). One problem associated with this approach is the need for additional addresses if MNs request a COA. This is not always a good idea considering the scarcity of IPv4 addresses.
- **Home agent (HA):** The HA provides several services for the MN and is located in the home network. The tunnel for packets toward the MN starts at the HA. The HA maintains a location registry, i.e., it is informed of the MN's location by the current COA. Three alternatives for the implementation of an HA exist.
  - The HA can be implemented on a router that is responsible for the home network. This is obviously the best position, because without optimizations to mobile IP, all packets for the MN have to go through the router anyway.
  - If changing the router's software is not possible, the HA could also be implemented on an arbitrary node in the subnet. One disadvantage of this solution is the double crossing of the router by the packet if the MN is in a foreign network. A packet for the MN comes in via the router; the HA sends it through the tunnel which again crosses the router.

- Finally, a home network is not necessary at all. The HA could be again on the 'router' but this time only acting as a manager for MNs belonging to a virtual home network. All MNs are always in a foreign network with this solution.

The example network in Figure 8.1 shows the following situation: A CN is connected via a router to the internet, as are the home network and the foreign network. The HA is implemented on the router connecting the home network with the internet, an FA is implemented on the router to the foreign network. The MN is currently in the foreign network. The tunnel for packets toward the MN starts at the HA and ends at the FA, for the FA has the COA in this example.

8.1.3 IP packet delivery

Figure 8.2 illustrates packet delivery to and from the MN using the example network of Figure 8.1. A correspondent node CN wants to send an IP packet to the MN. One of the requirements of mobile IP was to support hiding the mobility of the MN. CN does not need to know anything about the MN's current location and sends the packet as usual to the IP address of MN (step 1). This means that CN sends an IP packet with MN as a destination address and CN as a source address. The internet, not having information on the current location of MN, routes the packet to the router responsible for the home network of MN. This is done using the standard routing mechanisms of the internet.

The HA now intercepts the packet, knowing that MN is currently not in its home network. The packet is not forwarded into the subnet as usual, but encapsulated and tunnelled to the COA. A new header is put in front of the old IP header showing the COA as new destination and HA as source of the encapsulated packet (step 2). (Tunneling and encapsulation is described in more detail in section 8.1.6.) The foreign agent now decapsulates the packet, i.e., removes the additional header, and forwards the original packet with CN as source and MN as destination to the MN (step 3). Again, for the MN mobility is not visible. It receives the packet with the same sender and receiver address as it would have done in the home network.

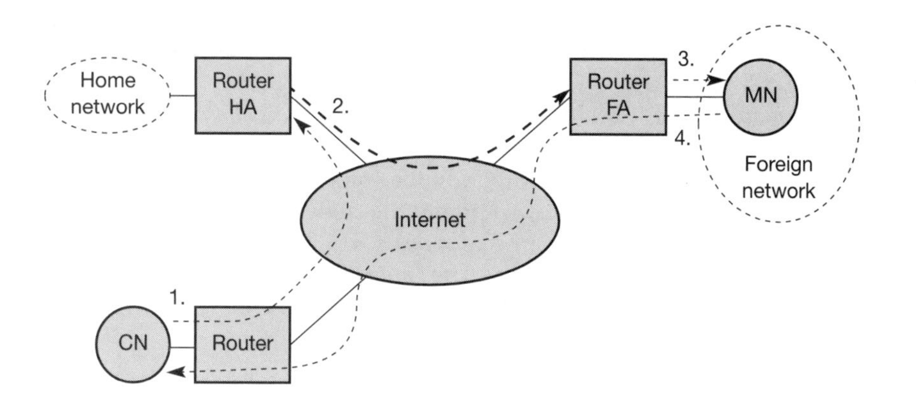

Figure 8.2
Packet delivery to and from the mobile node

At first glance, sending packets from the MN to the CN is much simpler; problems are discussed in section 8.1.8. The MN sends the packet as usual with its own fixed IP address as source and CN's address as destination (step 4). The router with the FA acts as default router and forwards the packet in the same way as it would do for any other node in the foreign network. As long as CN is a fixed node the remainder is in the fixed internet as usual. If CN were also a mobile node residing in a foreign network, the same mechanisms as described in steps 1 through 3 would apply now in the other direction.

The following sections present some additional mechanisms needed for mobile IP to work, some enhancements to the protocol, and some efficiency and security problems.

8.1.4 Agent discovery

One initial problem of an MN after moving is how to find a foreign agent. How does the MN discover that it has moved? For this purpose mobile IP describes two methods: agent advertisement and agent solicitation, which are in fact router discovery methods plus extensions.

8.1.4.1 Agent advertisement

For the first method, foreign agents and home agents advertise their presence periodically using special **agent advertisement** messages. These advertisement messages can be seen as a beacon broadcast into the subnet. For these advertisements Internet control message protocol (ICMP) messages according to RFC 1256 (Deering, 1991) are used with some mobility extensions. Routers in the fixed network implementing this standard also advertise their routing service periodically to the attached links.

The agent advertisement packet according to RFC 1256 with the extension for mobility is shown in Figure 8.3. The upper part represents the ICMP packet while the lower part is the extension needed for mobility. The fields necessary on lower layers for the agent advertisement are not shown in this figure. Clearly, mobile nodes must be reached with the appropriate link layer address. The TTL field of the IP packet is set to 1 for all advertisements to avoid forwarding them. The IP destination address according to standard router advertisements can be either set to 224.0.0.1, which is the multicast address for all systems on a link (Deering, 1989), or to the broadcast address 255.255.255.255.

The fields in the ICMP part are defined as follows. The **type** is set to 9, the **code** can be 0, if the agent also routes traffic from non-mobile nodes, or 16, if it does not route anything other than mobile traffic. Foreign agents are at least required to forward packets from the mobile node. The number of addresses advertised with this packet is in **#addresses** while the **addresses** themselves follow as shown. **Lifetime** denotes the length of time this advertisement is valid. **Preference** levels for each address help a node to choose the router that is the most eager one to get a new node.

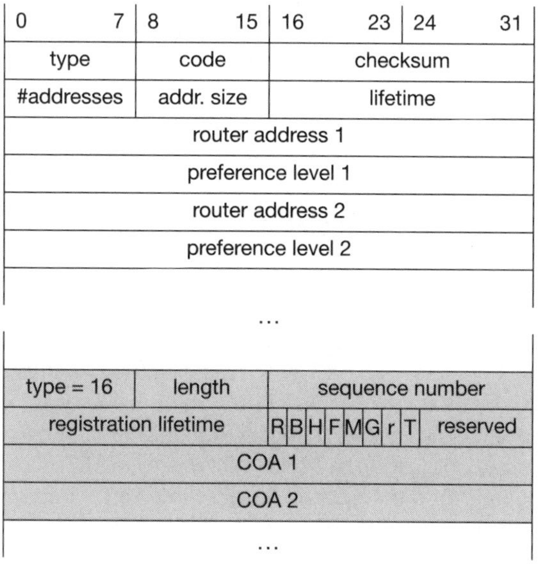

0 7	8 15	16 23	24 31
type	code	checksum	checksum
#addresses	addr. size	lifetime	lifetime
router address 1			
preference level 1			
router address 2			
preference level 2			
...			
type = 16	length	sequence number	sequence number
registration lifetime	registration lifetime	R B H F M G r T	reserved
COA 1			
COA 2			
...			

Figure 8.3
Agent advertisement packet (RFC 1256 + mobility extension)

The difference compared with standard ICMP advertisements is what happens after the router addresses. This extension for mobility has the following fields defined: **type** is set to 16, **length** depends on the number of COAs provided with the message and equals 6 + 4*(number of addresses). An agent shows the total number of advertisements sent since initialization in the **sequence number**. By the **registration lifetime** the agent can specify the maximum lifetime in seconds a node can request during registration as explained in section 8.1.5. The following bits specify the characteristics of an agent in detail. The **R** bit (registration) shows, if a registration with this agent is required even when using a colocated COA at the MN. If the agent is currently too busy to accept new registrations it can set the **B** bit. The following two bits denote if the agent offers services as a home agent (**H**) or foreign agent (**F**) on the link where the advertisement has been sent. Bits M and G specify the method of encapsulation used for the tunnel as explained in section 8.1.6. While IP-in-IP encapsulation is the mandatory standard, **M** can specify minimal encapsulation and **G** generic routing encapsulation. In the first version of mobile IP (RFC 2002) the **V** bit specified the use of header compression according to RFC 1144 (Jacobson, 1990). Now the field **r** at the same bit position is set to zero and must be ignored. The new field **T** indicates that reverse tunneling (see section 8.1.8) is supported by the FA. The following fields contain the **COAs** advertised. A foreign agent setting the F bit must advertise at least one COA. Further details and special extensions can be found in Perkins (1997) and RFC 3220. A mobile node in a subnet can now receive agent advertisements from either its home agent or a foreign agent. This is one way for the MN to discover its location.

8.1.4.2 Agent solicitation

If no agent advertisements are present or the inter-arrival time is too high, and an MN has not received a COA by other means, e.g., DHCP as discussed in section 8.2, the mobile node must send **agent solicitations**. These solicitations are again based on RFC 1256 for router solicitations. Care must be taken to ensure that these solicitation messages do not flood the network, but basically an MN can search for an FA endlessly sending out solicitation messages. Typically, a mobile node can send out three solicitations, one per second, as soon as it enters a new network. It should be noted that in highly dynamic wireless networks with moving MNs and probably with applications requiring continuous packet streams even one second intervals between solicitation messages might be too long. Before an MN even gets a new address many packets will be lost without additional mechanisms.

If a node does not receive an answer to its solicitations it must decrease the rate of solicitations exponentially to avoid flooding the network until it reaches a maximum interval between solicitations (typically one minute). Discovering a new agent can be done anytime, not just if the MN is not connected to one. Consider the case that an MN is looking for a better connection while still sending via the old path. This is the case while moving through several cells of different wireless networks.

After these steps of advertisements or solicitations the MN can now receive a COA, either one for an FA or a co-located COA. The MN knows its location (home network or foreign network) and the capabilities of the agent (if needed). The next step for the MN is the registration with the HA if the MN is in a foreign network as described in the following.

8.1.5 Registration

Having received a COA, the MN has to register with the HA. The main purpose of the registration is to inform the HA of the current location for correct forwarding of packets. Registration can be done in two different ways depending on the location of the COA.

- If the COA is at the FA, registration is done as illustrated in Figure 8.4 (left). The MN sends its registration request containing the COA (see Figure 8.5) to the FA which is forwarding the request to the HA. The HA now sets up a **mobility binding** containing the mobile node's home IP address and the current COA. Additionally, the mobility binding contains the lifetime of the registration which is negotiated during the registration process. Registration expires automatically after the lifetime and is deleted; so, an MN should reregister before expiration. This mechanism is necessary to avoid mobility bindings which are no longer used. After setting up the mobility binding, the HA sends a reply message back to the FA which forwards it to the MN.
- If the COA is co-located, registration can be simpler, as shown in Figure 8.4 (right). The MN may send the request directly to the HA and vice versa. This, by the way, is also the registration procedure for MNs returning to their home network. Here they also register directly with the HA. However, if the MN received an agent advertisement from the FA it should register via this FA if the R bit is set in the advertisement.

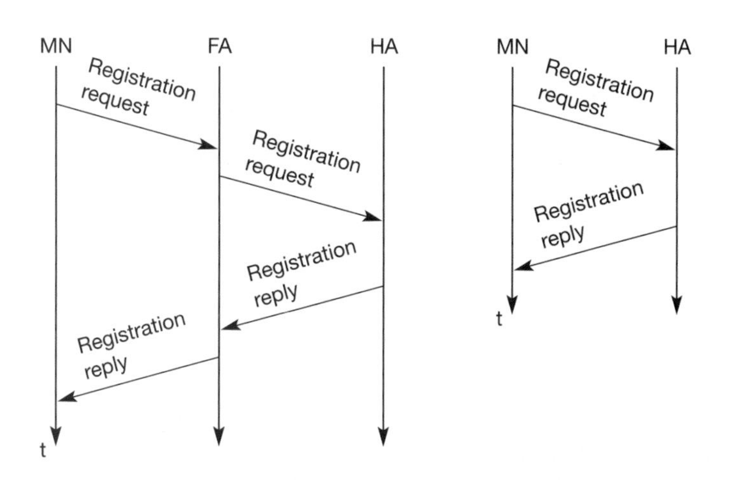

Figure 8.4 Registration of a mobile node via the FA or directly with the HA

0 – 7	8 – 15	16 – 23	24 – 31
type 1	S B D M G r T x	lifetime	
home address			
home agent			
COA			
identification			
extensions ...			

Figure 8.5 Registration request

UDP packets are used for **registration requests**. The IP source address of the packet is set to the interface address of the MN, the IP destination address is that of the FA or HA (depending on the location of the COA). The UDP destination port is set to 434. UDP is used because of low overheads and better performance compared to TCP in wireless environments (see chapter 9). The fields relevant for mobile IP registration requests follow as UDP data (see Figure 8.6). The fields are defined as follows.

The first field **type** is set to 1 for a registration request. With the **S** bit an MN can specify if it wants the HA to retain prior mobility bindings. This allows for simultaneous bindings. The following bits denote the requested behavior for packet forwarding. Setting the **B** bit generally indicates that an MN also wants to receive the broadcast packets which have been received by the HA in the home network. A more detailed description of how to filter broadcast messages which are not needed by the MN can be found in Perkins (1997). If an MN uses a co-located COA, it also takes care of the decapsulation at the tunnel endpoint. The **D** bit indicates this behavior. As already defined for agent advertisements, the following bits **M** and **G** denote the use of minimal encapsulation or generic routing encapsulation, respectively. **T** indicates reverse tunneling, **r** and **x** are set to zero.

Figure 8.6 Registration reply

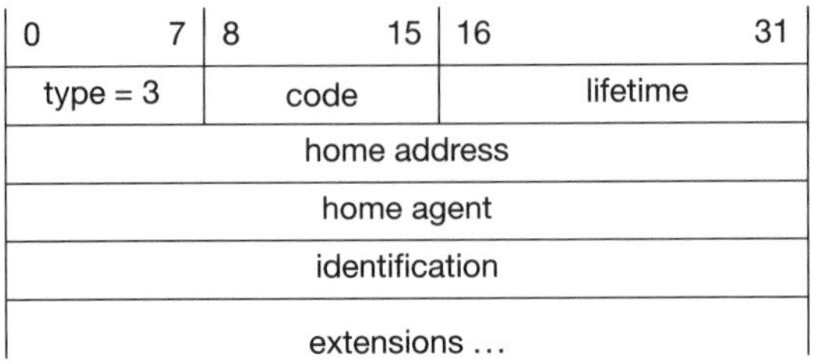

0 7	8 15	16 31
type = 3	code	lifetime
home address		
home agent		
identification		
extensions ...		

Lifetime denotes the validity of the registration in seconds. A value of zero indicates deregistration; all bits set indicates infinity. The **home address** is the fixed IP address of the MN, **home agent** is the IP address of the HA, and **COA** represents the tunnel endpoint. The 64 bit **identification** is generated by the MN to identify a request and match it with registration replies. This field is used for protection against replay attacks of registrations. The **extensions** must at least contain parameters for authentication.

A **registration reply**, which is conveyed in a UDP packet, contains a **type** field set to 3 and a **code** indicating the result of the registration request. Table 8.1 gives some example codes.

Table 8.1 Example registration reply codes

Registration	Code	Explanation
successful	0	registration accepted
	1	registration accepted, but simultaneous mobility bindings unsupported
denied by FA	65	administratively prohibited
	66	insufficient resources
	67	mobile node failed authentication
	68	home agent failed authentication
	69	requested lifetime too long
denied by HA	129	administratively prohibited
	130	insufficient resources
	131	mobile node failed authentication
	132	foreign agent failed authentication
	133	registration identification mismatch
	135	too many simultaneous mobility bindings

The **lifetime** field indicates how many seconds the registration is valid if it was successful. **Home address** and **home agent** are the addresses of the MN and the HA, respectively. The 64-bit **identification** is used to match registration requests with replies. The value is based on the identification field from the registration and the authentication method. Again, the **extensions** must at least contain parameters for authentication.

8.1.6 Tunneling and encapsulation

The following describes the mechanisms used for forwarding packets between the HA and the COA, as shown in Figure 8.2, step 2. A **tunnel** establishes a virtual pipe for data packets between a tunnel entry and a tunnel endpoint. Packets entering a tunnel are forwarded inside the tunnel and leave the tunnel unchanged. Tunneling, i.e., sending a packet through a tunnel, is achieved by using encapsulation.

Encapsulation is the mechanism of taking a packet consisting of packet header and data and putting it into the data part of a new packet. The reverse operation, taking a packet out of the data part of another packet, is called **decapsulation**. Encapsulation and decapsulation are the operations typically performed when a packet is transferred from a higher protocol layer to a lower layer or from a lower to a higher layer respectively. Here these functions are used within the same layer.

This mechanism is shown in Figure 8.7 and describes exactly what the HA at the tunnel entry does. The HA takes the original packet with the MN as destination, puts it into the data part of a new packet and sets the new IP header in such a way that the packet is routed to the COA. The new header is also called the **outer header** for obvious reasons. Additionally, there is an **inner header** which can be identical to the original header as this is the case for IP-in-IP encapsulation, or the inner header can be computed during encapsulation.

8.1.6.1 IP-in-IP encapsulation

There are different ways of performing the encapsulation needed for the tunnel between HA and COA. Mandatory for mobile IP is **IP-in-IP encapsulation** as specified in RFC 2003 (Perkins, 1996b). Figure 8.8 shows a packet inside the tunnel. The fields follow the standard specification of the IP protocol as defined

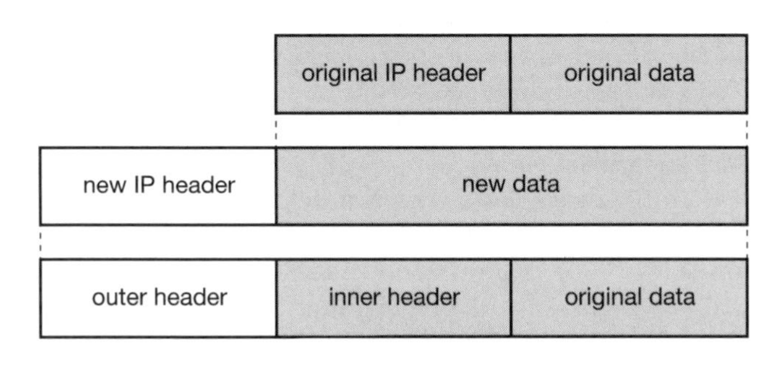

Figure 8.7
IP encapsulation

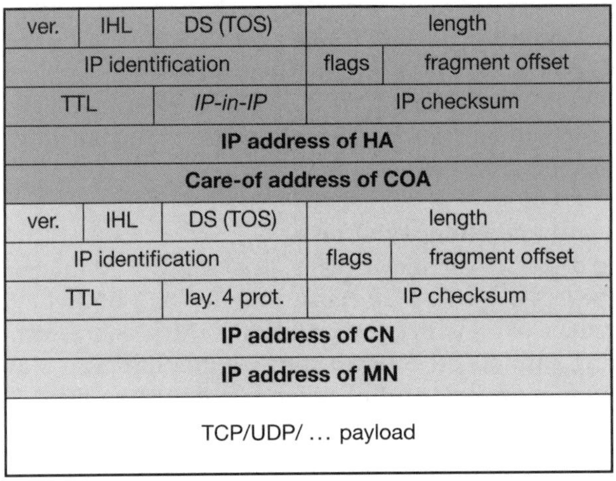

Figure 8.8
IP-in-IP encapsulation

ver.	IHL	DS (TOS)	length	
IP identification			flags	fragment offset
TTL		*IP-in-IP*	IP checksum	
IP address of HA				
Care-of address of COA				
ver.	IHL	DS (TOS)	length	
IP identification			flags	fragment offset
TTL		lay. 4 prot.	IP checksum	
IP address of CN				
IP address of MN				
TCP/UDP/ ... payload				

in RFC 791 (Postel, 1981) and the new interpretation of the former TOS, now DS field in the context of differentiated services (RFC 2474, Nichols, 1998). The fields of the outer header are set as follows. The version field **ver** is 4 for IP version 4, the internet header length (**IHL**) denotes the length of the outer header in 32 bit words. **DS(TOS)** is just copied from the inner header, the **length** field covers the complete encapsulated packet. The fields up to TTL have no special meaning for mobile IP and are set according to RFC 791. **TTL** must be high enough so the packet can reach the tunnel endpoint. The next field, here denoted with **IP-in-IP**, is the type of the protocol used in the IP payload. This field is set to 4, the protocol type for IPv4 because again an IPv4 packet follows after this outer header. IP **checksum** is calculated as usual. The next fields are the tunnel entry as source address (the **IP address of the HA**) and the tunnel exit point as destination address (the **COA**).

If no options follow the outer header, the inner header starts with the same fields as just explained. This header remains almost unchanged during encapsulation, thus showing the original sender CN and the receiver MN of the packet. The only change is TTL which is decremented by 1. This means that the whole tunnel is considered a single hop from the original packet's point of view. This is a very important feature of tunneling as it allows the MN to behave as if it were attached to the home network. No matter how many real hops the packet has to take in the tunnel, it is just one (logical) hop away for the MN. Finally, the payload follows the two headers.

8.1.6.2 Minimal encapsulation

As seen with IP-in-IP encapsulation, several fields are redundant. For example, TOS is just copied, fragmentation is often not needed etc. Therefore, **minimal encapsulation** (RFC 2004) as shown in Figure 8.9 is an optional encapsulation method for mobile IP (Perkins, 1996c). The tunnel entry point and endpoint are specified. In this case, the field for the type of the following header contains the

ver.	IHL	DS (TOS)	length	
IP identification			flags	fragment offset
TTL		*min. encap*	IP checksum	
IP address of HA				
care-of address of COA				
lay. 4 protoc.		S reserved	IP checksum	
IP address of MN				
original sender IP address (if S=1)				
TCP/UDP/ ... payload				

Figure 8.9
Minimal encapsulation

value 55 for the minimal encapsulation protocol. The inner header is different for minimal encapsulation. The type of the following protocol and the address of the MN are needed. If the **S** bit is set, the original sender address of the CN is included as omitting the source is quite often not an option. No field for fragmentation offset is left in the inner header and minimal encapsulation does not work with already fragmented packets.

8.1.6.3 Generic routing encapsulation

While IP-in-IP encapsulation and minimal encapsulation work only for IP, the following encapsulation scheme also supports other network layer protocols in addition to IP. **Generic routing encapsulation** (GRE) allows the encapsulation of packets of one protocol suite into the payload portion of a packet of another protocol suite (Hanks, 1994). Figure 8.10 shows this procedure. The packet of one protocol suite with the original packet header and data is taken and a new GRE header is prepended. Together this forms the new data part of the new packet. Finally, the header of the second protocol suite is put in front.

Figure 8.11 shows on the left side the fields of a packet inside the tunnel between home agent and COA using GRE as an encapsulation scheme according to RFC 1701. The outer header is the standard IP header with HA as source address and COA as destination address. The protocol type used in this outer IP

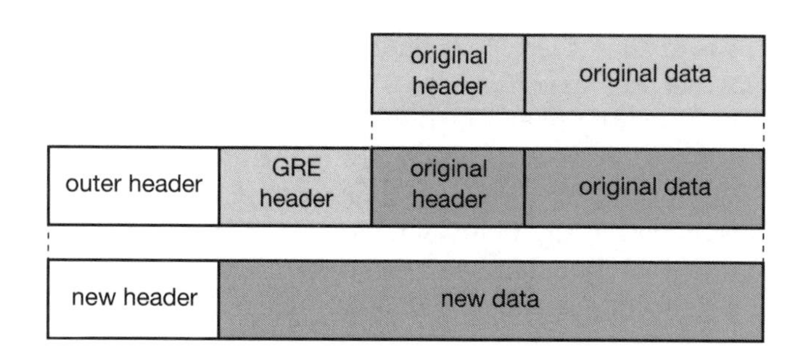

Figure 8.10
Generic routing encapsulation

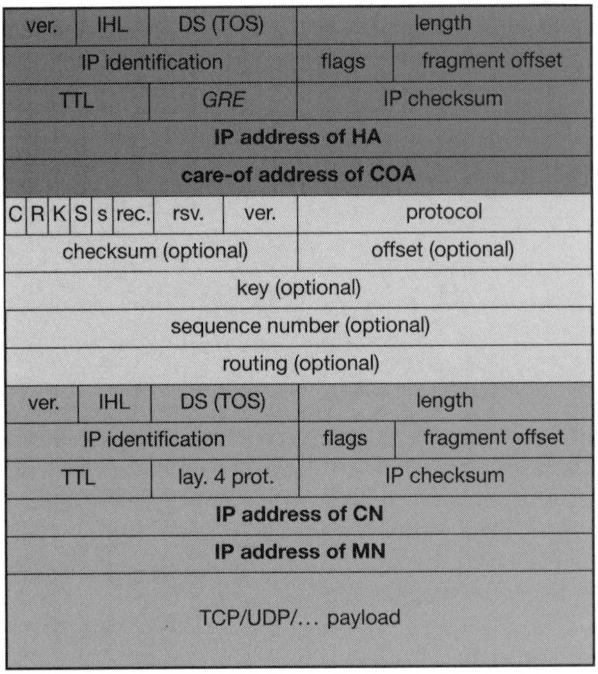

Figure 8.11
Protocol fields for GRE according to RFC 1701

ver.	IHL	DS (TOS)	length
IP identification		flags	fragment offset
TTL	*GRE*	IP checksum	
IP address of HA			
care-of address of COA			
C R K S s rec. rsv. ver.		protocol	
checksum (optional)		offset (optional)	
key (optional)			
sequence number (optional)			
routing (optional)			
ver.	IHL	DS (TOS)	length
IP identification		flags	fragment offset
TTL	lay. 4 prot.	IP checksum	
IP address of CN			
IP address of MN			
TCP/UDP/... payload			

header is 47 for GRE. The other fields of the outer packet, such as TTL and TOS, may be copied from the original IP header. However, the TTL must be decremented by 1 when the packet is decapsulated to prevent indefinite forwarding.

The GRE header starts with several flags indicating if certain fields are present or not. A minimal GRE header uses only 4 bytes; nevertheless, GRE is flexible enough to include several mechanisms in its header. The **C** bit indicates if the checksum field is present and contains valid information. If C is set, the **checksum** field contains a valid IP checksum of the GRE header and the payload. The **R** bit indicates if the offset and routing fields are present and contain valid information. The **offset** represents the offset in bytes for the first source **routing** entry. The routing field, if present, has a variable length and contains fields for source routing. If the C bit is set, the offset field is also present and, vice versa, if the R bit is set, the checksum field must be present. The only reason for this is to align the following fields to 4 bytes. The checksum field is valid only if C is set, and the offset field is valid only if R is set respectively.

GRE also offers a **key** field which may be used for authentication. If this field is present, the **K** bit is set. However, the authentication algorithms are not further specified by GRE. The sequence number bit **S** indicates if the **sequence** number field is present, if the s bit is set, strict source routing is used. Sequence numbers may be used by a decapsulator to restore packet order. This can be important, if a protocol guaranteeing in-order transmission is encapsulated and

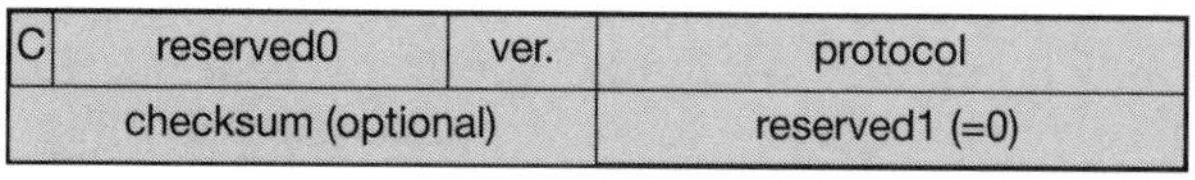

C	reserved0	ver.	protocol
checksum (optional)			reserved1 (=0)

Figure 8.12
Protocol fields for GRE according to RFC 2784

transferred using a protocol which does not guarantee in-order delivery, e.g., IP. Now the decapsulator at the tunnel exit must restore the sequence to maintain the characteristic of the protocol.

The **recursion control** field (rec.) is an important field that additionally distinguishes GRE from IP-in-IP and minimal encapsulation. This field represents a counter that shows the number of allowed recursive encapsulations. As soon as a packet arrives at an encapsulator it checks whether this field equals zero. If the field is not zero, additional encapsulation is allowed – the packet is encapsulated and the field decremented by one. Otherwise the packet will most likely be discarded. This mechanism prevents indefinite recursive encapsulation which might happen with the other schemes if tunnels are set up improperly (e.g., several tunnels forming a loop). The default value of this field should be 0, thus allowing only one level of encapsulation.

The following **reserved** fields must be zero and are ignored on reception. The **version** field contains 0 for the GRE version. The following 2 byte **protocol** field represents the protocol of the packet following the GRE header. Several values have been defined, e.g., 0×6558 for transparent Ethernet bridging using a GRE tunnel. In the case of a mobile IP tunnel, the protocol field contains 0×800 for IP.

The standard header of the original packet follows with the source address of the correspondent node and the destination address of the mobile node.

Figure 8.12 shows the simplified header of GRE following RFC 2784 (Farinacci, 2000), which is a more generalized version of GRE compared to RFC 1701. This version does not address mutual encapsulation and ignores several protocol-specific nuances on purpose. The field C indicates again if a checksum is present. The next 5 bits are set to zero, then 7 reserved bits follow. The **version** field contains the value zero. The **protocol** type, again, defines the protocol of the payload following RFC 3232 (Reynolds, 2002). If the flag C is set, then **checksum** field and a field called reserved1 follows. The latter field is constant zero set to zero follow. RFC 2784 deprecates several fields of RFC 1701, but can interoperate with RFC 1701-compliant implementations.

8.1.7 Optimizations

Imagine the following scenario. A Japanese and a German meet at a conference on Hawaii. Both want to use their laptops for exchanging data, both run mobile IP for mobility support. Now recall Figure 8.2 and think of the way the packets between both computers take.

If the Japanese sends a packet to the German, his computer sends the data to the HA of the German, i.e., from Hawaii to Germany. The HA in Germany now encapsulates the packets and tunnels them to the COA of the German laptop on Hawaii. This means that although the computers might be only meters away, the packets have to travel around the world! This inefficient behavior of a non-optimized mobile IP is called **triangular routing**. The triangle is made of the three segments, CN to HA, HA to COA/MN, and MN back to CN.

With the basic mobile IP protocol all packets to the MN have to go through the HA. This can cause unnecessary overheads for the network between CN and HA, but also between HA and COA, depending on the current location of the MN. As the example shows, latency can increase dramatically. This is particularly unfortunate if the MNs and HAs are separated by, e.g., transatlantic links.

One way to optimize the route is to inform the CN of the current location of the MN. The CN can learn the location by caching it in a **binding cache** which is a part of the local routing table for the CN. The appropriate entity to inform the CN of the location is the HA. The optimized mobile IP protocol needs four additional messages.

- **Binding request:** Any node that wants to know the current location of an MN can send a binding request to the HA. The HA can check if the MN has allowed dissemination of its current location. If the HA is allowed to reveal the location it sends back a binding update.
- **Binding update:** This message sent by the HA to CNs reveals the current location of an MN. The message contains the fixed IP address of the MN and the COA. The binding update can request an acknowledgement.
- **Binding acknowledgement:** If requested, a node returns this acknowledgement after receiving a binding update message.
- **Binding warning:** If a node decapsulates a packet for an MN, but it is not the current FA for this MN, this node sends a binding warning. The warning contains MN's home address and a target node address, i.e., the address of the node that has tried to send the packet to this MN. The recipient of the warning then knows that the target node could benefit from obtaining a fresh binding for the MN. The recipient can be the HA, so the HA should now send a binding update to the node that obviously has a wrong COA for the MN.

Figure 8.13 explains these additional four messages together with the case of an MN changing its FA. The CN can request the current location from the HA. If allowed by the MN, the HA returns the COA of the MN via an update message. The CN acknowledges this update message and stores the mobility binding. Now the CN can send its data directly to the current foreign agent FA_{old}. FA_{old} forwards the packets to the MN. This scenario shows a COA located at an FA. Encapsulation of data for tunneling to the COA is now done by the CN, not the HA.

The MN might now change its location and register with a new foreign agent, FA_{new}. This registration is also forwarded to the HA to update its location database. Furthermore, FA_{new} informs FA_{old} about the new registration of MN. MN's registration message contains the address of FA_{old} for this purpose. Passing this information is achieved via an update message, which is acknowledged by FA_{old}. Registration replies are not shown in this scenario. Without the informa-

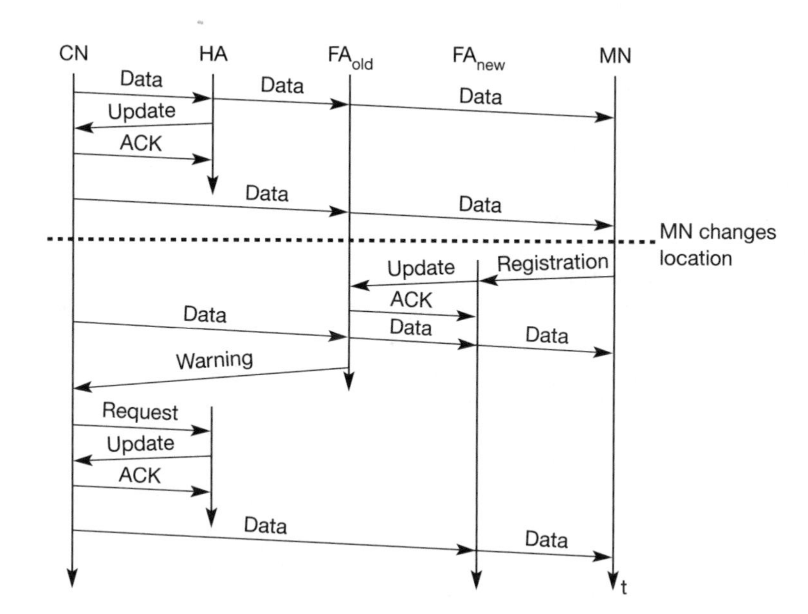

Figure 8.13
Change of the foreign agent with an optimized mobile IP

tion provided by the new FA, the old FA would not get to know anything about the new location of MN. In this case, CN does not know anything about the new location, so it still tunnels its packets for MN to the old FA, FA_{old}. This FA now notices packets with destination MN, but also knows that it is not the current FA of MN. FA_{old} might now forward these packets to the new COA of MN which is FA_{new} in this example. This forwarding of packets is another optimization of the basic Mobile IP providing **smooth handovers**. Without this optimization, all packets in transit would be lost while the MN moves from one FA to another. With TCP as the higher layer protocol this would result in severe performance degradation (see chapter 9).

To tell CN that it has a stale binding cache, FA_{old} sends, in this example, a binding warning message to CN. CN then requests a binding update. (The warning could also be directly sent to the HA triggering an update). The HA sends an update to inform the CN about the new location, which is acknowledged. Now CN can send its packets directly to FA_{new}, again avoiding triangular routing. Unfortunately, this optimization of mobile IP to avoid triangular routing causes several security problems (e.g., tunnel hijacking) as discussed in Montenegro (1998). Not all users of mobile communication systems want to reveal their current 'location' (in the sense of an IP subnet) to a communication partner.

8.1.8 Reverse tunneling

At first glance, the return path from the MN to the CN shown in Figure 8.2 looks quite simple. The MN can directly send its packets to the CN as in any other standard IP situation. The destination address in the packets is that of CN. But there are several severe problems associated with this simple solution.

- **Firewalls:** Almost all companies and many other institutions secure their internal networks (intranet) connected to the internet with the help of a firewall. All data to and from the intranet must pass through the firewall. Besides many other functions, firewalls can be set up to filter out malicious addresses from an administrator's point of view. Quite often firewalls only allow packets with topologically correct addresses to pass. This provides at least a first and simple protection against misconfigured systems of unknown addresses. However, MN still sends packets with its fixed IP address as source which is not topologically correct in a foreign network. Firewalls often filter packets coming from outside containing a source address from computers of the internal network. This avoids other computers that could use internal addresses and claim to be internal computers. However, this also implies that an MN cannot send a packet to a computer residing in its home network. Altogether, this means that not only does the destination address matter for forwarding IP packets, but also the source address due to security concerns. Further complications arise through the use of private addresses inside the intranet and the translation into global addresses when communicating with the internet. This **network address translation** (NAT, network address translator, RFC 3022, Srisuresh, 2001) is used by many companies to hide internal resources (routers, computers, printers etc.) and to use only some globally available addresses (Levkowetz, 2002, tries to solve the problems arising when using NAT together with mobile IP).
- **Multi-cast:** Reverse tunnels are needed for the MN to participate in a multicast group. While the nodes in the home network might participate in a multi-cast group, an MN in a foreign network cannot transmit multi-cast packets in a way that they emanate from its home network without a reverse tunnel. The foreign network might not even provide the technical infrastructure for multi-cast communication (multi-cast backbone, Mbone).
- **TTL:** Consider an MN sending packets with a certain TTL while still in its home network. The TTL might be low enough so that no packet is transmitted outside a certain region. If the MN now moves to a foreign network, this TTL might be too low for the packets to reach the same nodes as before. Mobile IP is no longer transparent if a user has to adjust the TTL while moving. A reverse tunnel is needed that represents only one hop, no matter how many hops are really needed from the foreign to the home network.

All these considerations led to RFC 2344 (Montenegro, 1998) defining reverse tunneling as an extension to mobile IP. The new RFC 3024 (Montenegro, 2001) renders RFC 2344 obsolete but comprises only some minor changes for the original standard. The RFC was designed backwards-compatible to mobile IP and defines topologically correct reverse tunneling as necessary to handle the problems described above. Reverse tunneling was added as an option to mobile IP in the new standard (RFC 3344).

Obviously, reverse tunneling now creates a triangular routing problem in the reverse direction. All packets from an MN to a CN go through the HA. RFC 3024 does not offer a solution for this reverse triangular routing, because it is not clear if the CN can decapsulate packets. Remember that mobile IP should work together with all traditional, non-mobile IP nodes. Therefore, one cannot assume that a CN is able to be a tunnel endpoint.

Reverse tunneling also raises several security issues which have not been really solved up to now. For example, tunnels starting in the private network of a company and reaching out into the internet could be hijacked and abused for sending packets through a firewall. It is not clear if companies would allow for setting up tunnels through a firewall without further checking of packets. It is more likely that a company will set up a special virtual network for visiting mobile nodes outside the firewall with full connectivity to the internet. This allows guests to use their mobile equipment, and at the same time, today's security standards are maintained. Initial architectures integrating mobility and security aspects within firewalls exist (Mink, 2000a and b).

8.1.9 IPv6

While mobile IP was originally designed for IP version 4, IP version 6 (Deering, 1998) makes life much easier. Several mechanisms that had to be specified separately for mobility support come free in IPv6 (Perkins, 1996d), (Johnson, 2002b). One issue is security with regard to authentication, which is now a required feature for all IPv6 nodes. No special mechanisms as add-ons are needed for securing mobile IP registration. Every IPv6 node masters address autoconfiguration – the mechanisms for acquiring a COA are already built in. Neighbor discovery as a mechanism mandatory for every node is also included in the specification; special foreign agents are no longer needed to advertise services. Combining the features of autoconfiguration and neighbor discovery means that every mobile node is able to create or obtain a topologically correct address for the current point of attachment.

Every IPv6 node can send binding updates to another node, so the MN can send its current COA directly to the CN and HA. These mechanisms are an integral part of IPv6. A soft handover is possible with IPv6. The MN sends its new COA to the old router servicing the MN at the old COA, and the old router encapsulates all incoming packets for the MN and forwards them to the new COA.

Altogether, mobile IP in IPv6 networks requires very few additional mechanisms of a CN, MN, and HA. The FA is not needed any more. A CN only has to be able to process binding updates, i.e., to create or to update an entry in the routing cache. The MN itself has to be able to decapsulate packets, to detect when it needs a new COA, and to determine when to send binding updates to the HA and CN. A HA must be able to encapsulate packets. However, IPv6 does not solve any firewall or privacy problems. Additional mechanisms on higher layers are needed for this.

8.1.10 IP micro-mobility support

Mobile IP exhibits several problems regarding the duration of handover and the scalability of the registration procedure. Assuming a large number of mobile devices changing networks quite frequently, a high load on the home agents as well as on the networks is generated by registration and binding update messages. IP micro-mobility protocols can complement mobile IP by offering fast and almost seamless handover control in limited geographical areas.

Consider a client arriving with his or her laptop at the customer's premises. The home agent only has to know an entry point to the customer's network, not the details within this network. The entry point acts as the current location. Changes in the location within the customer's network should be handled locally to minimize network traffic and to speed-up local handover. The basic underlying idea is the same for all micro-mobility protocols: Keep the frequent updates generated by local changes of the points of attachment away from the home network and only inform the home agent about major changes, i.e., changes of a region. In some sense all micro-mobility protocols establish a hierarchy. However, the debate is still going on if micro-mobility aspects should really be handled on the IP layer or if layer 2 is the better place for it. Layer 2 mobility support would comprise, e.g., the inter access point protocol (IAPP) of 802.11 WLANs (see chapter 7) or the mobility support mechanisms of mobile phone systems (see chapter 4).

The following presents three of the most prominent approaches, which should be seen neither as standards nor as final solutions of the micro-mobility problems. Campbell (2002) presents a comparison of the three approaches.

8.1.10.1 Cellular IP

Cellular IP (Valko, 1999), (Campbell, 2000) provides local handovers without renewed registration by instaling a single **cellular IP gateway (CIPGW)** for each domain, which acts to the outside world as a foreign agent (see Figure 8.14). Inside the cellular IP domain, all nodes collect routing information for accessing MNs based on the origin of packets sent by the MNs towards the CIPGW. Soft handovers are achieved by allowing simultaneous forwarding of packets destined for a mobile node along multiple paths. A mobile node moving between adjacent cells will temporarily be able to receive packets via both old and new **base stations (BS)** if this is supported by the lower protocol layers.

Concerning the manageability of cellular IP, it has to be noted that the approach has a simple and elegant architecture and is mostly self-configuring. However, mobile IP tunnels could be controlled more easily if the CIPGW was integrated into a firewall, but there are no detailed specifications in (Campbell, 2000) regarding such integration. Cellular IP requires changes to the basic mobile IP protocol and is not transparent to existing systems. The foreign network's routing tables are changed based on messages sent by mobile nodes. These should not be trusted blindly even if they have been authenticated. This could be exploited by systems in the foreign network for wiretapping packets

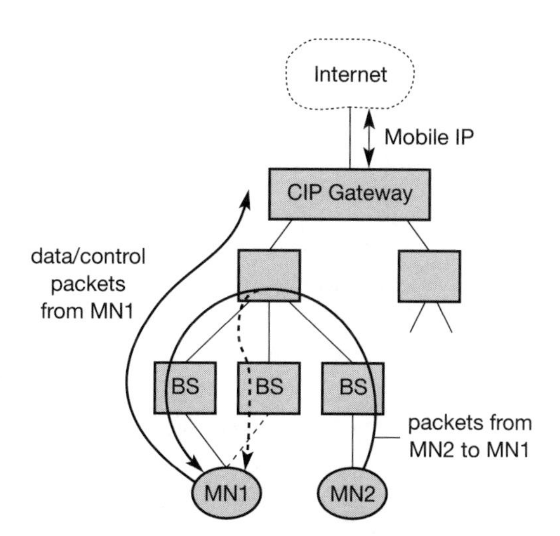

Figure 8.14
Basic architecture of cellular IP

destined for an MN by sending packets to the CIPGW with the source address set to the MN's address. In enterprise scenarios requiring basic communications security, this may not be acceptable.

Advantage

- Manageability: Cellular IP is mostly self-configuring, and integration of the CIPGW into a firewall would facilitate administration of mobility-related functionality. This is, however, not explicitly specified in (Campbell, 2000).

Disadvantages

- Efficiency: Additional network load is induced by forwarding packets on multiple paths.
- Transparency: Changes to MNs are required.
- Security: Routing tables are changed based on messages sent by mobile nodes. Additionally, all systems in the network can easily obtain a copy of all packets destined for an MN by sending packets with the MN's source address to the CIPGW.

8.1.10.2 Hawaii

HAWAII (Handoff-Aware Wireless Access Internet Infrastructure, Ramjee, 1999) tries to keep micro-mobility support as transparent as possible for both home agents and mobile nodes (which have to support route optimization). Its concrete goals are performance and reliability improvements and support for quality of service mechanisms. On entering an HAWAII domain, a mobile node obtains a co-located COA (see Figure 8.15, step 1) and registers with the HA (step 2). Additionally, when moving to another cell inside the foreign domain, the MN sends a registration request to the new base station as to a foreign agent

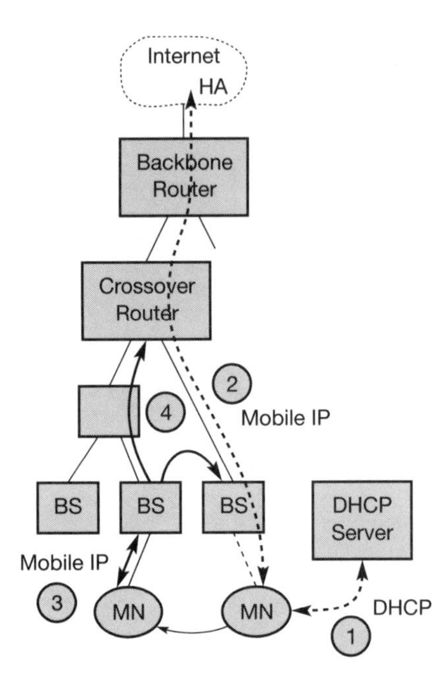

Figure 8.15
Basic architecture of HAWAII

(step 3), thus mixing the concepts of co-located COA and foreign agent COA. The base station intercepts the registration request and sends out a handoff update message, which reconfigures all routers on the paths from the old and new base station to the so-called crossover router (step 4). When routing has been reconfigured successfully, the base station sends a registration reply to the mobile node, again as if it were a foreign agent.

The use of challenge-response extensions for authenticating a mobile node is mandatory. In contrast to cellular IP, routing changes are always initiated by the foreign domain's infrastructure, and the corresponding messages could be authenticated, e.g., by means of an IPSec authentication header (AH; RFC 2402, Kent, 1998), reducing the risk of malicious rerouting of traffic initiated by bogus mobile hosts. However, this is not explicitly specified in Ramjee (1999). HAWAII claims to be mostly transparent to mobile nodes, but this claim has to be regarded with some caution as the requirement to support a co-located care-of-address as well as to interact with foreign agents could cause difficulties with some mobile nodes.

Advantages

- Security: Challenge-response extensions are mandatory. In contrast to Cellular IP, routing changes are always initiated by the foreign domain's infrastructure.
- Transparency: HAWAII is mostly transparent to mobile nodes.

Disadvantages

- Security: There are no provisions regarding the setup of IPSec tunnels.
- Implementation: No private address support is possible because of co-located COAs.

8.1.10.3 Hierarchical mobile IPv6 (HMIPv6)

As introducing hierarchies is the natural choice for handling micro-mobility issues, several proposals for a 'hierarchical' mobile IP exist. What follows is based on Soliman, (2002).

HMIPv6 provides micro-mobility support by installing a **mobility anchor point (MAP)**, which is responsible for a certain domain and acts as a local HA within this domain for visiting MNs (see Figure 8.16). The MAP receives all packets on behalf of the MN, encapsulates and forwards them directly to the MN's current address (link COA, **LCOA**). As long as an MN stays within the domain of a MAP, the globally visible COA (regional COA, **RCOA**) does not change. A MAP domain's boundaries are defined by the **access routers (AR)** advertising the MAP information to the attached MNs. A MAP assists with local handovers and maps RCOA to LCOA. MNs register their RCOA with the HA using a binding update. When a MN moves locally it must only register its new LCOA with its MAP. The RCOA stays unchanged. To support smooth handovers between MAP domains, an MN can send a binding update to its former MAP.

It should be mentioned as a security benefit that mobile nodes can be provided with some kind of limited location privacy because LCOAs on lower levels of the mobility hierarchy can be hidden from the outside world. However, this applies only to micro mobility, that is, as long as the mobile node rests in the same domain. A MN can also send a binding update to a CN who shares the same link. This reveals its location but optimizes packet flow (direct routing without going through the MAP). MNs can use their RCOA as source address. The extended mode of HMIPv6 supports both mobile nodes and mobile networks.

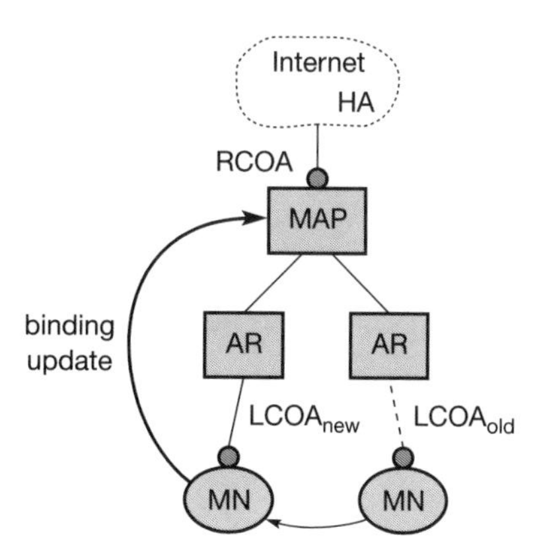

Figure 8.16 Basic architecture of hierarchical mobile IP

Advantages

- Security: MNs can have (limited) location privacy because LCOAs can be hidden.
- Efficiency: Direct routing between CNs sharing the same link is possible

Disadvantages

- Transparency: Additional infrastructure component (MAP).
- Security: Routing tables are changed based on messages sent by mobile nodes. This requires strong authentication and protection against denial of service attacks. Additional security functions might be necessary in MAPs

The main driving factors behind the three architectures presented here are efficiency, scalability, and seamless handover support. However, as security will be one of the key success factors of future mobile IP networks, first approaches adding this feature exist. (Mink 2000a and b.)

8.2 Dynamic host configuration protocol

The dynamic host configuration protocol (DHCP, RFC 2131, Drohms, 1997) is mainly used to simplifiy the installation and maintenance of networked computers. If a new computer is connected to a network, DHCP can provide it with all the necessary information for full system integration into the network, e.g., addresses of a DNS server and the default router, the subnet mask, the domain name, and an IP address. Providing an IP address, makes DHCP very attractive for mobile IP as a source of care-of-addresses. While the basic DHCP mechanisms are quite simple, many options are available as described in RFC 2132 (Alexander, 1997).

DHCP is based on a client/server model as shown in Figure 8.17. DHCP clients send a request to a server (DHCPDISCOVER in the example) to which the server responds. A client sends requests using MAC broadcasts to reach all devices in the LAN. A DHCP relay might be needed to forward requests across inter-working units to a DHCP server.

Figure 8.17 Basic DHCP configuration

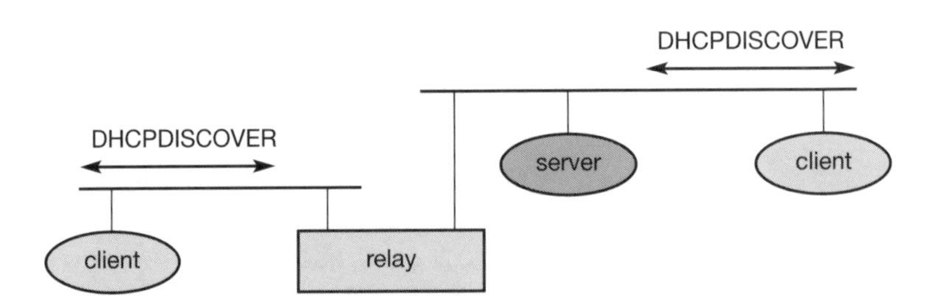

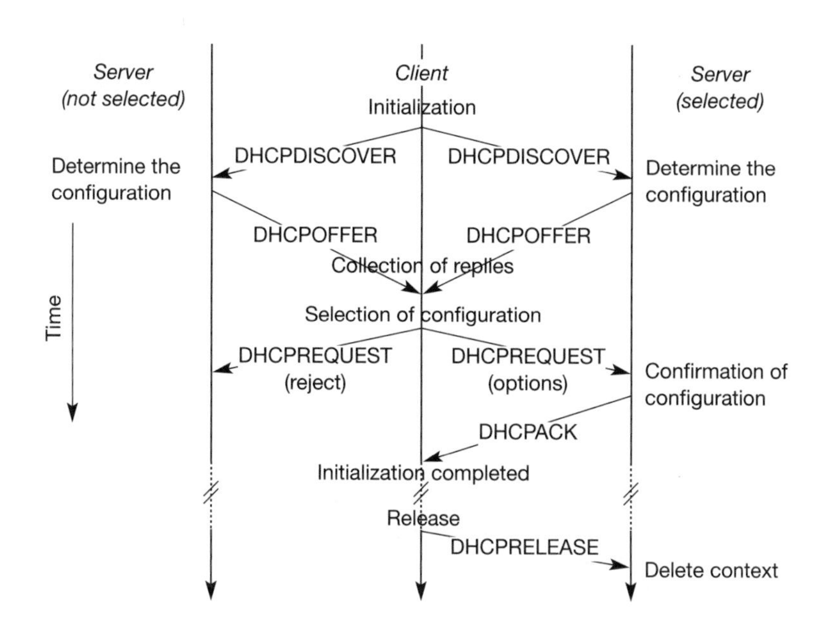

Figure 8.18 Client initialization via DHCP

A typical initialization of a DHCP client is shown in Figure 8.18. The figure shows one client and two servers. As described above, the client broadcasts a DHCPDISCOVER into the subnet. There might be a relay to forward this broadcast. In the case shown, two servers receive this broadcast and determine the configuration they can offer to the client. One example for this could be the checking of available IP addresses and choosing one for the client. Servers reply to the client's request with DHCPOFFER and offer a list of configuration parameters. The client can now choose one of the configurations offered. The client in turn replies to the servers, accepting one of the configurations and rejecting the others using DHCPREQUEST. If a server receives a DHCPREQUEST with a rejection, it can free the reserved configuration for other possible clients. The server with the configuration accepted by the client now confirms the configuration with DHCPACK. This completes the initialization phase.

If a client leaves a subnet, it should release the configuration received by the server using DHCPRELEASE. Now the server can free the context stored for the client and offer the configuration again. The configuration a client gets from a server is only leased for a certain amount of time, it has to be reconfirmed from time to time. Otherwise the server will free the configuration. This timeout of configuration helps in the case of crashed nodes or nodes moved away without releasing the context.

DHCP is a good candidate for supporting the acquisition of care-of-addresses for mobile nodes. The same holds for all other parameters needed, such as addresses of the default router, DNS servers, the timeserver etc. A DHCP server should be located in the subnet of the access point of the mobile node, or at least a DHCP relay should provide forwarding of the messages. RFC 3118

(Drohms, 2001) specifies authentication for DHCP messages which is needed to protect mobile nodes from malicious DHCP servers. Without authentication, the mobile node cannot trust a DHCP server, and the DHCP server cannot trust the mobile node.

8.3 Mobile ad-hoc networks

Mobility support described in sections 8.1 and 8.2 relies on the existence of at least some infrastructure. Mobile IP requires, e.g., a home agent, tunnels, and default routers. DHCP requires servers and broadcast capabilities of the medium reaching all participants or relays to servers. Cellular phone networks (see chapter 4) require base stations, infrastructure networks etc.

However, there may be several situations where users of a network cannot rely on an infrastructure, it is too expensive, or there is none at all. In these situations mobile ad-hoc networks are the only choice. It is important to note that this section focuses on so-called multi-hop ad-hoc networks when describing ad-hoc networking. The ad-hoc setting up of a connection with an infrastructure is not the main issue here. These networks should be mobile and use wireless communications. Examples for the use of such mobile, wireless, multi-hop ad-hoc networks, which are only called ad-hoc networks here for simplicity, are:

- **Instant infrastructure:** Unplanned meetings, spontaneous interpersonal communications etc. cannot rely on any infrastructure. Infrastructures need planning and administration. It would take too long to set up this kind of infrastructure; therefore, ad-hoc connectivity has to be set up.
- **Disaster relief:** Infrastructures typically break down in disaster areas. Hurricanes cut phone and power lines, floods destroy base stations, fires burn servers. Emergency teams can only rely on an infrastructure they can set up themselves. No forward planning can be done, and the set-up must be extremely fast and reliable. The same applies to many military activities, which is, to be honest, one of the major driving forces behind mobile ad-hoc networking research.
- **Remote areas:** Even if infrastructures could be planned ahead, it is sometimes too expensive to set up an infrastructure in sparsely populated areas. Depending on the communication pattern, ad-hoc networks or satellite infrastructures can be a solution.
- **Effectiveness:** Services provided by existing infrastructures might be too expensive for certain applications. If, for example, only connection-oriented cellular networks exist, but an application sends only a small status information every other minute, a cheaper ad-hoc packet-oriented network might be a better solution. Registration procedures might take too long, and communication overheads might be too high with existing networks. Application-tailored ad-hoc networks can offer a better solution.

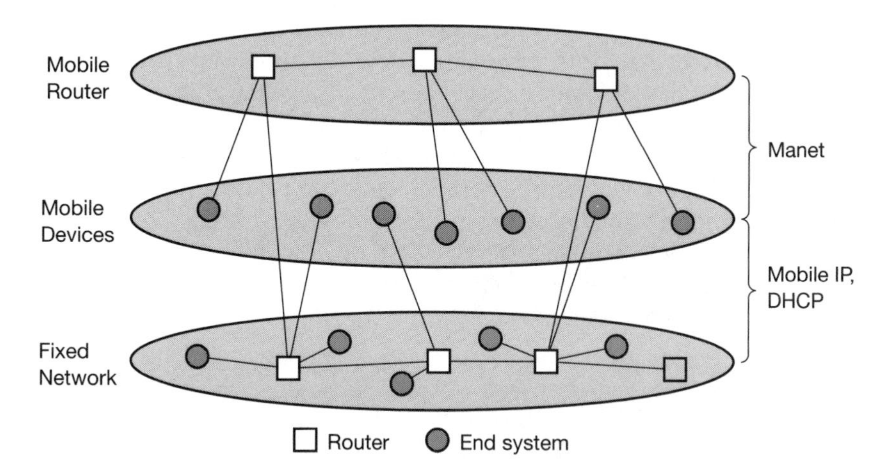

Figure 8.19
MANETs and mobile IP

Over the last few years ad-hoc networking has attracted a lot of research interest. This has led to creation of a working group at the IETF that is focussing on **mobile ad-hoc networking**, called **MANET** (MANET, 2002), (Corson, 1999). Figure 8.19 shows the relation of MANET to mobile IP and DHCP. While mobile IP and DHCP handle the connection of mobile devices to a fixed infrastructure, MANET comprises mobile routers, too. Mobile devices can be connected either directly with an infrastructure using Mobile IP for mobility support and DHCP as a source of many parameters, such as an IP address. MANET research is responsible for developing protocols and components to enable ad-hoc networking between mobile devices. It should be noted that the separation of end system and router is only a logical separation. Typically, mobile nodes in an ad-hoc scenario comprise routing and end system functionality.

The reason for having a special section about ad-hoc networks within a chapter about the network layer is that routing of data is one of the most difficult issues in ad-hoc networks. General routing problems are discussed in section 8.3.1 while the following sections give some examples for routing algorithms suited to ad-hoc networks. NB: routing functions sometimes exist in layer 2, not just in the network layer (layer 3) of the reference model. Bluetooth (see chapter 7), for example, offers forwarding/routing capabilities in layer 2 based on MAC addresses for ad-hoc networks.

One of the first ad-hoc wireless networks was the packet radio network started by ARPA in 1973. It allowed up to 138 nodes in the ad-hoc network and used IP packets for data transport. This made an easy connection possible to the ARPAnet, the starting point of today's Internet. Twenty radio channels between 1718.4–1840 MHz were used offering 100 or 400 kbit/s. The system used DSSS with 128 or 32 chips/bit.

A variant of distance vector routing was used in this ad-hoc network (Perlman, 1992). In this approach, each node sends a routing advertisement every 7.5 s. These advertisements contain a neighbor table with a list of link qualities to each neighbor. Each node updates the local routing table according to the distance vector algorithm based on these advertisements. Received packets also help to update the routing table. A sender now transmits a packet to its first hop neighbor using the local neighbor table. Each node forwards a packet received based on its own local neighbor table. Several enhancements to this simple scheme are needed to avoid routing loops and to reflect the possibly fast changing topology. The following sections discuss routing problems and enhanced routing mechanisms for ad-hoc networks in more detail. Perkins (2001a) comprises a collection of many routing protocols together with some initial performance considerations.

8.3.1 Routing

While in wireless networks with infrastructure support a base station always reaches all mobile nodes, this is not always the case in an ad-hoc network. A destination node might be out of range of a source node transmitting packets. Routing is needed to find a path between source and destination and to forward the packets appropriately. In wireless networks using an infrastructure, cells have been defined. Within a cell, the base station can reach all mobile nodes without routing via a broadcast. In the case of ad-hoc networks, each node must be able to forward data for other nodes. This creates many additional problems that are discussed in the following paragraphs.

Figure 8.20 gives a simple example of an ad-hoc network. At a certain time t_1 the network topology might look as illustrated on the left side of the figure. Five nodes, N_1 to N_5, are connected depending on the current transmission characteristics between them. In this snapshot of the network, N_4 can receive N_1 over a good link, but N_1 receives N_4 only via a weak link. Links do not necessarily have the same characteristics in both directions. The reasons for this are, e.g., different antenna characteristics or transmit power. N_1 cannot receive N_2 at all, N_2 receives a signal from N_1.

Figure 8.20
Example ad-hoc network

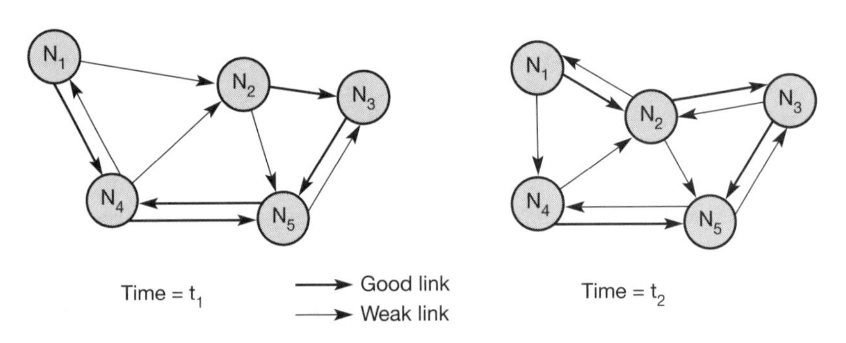

This situation can change quite fast as the snapshot at t_2 shows. N_1 cannot receive N_4 any longer, N_4 receives N_1 only via a weak link. But now N_1 has an asymmetric but bi-directional link to N_2 that did not exist before.

This very simple example already shows some fundamental differences between wired networks and ad-hoc wireless networks related to routing.

- **Asymmetric links:** Node A receives a signal from node B. But this does not tell us anything about the quality of the connection in reverse. B might receive nothing, have a weak link, or even have a better link than the reverse direction. Routing information collected for one direction is of almost no use for the other direction. However, many routing algorithms for wired networks rely on a symmetric scenario.
- **Redundant links:** Wired networks, too, have redundant links to survive link failures. However, there is only some redundancy in wired networks, which, additionally, are controlled by a network administrator. In ad-hoc networks nobody controls redundancy, so there might be many redundant links up to the extreme of a completely meshed topology. Routing algorithms for wired networks can handle some redundancy, but a high redundancy can cause a large computational overhead for routing table updates.
- **Interference:** In wired networks links exist only where a wire exists, and connections are planned by network administrators. This is not the case for wireless ad-hoc networks. Links come and go depending on transmission characteristics, one transmission might interfere with another, and nodes might overhear the transmissions of other nodes. Interference creates new problems by 'unplanned' links between nodes: if two close-by nodes forward two transmissions, they might interfere and destroy each other. On the other hand, interference might also help routing. A node can learn the topology with the help of packets it has overheard.
- **Dynamic topology:** The greatest problem for routing arises from the highly dynamic topology. The mobile nodes might move as shown in Figure 8.20 or medium characteristics might change. This results in frequent changes in topology, so snapshots are valid only for a very short period of time. In ad-hoc networks, routing tables must somehow reflect these frequent changes in topology, and routing algorithms have to be adapted. Routing algorithms used in wired networks would either react much too slowly or generate too many updates to reflect all changes in topology. Routing table updates in fixed networks, for example, take place every 30 seconds. This updating frequency might be too low to be useful for ad-hoc networks. Some algorithms rely on a complete picture of the whole network. While this works in wired networks where changes are rare, it fails completely in ad-hoc networks. The topology changes during the distribution of the 'current' snapshot of the network, rendering the snapshot useless.

Let us go back to the example network in Figure 8.20 and assume that node N_1 wants to send data to N_3 and needs an acknowledgement. If N_1 had a complete overview of the network at time t_1, which is not always the case in ad-hoc networks, it would choose the path N_1, N_2, N_3, for this requires only two hops (if we use hops as metric). Acknowledgements cannot take the same path, N_3 chooses N_3, N_5, N_4, N_1. This takes three hops and already shows that routing also strongly influences the function of higher layers. TCP, for example, makes round trip measurements assuming the same path in both directions. This is obviously wrong in the example shown, leading to misinterpretations of measurements and inefficiencies (see chapter 9).

Just a moment later, at time t_2, the topology has changed. Now N_3 cannot take the same path to send acknowledgements back to N_1, while N_1 can still take the old path to N_3. Although already more complicated than fixed networks, this example still assumes that nodes can have a complete insight into the current situation. The optimal knowledge for every node would be a description of the current connectivity between all nodes, the expected traffic flows, capacities of all links, delay of each link, and the computing and battery power of each node. While even in fixed networks traffic flows are not exactly predictable, for ad-hoc networks link capacities are additionally unknown. The capacity of each link can change from 0 to the maximum of the transmission technology used. In real ad-hoc networks no node knows all these factors, and establishing up-to-date snapshots of the network is almost impossible.

Ad-hoc networks using mobile nodes face additional problems due to hardware limitations. Using the standard routing protocols with periodic updates wastes battery power without sending any user data and disables sleep modes. Periodic updates waste bandwidth and these resources are already scarce for wireless links.

An additional problem is interference between two or more transmissions that do not use the same nodes for forwarding. If, for example, a second transmission from node N_4 to N_5 (see Figure 8.20) takes place at the same time as the transmission from N_1 to N_3, they could interfere. Interference could take place at N_2 which can receive signals from N_1 and N_4, or at N_5 receiving N_4 and N_2. If shielded correctly, there is no interference between two wires.

Considering all the additional difficulties in comparison to wired networks, the following observations concerning routing can be made for ad-hoc networks with moving nodes.

- Traditional routing algorithms known from wired networks will not work efficiently (e.g., distance vector algorithms such as RIP (Hendrik, 1988), (Malkin, 1998) converge much too slowly) or fail completely (e.g., link state algorithms such as OSPF (Moy, 1998) exchange complete pictures of the network). These algorithms have not been designed with a highly dynamic topology, asymmetric links, or interference in mind.

- Routing in wireless ad-hoc networks cannot rely on layer three knowledge alone. Information from lower layers concerning connectivity or interference can help routing algorithms to find a good path.
- Centralized approaches will not really work, because it takes too long to collect the current status and disseminate it again. Within this time the topology has already changed.
- Many nodes need routing capabilities. While there might be some without, at least one router has to be within the range of each node. Algorithms have to consider the limited battery power of these nodes.
- The notion of a connection with certain characteristics cannot work properly. Ad-hoc networks will be connectionless, because it is not possible to maintain a connection in a fast changing environment and to forward data following this connection. Nodes have to make local decisions for forwarding and send packets roughly toward the final destination.
- A last alternative to forward a packet across an unknown topology is flooding. This approach always works if the load is low, but it is very inefficient. A hop counter is needed in each packet to avoid looping, and the diameter of the ad-hoc network, i.e., the maximum number of hops, should be known. (The number of nodes can be used as an upper bound.)

Hierarchical clustering of nodes might help. If it is possible to identify certain groups of nodes belonging together, clusters can be established. While individual nodes might move faster, the whole cluster can be rather stationary. Routing between clusters might be simpler and less dynamic (see section 8.3.5.2).

The following sections give two examples for routing algorithms that were historically at the beginning of MANET research, DSDV and DSR, and useful metrics that are different from the usual hop counting. An overview of protocols follows. This is subdivided into the three categories: flat, hierarchical, and geographic-position-assisted routing based on Hong (2002).

8.3.2 Destination sequence distance vector

Destination sequence distance vector (DSDV) routing is an enhancement to distance vector routing for ad-hoc networks (Perkins, 1994). DSDV can be considered historically, however, an on-demand version (ad-hoc on-demand distance vector, AODV) is among the protocols currently discussed (see section 8.3.5). Distance vector routing is used as routing information protocol (RIP) in wired networks. It performs extremely poorly with certain network changes due to the count-to-infinity problem. Each node exchanges its neighbor table periodically with its neighbors. Changes at one node in the network propagate slowly through the network (step-by-step with every exchange). The strategies to avoid this problem which are used in fixed networks (poisoned-reverse/split-horizon (Perlman, 1992)) do not help in the case of wireless ad-hoc networks, due to the rapidly changing topology. This might create loops or unreachable regions within the network.

DSDV now adds two things to the distance vector algorithm:

- **Sequence numbers:** Each routing advertisement comes with a sequence number. Within ad-hoc networks, advertisements may propagate along many paths. Sequence numbers help to apply the advertisements in correct order. This avoids the loops that are likely with the unchanged distance vector algorithm.
- **Damping:** Transient changes in topology that are of short duration should not destabilize the routing mechanisms. Advertisements containing changes in the topology currently stored are therefore not disseminated further. A node waits with dissemination if these changes are probably unstable. Waiting time depends on the time between the first and the best announcement of a path to a certain destination.

The routing table for N_1 in Figure 8.20 would be as shown in Table 8.2.

For each node N_1 stores the next hop toward this node, the metric (here number of hops), the sequence number of the last advertisement for this node, and the time at which the path has been installed first. The table contains flags and a settling time helping to decide when the path can be assumed stable. Router advertisements from N_1 now contain data from the first, third, and fourth column: destination address, metric, and sequence number. Besides being loop-free at all times, DSDV has low memory requirements and a quick convergence via triggered updates.

8.3.3 Dynamic source routing

Imagine what happens in an ad-hoc network where nodes exchange packets from time to time, i.e., the network is only lightly loaded, and DSDV or one of the traditional distance vector or link state algorithms is used for updating routing tables. Although only some user data has to be transmitted, the nodes exchange routing information to keep track of the topology. These algorithms maintain routes between all nodes, although there may currently be no data exchange at all. This causes unnecessary traffic and prevents nodes from saving battery power.

Table 8.2 Part of a routing table for DSDV

Destination	Next hop	Metric	Sequence no.	Instal time
N_1	N_1	0	S_1–321	T_4–001
N_2	N_2	1	S_2–218	T_4–001
N_3	N_2	2	S_3–043	T_4–002
N_4	N_4	1	S_4–092	T_4–001
N_5	N_4	2	S_5–163	T_4–002

Dynamic source routing (DSR), therefore, divides the task of routing into two separate problems (Johnson, 1996), (Johnson, 2002a):

- **Route discovery:** A node only tries to discover a route to a destination if it has to send something to this destination and there is currently no known route.
- **Route maintenance:** If a node is continuously sending packets via a route, it has to make sure that the route is held upright. As soon as a node detects problems with the current route, it has to find an alternative.

The basic principle of source routing is also used in fixed networks, e.g. token rings. Dynamic source routing eliminates all periodic routing updates and works as follows. If a node needs to discover a route, it broadcasts a route request with a unique identifier and the destination address as parameters. Any node that receives a route request does the following.

- If the node has already received the request (which is identified using the unique identifier), it drops the request packet.
- If the node recognizes its own address as the destination, the request has reached its target.
- Otherwise, the node appends its own address to a list of traversed hops in the packet and broadcasts this updated route request.

Using this approach, the route request collects a list of addresses representing a possible path on its way towards the destination. As soon as the request reaches the destination, it can return the request packet containing the list to the receiver using this list in reverse order. One condition for this is that the links work bi-directionally. If this is not the case, and the destination node does not currently maintain a route back to the initiator of the request, it has to start a route discovery by itself. The destination may receive several lists containing different paths from the initiator. It could return the best path, the first path, or several paths to offer the initiator a choice.

Applying route discovery to the example in Figure 8.20 for a route from N_1 to N_3 at time t_1 results in the following.

- N_1 broadcasts the request ((N_1), id = 42, target = N_3), N_2 and N_4 receive this request.
- N_2 then broadcasts ((N_1, N_2), id = 42, target = N_3), N_4 broadcasts ((N_1, N_4), id = 42, target = N_3). N_3 and N_5 receive N_2's broadcast, N_1, N_2, and N_5 receive N_4's broadcast.
- N_3 recognizes itself as target, N_5 broadcasts ((N_1, N_2, N_5), id = 42, target = N_3). N_3 and N_4 receive N_5's broadcast. N_1, N_2, and N_5 drop N_4's broadcast packet, because they all recognize an already received route request (and N_2's broadcast reached N_5 before N_4's did).

- N_4 drops N_5's broadcast, N_3 recognizes (N_1, N_2, N_5) as an alternate, but longer route.
- N_3 now has to return the path (N_1, N_2, N_3) to N_1. This is simple assuming symmetric links working in both directions. N_3 can forward the information using the list in reverse order.

The assumption of bi-directional links holds for many ad-hoc networks. However, if links are not bi-directional, the scenario gets more complicated. The algorithm has to be applied again, in the reverse direction if the target does not maintain a current path to the source of the route request.

- N_3 has to broadcast a route request ((N_3), id = 17, target = N_1). Only N_5 receives this request.
- N_5 now broadcasts ((N_3, N_5), id = 17, target = N_1), N_3 and N_4 receive the broadcast.
- N_3 drops the request because it recognizes an already known id. N_4 broadcasts ((N_3, N_5, N_4), id = 17, target = N_1), N_5, N_2, and N_1 receive the broadcast.
- N_5 drops the request packet, N_1 recognizes itself as target, and N_2 broadcasts ((N_3, N_5, N_4, N_2), id = 17, target = N_1). N_3 and N_5 receive N_2's broadcast.
- N_3 and N_5 drop the request packet.

Now N_3 holds the list for a path from N_1 to N_3, (N_1, N_2, N_3), and N_1 knows the path from N_3 to N_1, (N_3, N_5, N_4, N_1). But N_1 still does not know how to send data to N_3! The only solution is to send the list (N_1, N_2, N_3) with the broadcasts initiated by N_3 in the reverse direction. This example shows clearly how much simpler routing can be if links are symmetrical.

The basic algorithm for route discovery can be optimized in many ways.

- To avoid too many broadcasts, each route request could contain a counter. Every node rebroadcasting the request increments the counter by one. Knowing the maximum network diameter (take the number of nodes if nothing else is known), nodes can drop a request if the counter reaches this number.
- A node can cache path fragments from recent requests. These fragments can now be used to answer other route requests much faster (if they still reflect the topology!).
- A node can also update this cache from packet headers while forwarding other packets.
- If a node overhears transmissions from other nodes, it can also use this information for shortening routes.

After a route has been discovered, it has to be maintained for as long as the node sends packets along this route. Depending on layer two mechanisms, different approaches can be taken:

- If the link layer uses an acknowledgement (as, for example, IEEE 802.11) the node can interpret this acknowledgement as an intact route.
- If possible, the node could also listen to the next node forwarding the packet, so getting a passive acknowledgement.
- A node could request an explicit acknowledgement.

Again, this situation is complicated if links are not bi-directional. If a node detects connectivity problems, it has to inform the sender of a packet, initiating a new route discovery starting from the sender. Alternatively, the node could try to discover a new route by itself.

Although dynamic source routing offers benefits compared to other algorithms by being much more bandwidth efficient, problems arise if the topology is highly dynamic and links are asymmetrical.

8.3.4 Alternative metrics

The examples shown in this chapter typically use the number of hops as routing metric. Although very simple, especially in wireless ad-hoc networks, this is not always the best choice. Even for fixed networks, e.g., bandwidth can also be a factor for the routing metric. Due to the varying link quality and the fact that different transmissions can interfere, other metrics can be more useful.

One other metric, called **least interference routing** (LIR), takes possible interference into account. Figure 8.21 shows an ad-hoc network topology. Sender S_1 wants to send a packet to receiver R_1, S_2 to R_2. Using the hop count as metric, S_1 could choose three different paths with three hops, which is also the minimum. Possible paths are (S_1, N_3, N_4, R_1), (S_1, N_3, N_2, R_1), and (S_1, N_1, N_2, R_1). S_2 would choose the only available path with only three hops (S_2, N_5, N_6, R_2). Taking interference into account, this picture changes. To calculate the possible

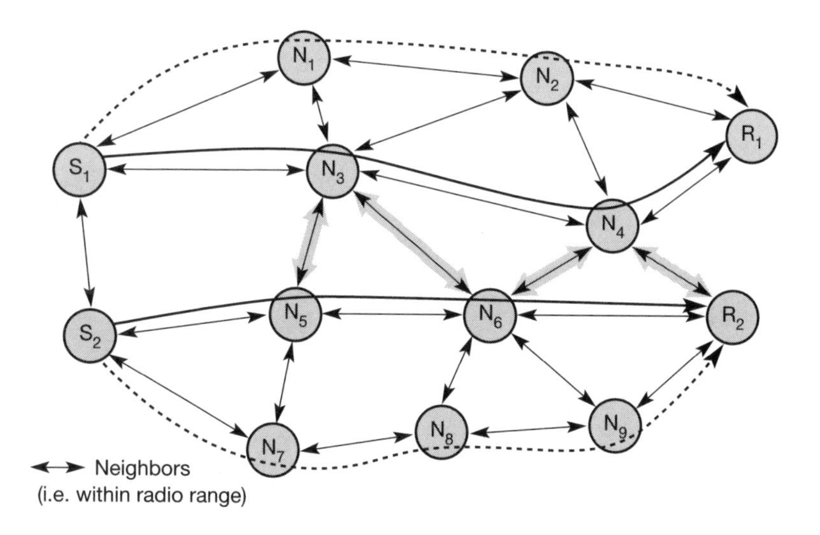

Figure 8.21
Example for least interference routing

interference of a path, each node calculates its possible interference (interference is defined here as the number of neighbors that can overhear a transmission). Every node only needs local information to compute its interference.

In this example, the interference of node N_3 is 6, that of node N_4 is 5 etc. Calculating the costs of possible paths between S_1 and R_1 results in the following:

$C1 = cost(S_1, N_3, N_4, R_1) = 16,$

$C_2 = cost(S_1, N_3, N_2, R_1) = 15,$

and $C_3 = cost(S_1, N_1, N_2, R_1) = 12.$

All three paths have the same number of hops, but the last path has the lowest cost due to interference. Thus, S_1 chooses (S_1, N_1, N_2, R_1). S_2 also computes the cost of different paths, examples are $C_4 = cost(S_2, N_5, N_6, R_2) = 16$ and $C_5 = cost(S_2, N_7, N_8, N_9, R_2) = 15$. S_2 would, therefore, choose the path $(S_2, N_7, N_8, N_9, R_2)$, although this path has one hop more than the first one.

With both transmissions taking place simultaneously, there would have been interference between them as shown in Figure 8.21. In this case, least interference routing helped to avoid interference. Taking only local decisions and not knowing what paths other transmissions take, this scheme can just lower the probability of interference. Interference can only be avoided if all senders know of all other transmissions (and the whole routing topology) and base routing on this knowledge.

Routing can take several metrics into account at the same time and weigh them. Metrics could be the number of hops h, interference i, reliability r, error rate e etc. The cost of a path could then be determined as:

$$cost = \alpha h + \beta i + \gamma r + \delta e + \ldots$$

It is not at all easy (if even possible) to choose the weights $\alpha, \beta, \gamma, \delta, \ldots$ to achieve the desired routing behavior.

8.3.5 Overview of ad-hoc routing protocols

As already mentioned, ad-hoc networking has attracted a lot of research over the last few years. This has led to the development of many new routing algorithms. They all come with special pros and cons (Royer, 1999), (Perkins, 2001a). Hong (2002) separates them into three categories: flat routing, hierarchical routing, and geographic-position-assisted routing.

8.3.5.1 Flat ad-hoc routing

Flat ad-hoc routing protocols comprise those protocols that do not set up hierarchies with clusters of nodes, special nodes acting as the head of a cluster, or different routing algorithms inside or outside certain regions. All nodes in this approach play an equal role in routing. The addressing scheme is flat.

This category again falls into two subcategories: proactive and reactive protocols. **Proactive protocols** set up tables required for routing regardless of any traffic that would require routing functionality. DSDV, as presented in section 8.3.2 is a classic member of this group. Many protocols belonging to this group are based on a link-state algorithm as known from fixed networks. Link-state algorithms flood their information about neighbors periodically or event triggered (Kurose, 2003). In mobile ad-hoc environments this method exhibits severe drawbacks: either updating takes place often enough to reflect the actual configuration of the network or it tries to minimize network load. Both goals cannot be achieved at the same time without additional mechanisms. **Fisheye state routing** (FSR, Pei, 2000) and **fuzzy sighted link-state** (FSLS, Santivanez, 2001) attack this problem by making the update period dependent on the distance to a certain hop. Routing entries corresponding to a faraway destination are propagated with lower frequency than those corresponding to nearby destinations. The result are routing tables that reflect the proximity of a node very precisely, while imprecise entries may exist for nodes further away. Other link-state protocols that try to reduce the traffic caused by link-state information dissemination are **topology broadcast based on reverse path forwarding** (TBRPF, Ogier, 2002) and **optimized link-state routing** (OLSR, Clausen, 2002). A general **advantage** of proactive protocols is that they can give QoS guarantees related to connection set-up, latency or other real-time requirements. As long as the topology does not change too fast, the routing tables reflect the current topology with a certain precision. The propagation characteristics (delay, bandwidth etc.) of a certain path between a sender and a receiver are already known before a data packet is sent. A big **disadvantage** of proactive schemes are their overheads in lightly loaded networks. Independent of any real communication the algorithm continuously updates the routing tables. This generates a lot of unnecessary traffic and drains the batteries of mobile devices.

Reactive protocols try to avoid this problem by setting up a path between sender and receiver only if a communication is waiting. The two most prominent members of this group are **dynamic source routing** (DSR, Johnson, 1996), as presented in section 8.3.3, and **ad-hoc on-demand distance vector** (AODV, Perkins, 2001a), an on-demand version of DSDV. AODV acquires and maintains routes only on demand like DSR does. A comparison of both protocols is given in Perkins (2001b), while Maltz (2001) gives some actual measurements done with DSR. Both protocols, DSR and AODV, are the leading candidates for standardization in the IETF. However, up to now there seems to be no clear winner. A dozen more reactive protocols already exist (Hong, 2002).

A clear **advantage** of on-demand protocols is scalability as long as there is only light traffic and low mobility. Mobile devices can utilize longer low-power periods as they only have to wake up for data transmission or route discovery. However, these protocols also exhibit **disadvantages**. The initial search latency may degrade the performance of interactive applications and the quality of a path is not known *a priori*. Route caching, a mechanism typically employed by on-demand protocols, proves useless in high mobility situations as routes change too frequently.

8.3.5.2 Hierarchical ad-hoc routing

Algorithms such as DSDV, AODV, and DSR only work for a smaller number of nodes and depend heavily on the mobility of nodes. For larger networks, clustering of nodes and using different routing algorithms between and within clusters can be a scalable and efficient solution. The motivation behind this approach is the locality property, meaning that if a cluster can be established, nodes typically remain within a cluster, only some change clusters. If the topology within a cluster changes, only nodes of the cluster have to be informed. Nodes of other clusters only need to know how to reach the cluster. The approach basically hides all the small details in clusters which are further away.

From time to time each node needs to get some information about the topology. Again, updates from clusters further away will be sent out less frequently compared to local updates. Clusters can be combined to form super clusters etc., building up a larger hierarchy. Using this approach, one or more nodes can act as clusterheads, representing a router for all traffic to/from the cluster. All nodes within the cluster and all other clusterheads use these as gateway for the cluster. Figure 8.22 shows an ad-hoc network with interconnection to the internet via a base station. This base station transfers data to and from the cluster heads. In this example, one cluster head also acts as head of the super cluster, routing traffic to and from the super cluster. Different routing protocols may be used inside and outside clusters.

Clusterhead-Gateway Switch Routing (CGSR, Chiang, 1997) is a typical representative of hierarchical routing algorithms based on distance vector (DV) routing (Kurose, 2003). Compared to DV protocols, the hierarchy helps to reduce routing tables tremendously. However, it might be difficult to maintain

Figure 8.22
Building hierarchies in ad-hoc networks

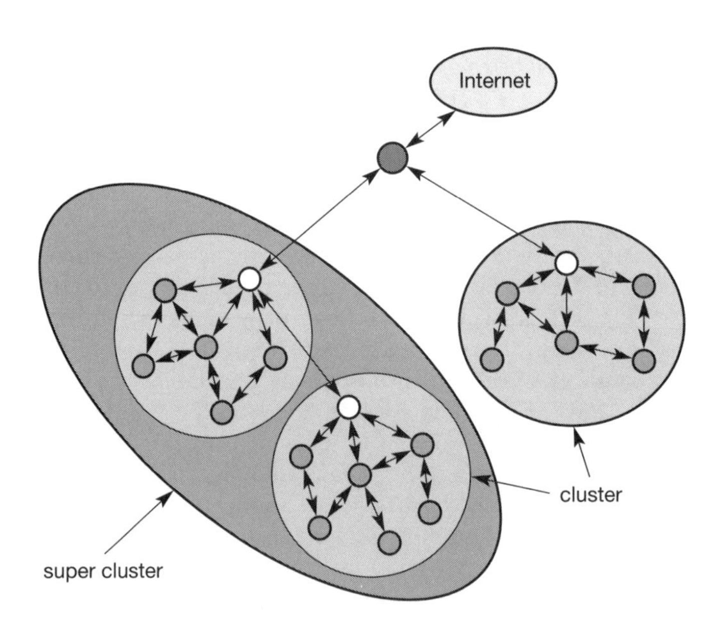

the cluster structure in a highly mobile environment. An algorithm based on the link-state (LS) principle is **hierarchical state routing** (HSR, Pei, 1999). This applies the principle of clustering recursively, creating multiple levels of clusters and clusters of clusters etc. This recursion is also reflected in a hierarchical addressing scheme. A typical hybrid hierarchical routing protocol is the **zone routing protocol** (ZRP, Haas, 2001). Each node using ZRP has a predefined zone with the node as the center. The zone comprises all other nodes within a certain hop-limit. Proactive routing is applied within the zone, while on-demand routing is used outside the zone.

Due to the established hierarchy, HSR and CGSR force the traffic to go through certain nodes which may be a bottleneck and which may lead to sub-optimal paths. Additionally, maintaining clusters or a hierarchy of clusters causes additional overheads. ZRP faces the problem of flat on-demand schemes as soon as the network size increases as many destinations are then outside the zone.

8.3.5.3 Geographic-position-assisted ad-hoc routing

If mobile nodes know their geographical position this can be used for routing purposes. This improves the overall performance of routing algorithms if geographical proximity also means radio proximity (which is typically, but not always, the case – just think of obstacles between two close-by nodes). One way to acquire position information is via the global positioning system (GPS). Mauve (2001) gives an overview of several position-based ad-hoc routing protocols.

GeoCast (Navas, 1997) allows messages to be sent to all nodes in a specific region. This is done using addresses based on geographic information instead of logical numbers. Additionally, a hierarchy of geographical routers can be employed which are responsible for regions of different scale. The **location-aided routing** protocol (LAR, Ko, 2000) is similar to DSR, but limits route discovery to certain geographical regions. Another protocol that is based on location information is **greedy perimeter stateless routing** (GPSR, Karp, 2000). This uses only the location information of neighbors that are exchanged via periodic beacon messages or via piggybacking in data packets. The main scheme of the protocol, which is the greedy part, is quite simple. Packets are always forwarded to the neighbor that is geographically closest to the destination. Additional mechanisms are applied if a dead end is reached (no neighbor is closer to the destination than the node currently holding the data packet to be forwarded).

8.4 Summary

Mobility support on the network layer is of special importance, as the network layer holds together the huge internet with the common protocol IP. Although based on possibly different wireless or wired technologies, all nodes of the network should be able to communicate. Mobile IP (an extension of the classical IP) has been designed to enable mobility in the internet without changing existing wired systems. However, mobile IP leaves some points unresolved, especially when it comes to security, efficiency of the packet flow, and support for quality of service. Some of these issues are addressed in IP version 6 making working with mobility much simpler. Many of these issues are still unresolved. Much has been done over the last years to support micro-mobility, and to support a seamless, or at least very fast, handover (Seamoby, 2002). But there are even some more fundamental questions: is the network layer the right place for fast handover or micro-mobility? Some researchers argue that, e.g., mobile phone systems or layer 2 do a better job here.

DHCP offers a fully automatic mechanism for a node to acquire all the necessary information to be integrated into a network, supporting installation of new computers and the integration of mobile computers into networks. DHCP will be a major source of care-of-addresses needed for mobile IP.

Finally, ad-hoc networks offer a completely new way of setting up mobile communications if no infrastructure is available. In these networks routing is a major topic, because there is no base station that can reach all nodes via broadcast as in cellular networks. Traditional routing algorithms do not work at all well in the highly dynamic environment of ad-hoc networks, so extensions of existing or completely new algorithms have to be applied. For larger groups of nodes only hierarchical approaches solve the routing problem, flat algorithms such as DSR or AODV do not scale well. An important difference in wireless networks is the knowledge required about layer 2 characteristics. Information about interference and acknowledgements can help in finding a good route. Location information can further optimize routing. Lu (2001) comprises several articles presenting challenges in mobile ad-hoc networking, such as higher layer performance, self-organization, power consumption, and quality of service. Zhou (1999) presents special security features for mobile ad-hoc networks. These networks cannot rely on security mechanisms provided by an infrastructure (e.g., authentication systems of mobile phone systems).

There are also critical positions towards mobile multi-hop ad-hoc networks that do not foresee applications outside the military area. For almost all civil applications, such as electronic classrooms, meeting points etc. an infrastructure is available. If ad-hoc communication is needed without an infrastructure this is typically not a multi-hop scenario but, e.g., a spontaneous exchange of data between several devices within broadcast range. The real demand for civil multi-hop ad-hoc networks has still to be shown. A premier candidate for such a scenario could be networks between cars on a highway.

The **network mobility (nemo)** working group at the IETF looks at the mobility of an entire network. This network is viewed as a single unit, which can change its point of attachment to the fixed network. One or more mobile routers perform data forwarding within this mobile network. One basic assumption of this approach is the transparency of mobility to the end-systems within the mobile network. A typical example for such a network is an onboard network of a car or a train. Similar to mobile IP the mobile routers could be connected to a home agent via a bidirectional tunnel to enable permanent IP addresses for end systems.

However, many very interesting questions of a mobile network layer are still open, new questions arise every day, and no protocol emerges as the clear winner. Altogether, the network layer remains an open and very interesting research field.

8.5 Review exercises

1. Recall routing in fixed IP networks (Kurose, 2003). Name the consequences and problems of using IP together with the standard routing protocols for mobile communications.
2. What could be quick 'solutions' and why don't they work?
3. Name the requirements for a mobile IP and justify them. Does mobile IP fulfill them all?
4. List the entities of mobile IP and describe data transfer from a mobile node to a fixed node and vice versa. Why and where is encapsulation needed?
5. How does registration on layer 3 of a mobile node work?
6. Show the steps required for a handover from one foreign agent to another foreign agent including layer 2 and layer 3.
7. Explain packet flow if two mobile nodes communicate and both are in foreign networks. What additional routes do packets take if reverse tunneling is required?
8. Explain how tunneling works in general and especially for mobile IP using IP-in-IP, minimal, and generic routing encapsulation, respectively. Discuss the advantages and disadvantages of these three methods.
9. Name the inefficiencies of mobile IP regarding data forwarding from a correspondent node to a mobile node. What are optimizations and what additional problems do they cause?
10. What advantages does the use of IPv6 offer for mobility? Where are the entities of mobile IP now?
11. What are general problems of mobile IP regarding security and support of quality of service?
12. What is the basic purpose of DHCP? Name the entities of DHCP.
13. How can DHCP be used for mobility and support of mobile IP?
14. Name the main differences between multi-hop ad-hoc networks and other networks. What advantages do these ad-hoc networks offer?
15. Why is routing in multi-hop ad-hoc networks complicated, what are the special challenges?
16. Recall the distance vector and link state routing algorithms for fixed networks. Why are both difficult to use in multi-hop ad-hoc networks?

17 What are the differences between AODV and the standard distance vector algorithm? Why are extensions needed?

18 How does dynamic source routing handle routing? What is the motivation behind dynamic source routing compared to other routing algorithms from fixed networks?

19 How does the symmetry of wireless links influence the routing algorithms proposed?

20 Why are special protocols for the support of micro mobility on the network layer needed?

21 What are the benefits of location information for routing in ad-hoc networks, which problems arise?

22 Think of ad-hoc networks with fast moving nodes, e.g., cars in a city. What problems arise even for the routing algorithms adapted to ad-hoc networks? What is the situation on highways?

8.6 References

Alexander, S., Drohms, R. (1997) *DHCP options and BOOTP vendor extensions*, RFC 2132.

Brewer, E.A., Katz, R.H., Chawathe, Y., Gribble, S.D., Hodes, T., Nguyen, G., Stemm, M., Henderson, T., Amit, E., Balakrishnan, H., Fox, A., Padmanabhan, V., Seshan, S. (1998) 'A network architecture for heterogeneous mobile computing,' *IEEE Personal Communications*, 5(5).

Campbell, A., Gomez, J., Kim, S., Valko, A., Wan, C.-Y., Turanyi, Z. (2000) 'Design, implementation and evaluation of Cellular IP,' *IEEE Personal Communications*, 7(4).

Campbell, A., Gomez, J., Kim, S., Wan, C.-Y. (2002) 'Comparison of IP Micromobility Protocols,' *IEEE Wireless Communications*, 9(1).

Chiang, C.-C., Gerla, M. (1997) *Routing and Multicast in Multihop, Mobile Wireless Networks*, proc. IEEE ICUPC 1997, San Diego.

Clausen, T., Jacquet, P., Laouiti, A., Minet, P, Muhlethaler, P., Qayyum, A., Viennot, L. (2002) *Optimized Link State Routing Protocol*, draft-ietf-manet-olsr-07.txt, (work in progress).

Corson, S., Macker, J. (1999) *Mobile ad-hoc Networking (MANET): Routing Protocol Performance Issues and Evaluation Considerations*, RFC 2501.

Deering, S. (1989) *Host extensions for IP multicasting*, RFC 1112, updated by RFC 2236.

Deering, S. (1991) *ICMP router discovery messages*, RFC 1256.

Deering, S., Hinden, R. (1998) Internet Protocol, version 6 (IPv6) Specification, RFC 2460.

Drohms, R. (1997) *Dynamic Host Configuration Protocol*, RFC 2131.

Drohms, R., Arbaugh, W. (2001) *Authentication for DHCP Messages*, RFC 3118.

Farinacci, D., Li, T., Hanks, S., Meyer, D., Traina, P. (2000) *Generic Routing Encapsulation (GRE)*, RFC 2784.

Fuller, V.; Li, T., Yu, J., Varadhan, K. (1993) *Classless Inter-Domain Routing (CIDR): an Address Assignment and Aggregation Strategy*, RFC 1519.

Haas, Z., Pearlman, M. (2001) 'The Performance of Query Control Schemes for the Zone Routing Protocol,' *ACM/IEEE Transactions on Networking*, 9(4).

Hanks, S., Li, T., Farinacci, D., Traina, P. (1994) *Generic Routing Encapsulation (GRE)*, RFC 1701.

Hendrick, C. (1988) *Routing Information Protocol*, RFC 1058, updated by RFC 2453 (RIP v2).

Hong, X., Xu, K., Gerla, M. (2002) 'Scalable Routing Protocols for Mobile ad-hoc Networks,' *IEEE Network*, 16(4).

Jacobson, V. (1990) *Compressing TCP/IP headers for low-speed serial links*, RFC 1144.

Johnson, D., Maltz, D. (1996) 'Dynamic source routing in ad-hoc wireless networks,' *Mobile Computing* (eds. Imielinski, Korth). Kluwer Academic Publishers.

Johnson, D., Maltz, D., Hu, Y.-C., Jetcheva, J. (2002a) *The dynamic source routing protocol for mobile ad hoc networks*, draft-ietf-manet-dsr-07.txt (work in progress).

Johnson, D., Perkins, C., Arkko, J. (2002b) *Mobility Support in IPv6*, draft-ietf-mobileip-ipv6-18.txt (work in progress).

Karp, B., Kung, H. (2000) *GPSR: Greedy Perimeter Stateless Routing for Wireless Networks*, proc. sixth Annual International Conference on Mobile Computing and Networking (MobiCom 2000), Boston, USA.

Kent, S., Atkinson, R. (1998) *IP Authentication Header*, RFC 2402.

Ko, Y.-B., Vaidya, N. (2000) 'Location-aided Routing (LAR) in Mobile Ad hoc Networks,' *ACM/Baltzer WINET Journal*, 6(4).

Kurose, J. F., Ross, K. (2003) Computer Networking – *A top-down approach featuring the Internet*. Addison-Wesley.

Levkowetz, H., Vaarala, S. (2002) *Mobile IP NAT/NAPT Traversal using UDP Tunnelling*, draft-ietf-mobileip-nat-traversal-05.txt (work in progress).

Lu, W., Giordano, S. (2001) 'Challenges in Mobile Ad Hoc Networking,' collection of articles, *IEEE Communications Magazine*, 39(6).

Malkin, G. (1998) *RIP version 2*, RFC 2453.

Maltz, D., Broch, J., Johnson, D. (2001) 'Lessons from a Full-Scale Multihop Wireless Ad Hoc Network Testbed,' *IEEE Personal Communications*, 8(1).

MANET (2002) Mobile Ad-hoc Networks, http://www.ietf.org/html.charters/manet-charter.html.

Mauve, M., Widmer, J., Hartenstein, H. (2001) 'A Survey on Position-Based Routing in Mobile Ad Hoc Networks,' *IEEE Network*, 15(6).

Mink, S., Pählke, F, Schäfer, G., Schiller, J. (2000a) *FATIMA: A Firewall-Aware Transparent Internet Mobility Architecture*, IEEE International Symposium on Computerts and Communication, ISCC 2000, Antibes, France.

Mink, S., Pählke, F., Schäfer, G., Schiller, J. (2000b) *Towards Secure Mobility Support for IP Networks*, IFIP International Conference on Communication Technologies, ICCT, Beijing, China.

Montenegro, G. (1998) *Reverse Tunneling for Mobile IP*, RFC 2344.

Montenegro, G. (2001) *Reverse Tunneling for Mobile IP*, RFC 3024.

Moy, J. (1998) *OSPF version 2*, RFC 2328.

Navas, J., Imielinski, T. (1997) *Geographic Addressing and Routing, proc.* Third ACM/IEEE International Conference on Mobile Computing and Networking, MobiCom'97, Budapest, Hungary.

Nichols, K., Blake, S., Baker, F., Black, D. (1998) *Definition of the Differentiated Services Field (DS Field) in the IPv4 and IPv6 Headers*, RFC 2474.

Ogier, R., Templin, F., Bellur, B., Lewis, M. (2002) *Topology Broadcast based on Reverse-Path Forwarding (TBRPF)*, draft-ietf-manet-tbrpf-05.txt (work in progress).

Pei, G., Gerla, M., Hong, X., Chiang, C.C. (1999) *A Wireless Hierarchical Routing Protocol with Group Mobility*, proc. IEEE WCNC'99, New Orleans, USA.

Pei, G., Gerla, M., Chen, T.W. (2000) *Fisheye State Routing: A Routing Scheme for Ad Hoc Wireless Networks*,' proc. ICC 2000, New Orleans, USA.

Perkins, C., Bhagwat, P. (1994) *Highly dynamic Destination-Sequenced Distance Vector routing (DSDV) for mobile computers*, proc. ACM SIGCOMM '94, London, UK.

Perkins, C. (1996a) *IP Mobility Support*, RFC 2002.

Perkins, C. (1996b) *IP Encapsulation within IP*, RFC 2003.

Perkins, C. (1996c) *Minimal Encapsulation within IP*, RFC 2004.

Perkins, C., Johnson, D.B. (1996d) *Mobility support in IPv6*, proc. ACM Mobicom 96.

Perkins, C. (1997) *Mobile IP: Design Principles and Practice*. Addison-Wesley.

Perkins, C. (2001a) *Ad Hoc Networking*. Addison-Wesley.

Perkins, C., Royer, E., Das, S., Marina, M. (2001b) 'Performance Comparison of Two On-Demand Routing Protocols for Ad Hoc Networks,' *IEEE Personal Communications*, 8(1).

Perkins, C. (2002) *IP Mobility Support for IPv4*, RFC 3344.

Perlman, R. (1992) *Interconnections: Bridges and Routers*. Addison-Wesley.

Postel, J.B. (1981) *Internet Protocol*, RFC 791.

Ramjee, R., La-Porta, T., Thuel, S., Varadhan, K., Wang, S. (1999) *HAWAII: a domain based approach for supporting mobility in wide-area wireless networks*, proc. International Conference on Network Protocols, ICNP'99, Toronto, Canada.

Reynolds, J. (2002) *Assigned Numbers: RFC 1700 is Replaced by an On-line Database*, RFC 3232, http://www.iana.org/.

Royer, E., Toh, C.K. (1999) 'A Review of Current Routing Protocols for Ad-Hoc Mobile Wireless Networks,' *IEEE Personal Communications*, 6(2).

Santivanez, C., Ramanathan, R., Stavrakakis, I. (2001) *Making Link-State Routing Scale for Ad Hoc Networks*, proc. ACM International Symposium on Mobile Ad Hoc Networking & Computing (MobiHOC 2001), Long Beach, USA.

Seamoby (2002) *Context Transfer, Handoff Candidate Discovery, and Dormant Mode Host Alerting (seamoby)*, http://www.ietf.org/html.charters/seamoby-charter.html.

Soliman, H.; Castelluccia, C., El-Malki, K., Bellier, L. (2002) *Hierarchical MIPv6 mobility management (HMIPv6)*, draft-ietf-mobileip-hmipv6-06.txt (work in progress).

Solomon, J. D. (1998) *Mobile IP – The Internet Unplugged*. Prentice Hall.

Srisuresh, P., Egevang, K. (2001) *Traditional IP Network Address Translator (Traditional NAT)*, RFC 3022.

Stevens, W. R. (1994) *TCP/IP Illustrated, Volume 1: The Protocols*. Addison Wesley.

Valko, A. (1999) 'Cellular IP – a new approach of Internet host mobility,' *ACM Computer Communication Reviews*, January.

Vixie, P., Thomson, S., Rekhter, Y., Bound, J. (1997) *Dynamic Updates in the Domain Name System (DNS UPDATE)*, RFC 2136, updated by RFC 3007.

Zhou, L., Haas, Z. (1999) 'Securing Ad Hoc Networks,' IEEE Network, 13(6).

9 Mobile transport layer

Supporting mobility only on lower layers up to the network layer is not enough to provide mobility support for applications. Most applications rely on a transport layer, such as TCP (transmission control protocol) or UDP (user datagram protocol) in the case of the internet. Two functions of the transport layer in the internet are checksumming over user data and multiplexing/demultiplexing of data from/to applications. While the network layer only addresses a host, ports in UDP or TCP allow dedicated applications to be addressed. The connectionless UDP does not offer much more than this addressing, so, the following concentrates on TCP. While UDP is connectionless and does not give certain guarantees about reliable data delivery, TCP is much more complex and, needs special mechanisms to be useful in mobile environments. Mobility support in IP (such as mobile IP) is already enough for UDP to work.

The main difference between UDP and TCP is that TCP offers connections between two applications. Within a connection TCP can give certain guarantees, such as in-order delivery or reliable data transmission using retransmission techniques. TCP has built-in mechanisms to behave in a 'network friendly' manner. If, for example, TCP encounters packet loss, it assumes network internal congestion and slows down the transmission rate. This is one of the main reasons to stay with protocols like TCP. One key requirement for new developments in the internet is 'TCP friendliness'. UDP requires that applications handle reliability, in-order delivery etc. UDP does not behave in a network friendly manner, i.e., does not pull back in case of congestion and continues to send packets into an already congested network.

The following section gives an overview of mechanisms within TCP that play an important role when using TCP for mobility. The main problem with many mechanisms is that they have been designed for situations that are completely different from those in mobile networks. Based on these problems, which can lead to a complete breakdown of TCP traffic, a set of solutions has been developed (Xylomenos, 2001). Several classical solutions are presented in sections 9.2.1 to 9.2.7; each solution has its specific strengths and weaknesses. Section 9.3 presents current efforts to adapt TCP to emerging 3G networks, while section 9.4 discusses current findings for performance enhancing proxies (PEP) in general.

9.1 Traditional TCP

This section highlights several mechanisms of the transmission control protocol (TCP) (Postel, 1981) that influence the efficiency of TCP in a mobile environment. A very detailed presentation of TCP is given in Stevens (1994).

9.1.1 Congestion control

A transport layer protocol such as TCP has been designed for fixed networks with fixed end-systems. Data transmission takes place using network adapters, fiber optics, copper wires, special hardware for routers etc. This hardware typically works without introducing transmission errors. If the software is mature enough, it will not drop packets or flip bits, so if a packet on its way from a sender to a receiver is lost in a fixed network, it is not because of hardware or software errors. The probable reason for a packet loss in a fixed network is a temporary overload some point in the transmission path, i.e., a state of congestion at a node.

Congestion may appear from time to time even in carefully designed networks. The packet buffers of a router are filled and the router cannot forward the packets fast enough because the sum of the input rates of packets destined for one output link is higher than the capacity of the output link. The only thing a router can do in this situation is to drop packets. A dropped packet is lost for the transmission, and the receiver notices a gap in the packet stream. Now the receiver does not directly tell the sender which packet is missing, but continues to acknowledge all in-sequence packets up to the missing one.

The sender notices the missing acknowledgement for the lost packet and assumes a packet loss due to congestion. Retransmitting the missing packet and continuing at full sending rate would now be unwise, as this might only increase the congestion. Although it is not guaranteed that all packets of the TCP connection take the same way through the network, this assumption holds for most of the packets. To mitigate congestion, TCP slows down the transmission rate dramatically. All other TCP connections experiencing the same congestion do exactly the same so the congestion is soon resolved. This cooperation of TCP connections in the internet is one of the main reasons for its survival as it is today. Using UDP is not a solution, because the throughput is higher compared to a TCP connection just at the beginning. As soon as everyone uses UDP, this advantage disappears. After that, congestion is standard and data transmission quality is unpredictable. Even under heavy load, TCP guarantees at least sharing of the bandwidth.

9.1.2 Slow start

TCP's reaction to a missing acknowledgement is quite drastic, but it is necessary to get rid of congestion quickly. The behavior TCP shows after the detection of congestion is called **slow start** (Kurose, 2003).

The sender always calculates a **congestion window** for a receiver. The start size of the congestion window is one segment (TCP packet). The sender sends one packet and waits for acknowledgement. If this acknowledgement arrives, the sender increases the congestion window by one, now sending two packets (congestion window = 2). After arrival of the two corresponding acknowledgements, the sender again adds 2 to the congestion window, one for each of the acknowledgements. Now the congestion window equals 4. This scheme doubles the congestion window every time the acknowledgements come back, which takes one round trip time (RTT). This is called the exponential growth of the congestion window in the slow start mechanism.

It is too dangerous to double the congestion window each time because the steps might become too large. The exponential growth stops at the **congestion threshold**. As soon as the congestion window reaches the congestion threshold, further increase of the transmission rate is only linear by adding 1 to the congestion window each time the acknowledgements come back.

Linear increase continues until a time-out at the sender occurs due to a missing acknowledgement, or until the sender detects a gap in transmitted data because of continuous acknowledgements for the same packet. In either case the sender sets the congestion threshold to half of the current congestion window. The congestion window itself is set to one segment and the sender starts sending a single segment. The exponential growth (as described above) starts once more up to the new congestion threshold, then the window grows in linear fashion.

9.1.3 Fast retransmit/fast recovery

Two things lead to a reduction of the congestion threshold. One is a sender receiving continuous acknowledgements for the same packet. This informs the sender of two things. One is that the receiver got all packets up to the acknowledged packet in sequence. In TCP, a receiver sends acknowledgements only if it receives any packets from the sender. Receiving acknowledgements from a receiver also shows that the receiver continuously receives something from the sender. The gap in the packet stream is not due to severe congestion, but a simple packet loss due to a transmission error. The sender can now retransmit the missing packet(s) before the timer expires. This behavior is called **fast retransmit** (Kurose, 2003).

The receipt of acknowledgements shows that there is no congestion to justify a slow start. The sender can continue with the current congestion window. The sender performs a **fast recovery** from the packet loss. This mechanism can improve the efficiency of TCP dramatically.

The other reason for activating slow start is a time-out due to a missing acknowledgement. TCP using fast retransmit/fast recovery interprets this congestion in the network and activates the slow start mechanism.

9.1.4 Implications on mobility

While slow start is one of the most useful mechanisms in fixed networks, it drastically decreases the efficiency of TCP if used together with mobile receivers or senders. The reason for this is the use of slow start under the wrong assumptions. From a missing acknowledgement, TCP concludes a congestion situation. While this may also happen in networks with mobile and wireless end-systems, it is not the main reason for packet loss.

Error rates on wireless links are orders of magnitude higher compared to fixed fiber or copper links. Packet loss is much more common and cannot always be compensated for by layer 2 retransmissions (ARQ) or error correction (FEC). Trying to retransmit on layer 2 could, for example, trigger TCP retransmission if it takes too long. Layer 2 now faces the problem of transmitting the same packet twice over a bad link. Detecting these duplicates on layer 2 is not an option, because more and more connections use end-to-end encryption, making it impossible to look at the packet.

Mobility itself can cause packet loss. There are many situations where a soft handover from one access point to another is not possible for a mobile end-system. For example, when using mobile IP, there could still be some packets in transit to the old foreign agent while the mobile node moves to the new foreign agent. The old foreign agent may not be able to forward those packets to the new foreign agent or even buffer the packets if disconnection of the mobile node takes too long. This packet loss has nothing to do with wireless access but is caused by the problems of rerouting traffic.

The TCP mechanism detecting missing acknowledgements via time-outs and concluding packet loss due to congestion cannot distinguish between the different causes. This is a fundamental design problem in TCP: An error control mechanism (missing acknowledgement due to a transmission error) is misused for congestion control (missing acknowledgement due to network overload). In both cases packets are lost (either due to invalid checksums or to dropping in routers). However, the reasons are completely different. TCP cannot distinguish between these two different reasons. Explicit congestion notification (ECN) mechanisms are currently discussed and some recommendations have been already given (RFC 3168, Ramakrishnan, 2001). However, RFC 3155 (Dawkins, 2001b) states that ECN cannot be used as surrogate for explicit transmission error notification. Standard TCP reacts with slow start if acknowledgements are missing, which does not help in the case of transmission errors over wireless links and which does not really help during handover. This behavior results in a severe performance degradation of an unchanged TCP if used together with wireless links or mobile nodes.

However, one cannot change TCP completely just to support mobile users or wireless links. The same arguments that were given to keep IP unchanged also apply to TCP. The installed base of computers using TCP is too large to be changed and, more important, mechanisms such as slow start keep the internet operable. Every enhancement to TCP, therefore, has to remain compatible with the standard TCP and must not jeopardize the cautious behavior of TCP in case of congestion. The following sections present some classical solutions before discussing current TCP tuning recommendations.

9.2 Classical TCP improvements

Together with the introduction of WLANs in the mid-nineties several research projects were started with the goal to increase TCP's performance in wireless and mobile environments.

9.2.1 Indirect TCP

Two competing insights led to the development of indirect TCP (I-TCP) (Bakre, 1995). One is that TCP performs poorly together with wireless links; the other is that TCP within the fixed network cannot be changed. I-TCP segments a TCP connection into a fixed part and a wireless part. Figure 9.1 shows an example with a mobile host connected via a wireless link and an access point to the 'wired' internet where the correspondent host resides. The correspondent node could also use wireless access. The following would then also be applied to the access link of the correspondent host.

Standard TCP is used between the fixed computer and the access point. No computer in the internet recognizes any changes to TCP. Instead of the mobile host, the access point now terminates the standard TCP connection, acting as a proxy. This means that the access point is now seen as the mobile host for the fixed host and as the fixed host for the mobile host. Between the access point and the mobile host, a special TCP, adapted to wireless links, is used. However, changing TCP for the wireless link is not a requirement. Even an unchanged TCP can benefit from the much shorter round trip time, starting retransmission much faster. A good place for segmenting the connection between mobile host and correspondent host is at the foreign agent of mobile IP (see chapter 8). The foreign agent controls the mobility of the mobile host anyway and can also hand over the connection to the next foreign agent when the mobile host

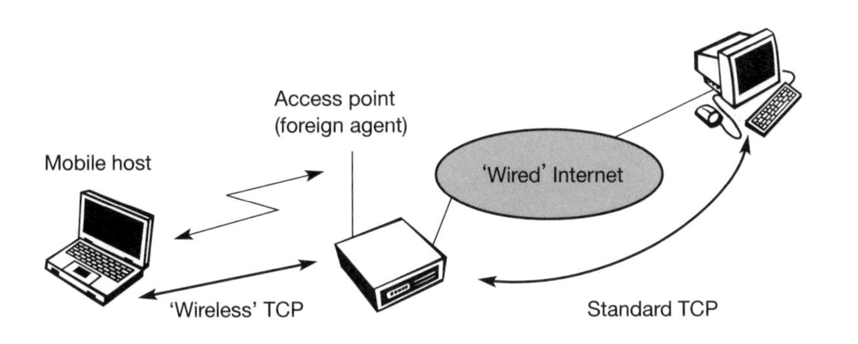

Figure 9.1
Indirect TCP segments a TCP connection into two parts

moves on. However, one can also imagine separating the TCP connections at a special server, e.g., at the entry point to a mobile phone network (e.g., IWF in GSM, GGSN in GPRS).

The correspondent host in the fixed network does not notice the wireless link or the segmentation of the connection. The foreign agent acts as a proxy and relays all data in both directions. If the correspondent host sends a packet, the foreign agent acknowledges this packet and tries to forward the packet to the mobile host. If the mobile host receives the packet, it acknowledges the packet. However, this acknowledgement is only used by the foreign agent. If a packet is lost on the wireless link due to a transmission error, the correspondent host would not notice this. In this case, the foreign agent tries to retransmit this packet locally to maintain reliable data transport.

Similarly, if the mobile host sends a packet, the foreign agent acknowledges this packet and tries to forward it to the correspondent host. If the packet is lost on the wireless link, the mobile hosts notice this much faster due to the lower round trip time and can directly retransmit the packet. Packet loss in the wired network is now handled by the foreign agent.

I-TCP requires several actions as soon as a handover takes place. As Figure 9.2 demonstrates, not only the packets have to be redirected using, e.g., mobile IP. In the example shown, the access point acts as a proxy buffering packets for retransmission. After the handover, the old proxy must forward buffered data to the new proxy because it has already acknowledged the data. As explained in chapter 8, after registration with the new foreign agent, this new foreign agent can inform the old one about its location to enable packet forwarding. Besides buffer content, the sockets of the proxy, too, must migrate to the new foreign agent located in the access point. The socket reflects the current state of the TCP connection, i.e., sequence number, addresses, ports etc. No new connection may be established for the mobile host, and the correspondent host must not see any changes in connection state.

Figure 9.2
Socket and state migration after handover of a mobile host

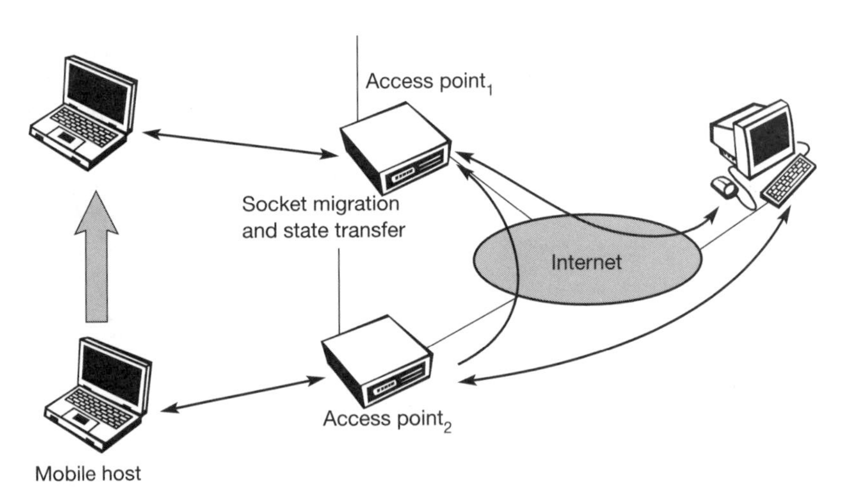

There are several advantages with I-TCP:

- I-TCP does not require any changes in the TCP protocol as used by the hosts in the fixed network or other hosts in a wireless network that do not use this optimization. All current optimizations for TCP still work between the foreign agent and the correspondent host.
- Due to the strict partitioning into two connections, transmission errors on the wireless link, i.e., lost packets, cannot propagate into the fixed network. Without partitioning, retransmission of lost packets would take place between mobile host and correspondent host across the whole network. Now only packets in sequence, without gaps leave the foreign agent.
- It is always dangerous to introduce new mechanisms into a huge network such as the internet without knowing exactly how they will behave. However, new mechanisms are needed to improve TCP performance (e.g., disabling slow start under certain circumstances), but with I-TCP only between the mobile host and the foreign agent. Different solutions can be tested or used at the same time without jeopardizing the stability of the internet. Furthermore, optimizing of these new mechanisms is quite simple because they only cover one single hop.
- The authors assume that the short delay between the mobile host and foreign agent could be determined and was independent of other traffic streams. An optimized TCP could use precise time-outs to guarantee retransmission as fast as possible. Even standard TCP could benefit from the short round trip time, so recovering faster from packet loss. Delay is much higher in a typical wide area wireless network than in wired networks due to FEC and MAC. GSM has a delay of up to 100 ms circuit switched, 200 ms and more packet switched (depending on packet size and current traffic). This is even higher than the delay on transatlantic links.
- Partitioning into two connections also allows the use of a different transport layer protocol between the foreign agent and the mobile host or the use of compressed headers etc. The foreign agent can now act as a gateway to translate between the different protocols.

But the idea of segmentation in I-TCP also comes with some **disadvantages**:

- The loss of the end-to-end semantics of TCP might cause problems if the foreign agent partitioning the TCP connection crashes. If a sender receives an acknowledgement, it assumes that the receiver got the packet. Receiving an acknowledgement now only means (for the mobile host and a correspondent host) that the foreign agent received the packet. The correspondent node does not know anything about the partitioning, so a crashing access node may also crash applications running on the correspondent node assuming reliable end-to-end delivery.

- In practical use, increased handover latency may be much more problematic. All packets sent by the correspondent host are buffered by the foreign agent besides forwarding them to the mobile host (if the TCP connection is split at the foreign agent). The foreign agent removes a packet from the buffer as soon as the appropriate acknowledgement arrives. If the mobile host now performs a handover to another foreign agent, it takes a while before the old foreign agent can forward the buffered data to the new foreign agent. During this time more packets may arrive. All these packets have to be forwarded to the new foreign agent first, before it can start forwarding the new packets redirected to it.
- The foreign agent must be a trusted entity because the TCP connections end at this point. If users apply end-to-end encryption, e.g., according to RFC 2401 (Kent, 1998a), the foreign agent has to be integrated into all security mechanisms.

9.2.2 Snooping TCP

One of the drawbacks of I-TCP is the segmentation of the single TCP connection into two TCP connections. This loses the original end-to-end TCP semantic. The following TCP enhancement works completely transparently and leaves the TCP end-to-end connection intact. The main function of the enhancement is to buffer data close to the mobile host to perform fast local retransmission in case of packet loss. A good place for the enhancement of TCP could be the foreign agent in the Mobile IP context (see Figure 9.3).

In this approach, the foreign agent buffers all packets with **destination mobile host** and additionally 'snoops' the packet flow in both directions to recognize acknowledgements (Balakrishnan, 1995), (Brewer, 1998). The reason for buffering packets toward the mobile node is to enable the foreign agent to perform a local retransmission in case of packet loss on the wireless link. The foreign agent buffers every packet until it receives an acknowledgement from the mobile host. If the foreign agent does not receive an acknowledgement from the mobile host within a certain amount of time, either the packet or the acknowledgement has been lost. Alternatively, the foreign agent could receive a duplicate ACK which also shows the loss of a packet. Now the foreign agent

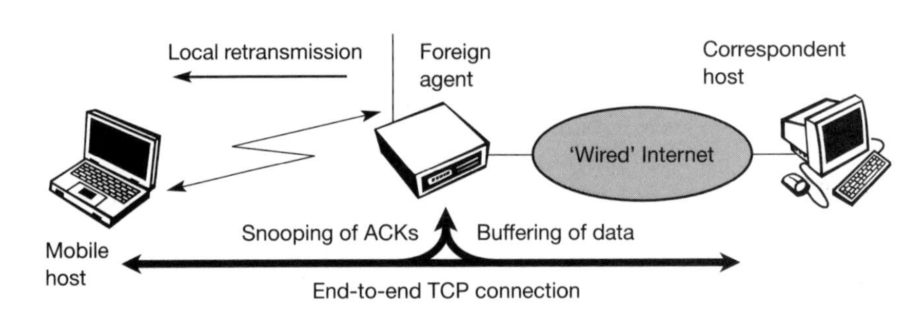

Figure 9.3 Snooping TCP as a transparent TCP extension

retransmits the packet directly from the buffer, performing a much faster retransmission compared to the correspondent host. The time out for acknowledgements can be much shorter, because it reflects only the delay of one hop plus processing time.

To remain transparent, the foreign agent must not acknowledge data to the correspondent host. This would make the correspondent host believe that the mobile host had received the data and would violate the end-to-end semantic in case of a foreign agent failure. However, the foreign agent can filter the duplicate acknowledgements to avoid unnecessary retransmissions of data from the correspondent host. If the foreign agent now crashes, the time-out of the correspondent host still works and triggers a retransmission. The foreign agent may discard duplicates of packets already retransmitted locally and acknowledged by the mobile host. This avoids unnecessary traffic on the wireless link.

Data transfer from the mobile host with **destination correspondent host** works as follows. The foreign agent snoops into the packet stream to detect gaps in the sequence numbers of TCP. As soon as the foreign agent detects a missing packet, it returns a negative acknowledgement (NACK) to the mobile host. The mobile host can now retransmit the missing packet immediately. Reordering of packets is done automatically at the correspondent host by TCP.

Extending the functions of a foreign agent with a 'snooping' TCP has several **advantages**:

- The end-to-end TCP semantic is preserved. No matter at what time the foreign agent crashes (if this is the location of the buffering and snooping mechanisms), neither the correspondent host nor the mobile host have an inconsistent view of the TCP connection as is possible with I-TCP. The approach automatically falls back to standard TCP if the enhancements stop working.
- The correspondent host does not need to be changed; most of the enhancements are in the foreign agent. Supporting only the packet stream from the correspondent host to the mobile host does not even require changes in the mobile host.
- It does not need a handover of state as soon as the mobile host moves to another foreign agent. Assume there might still be data in the buffer not transferred to the next foreign agent. All that happens is a time-out at the correspondent host and retransmission of the packets, possibly already to the new care-of address.
- It does not matter if the next foreign agent uses the enhancement or not. If not, the approach automatically falls back to the standard solution. This is one of the problems of I-TCP, since the old foreign agent may have already signaled the correct receipt of data via acknowledgements to the correspondent host and now has to transfer these packets to the mobile host via the new foreign agent.

However, the simplicity of the scheme also results in some **disadvantages:**

- Snooping TCP does not isolate the behavior of the wireless link as well as I-TCP. Assume, for example, that it takes some time until the foreign agent can successfully retransmit a packet from its buffer due to problems on the wireless link (congestion, interference). Although the time-out in the foreign agent may be much shorter than the one of the correspondent host, after a while the time-out in the correspondent host triggers a retransmission. The problems on the wireless link are now also visible for the correspondent host and not fully isolated. The quality of the isolation, which snooping TCP offers, strongly depends on the quality of the wireless link, time-out values, and further traffic characteristics. It is problematic that the wireless link exhibits very high delays compared to the wired link due to error correction on layer 2 (factor 10 and more higher). This is similar to I-TCP. If this is the case, the timers in the foreign agent and the correspondent host are almost equal and the approach is almost ineffective.
- Using negative acknowledgements between the foreign agent and the mobile host assumes additional mechanisms on the mobile host. This approach is no longer transparent for arbitrary mobile hosts.
- All efforts for snooping and buffering data may be useless if certain encryption schemes are applied end-to-end between the correspondent host and mobile host. Using IP encapsulation security payload (RFC 2406, (Kent, 1998b)) the TCP protocol header will be encrypted – snooping on the sequence numbers will no longer work. Retransmitting data from the foreign agent may not work because many security schemes prevent replay attacks – retransmitting data from the foreign agent may be misinterpreted as replay. Encrypting end-to-end is the way many applications work so it is not clear how this scheme could be used in the future. If encryption is used above the transport layer (e.g., SSL/TLS) snooping TCP can be used.

9.2.3 Mobile TCP

Dropping packets due to a handover or higher bit error rates is not the only phenomenon of wireless links and mobility – the occurrence of lengthy and/or frequent disconnections is another problem. Quite often mobile users cannot connect at all. One example is islands of wireless LANs inside buildings but no coverage of the whole campus. What happens to standard TCP in the case of disconnection?

A TCP sender tries to retransmit data controlled by a retransmission timer that doubles with each unsuccessful retransmission attempt, up to a maximum of one minute (the initial value depends on the round trip time). This means that the sender tries to retransmit an unacknowledged packet every minute and will give up after 12 retransmissions. What happens if connectivity is back earlier than this? No data is successfully transmitted for a period of one minute! The retransmission time-out is still valid and the sender has to wait. The sender also goes into slow-start because it assumes congestion.

What happens in the case of I-TCP if the mobile is disconnected? The proxy has to buffer more and more data, so the longer the period of disconnection, the more buffer is needed. If a handover follows the disconnection, which is typical, even more state has to be transferred to the new proxy. The snooping approach also suffers from being disconnected. The mobile will not be able to send ACKs so, snooping cannot help in this situation.

The **M-TCP (mobile TCP)**[1] approach has the same goals as I-TCP and snooping TCP: to prevent the sender window from shrinking if bit errors or disconnection but not congestion cause current problems. M-TCP wants to improve overall throughput, to lower the delay, to maintain end-to-end semantics of TCP, and to provide a more efficient handover. Additionally, M-TCP is especially adapted to the problems arising from lengthy or frequent disconnections (Brown, 1997).

M-TCP splits the TCP connection into two parts as I-TCP does. An unmodified TCP is used on the standard host-**supervisory host (SH)** connection, while an optimized TCP is used on the SH-MH connection. The supervisory host is responsible for exchanging data between both parts similar to the proxy in I-TCP (see Figure 9.1). The M-TCP approach assumes a relatively low bit error rate on the wireless link. Therefore, it does not perform caching/retransmission of data via the SH. If a packet is lost on the wireless link, it has to be retransmitted by the original sender. This maintains the TCP end-to-end semantics.

The SH monitors all packets sent to the MH and ACKs returned from the MH. If the SH does not receive an ACK for some time, it assumes that the MH is disconnected. It then chokes the sender by setting the sender's window size to 0. Setting the window size to 0 forces the sender to go into **persistent mode**, i.e., the state of the sender will not change no matter how long the receiver is disconnected. This means that the sender will not try to retransmit data. As soon as the SH (either the old SH or a new SH) detects connectivity again, it reopens the window of the sender to the old value. The sender can continue sending at full speed. This mechanism does not require changes to the sender's TCP.

The wireless side uses an adapted TCP that can recover from packet loss much faster. This modified TCP does not use slow start, thus, M-TCP needs a **bandwidth manager** to implement fair sharing over the wireless link.

The **advantages** of M-TCP are the following:

- It maintains the TCP end-to-end semantics. The SH does not send any ACK itself but forwards the ACKs from the MH.
- If the MH is disconnected, it avoids useless retransmissions, slow starts or breaking connections by simply shrinking the sender's window to 0.

---

1 The reader should be aware that mobile TCP does not have the same status as mobile IP, which is an internet RFC.

- Since it does not buffer data in the SH as I-TCP does, it is not necessary to forward buffers to a new SH. Lost packets will be automatically retransmitted to the new SH.

The lack of buffers and changing TCP on the wireless part also has some **disadvantages:**

- As the SH does not act as proxy as in I-TCP, packet loss on the wireless link due to bit errors is propagated to the sender. M-TCP assumes low bit error rates, which is not always a valid assumption.
- A modified TCP on the wireless link not only requires modifications to the MH protocol software but also new network elements like the bandwidth manager.

9.2.4 Fast retransmit/fast recovery

As described in section 9.1.4, moving to a new foreign agent can cause packet loss or time out at mobile hosts or corresponding hosts. TCP concludes congestion and goes into slow start, although there is no congestion. Section 9.1.3 showed the mechanisms of fast recovery/fast retransmit a host can use after receiving duplicate acknowledgements, thus concluding a packet loss without congestion.

The idea presented by Caceres (1995) is to artificially force the fast retransmit behavior on the mobile host and correspondent host side. As soon as the mobile host registers at a new foreign agent using mobile IP, it starts sending duplicated acknowledgements to correspondent hosts. The proposal is to send three duplicates. This forces the corresponding host to go into fast retransmit mode and not to start slow start, i.e., the correspondent host continues to send with the same rate it did before the mobile host moved to another foreign agent.

As the mobile host may also go into slow start after moving to a new foreign agent, this approach additionally puts the mobile host into fast retransmit. The mobile host retransmits all unacknowledged packets using the current congestion window size without going into slow start.

The **advantage** of this approach is its simplicity. Only minor changes in the mobile host's software already result in a performance increase. No foreign agent or correspondent host has to be changed.

The main **disadvantage** of this scheme is the insufficient isolation of packet losses. Forcing fast retransmission increases the efficiency, but retransmitted packets still have to cross the whole network between correspondent host and mobile host. If the handover from one foreign agent to another takes a longer time, the correspondent host will have already started retransmission. The approach focuses on loss due to handover: packet loss due to problems on the wireless link is not considered. This approach requires more cooperation between the mobile IP and TCP layer making it harder to change one without influencing the other.

9.2.5 Transmission/time-out freezing

While the approaches presented so far can handle short interruptions of the connection, either due to handover or transmission errors on the wireless link, some were designed for longer interruptions of transmission. Examples are the use of mobile hosts in a car driving into a tunnel, which loses its connection to, e.g., a satellite (however, many tunnels and subways provide connectivity via a mobile phone), or a user moving into a cell with no capacity left over. In this case, the mobile phone system will interrupt the connection. The reaction of TCP, even with the enhancements of above, would be a disconnection after a time out.

Quite often, the MAC layer has already noticed connection problems, before the connection is actually interrupted from a TCP point of view. Additionally, the MAC layer knows the real reason for the interruption and does not assume congestion, as TCP would. The MAC layer can inform the TCP layer of an upcoming loss of connection or that the current interruption is not caused by congestion. TCP can now stop sending and 'freezes' the current state of its congestion window and further timers. If the MAC layer notices the upcoming interruption early enough, both the mobile and correspondent host can be informed. With a fast interruption of the wireless link, additional mechanisms in the access point are needed to inform the correspondent host of the reason for interruption. Otherwise, the correspondent host goes into slow start assuming congestion and finally breaks the connection.

As soon as the MAC layer detects connectivity again, it signals TCP that it can resume operation at exactly the same point where it had been forced to stop. For TCP time simply does not advance, so no timers expire.

The **advantage** of this approach is that it offers a way to resume TCP connections even after longer interruptions of the connection. It is independent of any other TCP mechanism, such as acknowledgements or sequence numbers, so it can be used together with encrypted data. However, this scheme has some severe **disadvantages**. Not only does the software on the mobile host have to be changed, to be more effective the correspondent host cannot remain unchanged. All mechanisms rely on the capability of the MAC layer to detect future interruptions. Freezing the state of TCP does not help in case of some encryption schemes that use time-dependent random numbers. These schemes need resynchronization after interruption.

9.2.6 Selective retransmission

A very useful extension of TCP is the use of selective retransmission. TCP acknowledgements are cumulative, i.e., they acknowledge in-order receipt of packets up to a certain packet. If a single packet is lost, the sender has to retransmit everything starting from the lost packet (go-back-n retransmission). This obviously wastes bandwidth, not just in the case of a mobile network, but for any network (particularly those with a high path capacity, i.e., bandwidth-delay-product).

Using RFC 2018 (Mathis, 1996), TCP can indirectly request a selective retransmission of packets. The receiver can acknowledge single packets, not only trains of in-sequence packets. The sender can now determine precisely which packet is needed and can retransmit it.

The **advantage** of this approach is obvious: a sender retransmits only the lost packets. This lowers bandwidth requirements and is extremely helpful in slow wireless links. The gain in efficiency is not restricted to wireless links and mobile environments. Using selective retransmission is also beneficial in all other networks. However, there might be the minor **disadvantage** of more complex software on the receiver side, because now more buffer is necessary to resequence data and to wait for gaps to be filled. But while memory sizes and CPU performance permanently increase, the bandwidth of the air interface remains almost the same. Therefore, the higher complexity is no real disadvantage any longer as it was in the early days of TCP.

9.2.7 Transaction-oriented TCP

Assume an application running on the mobile host that sends a short request to a server from time to time, which responds with a short message. If the application requires reliable transport of the packets, it may use TCP (many applications of this kind use UDP and solve reliability on a higher, application-oriented layer).

Using TCP now requires several packets over the wireless link. First, TCP uses a three-way handshake to establish the connection. At least one additional packet is usually needed for transmission of the request, and requires three more packets to close the connection via a three-way handshake. Assuming connections with a lot of traffic or with a long duration, this overhead is minimal. But in an example of only one data packet, TCP may need seven packets altogether. Figure 9.4 shows an example for the overhead introduced by using TCP over GPRS in a web scenario. Web services are based on HTTP which requires a reliable transport system. In the internet, TCP is used for this purpose. Before a

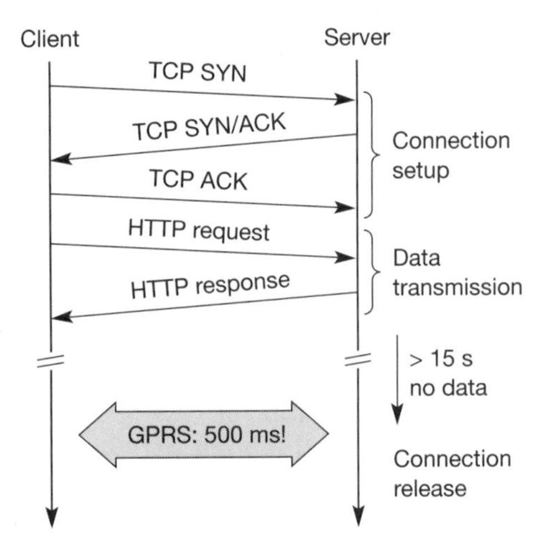

Figure 9.4
Example TCP connection setup overhead

HTTP request can be transmitted the TCP connection has to be established. This already requires three messages. If GPRS is used as wide area transport system, one-way delays of 500 ms and more are quite common. The setup of a TCP connection already takes far more than a second.

This led to the development of a transaction-oriented TCP (T/TCP, RFC 1644 (Braden, 1994)). T/TCP can combine packets for connection establishment and connection release with user data packets. This can reduce the number of packets down to two instead of seven. Similar considerations led to the development of a transaction service in WAP (see chapter 10).

The obvious **advantage** for certain applications is the reduction in the overhead which standard TCP has for connection setup and connection release. However, T/TCP is not the original TCP anymore, so it requires changes in the mobile host and all correspondent hosts, which is a major **disadvantage**. This solution no longer hides mobility. Furthermore, T/TCP exhibits several security problems (de Vivo, 1999).

Table 9.1 Overview of classical enhancements to TCP for mobility

Approach	Mechanism	Advantages	Disadvantages
Indirect TCP	Splits TCP connection into two connections	Isolation of wireless link, simple	Loss of TCP semantics, higher latency at handover, security problems
Snooping TCP	Snoops data and acknowledgements, local retransmission	Transparent for end-to-end connection, MAC integration possible	Insufficient isolation of wireless link, security problems
M-TCP	Splits TCP connection, chokes sender via window size	Maintains end-to-end semantics, handles long term and frequent disconnections	Bad isolation of wireless link, processing overhead due to bandwidth management, security problems
Fast retransmit/ fast recovery	Avoids slow-start after roaming	Simple and efficient	Mixed layers, not transparent
Transmission/ time-out freezing	Freezes TCP state at disconnection, resumes after reconnection	Independent of content, works for longer interruptions	Changes in TCP required, MAC dependent
Selective retransmission	Retransmits only lost data	Very efficient	Slightly more complex receiver software, more buffer space needed
Transaction-oriented TCP	Combines connection setup/release and data transmission	Efficient for certain applications	Changes in TCP required, not transparent, security problems

Table 9.1 shows an overview of the classical mechanisms presented together with some advantages and disadvantages. The approaches are not all exclusive, but can be combined. Selective retransmission, for example, can be used together with the others and can even be applied to fixed networks.

An additional scheme that can be used to reduce TCP overhead is **header compression** (Degermark, 1997). Using tunneling schemes as in mobile IP (see section 8.1) together with TCP, results in protocol headers of 60 byte in case of IPv4 and 100 byte for IPv6 due to the larger addresses. Many fields in the IP and TCP header remain unchanged for every packet. Only just transmitting the differences is often sufficient. Especially delay sensitive applications like, e.g., interactive games, which have small packets benefit from small headers. However, header compression experiences difficulties when error rates are high due to the loss of the common context between sender and receiver.

With the new possibilities of wireless wide area networks (WWAN) and their tremendous success, the focus of research has shifted more and more towards these 2.5G/3G networks. Up to now there are no final solutions to the problems arising when TCP is used in WWANs. However, some guidelines do exist.

9.3 TCP over 2.5/3G wireless networks

The current internet draft for TCP over 2.5G/3G wireless networks (Inamura, 2002) describes a profile for optimizing TCP over today's and tomorrow's wireless WANs such as GSM/GPRS, UMTS, or cdma2000. The configuration optimizations recommended in this draft can be found in most of today's TCP implementations so this draft does not require an update of millions of TCP stacks. The focus on 2.5G/3G for transport of internet data is important as already more than 1 billion people use mobile phones and it is obvious that the mobile phone systems will also be used to transport arbitrary internet data.

The following characteristics have to be considered when deploying applications over 2.5G/3G wireless links:

- **Data rates:** While typical data rates of today's 2.5G systems are 10–20 kbit/s uplink and 20–50 kbit/s downlink, 3G and future 2.5G systems will initially offer data rates around 64 kbit/s uplink and 115–384 kbit/s downlink. Typically, data rates are asymmetric as it is expected that users will download more data compared to uploading. Uploading is limited by the limited battery power. In cellular networks, asymmetry does not exceed 3–6 times, however, considering broadcast systems as additional distribution media (digital radio, satellite systems), asymmetry may reach a factor of 1,000. Serious problems that may reduce throughput dramatically are bandwidth oscillations due to dynamic resource sharing. To support multiple users within a radio cell, a scheduler may have to repeatedly allocate and deallocate resources for each user. This may lead to a periodic allocation and release of a high-speed channel.

- **Latency:** All wireless systems comprise elaborated algorithms for error correction and protection, such as forward error correction (FEC), check summing, and interleaving. FEC and interleaving let the round trip time (RTT) grow to several hundred milliseconds up to some seconds. The current GPRS standard specifies an average delay of less than two seconds for the transport class with the highest quality (see chapter 4).
- **Jitter:** Wireless systems suffer from large delay variations or 'delay spikes'. Reasons for sudden increase in the latency are: link outages due to temporal loss of radio coverage, blocking due to high-priority traffic, or handovers. Handovers are quite often only virtually seamless with outages reaching from some 10 ms (handover in GSM systems) to several seconds (intersystem handover, e.g., from a WLAN to a cellular system using Mobile IP without using additional mechanisms such as multicasting data to multiple access points).
- **Packet loss:** Packets might be lost during handovers or due to corruption. Thanks to link-level retransmissions the loss rates of 2.5G/3G systems due to corruption are relatively low (but still orders of magnitude higher than, e.g., fiber connections!). However, recovery at the link layer appears as jitter to the higher layers.

Based on these characteristics, (Inamura, 2002) suggests the following configuration **parameters** to adapt TCP to wireless environments:

- **Large windows:** TCP should support large enough window sizes based on the bandwidth delay product experienced in wireless systems. With the help of the windows scale option (RFC 1323) and larger buffer sizes this can be accomplished (typical buffer size settings of 16 kbyte are not enough). A larger initial window (more than the typical one segment) of 2 to 4 segments may increase performance particularly for short transmissions (a few segments in total).
- **Limited transmit:** This mechanism, defined in RFC 3042 (Allman, 2001) is an extension of Fast Retransmission/Fast Recovery (Caceres, 1995) and is particularly useful when small amounts of data are to be transmitted (standard for, e.g., web service requests).
- **Large MTU:** The larger the MTU (Maximum Transfer Unit) the faster TCP increases the congestion window. Link layers fragment PDUs for transmission anyway according to their needs and large MTUs may be used to increase performance. MTU path discovery according to RFC 1191 (IPv4) or RFC 1981 (IPv6) should be used to employ larger segment sizes instead of assuming the small default MTU.

- **Selective Acknowledgement (SACK):** SACK (RFC 2018) allows the selective retransmission of packets and is almost always beneficial compared to the standard cumulative scheme.
- **Explicit Congestion Notification (ECN):** ECN as defined in RFC 3168 (Ramakrishnan, 2001) allows a receiver to inform a sender of congestion in the network by setting the ECN-Echo flag on receiving an IP packet that has experienced congestion. This mechanism makes it easier to distinguish packet loss due to transmission errors from packet loss due to congestion. However, this can only be achieved when ECN capable routers are deployed in the network.
- **Timestamp:** TCP connections with large windows may benefit from more frequent RTT samples provided with timestamps by adapting quicker to changing network conditions. With the help of timestamps higher delay spikes can be tolerated by TCP without experiencing a spurious timeout. The effect of bandwidth oscillation is also reduced.
- **No header compression:** As the TCP header compression mechanism according to RFC 1144 does not perform well in the presence of packet losses this mechanism should not be used. Header compression according to RFC 2507 or RFC 1144 is not compatible with TCP options such as SACK or timestamps.

It is important to note that although these recommendations are still at the draft-stage, they are already used in i-mode running over FOMA as deployed in Japan and are part of the WAP 2.0 standard (aka TCP with wireless profile).

9.4 Performance enhancing proxies

RFC 3135 'Performance Enhancing Proxies Intended to Mitigate Link-Related Degradations' lists many proxy architectures that can also be beneficial for wireless and mobile internet access (Border, 2001). Some initial proxy approaches, such as snooping TCP and indirect TCP have already been discussed. In principle, proxies can be placed on any layer in a communication system. However, the approaches discussed in RFC 3135 are located in the transport and application layer. One of the key features of a proxy is its transparency with respect to the end systems, the applications and the users.

Transport layer proxies are typically used for local retransmissions, local acknowledgements, TCP acknowledgement filtering or acknowledgement handling in general. Application level proxies can be used for content filtering, content-aware compression, picture downscaling etc. Prominent examples are internet/WAP gateways making at least some of the standard web content accessible from WAP devices (see chapter 10). Figure 9.5 shows the general architecture of a wireless system connected via a proxy with the internet.

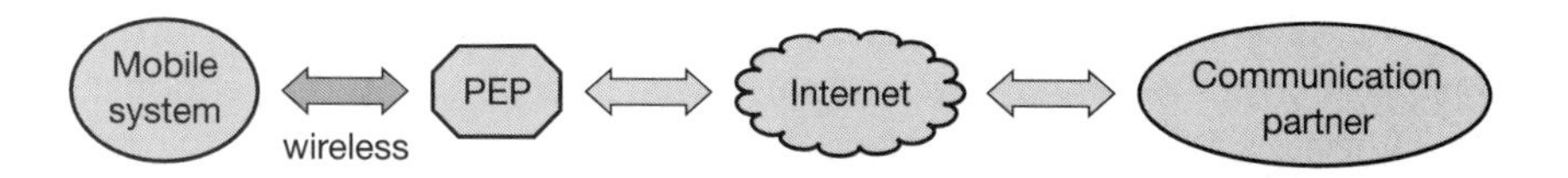

Figure 9.5
Performance enhancing proxy

However, all proxies share a common problem as they break the end-to-end semantics of a connection. According to RFC 3135, the most detrimental negative implication of breaking the end-to-end semantics is that it disables end-to-end use of IP security (RFC 2401). Using IP security with ESP (encapsulation security payload) the major part of the IP packet including the TCP header and application data is encrypted so is not accessible for a proxy. For any application one has to choose between using a performance enhancing proxy and using IP security. This is a killer criterion in any commercial environment as the only 'solution' would mean the integration of the proxy into the security association between the end systems. Typically this is not feasible as the proxy does not belong to the same organisation as the mobile node and the corresponding node.

9.5 Summary

This chapter introduced the problems of TCP as a connection-oriented protocol in a mobile environment. The basic assumptions while designing TCP have been completely different from the reality of using mobile hosts. The mechanisms of TCP that make the protocol network-friendly and keep the internet together, cause severe efficiency problems.

TCP assumes a network congestion if acknowledgements do not arrive in time. However, wireless links have much higher error rates compared to, e.g., a twisted pair or fiber optics, that way causing higher packet loss rates. The link layer may try to correct many of those errors which can hide link layer characteristics. This quite often leads to unwanted high delays or jitter. Link layer error correction should therefore be used application dependent. Mobility itself, i.e., the handover between different access points, can cause packet loss without any congestion in the network. In either case, TCP goes into a slow start state reducing its sending rate drastically.

Several classic solutions have been presented which have tried to increase the efficiency of TCP in mobile and wireless environments. This chapter showed current considerations related to TCP over 2.5G/3G networks. Besides the failure of performance enhancing proxies in an IP security enhanced network, additional issues are still open. RFC 3150 'End-to-end performance implications of **slow links**' gives recommendations for networks where hosts can saturate the available bandwidth (Dawkins, 2001a). It is recommended here, among others, that header compression following RFC 1144 or RFC 2507 should be used. It is also suggested that the timestamp option is turned off. These recommendations contrast with the 2.5G/3G recommendations described above if these links are considered slow in the sense of RFC 3150. RFC 3150 sees smaller MTU sizes as useful for slow links with lossy characteristics.

RFC 3155 'End-to-end performance implications of **links with errors**' discusses the implications of the use of wireless links for internet access on the performance of TCP (Dawkins, 2001b). Among others, it is stated that it is not possible to use the explicit congestion notification (RFC 2481) as a surrogate for explicit transmission error notification. Such a mechanism is still lacking in the internet.

It is easy to see that it is not easy to adjust TCP behavior according to the current environment. Users may roam between WLANs, 2.5G/3G cellular systems and other wireless/wired technologies. Each technology may exhibit a special behavior which can be classified as 'link with error', 'slow link' etc. Without permanent adaptation, TCP's performance will be poor as will the performance of all protocols built on top of TCP (such as HTTP, SOAP). All the problems related to the relatively high connection set-up time due to a three-way handshake still remain if a stream-oriented protocol such as TCP is used in a transaction-oriented manner. Very short lived connections and TCP still do not go together very well.

An unchanged TCP faces even more problems when used over satellite links or in general links to a spacecraft (ranging from an LEO to interplanetary deep-space probes). The main problems are the extremely high RTT, error-prone links, limited link capacity, intermittent connectivity, and asymmetric channels (up to 1,000:1). Asymmetric channels with, for example, a high bandwidth from the spacecraft to ground control, limit throughput due to the limited capacity for the acknowledgements on the return path. (Durst, 1997) presents a set of TCP enhancements, primarily a **selective negative acknowledgement (SNACK)** option, that adapt TCP to the requirements in space communication. The set of protocols developed for space communication is known as **space communications protocol standards (SCPS)**, the extended TCP is called **SCPS-transport protocol (SCPS-TP)**. RFC 2488 (Allman, 1999a) specifies the best current practise for enhancing TCP over satellite channels using standard mechanisms already available in TCP. Choosing the right parameter settings enables TCP to more effectively utilize the available capacity of the network path.

Many questions on the transport layer are still unsolved. Parameters like RTT are difficult to estimate due to high jitter. This influences many time-out values in TCP like the retransmission timer. For an initial estimation of **TCP's performance** the following formulas can be used (Karn, 2002). Both formulas assume long running connections, large enough receiver windows, and Reno TCP according to RFC 2581 (Allman, 1999b). The upper bound on the bandwidth (BW) of a TCP connection is given by $BW = \frac{0.93 \cdot MSS}{RTT \cdot \sqrt{p}}$ (Mathis, 1997). RTT is the average end-to-end round trip time of the TCP connection. The maximum segment size (MSS) is the segment size being used by the TCP connection. p denotes the packet loss probability for the path.

This simple formula neglects retransmissions due to errors. If error rate is above one per cent these retransmissions have to be considered. This leads to a more complicated formula:

$$BW = \frac{MSS}{RTT \cdot \sqrt{1.33p} + RTO \cdot p \cdot (1 + 32 \cdot p^2) \cdot \min(1, 3\sqrt{0.75p})}$$

(Padhye, 1998).

This formula also integrates the retransmission timeout (*RTO*), which TCP bases on the RTT. Typically, the simplification *RTO* = 5 *RTT* can be made. For short living connections (less than 10 packets) TCP performance is completely driven by the TCP slow start algorithm without additional enhancements.

To make things even more complicated, the reader may think of using TCP over ad-hoc networks as described in chapter 8. Again, lossy channels and mobility may lead TCP to idle states. De Oliveira (2001) gives an overview of several approaches and points out their premature state with respect to scalability and to security issues.

9.6 Review exercises

1. Compare the different types of transmission errors that can occur in wireless and wired networks. What additional role does mobility play?
2. What is the reaction of standard TCP in case of packet loss? In what situation does this reaction make sense and why is it quite often problematic in the case of wireless networks and mobility?
3. Can the problems using TCP be solved by replacing TCP with UDP? Where could this be useful and why is it quite often dangerous for network stability?
4. How and why does I-TCP isolate problems on the wireless link? What are the main drawbacks of this solution?
5. Show the interaction of mobile IP with standard TCP. Draw the packet flow from a fixed host to a mobile host via a foreign agent. Then a handover takes place. What are the following actions of mobile IP and how does TCP react?
6. Now show the required steps during handover for a solution with a PEP. What are the state and function of foreign agents, home agents, correspondent host, mobile host, PEP and care-of-address before, during, and after handover? What information has to be transferred to which entity to maintain consistency for the TCP connection?
7. What are the influences of encryption on the proposed schemes? Consider for example IP security that can encrypt the payload, i.e., the TCP packet.
8. Name further optimizations of TCP regarding the protocol overhead which are important especially for narrow band connections. Which problems may occur?

9 Assume a fixed internet connection with a round trip time of 20 ms and an error rate of 10^{-10}. Calculate the upper bound on TCP's bandwidth for a maximum segment size of 1,000 byte. Now two different wireless access networks are added. A WLAN with 2 ms additional one-way delay and an error rate of 10^{-3}, and a GPRS network with an additional RTT of 2 s and an error rate of 10^{-7}. Redo the calculation ignoring the fixed network's error rate. Compare these results with the ones derived from the second formula (use RTO = 5 RTT). Why are some results not realistic?

10 Why does the link speed not appear in the formulas presented to estimate TCP's throughput? What is wrong if the estimated bandwidth is higher than the link speed?

9.7 References

Allman, M., Glover, D., Sanchez, L. (1999a) *Enhancing TCP Over Satellite Channels using Standard Mechanisms*, RFC 2488.

Allman, M., Paxson, V., Stevens, W. (1999b) *TCP Congestion Control*, RFC 2581.

Allman, M., Balakrishnan, H., Floyd, S. (2001) *Enhancing TCP's Loss Recovery Using Limited Transmit*, RFC 3042.

Bakre, A., Badrinath, B. (1995) *I-TCP: Indirect TCP for mobile hosts*, proc. Fifteenth International Conference on Distributed Computing Systems (ICDCS), Vancouver, Canada.

Balakrishnan, H., Seshan, S., Katz, R.H. (1995) 'Improving reliable transport and handoff performance in cellular wireless networks,' *Wireless Networks*, J.C. Baltzer, no. 1.

Border, J., Kojo, M., Griner, J., Montenegro, G., Shelby, Z. (2001) *Performance Enhancing Proxies Intended to Mitigate Link-Related Degradations*, RFC 3135.

Braden, R. (1994) *T-TCP – TCP extensions for transactions functional specification*, RFC 1644.

Brewer, E.A., Katz, R.H., Chawathe, Y., Gribble, S.D., Hodes, T., Nguyen, G., Stemm, M., Henderson, T., Amit, E., Balakrishnan, H., Fox, A., Padmanabhan, V., Seshan, S. (1998) 'A network architecture for heterogeneous mobile computing,' *IEEE Personal Communications*, 5(5).

Brown, K., Singh, S. (1997) 'M-TCP: TCP for mobile cellular networks,' *ACM Computer Communications Review*, 27(5).

Caceres, R., Iftode, L. (1995) 'Improving the performance of reliable transport protocols in mobile computing environments,' *IEEE Journal on Selected Areas in Communications*, 13(5).

Dawkins, S., Kojo, M., Magret, V. (2001a) *End-to-end Performance Implications of Slow Links*, RFC 3150.

Dawkins, S., Montenegro, G., Kojo, M., Magret, V., Vaidya, N. (2001b) *End-to-end Performance Implications of Links with Errors*, RFC 3155.

De Oliveira, R., Braun, T. (2002) 'TCP in Wireless Mobile Ad-Hoc Networks,' *Technical Report*, University of Berne, TR-IAM-02-003.

De Vivo, M., de Vivo, O., Koeneke, G., Isern, G. (1999) 'Internet Vulnerabilities Related to TCP/IP and T/TCP,' *ACM Computer Communication Review*, 29(1).

Degermark, M., Engan, M., Nordgren, B., Pink, S. (1997) 'Low-loss TCP/IP header compression for wireless networks,' *Wireless Networks*, J.C. Baltzer, no. 3.

Durst, R.C., Miller, G.J., Travis, E.J. (1997) 'TCP extensions for space communications,' *Wireless Networks*, J.C. Baltzer, no. 3.

Inamura, H., Montenegro, G., Ludwig, R., Gurtov, A., Khafizov, F. (2002) *TCP over Second (2.5G) and Third (3G) Generation Wireless Networks*, draft-ietf-pilc-2.5g3g-09.txt, (work in progress).

Karn, P. (2002) *Advice for Internet Subnetwork Designers*, draft.-ietf-pilc-link-design-12.txt, (work in progress).

Kent, S., Atkinson, R. (1998a) *Security Architecture for the Internet Protocol*, RFC 2401, updated by RFC 3168.

Kent, S., Atkinson, R. (1998b) *IP Encapsulating Security Payload (ESP)*, RFC 2406.

Kurose, J.F., Ross, K. (2003) *Computer Networking – A top-down approach featuring the Internet*. Addison-Wesley.

Mathis, M., Mahdavi, J., Floyd, S., Romanow, A. (1996) *TCP selective acknowledgement options*, RFC 2018.

Mathis, M., Semke, J., Mahdavi, J., Ott, T. (1997) 'The Macroscopic Behavior of the TCP Congestion Avoidance Algorithm,' *Computer Communication Review*, 27(3).

Padhye, J., Firoiu, V., Towsley, D., Kurose, J. (1998) 'Modeling TCP Throughput: a Simple Model and its Empirical Validation,' *UMASS CMPSCI Technical Report*, TR98-008.

Postel, J. (1981) *Transmission Control Protocol*, RFC 793.

Ramakrishnan, K., Floyd, S., Black, D. (2001) *The Addition of Explicit Congestion Notification (ECN) to IP*, RFC 3168.

Stevens, W. R. (1994) *TCP/IP Illustrated, Volume 1: The Protocols*. Addison-Wesley Longman.

Xylomenos, G., Polyzos, G., Mähönen, P., Saaranen, M. (2001) 'TCP Performance Issues over Wireless Links,' *IEEE Communications Magazine*, 39(4).

Support for mobility 10

Transferring data from a sender to a single receiver or many receivers is not enough. Only applications make a communication network useful. However, to use well-known applications from fixed networks, some additional components are needed in a mobile and wireless communication system. Examples are file systems, databases, security, accounting and billing mechanisms. As mobile devices have limited energy resources, power consumption is an important issue.

This chapter focuses on two aspects, file systems/file synchronization and access to the **world wide web (www)**. Some years ago, many research projects dealt with the problems of **distributed file systems**. Some focused on the support of mobile devices, low bandwidth wireless links, and disconnected operation. The main problem for distributed, loosely coupled file systems is the maintenance of consistency. Are all views on the file system the same? What happens if a disconnected user changes data? When and how should the system propagate changes to a user? Section 10.1 discusses several problems and presents some research projects, while section 10.5 focuses on a new framework for synchronization, SyncML.

However, the success of the www shifted the focus of many projects. A lot of research effort was, and still is, put into the support of web browsing for mobile users, as the web is the application driving the internet. Section 10.2 explains some basic properties of the web and presents the hypertext transfer protocol (HTTP) and hypertext markup language (HTML) in a short overview. For this section, it is important to demonstrate the fundamental problems with HTTP and HTML if used in a mobile network with only low-bandwidth wireless access. The web has been designed for conventional computers and fixed networks. Several new system architectures try to alleviate these problems. These architectures are also good examples for client/server scenarios in wireless environments.

Section 10.3 which presents the **wireless application protocol (WAP)** version 1.x is the main part of the chapter. WAP is a common effort of many companies and organizations to set up a framework for wireless and mobile web access using many different transport systems. Examples are GSM, GPRS, and UMTS as presented in chapter 4. WAP integrates several communication layers for security mechanisms, transaction-oriented protocols, and application support. In the current www, these features are not an integral part but add-ons.

WAP combines the telephone network and the internet by integrating telephony applications into the web using its own wireless markup language (WML) and scripting language (WMLScript).

WAP, as introduced some years ago, was not a commercial success. Aside from errors in marketing WAP (announced as "Internet on the mobile phone" which WAP 1.x is not at all), the fatal combination of an interactive application (web browsing) with a connection-oriented transport system (typically GSM CSD) was one reason for the failure. I-mode, introduced in 1999 in Japan, was a big commercial success. Section 10.4 gives a quick overview of i-mode and reasons for the success.

Finally, section 10.6 presents the architecture of WAP version 2.0. This version combines the architecture and protocols of WAP 1.x with protocols and content formats known from the internet. This also reflects the fact that devices become more powerful over time so can handle more complex protocols and content formats. However, the reader should be aware that new applications and mechanisms are still evolving.

10.1 File systems

The general goal of a file system is to support efficient, transparent, and consistent access to files, no matter where the client requesting files or the server(s) offering files are located. **Efficiency** is of special importance for wireless systems as the bandwidth is low so the protocol overhead and updating operations etc. should be kept at a minimum. **Transparency** addresses the problems of location-dependent views on a file system. To support mobility, the file system should provide identical views on directories, file names, access rights etc., independent of the current location. The main problem is **consistency** as section 10.1.1 illustrates in more detail.

General problems are the limited resources on portable devices and the low bandwidth of the wireless access. File systems cannot rely on large caches in the end-system or perform many updates via the wireless link. Portable devices may also be disconnected for a longer period. Hardware and software components of portable devices often do not follow standard computer architectures or operating systems. Mobile phones, PDAs, and other devices have their own operating system, hardware, and application software. Portable devices are not as reliable as desktop systems or traditional file servers.

Standard file systems like the network file system (NFS) are very inefficient and almost unusable in a mobile and wireless environment (Honeyman, 1995). Traditional file systems do not expect disconnection, low bandwidth connections, and high latencies. To support disconnected operation, the portable device may replicate files or single objects. This can be done in advance by prefetching or while fetching data (caching). The main problem is consistency of the copy with the original data. The following section presents some more problems and solutions regarding consistency.

10.1.1 Consistency

The basic problem for distributed file systems that allow replication of data for performance reasons is the consistency of replicated objects (files, parts of files, parts of a data structure etc.). What happens, for example, if two portable devices hold copies of the same object, then one device changes the value of the object and after that, both devices read the value? Without further mechanisms, one portable device reads an old value.

To avoid inconsistencies many traditional systems apply mechanisms to maintain a permanent consistent view for all users of a file system. This **strong consistency** is achieved by atomic updates similar to database systems. A writer of an object locks the object, changes the object, and unlocks the object after the change. If an object is locked, no other device can write the object. Cached objects are invalidated after a change. Maintaining strong consistency is not only very expensive in terms of exchanging updates via the wireless link, but is also sometimes impossible. Assume a temporarily disconnected device with several objects in its cache. It is impossible to update the objects or invalidate them. Locking the cached objects may not be visible to other users.

One solution is to forbid access to disconnected objects. This would prohibit any real application based on the file system. Mobile systems have to use a **weak consistency** model for file systems. Weak consistency implies certain periods of inconsistency that have to be tolerated for performance reasons. However, the overall file system should remain consistent so conflict resolution strategies are needed for reintegration. **Reintegration** is the process of merging objects from different users resulting in one consistent file system. A user could hold a copy of an object, disconnect from the network, change the object, and reconnect again. The changed object must then be reintegrated. A **conflict** may occur, e.g., if an object has been changed by two users working with two copies. During reintegration the file system may notice that both copies differ, the conflict resolution strategy has to decide which copy to use or how to proceed. The system may detect conflicts based on time stamps, version numbering, hash values, content comparison etc.

Assume, for example, that several people are writing an article. Each person is working on one section using his or her own laptop. As long as everyone stays within his or her section, reintegration is simple. As soon as one person makes a copy of another section and starts making changes, reintegration becomes difficult and is content-dependent. The examples in the following sections show different solutions for file systems. These solutions vary in the granularity of caching and pre-fetching (files, directories, sub-trees, disk partitions), in the location of mobility support (fixed network and/or mobile computer), and in their conflict resolution strategies.

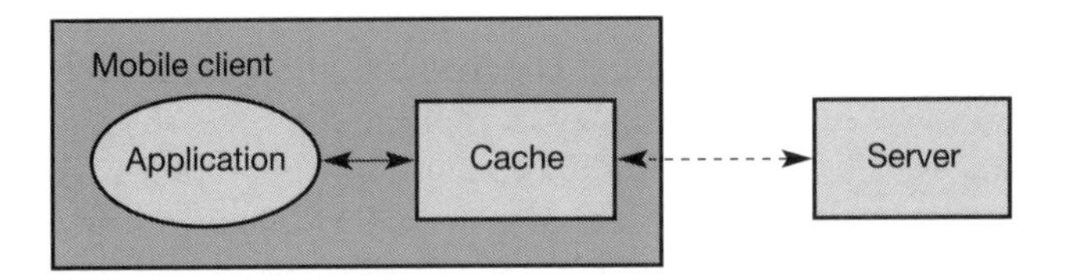

Figure 10.1
Application, cache, and server in Coda

10.1.2 Coda

The predecessor of many distributed file systems that can be used for mobile operation is the Andrew file system (AFS, (Howard, 1988)). Coda is the successor of AFS and offers two different types of replication: server replication and caching on clients. Disconnected clients work only on the cache, i.e., applications use only cached replicated files. Figure 10.1 shows the cache between an application and the server. Coda is a transparent extension of the client's cache manager. This very general architecture is valid for most of today's mobile systems that utilise a cache.

To provide all the necessary files for disconnected work, Coda offers extensive mechanisms for pre-fetching of files while still connected, called **hoarding** (Kistler, 1992). If the client is connected to the server with a strong connection (see Figure 10.2), hoarding transparently pre-fetches files currently used. This automatic data collection is necessary for it is impossible for a standard user to know all the files currently used. While standard programs and application data may be familiar to a user, he or she typically does not know anything about the numerous small system files needed in addition (e.g., profiles, shared libraries, drivers, fonts).

A user can pre-determine a list of files, which Coda should explicitly pre-fetch. Additionally, a user can assign priorities to certain programs. Coda now decides on the current cache content using the list and a least-recently-used (LRU) strategy.

As soon as the client is disconnected, applications work on the replicates (see Figure 10.2, **emulating**). Coda follows an optimistic approach and allows read and write access to all files. The system keeps a record of changed files, but does not maintain a history of changes for each file. The cache always has only one

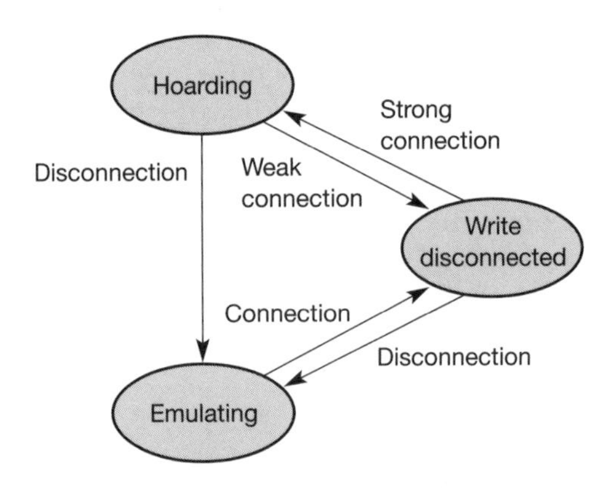

Figure 10.2
States of a client in Coda

replicate (possibly changed). After reconnection, Coda compares the replicates with the files on the server as described in Kistler (1992). If Coda notices that two different users have changed a file, reintegration of this file fails and Coda saves the changed file as a copy on the server to allow for manual reintegration.

The optimistic approach of Coda is very coarse grained, working on whole files. The success of Coda relies on the fact that files in UNIX are seldom written by more than one user. Most files are just read, only some files are changed. Experiences with Coda showed that only 0.72 per cent of all file accesses resulted in write conflicts (Satyanarayanan, 1993). Considering only user files this is reduced to 0.3 per cent. However, this low conflict rate is not applicable to arbitrary shared files as used in, e.g., computer-supported cooperative work (CSCW). The tool application specific resolver (ASR) was developed to automate conflict resolution after failed reintegration (Kumar, 1993). A general problem with these tools is that they can only work after the fact. This means that the tools have to reconstruct a history of changes based on the replicate because Coda does not record every single change.

Another problem of Coda is the definition of a conflict. Coda detects only write conflicts, i.e., if two or more users change a file. Now consider two files f_1 and f_2. One client uses values from files f_1 and f_2 to calculate something and stores the result in file f_1. The other client uses values from files f_1 and f_2 to calculate something else and stores the result in file f_2. Coda would not detect any problem during reintegration of the files. However, the results may not reflect the correct values based on the files. The order of execution plays an important role. To solve this problem, a simple transaction mechanism was introduced into Coda as an option, the so-called isolation-only transactions (IOT, (Lu, 1994)). IOT allows grouping certain operations and checks them for serial execution.

While in the beginning Coda simply distinguished the two states "hoarding" while connected and "emulating" while disconnected, the loosely connected state **write disconnected** was later integrated, (see Figure 10.2, (Mummert, 1995)). If a client is only weakly connected, Coda decides if it is worthwhile to fetch a file via this connection or to let the user wait until a better connection is available. In other words, Coda models the patience of a user and weighs it against the cost of fetching the file required by the user.

Figure 10.2 illustrates the three states of a client in Coda. The client only performs hoarding while a strong connection to the server exists. If the connection breaks completely, the client goes into emulating and uses only the cached replicates. If the client loses the strong connection and only a weak connection remains, it does not perform hoarding, but decides if it should fetch the file in case of a cache miss considering user patience and file type. The weak connection, however, is not used for reintegration of files.

10.1.3 Little Work

The distributed file system Little Work is, like Coda, an extension of AFS (Huston, 1993), (Honeyman, 1995). Little Work only requires changes to the cache manager of the client and detects write conflicts during reintegration. Little Work has no specific tools for reintegration and offers no transaction service.

However, Little Work uses more client states to maintain consistency.

- **Connected:** The operation of the client is normal, i.e., no special mechanisms from Little Work are required. This mode needs a continuous high bandwidth as available in typical office environments using, e.g., a WLAN.
- **Partially connected:** If a client has only a lower bandwidth connection, but still has the possibility to communicate continuously, it is referred to as partially connected. Examples for this type of network are packet radio networks. These networks typically charge based on the amount of traffic and not based on the duration of a connection. This client state allows to use cache consistency protocols similar to the normal state, but with a delayed write to the server to lower communication cost if the client changes the file again. This helps to avoid consistency problems, although no high-bandwidth connection is available.
- **Fetch only:** If the only network available offers connections on demand, the client goes into the fetch only state. Networks of this type are cellular networks such as GSM with costs per call. The client uses the replicates in the cache in an optimistic way, but fetches files via the communication link if they are not available in the cache. This enables a user to access all files of the server, but this also tries to minimize communication by working on replicates and reintegrate after reconnection using a continuously high bandwidth link.
- **Disconnected:** Without any network, the client is disconnected. Little Work now aborts if a cache mis-occurs, otherwise replicates are used.

10.1.4 Ficus

Ficus is a distributed file system, which is not based on a client/server approach (Popek, 1990), (Heidemann, 1992). Ficus allows the optimistic use of replicates, detects write conflicts, and solves conflicts on directories. Ficus uses so-called **gossip protocols**, an idea many other systems took over later. A mobile computer does not necessarily need to have a direct connection to a server. With the help of other mobile computers, it can propagate updates through the network until it reaches a fixed network and the server. Thus, changes on files propagate through the network step-by-step. Ficus tries to minimize the exchange of files that are valid only for a short time, e.g. temporary files. A critical issue for gossip protocols is how fast they propagate to the client that needs this information and how much unnecessary traffic it causes to propagate information to clients that are not interested.

10.1.5 MIo-NFS

The system mobile integration of NFS (MIo–NFS) is an extension of the Network File System (NFS, (Guedes, 1995)). In contrast to many other systems, MIo-NFS uses a pessimistic approach with tokens controling access to files. Only the token-holder for a specific file may change this file, so MIo-NFS avoids write conflicts. Read/write conflicts as discussed in section 10.1.2 cannot be avoided. MIo-NFS supports three different modes:

- **Connected:** The server handles all access to files as usual.
- **Loosely connected:** Clients use local replicates, exchange tokens over the network, and update files via the network.
- **Disconnected:** The client uses only local replicates. Writing is only allowed if the client is token-holder.

10.1.6 Rover

Compared to Coda, the Rover platform uses another approach to support mobility (Joseph, 1997a and 1997b). Instead of adapting existing applications for mobile devices, Rover provides a platform for developing new, mobility aware applications. Two new components have been introduced in Rover. **Relocatable dynamic objects** are objects that can be dynamically loaded into a client computer from a server (or vice-versa) to reduce client-server communication. A trade-off between transferring objects and transferring only data for objects has to be found. If a client needs an object quite often, it makes sense to migrate the object. Object migration for a single access, on the other hand, creates too much overhead. **Queued remote procedure calls** allow for non-blocking RPCs even when a host is disconnected. Requests and responses are exchanged as soon as a connection is available again. Conflict resolution is done in the server and is application specific.

Some more platforms for mobile computing were in the late nineties developed (e.g., MobiWare (Angin, 1998), a mobile middleware environment using CORBA and Java). However, while some ideas of the systems described in these sections have been integrated into commercial products, none of the above systems is in use everyday. The focus of research has shifted more and more towards the www.

10.2 World wide web

This section discusses some problems that web applications encounter when used in a mobile and wireless environment. The reader should be familiar with the basic concepts of the world wide web, its protocols (HTTP) and language (HTML). Sections 10.3 and 10.6 present a complete framework, the wireless application protocol (WAP), that handles many of the problems discussed here and, thus, this section serves as a basis for this framework. The approaches mentioned in this section are only discussed briefly in favor of a broader presentation of the WAP framework. The first two subsections give short overviews of HTTP and HTML together with their problems in wireless environments. Some approaches to improve HTML and HTTP are presented, most of them proprietary. The last subsection introduces different system architectures used for web access, each trying to improve the classic client/server scenario.

10.2.1 Hypertext transfer protocol

The **hypertext transfer protocol (HTTP)** is a stateless, lightweight, application-level protocol for data transfer between servers and clients. The first version, **HTTP/1.0** (Berners-Lee, 1996), never became a formal standard due to too many variant implementations. **HTTP/1.1** is the standard currently used by most implementations (Fielding, 1999). Krishnamurthy (1999) lists the key differences between the two versions. An HTTP **transaction** consists of an HTTP **request** issued by a client and an HTTP **response** from a server. Stateless means that all HTTP transactions are independent of each other. HTTP does not 'remember' any transaction, request, or response. This results in a very simple implementation without the need for complex state machines.

A simple request might proceed as follows. `GET` requests the source following next, here / indicates the index file in the web root directory (index.html). Additionally, the protocol `HTTP` and `version 1.1` is indicated. As this is not an introduction into HTTP, the reader is referred to the extensive literature about the web and its protocols. Everyone can try this, just send the following to port 80 of your web server (using, e.g., telnet):

```
GET / HTTP/1.1
Host: www.inf.fu-berlin.de
```

The server might answer with something similar to the following (the response):

```
HTTP/1.1 200 OK
Date: Wed, 30 Oct 2002 19:44:26 GMT
Server: Apache/1.3.12 (Unix) mod_perl/1.24
Last-Modified: Wed, 30 Oct 2002 13:16:31 GMT
ETag: "2d8190-2322-3dbfdbaf"
Accept-Ranges: bytes
Content-Length: 8994
Content-Type: text/html

<DOCTYPE HTML PUBLIC "-//W3C//DTD HTML 4.01
  Transitional//EN">
<html>
  <head>
```

```
    <title>FU-Berlin: Institut f&uuml;r
      Informatik</TITLE>
    <base href="http://www.inf.fu-berlin.de">
    <link rel="stylesheet" type="text/css"
     href="http://www.inf.fu-
       berlin.de/styles/homepage.css">
    <!--script language="JavaScript" src="fuinf.js"-->
    <!--/script-->
  </head>


  <body onResize="self.location.reload();">
...
```

The first line contains the status code (200) which shows that everything was ok (not the OK in the plain text, this is also sent with error codes indicating only that everything works). The HTTP header follows with information about date, time, server version, connection information, and type of the following content (the body of the response). Here the content is the (truncated) HTML code of the web page index.html in the root directory of the web server.

HTTP assumes a reliable underlying protocol; typically, TCP is used in the internet. While HTTP/1.0 establishes a new connection for each request, version 1.1 keeps the connection alive for multiple requests. Try using `GET / HTTP/1.0` in the above example (without the second line Host: ...). The answer will almost be the same. However, now the connection to the server is closed at once (and the line `Connection: close` is inserted into the HTTP header). Some more enhancements have been integrated into HTTP version 1.1 as explained in section 10.2.3.

Without these enhancements this means that if a web page contains five icons, two pictures, and some text, altogether eight TCP connections will be established with version 1.0 – one for the pages itself including the text, five for fetching the icons, and two for the pictures. The typical **request method** of HTTP is `GET` as already shown, which returns the requested resource. This `GET` can become conditional if an `If-Modified-Since` is added to the header, which allows for fetching newer content only. HTTP additionally makes it possible to request only the header without a body using the `HEAD` request. If a client wants to provide data to a function on a server, it can use the `POST` method.

The server may answer with different **status codes**. An example is the "200" from above indicating that the request has been accepted. A server can redirect a client to another location, it can show that user authentication is needed for a certain resource, that it refuses to fulfill the request, or that it is currently unable to handle the request.

HTTP/1.0 supported only simple **caching** mechanisms. Caching is useful to avoid unnecessary retransmissions of content that has not changed since the last access. Caches may be located anywhere between a server and a client.

Typically, each client maintains a cache locally to minimize delay when jumping back and forth on web pages. Caches can also exist for a whole company, university, region etc. The same pages will be accessed by many people, so, it makes sense to cache those pages closer to the clients. Different header information supports caching. For example, one can assign an expiry date to a page. This means that an application must not cache this page beyond expiration. A no-cache entry in the header disables caching in version 1.0 altogether. This may be useful for pages with dynamic content. Additional information regarding early caching mechanisms in HTTP can be found in Berners-Lee (1996).

HTTP (in particular version 1.0) causes **many problems** already in fixed networks but even more in wireless networks.

- **Bandwidth and delay:** HTTP has not been designed with low bandwidth/high delay connections in mind. The original environment has been networked with workstations running TCP/IP over wired networks with some Mbit/s bandwidth. HTTP protocol headers are quite large and redundant. Many information fields are transferred repeatedly with each request because HTTP is stateless. Headers are readable for humans and transferred in plain ASCII. Servers transfer content uncompressed, i.e., if applications do not compress content (as is the case for GIF or JPEG coded images), the server will not perform any compression. As TCP connections are typically used for each item on a web page (icons, images etc.), a huge overhead comes with each item in HTTP/1.0. Think of a 50-byte large icon, then a TCP connection has to be established including a three-way-handshake, data transmission, and reliable disconnection. As pointed out in chapter 9, this may imply seven PDUs exchanged between client and server! TCP has not been designed for this transaction like request/response scheme with only some data exchanged. As also shown in chapter 9, the slow-start mechanism built into TCP can cause additional problems. TCP may be too cautious in the beginning of a transmission, but before it can utilize the available bandwidth, the transmission is over. In other words, TCP never leaves the slow-start that way, causing unnecessary high delay. Another problem is caused by the DNS look-up, necessary for many items on a web page, reducing bandwidth and increasing the delay even further. Each time a browser reads a hyperlink reference to a new sever it has to resolve the logical name into an IP address before fetching the item from the server. This requires an additional request to a DNS server over the wireless link adding a round-trip time to the delay.
- **Caching:** Although useful in many cases, caching is quite often disabled by content providers. Many companies want to place advertisements on web pages and need feedback, e.g., through the number of clicks on a page to estimate the number of potential customers. With a cache between a server and a client, companies cannot get realistic feedback. Either caches need additional mechanisms to create usage profiles or caching is disabled from

the beginning using the `no-cache` keyword in the HTTP/1.0 header. Version 1.1 provides more detailed caching mechanisms. Network providers need someone to pay for pages and follow this no-caching requirement from their customers. Users suffer by downloading the same content repeatedly from the server. Many present-day pages contain dynamic objects that cannot be cached. Examples are: access counters, time, date, or other customized items. This content changes over time or for each access; sometimes at least a part of a page is static and can be cached. Many of today's companies generate customized pages on demand (via CGI, ASP etc.). It is not possible to save a bookmark to a point further down the link hierarchy. Instead, a user always has to enter the company's pages from the home page. Customization is saved in cookies.[1] This more or less prevents any caching because the names of links are also generated dynamically and the caching algorithms cannot detect access to the same content if the links differ. The homepages of companies are often created dynamically depending on the type of browser, client hardware, client location etc. Even if a cache could store some static content, it is often impossible to merge this with the dynamic remainder of a page. Mobility quite often inhibits caching because the ways of accessing web servers change over time due to changing access points. Caches at entry points of mobile networks may save some bandwidth and time. Many security mechanisms also inhibit caching. Authentication is often between a client and a server, not between a client and its cache. Keys for authentication have an associated time-out after which they are not valid anymore. Caching content for this type of secured transactions is useless.

- **POSTing:** Sending content from a client to a server can cause additional problems if the client is currently disconnected. The `POST` request cannot be fulfilled in a disconnected state, so a server could be simulated by accepting the posting via an additional process. However, this clearly causes additional problems, e.g., if the real server does not accept the posting or if the server cannot accept the deferred posting.

10.2.2 Hypertext markup language

HTML is broadly used to describe the content of web pages in the world wide web (Raggett, 1998). No matter which version is used, they all share common properties: HTML was designed for standard desktop computers connected to the internet with a fixed wire. These computers share common properties, such as a relatively high performance (especially when compared to handheld devices), a color high-resolution display (24 bit true color, 1,200 × 1,024 pixels is standard), mouse, sound system, and large hard disks.

---

1 Usually, a cookie is represented as an entry in a file that stores user-specific information for web servers on the client side. A company can store information in a cookie and retrieve this information as soon as the user visits the company's web pages again.

What do standard handheld devices offer? Due to restrictions in power consumption and form factor (they should still be "handheld"), these devices have rather small displays, some still only black and white, with a low resolution (e.g., 320 × 240), very limited user interfaces (touch screens, soft keyboards, voice commands etc.), and low performance CPUs (compared to desktops).

The network connection of desktop computers often consists of 100 Mbit/s LANs, some 5 Mbit/s DSL connections, or at least a 64 kbit/s ISDN connection. Round-trip delays are in the range of some ms, probably a few 100 ms in transatlantic links. What do today's wireless connections offer in the wide area? 10 kbit/s for standard GSM, 50 kbit/s with GPRS, and about 120 kbit/s with, e.g., UMTS. Round-trip delays are often in the range of some seconds.

Web pages using the current HTML often ignore these differences in end-systems. Pages are designed primarily for a nice presentation of content, not for efficient transfer of this content. HTML itself offers almost no way of optimizing pages for different clients or different transmission technologies. However, HTML is not the biggest problem when accessing web pages from wireless handheld clients.

Almost all of today's web pages, especially those of companies, are 'enriched' with special 'features', some using HTML, some not. These features include animated GIFs, Java Applets, Frames, ActiveX controls, multi-media content following different proprietary formats etc. Some of them can be interpreted directly by the client's browser, some need a special **plug-in**. These additional content formats cause several problems. First of all, appropriate plug-ins are often only available for the most common computer platforms, not for those many handheld devices, each with its own operating system. If a plug-in was available, the browser would still have the problem of displaying, e.g., a true-color video on a small black and white display, or displaying a GIF with many 'clickable' areas etc. Many web pages use exactly these GIFs for navigation, the user just has to click in the right area. But what if those GIFs cannot be displayed?

The approaches using content distillation or semantic compression might work with HTML, but those many additional plug-ins each need their own mechanism to translate them into a useful format for a wireless device with limited capabilities. Without additional mechanisms and a more integrated approach, large high-resolution pictures would be transferred to a mobile phone with a low-resolution display causing high costs, because the user does not exactly know the consequences of following a link. Web pages typically ignore the heterogeneity of end-systems altogether.

10.2.3 Some approaches that might help wireless access

The problems with HTTP and HTML are well known and have encouraged many different proprietary and standardized solutions (or better partial solutions). Some of the efforts are:

- **Image scaling:** If a page contains a true color, high-resolution picture, this picture can be scaled down to fewer colors, lower resolution, or to just the title of the picture. The user can then decide to download the picture separately. Clipping, zooming, or detail studies can be offered to users if they are interested in a part of the picture.
- **Content transformation:** Many documents are only available in certain formats, e.g. Postscript or portable document format (PDF) (Adobe, 2002). Before transmitting such documents to a client without the appropriate reader, a special converter could translate this document into plain text (e.g. Fox, 1996a).
- **Content extraction/semantic compression:** Besides transforming the content, e.g., headlines or keywords could be extracted from a document and presented to a user (e.g. Bickmore, 1997). The user could then decide to download more information relating to a certain headline or keyword. An abstract from some given text could be automatically generated. This semantic compression is quite difficult for arbitrary text. Extracting headlines is simpler, but sometimes useless if HTML headlines are used for layout purposes and not for structuring a document.
- **Special languages and protocols:** Other approaches try to replace HTML and HTTP with other languages and protocols better adapted to a wireless environment. Early examples are the handheld device transport protocol (HDTP) and the handheld device markup language (HDML) from Unwired Planet (King, 1997; the company was renamed to Phone.com, merged with Software.com and formed the company Openwave (2002) in November 2000 as dotcom in a name was no longer en vogue ...). Ideas from these proprietary solutions have been integrated into a broader approach (wireless application protocol) and will be discussed in the next section.
- **Push technologies:** Instead of pulling content from a server, the server could also push content to a client. This avoids the overhead of setting up connections for each item, but is only useful for some content, e.g. news, weather information, road conditions, where users do not have to interact much.

Typically, many of these enhancements will be placed in the fixed network integrated either into the server or into a gateway between the fixed and the mobile network. These application gateways are already used to provide www content to users with mobile phones and comprise entities for compression, filtering, content extraction, and automatic adaptation to network characteristics. However, many proprietary approaches typically require enhancements to standard browsers and cannot really handle the broad range of heterogeneous devices. The standard transfer protocol for web content HTTP has also been improved.

HTTP version 1.1 (RFC 2616 (Fielding, 1999)) offers several improvements:

- **Connection re-use:** Clients and servers can use the same TCP connection for several requests and responses (persistent connections, see the simple example above). Persistent connections are default in 1.1 (version 1.0 could use the keep-alive option). A client may send multiple requests at the beginning of a session, and the server can send all responses in the same order (pipelining). This avoids waiting for a response before the next request may be transmitted. Considering the high RTTs of wireless connections, pipelining improves performance dramatically.
- **Caching enhancements:** A cache may now also store cacheable responses to reduce response time and bandwidth for future, equivalent responses. Caching tries to achieve semantic transparency, i.e., a cache should not affect client or server besides increasing the performance. The correctness of cached entries has been enhanced. To fetch the most up-to-date version of an item, the item can be revalidated with the origin server, the entry can be considered as fresh enough, a warning can be included if the freshness has been violated, it can be shown that the item has not been modified etc. Web pages can contain further information about cacheability and semantic transparency. A special tag allows for the identification of content and helps to determine if two different URIs map to the same content. Several more tags determine if content is cacheable, cacheable in private caches only etc. Altogether, HTTP/1.1 defines a large set of cache-control directives.
- **Bandwidth optimization:** HTTP/1.1 supports not only compression, but also the negotiation of compression parameters and different compression styles (hop-by-hop or end-to-end). It allows for partial transmission of objects. For example, first the initial part of an image is read to determine its geometry (useful for the page layout before the whole picture is loaded). Partial transmissions can also be used to recover from network failure (partial cache contents can be completed to a full response).
- **Security:** HTTP/1.1 comprises further mechanisms to check message integrity and to authenticate clients, proxies, and servers.

Some kind of state can be introduced into the stateless behavior of HTTP by using **cookies** (Kristol, 2000). Cookies can set up a long-term "session" by storing state upon request. When a server asks to store a cookie on the client's side, this "starts" the "session". Depending on server requirements, a cookie may reflect the current state of browsing, client capabilities, user profiles etc. A session is "resumed" by returning a cookie to a server. Cookies may have additional attributes, such as a maximum age. However, this cookie mechanism is not really integrated into HTTP and cannot replace real sessions with mechanisms to suspend the session upon user request, to set-back to a certain state etc. Many users feel uncomfortable using cookies because it is not obvious what they store and what they reveal to servers.

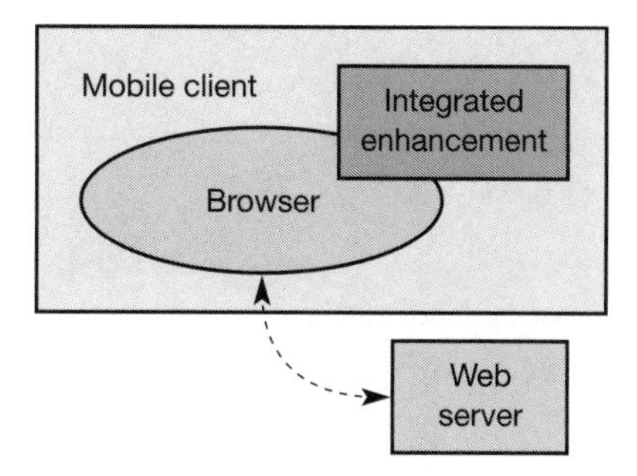

Figure 10.3 Integrated browser enhancement

10.2.4 System architecture

The classic underlying system architecture of the www is a client/server system. The client, a web browser running as an application on a computer, requests content from a server, the web server running on another computer. Without any enhancements, each click on a hyperlink initiates the transfer of the content the link points to (and possibly much more if the page contains further references – the browser fetches them automatically using one or more TCP connections). The browser uses the HTTP protocol for content transfer (see section 10.2.1). Web pages are described using HTML (see section 10.2.2) and many more (proprietary) formats.

Caching is a major topic in the web client/server scenario. While caching is also useful for wired computers because it reduces the delay of displaying previously accessed pages, it is the only way of supporting (partially) disconnected web browsers. Especially on mobile, wireless clients, network connections can be disrupted or quite often be of bad quality. The first enhancement was the integration of caching into web browsers. This is standard for all of today's browsers (e.g. Netscape, (2002), Microsoft, (2002)). Figure 10.3 shows a mobile client with a web browser running. This browser has an integrated caching mechanism as enhancement. This cache does not perform automatic pre-fetching of pages but stores already transferred content up to a certain limit. A user can then go "offline" and still browse through the cached content (pages, pictures, multi-media objects etc.). Caching strategies are very simple. The user can, for example, determine if a check for updating is performed every time he or she accesses a page, only after restarting the browser, or never (i.e., a page has to be refreshed manually).

Figure 10.4 shows an architecture for an early approach to enhance web access for mobile clients. The initial WebWhacker, for example, is a companion application for the browser that supports pre-fetching of content, caching and disconnected service (Blue Squirrel, 2002). However, this approach is not transparent for a browser as there are now two different ways of accessing content (one directly to the web server, one via the additional application).

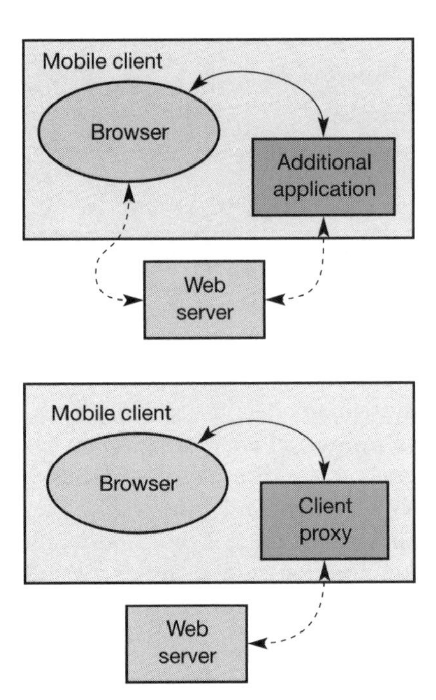

Figure 10.4
Additional application supporting browsing

Figure 10.5
Client proxy as browser support

The typical enhancements for web browsing act as a transparent proxy as shown in Figure 10.5. The browser accesses the web server through the client proxy, i.e., the proxy acts as server for the browser and as client for the web server. The proxy can now pre-fetch and cache content according to many strategies. As soon as the client is disconnected, the proxy serves the content. Many approaches follow this scheme which is independent of the browser so allows other developments (e.g., Caubweb (LoVerso, 1997), TeleWeb (Schilit, 1996), Weblicator/Domino Offline Services (Lotus, 2002), WebWhacker (Blue Squirrel, 2002)).

Example strategies for pre-fetching could be: all pages the current pages points to, all pages including those the pre-fetched pages point to (down to a certain level), pages but no pictures, all pages with the same keyword on the same server etc.

A proxy can also support a mobile client on the network side (see Figure 10.6). This network proxy can perform adaptive content transformation (e.g., semantic compression, headline extraction etc, see section 10.2.3) or pre-fetch and cache content. Pre-fetching and caching is useful in a wireless environment with higher error probability. Similar to the enhancements, for example, I-TCP achieves (see chapter 9), splitting web access into a mobile and fixed part can improve overall system performance. For the web server the network proxy acts like any fixed browser with wired access. Disconnection of the mobile client does not influence the web server. Examples for this approach are TranSend (Fox, 1996a, 1996b), Digestor (Bickmore, 1997).

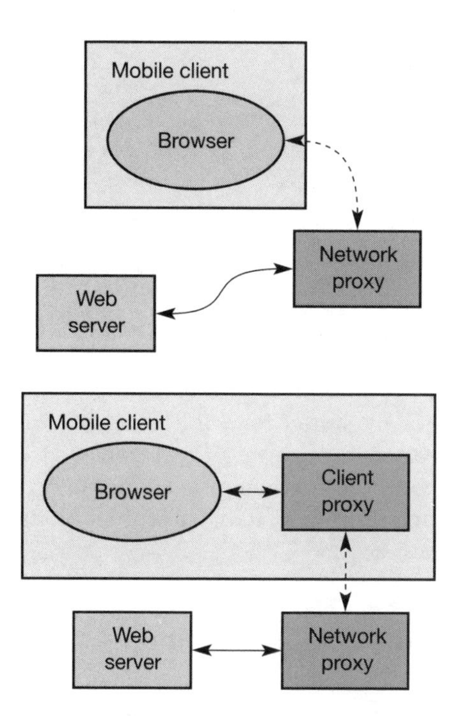

Figure 10.6
Network proxy as browser support

Figure 10.7
Client and network proxy as browser support

The benefits of client and network proxies can be combined, which results in a system architecture as illustrated in Figure 10.7. An example for this approach is WebExpress (Housel, 1996), (Floyd, 1998). Client proxy and network proxy can now interact better in pre-fetching and caching of data. The client proxy could inform, for example, the network proxy about user behavior, the network proxy can then pre-fetch pages according to this information. The whole approach is still transparent to the web server and the client browser.

You can even go one step further and implement a specialized network subsystem as shown in Figure 10.8. This solution has the same benefits as the previous one but now, content transfer can be further optimized. Examples are on line compression and replacement of transfer protocols, such as HTTP and TCP, with protocols better adapted to the mobility and wireless access of the client.

One example for such a system is Mowgli (Liljeberg, 1995), (Liljeberg, 1996). This system supports web access over cellular telephone networks, i.e., networks with low bandwidth and relatively high delay. The system not only replaces transport protocols but also performs additional content transformation needed for mobile phones. The browser still uses HTTP to the client proxy. The client proxy then uses a specialized transport service, the Mowgli data channel service, to the network proxy. Standard protocols are used to the web servers. Client and network proxy exchange their messages over long-lived Mowgli connections. This avoids TCP's slow start and the one TCP connection per HTTP request behavior of HTTP/1.0.

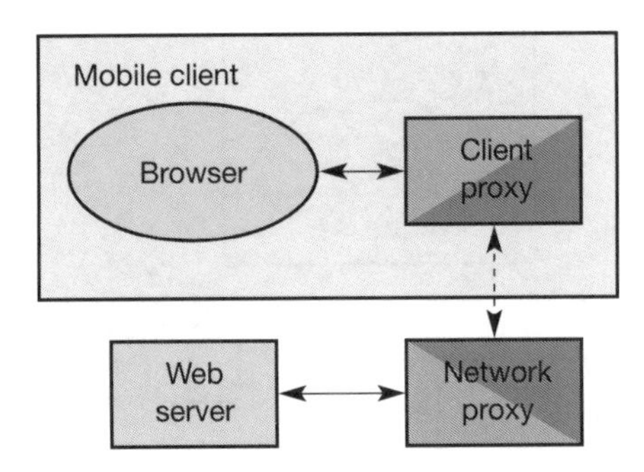

Figure 10.8 Client and network proxy with special transmission protocol

Many other enhancements are possible. Examples are server extensions to provide content especially suited for wireless access and mobile, handheld clients. The following section presents a framework that includes many of the ideas discussed in the previous subsections: enhancements to HTML, support of different system architectures, and transfer protocols adapted to the requirements or mobile, wireless access.

10.3 Wireless application protocol (version 1.x)

The growth of the internet, internet applications, and mobile communications led to many early proprietary solutions providing internet services for mobile, wireless devices. Some of the problems these partial solutions face were discussed in section 10.2 because the world wide web is the most important and fastest growing internet application. To avoid many islands of incompatible solutions, e.g. special solutions for GSM, IS-136, or certain manufacturers, the **wireless application protocol forum (WAP Forum)** was founded in June 1997 by Ericsson, Motorola, Nokia, and Unwired Planet (renamed to Phone.Com, renamed to Openwave; WAP Forum, 2000a). In summer 2002, the WAP forum together with the open mobile architecture forum and the SyncML initiative formed the **open mobile alliance** (OMA, 2002). OMA cooperates with many other standardization bodies, such as ETSI (2002), IETF (2002), 3GPP (2002). As this section describes version 1.x of WAP and the standards are still known as WAP Forum standards, this name was kept throughout the description of WAP.

The basic objectives of the WAP Forum and now of the OMA are to bring diverse internet content (e.g., web pages, push services) and other data services (e.g., stock quotes) to digital cellular phones and other wireless, mobile terminals (e.g., PDAs, laptops). Moreover, a protocol suite should enable global wireless communication across different wireless network technologies, e.g., GSM, CDPD, UMTS etc. The forum is embracing and extending existing standards and technologies of the internet wherever possible and is creating a framework for the development of contents and applications that scale across a very wide range of wireless bearer networks and wireless device types.

All solutions must be:

- **interoperable**, i.e., allowing terminals and software from different vendors to communicate with networks from different providers;
- **scaleable**, i.e., protocols and services should scale with customer needs and number of customers;
- **efficient**, i.e., provision of QoS suited to the characteristics of the wireless and mobile networks;
- **reliable**, i.e., provision of a consistent and predictable platform for deploying services; and
- **secure**, i.e., preservation of the integrity of user data, protection of devices and services from security problems.

The WAP Forum published its first set of specifications in April 1998, version 1.0, already covered many aspects of the whole architecture. Versions 1.1 (May 1999) and 1.2 (November 1999) followed. In June 2000 version 1.2.1 was released (WAP Forum, 2000a–r). This set of specifications forms the basis for the following sections and is used by most of today's WAP-enabled mobile phones. Section 10.3.1 presents the overall architecture of WAP 1.x and compares the WAP standardization with existing internet protocols and applications; sections 10.3.2 to 10.3.11 discuss the components of the WAP architecture, while section, 10.3.12, presents example configurations. All specifications are available from OMA (2002). Singhal (2001) presents a comprehensive and detailed overview of WAP 1.x. Section 10.6 presents the new WAP architecture, WAP 2.0 that was standardized in 2001.

10.3.1 Architecture

Figure 10.9 gives an overview of the WAP architecture, its protocols and components, and compares this architecture with the typical internet architecture when using the world wide web. This comparison is often cited by the WAP Forum and it helps to understand the architecture (WAP Forum, 2000a). This comparison can be misleading as not all components and protocols shown at the same layer are comparable (Khare, 1999). For consistency reasons with the existing specification, the following stays with the model as shown in Figure 10.9.

The basis for transmission of data is formed by different **bearer services**. WAP does not specify bearer services, but uses existing data services and will integrate further services. Examples are message services, such as short message service (SMS) of GSM, circuit-switched data, such as high-speed circuit switched data (HSCSD) in GSM, or packet switched data, such as general packet radio service (GPRS) in GSM. Many other bearers are supported, such as CDPD, IS-136, PHS. No special interface has been specified between the bearer service and the next higher layer, the **transport layer** with its **wireless datagram protocol (WDP)** and the additional **wireless control message protocol (WCMP)**, because the adaptation of

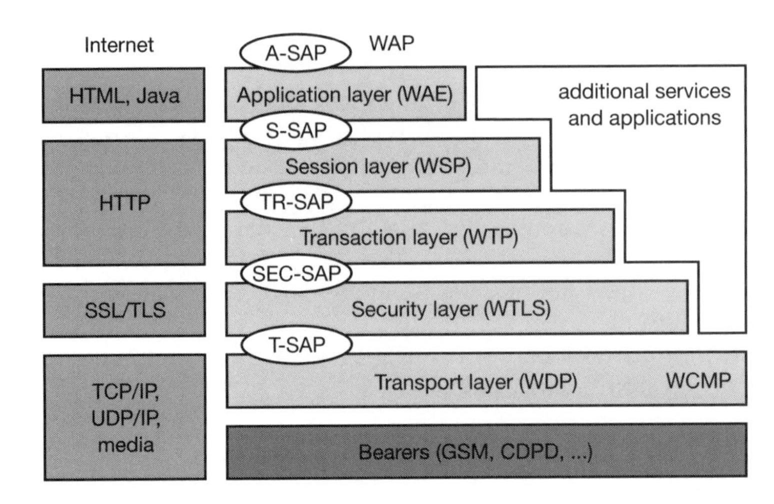

Figure 10.9 Components and interface of the WAP 1.x architecture

these protocols are bearer-specific (WAP Forum, 2000u). The transport layer offers a bearer independent, consistent datagram-oriented service to the higher layers of the WAP architecture. Communication is done transparently over one of the available bearer services. The **transport layer service access point (T-SAP)** is the common interface to be used by higher layers independent of the underlying network. WDP and WCMP are discussed in more detail in section 10.3.2.

The next higher layer, the **security layer** with its **wireless transport layer security** protocol **WTLS** offers its service at the **security SAP (SEC-SAP)**. WTLS is based on the transport layer security (TLS, formerly SSL, secure sockets layer) already known from the www. WTLS has been optimized for use in wireless networks with narrow-band channels. It can offer data integrity, privacy, authentication, and (some) denial-of-service protection. It is presented in section 10.3.3.

The WAP **transaction layer** with its **wireless transaction protocol (WTP)** offers a lightweight transaction service at the **transaction SAP (TR-SAP)**. This service efficiently provides reliable or unreliable requests and asynchronous transactions as explained in section 10.3.4. Tightly coupled to this layer is the next higher layer, if used for connection-oriented service as described in section 10.3.5. The **session layer** with the **wireless session protocol (WSP)** currently offers two services at the **session-SAP (S-SAP)**, one connection-oriented and one connectionless if used directly on top of WDP. A special service for browsing the web (WSP/B) has been defined that offers HTTP/1.1 functionality, long-lived session state, session suspend and resume, session migration and other features needed for wireless mobile access to the web.

Finally the **application layer** with the **wireless application environment (WAE)** offers a framework for the integration of different www and mobile telephony applications. It offers many protocols and services with special service access points as described in sections 10.3.6–10.3.11. The main issues here are scripting languages, special markup languages, interfaces to telephony applications, and many content formats adapted to the special requirements of small, handheld, wireless devices.

Figure 10.9 not only shows the overall WAP architecture, but also its relation to the traditional internet architecture for www applications. The WAP transport layer together with the bearers can be (roughly) compared to the services offered by TCP or UDP over IP and different media in the internet. If a bearer in the WAP architecture already offers IP services (e.g., GPRS, CDPD) then UDP is used as WDP. The TLS/SSL layer of the internet has also been adopted for the WAP architecture with some changes required for optimization. The functionality of the session and transaction layer can roughly be compared with the role of HTTP in the web architecture. However, HTTP does not offer all the additional mechanisms needed for efficient wireless, mobile access (e.g., session migration, suspend/resume). Finally, the application layer offers similar features as HTML and Java. Again, special formats and features optimized for the wireless scenario have been defined and telephony access has been added.

WAP does not always force all applications to use the whole protocol architecture. Applications can use only a part of the architecture as shown in Figure 10.9. For example, this means that, if an application does not require security but needs the reliable transport of data, it can **directly** use a service of the transaction layer. Simple applications can directly use WDP.

Different scenarios are possible for the integration of WAP components into existing wireless and fixed networks (see Figure 10.10). On the left side, different fixed networks, such as the traditional internet and the public switched telephone network (PSTN), are shown. One cannot change protocols and services of these existing networks so several new elements will be implemented between these networks and the WAP-enabled wireless, mobile devices in a wireless network on the right-hand side.

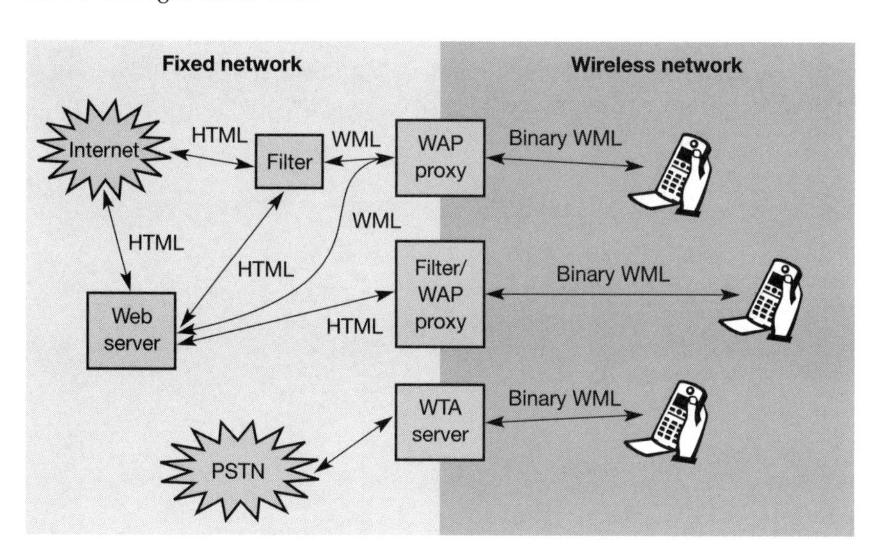

Figure 10.10
Examples for the integration of WAP components

The current www in the internet offers web pages with the help of HTML and web servers. To be able to browse these pages or additional pages with handheld devices, a wireless markup language (WML) has been defined in WAP. Special filters within the fixed network can now translate HTML into WML, web servers can already provide pages in WML, or the gateways between the fixed and wireless network can translate HTML into WML. These gateways not only filter pages but also act as proxies for web access, as explained in the following sections. WML is additionally converted into binary WML for more efficient transmission.

In a similar way, a special gateway can be implemented to access traditional telephony services via binary WML. This wireless telephony application (WTA) server translates, e.g., signaling of the telephone network (incoming call etc.) into WML events displayed at the handheld device. It is important to notice the integrated view for the wireless client of all different services, telephony and web, via the WAE (see section 10.3.6).

10.3.2 Wireless datagram protocol

The **wireless datagram protocol (WDP)** operates on top of many different bearer services capable of carrying data. At the T-SAP WDP offers a consistent datagram transport service independent of the underlying bearer (WAP Forum, 2000b). To offer this consistent service, the adaptation needed in the transport layer can differ depending on the services of the bearer. The closer the bearer service is to IP, the smaller the adaptation can be. If the bearer already offers IP services, UDP (Postel, 1980) is used as WDP. WDP offers more or less the same services as UDP.

WDP offers **source** and **destination port numbers** used for multiplexing and demultiplexing of data respectively. The service primitive to send a datagram is **T-DUnitdata.req** with the **destination address (DA), destination port (DP), Source address (SA), source port (SP)**, and **user data (UD)** as mandatory parameters (see Figure 10.11). Destination and source address are unique addresses for the receiver and sender of the user data. These could be MSISDNs (i.e., a telephone number), IP addresses, or any other unique identifiers. The **T-DUnitdata.ind** service primitive indicates the reception of data. Here destination address and port are only optional parameters.

Figure 10.11
WDP service primitives

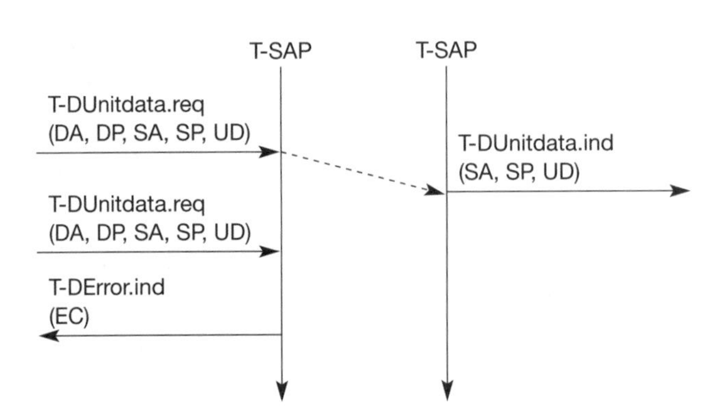

If a higher layer requests a service the WDP cannot fulfill, this error is indicated with the **T-DError.ind** service primitive as shown in Figure 10.11. An **error code (EC)** is returned indicating the reason for the error to the higher layer. WDP is not allowed to use this primitive to indicate problems with the bearer service. It is only allowed to use the primitive to indicate local problems, such as a user data size that is too large.

If any errors happen when WDP datagrams are sent from one WDP entity to another (e.g. the destination is unreachable, no application is listening to the specified destination port etc.), the **wireless control message protocol (WCMP)** provides error handling mechanisms for WDP (WAP Forum, 2000r) and should therefore be implemented. WCMP contains control messages that resemble the internet control message protocol (ICMP (Postel, 1981b) for IPv4, (Conta, 1998) for IPv6) messages and can also be used for diagnostic and informational purposes. WCMP can be used by WDP nodes and gateways to report errors. However, WCMP error messages must not be sent as response to other WCMP error messages. In IP-based networks, ICMP will be used as WCMP (e.g., CDPD, GPRS). Typical WCMP messages are **destination unreachable** (route, port, address unreachable), **parameter problem** (errors in the packet header), **message too big**, **reassembly failure**, or **echo request/reply**.

An additional **WDP management entity** supports WDP and provides information about changes in the environment, which may influence the correct operation of WDP. Important information is the current configuration of the device, currently available bearer services, processing and memory resources etc. Design and implementation of this management component is considered vendor-specific and is outside the scope of WAP.

If the bearer already offers IP transmission, WDP (i.e., UDP in this case) relies on the segmentation (called fragmentation in the IP context) and reassembly capabilities of the IP layer as specified in (Postel, 1981a). Otherwise, WDP has to include these capabilities, which is, e.g., necessary for the GSM SMS. The WAP specification provides many more adaptations to almost all bearer services currently available or planned for the future (WAP Forum, 2000q), (WAP Forum, 2000b).

10.3.3 Wireless transport layer security

If requested by an application, a security service, the **wireless transport layer security (WTLS)**, can be integrated into the WAP architecture on top of WDP as specified in (WAP Forum, 2000c). WTLS can provide different levels of security (for privacy, data integrity, and authentication) and has been optimized for low bandwidth, high-delay bearer networks. WTLS takes into account the low processing power and very limited memory capacity of the mobile devices for cryptographic algorithms. WTLS supports datagram and connection-oriented transport layer protocols. New compared to, e.g. GSM, is the security relation between two peers and not only between the mobile device and the base station (see chapter 4). WTLS took over many features and mechanisms from TLS (formerly SSL, secure sockets layer (Dierks, 1999)), but it has an optimized handshaking between the peers.

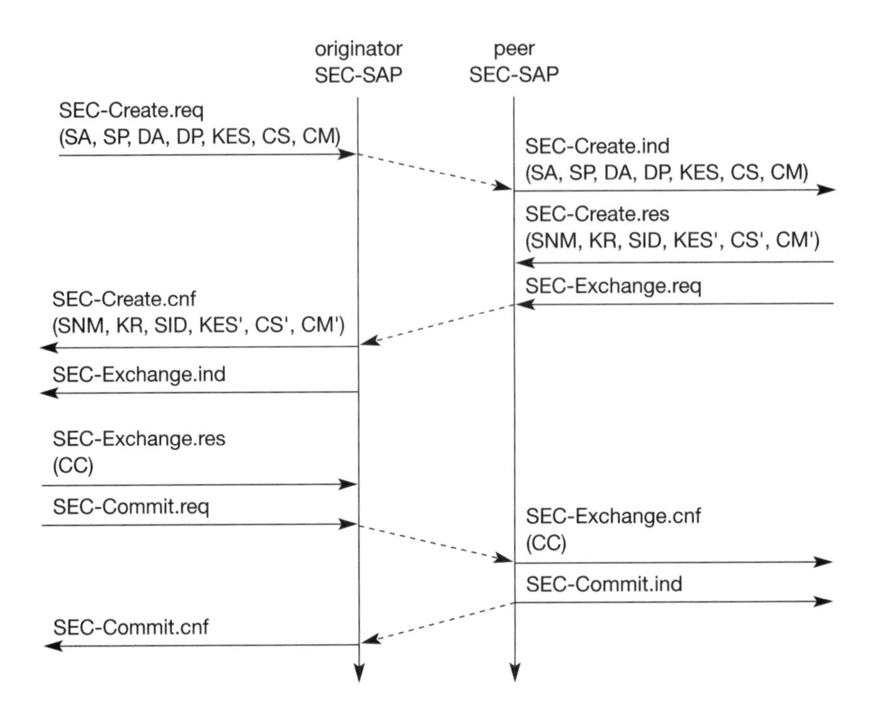

Figure 10.12 WTLS establishing a secure session

Before data can be exchanged via WTLS, a secure session has to be established. This session establishment consists of several steps: Figure 10.12 illustrates the sequence of service primitives needed for a so-called 'full handshake' (several optimizations are possible). The originator and the peer of the secure session can both interrupt session establishment any time, e.g., if the parameters proposed are not acceptable.

The first step is to initiate the session with the **SEC-Create** primitive. Parameters are **source address (SA), source port (SP)** of the originator, **destination address (DA), destination port (DP)** of the peer. The originator proposes **a key exchange suite (KES)** (e.g., RSA (Rivest, 1978), DH (Diffie, 1976), ECC (Certicom, 2002)), a **cipher suite (CS)** (e.g., DES, IDEA (Schneier, 1996), and **a compression method (CM)** (currently not further specified). The peer answers with parameters for the **sequence number mode (SNM)**, the **key refresh** cycle (KR) (i.e., how often keys are refreshed within this secure session), the **session identifier (SID)** (which is unique with each peer), and the selected **key exchange suite (KES′), cipher suite (CS′), compression method (CM′)**. The peer also issues a **SEC-Exchange primitive**. This indicates that the peer wishes to perform public-key authentication with the client, i.e., the peer requests a **client certificate** (CC) from the originator.

The first step of the secure session creation, the negotiation of the security parameters and suites, is indicated on the originator's side, followed by the request for a certificate. The originator answers with its certificate and issues a **SEC-Commit.req** primitive. This primitive indicates that the handshake is com-

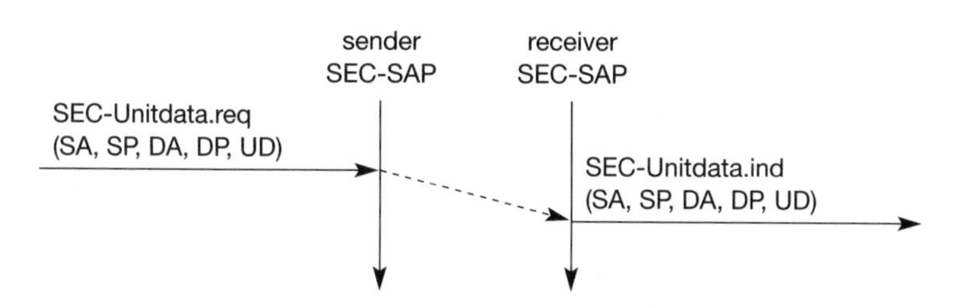

Figure 10.13
WTLS datagram transfer

pleted for the originator's side and that the originator now wants to switch into the newly negotiated connection state. The certificate is delivered to the peer side and the SEC-Commit is indicated. The WTLS layer of the peer sends back a confirmation to the originator. This concludes the full handshake for secure session setup.

After setting up a secure connection between two peers, user data can be exchanged. This is done using the simple **SEC-Unitdata** primitive as shown in Figure 10.13. SEC-Unitdata has exactly the same function as T-DUnitdata on the WDP layer, namely it transfers a datagram between a sender and a receiver. This data transfer is still unreliable, but is now secure. This shows that WTLS can be easily plugged into the protocol stack on top of WDP. The higher layers simply use SEC-Unitdata instead of T-DUnitdata. The parameters are the same here: **source address (SA)**, **source port (SP)**, **destination address (DA)**, **destination port (DP)**, and **user data (UD)**.

This section will not discuss the security-related features of WTLS or the pros and cons of different encryption algorithms. The reader is referred to the specification (WAP Forum, 2000c) and excellent cryptography literature e.g., (Schneier, 1996), (Kaufman, 1995).

Although WTLS allows for different encryption mechanisms with different key lengths, it is quite clear that due to computing power on the handheld devices the encryption provided cannot be very strong. If applications require stronger security, it is up to an application or a user to apply stronger encryption on top of the whole protocol stack and use WTLS as a basic security level only. Many programs are available for this purpose. It is important to note that the security association in WTLS exists between the mobile WAP-enabled device and a WAP server or WAP gateway only. If an application accesses another server via the gateway, additional mechanisms are needed for end-to-end security. If for example a user accesses his or her bank account using WAP, the WTLS security association typically ends at the WAP gateway inside the network operator's domain. The bank and user will want to apply additional security mechanisms in this scenario.

Future work in the WTLS layer comprises consistent support for application level security (e.g. digital signatures) and different implementation classes with different capabilities to select from.

10.3.4 Wireless transaction protocol

The **wireless transaction protocol (WTP)** is on top of either WDP or, if security is required, WTLS (WAP Forum, 2000d). WTP has been designed to run on very thin clients, such as mobile phones. WTP offers several advantages to higher layers, including an improved reliability over datagram services, improved efficiency over connection-oriented services, and support for transaction-oriented services such as web browsing. In this context, a transaction is defined as a request with its response, e.g. for a web page.

WTP offers many features to the higher layers. The basis is formed from three **classes of transaction service** as explained in the following paragraphs. Class 0 provides unreliable message transfer without any result message. Classes 1 and 2 provide reliable message transfer, class 1 without, class 2 with, exactly one reliable result message (the typical request/response case). WTP achieves reliability using **duplicate removal, retransmission, acknowledgements** and unique **transaction identifiers**. No WTP-class requires any connection set-up or tear-down phase. This avoids unnecessary overhead on the communication link. WTP allows for **asynchronous transactions, abort of transactions, concatenation of messages**, and can **report success or failure** of reliable messages (e.g., a server cannot handle the request).

To be consistent with the specification, in the following the term **initiator** is used for a WTP entity initiating a transaction (aka client), and the term **responder** for the WTP entity responding to a transaction (aka server). The three service primitives offered by WTP are **TR-Invoke** to initiate a new transaction, **TR-Result** to send back the result of a previously initiated transaction, and **TR-Abort** to abort an existing transaction. The PDUs exchanged between two WTP entities for normal transactions are the **invoke PDU, ack PDU**, and **result PDU**. The use of the service primitives, the PDUs, and the associated parameters with the classes of transaction service will be explained in the following sections.

A special feature of WTP is its ability to provide a **user acknowledgement** or, alternatively, an **automatic acknowledgement** by the WTP entity. If user acknowledgement is required, a WTP user has to confirm every message received by a WTP entity. A user acknowledgement provides a stronger version of a confirmed service because it guarantees that the response comes from the user of the WTP and not the WTP entity itself.

10.3.4.1 WTP class 0

Class 0 offers an unreliable transaction service without a result message. The transaction is stateless and cannot be aborted. The service is requested with the **TR-Invoke.req** primitive as shown in Figure 10.14. Parameters are the **source address (SA), source port (SP), destination address (DA), destination port (DP)** as already explained in section 10.3.2. Additionally, with the **A** flag the user of this service can determine, if the responder WTP entity should generate an **acknowledgement** or if a user acknowledgement should be used. The WTP layer will transmit the **user data (UD)** transparently to its destination. The class type **C** indicates here class 0. Finally, the transaction **handle H** provides a simple index to uniquely identify the transaction and is an alias for the tuple (SA, SP, DA, DP), i.e., a socket pair, with only local significance.

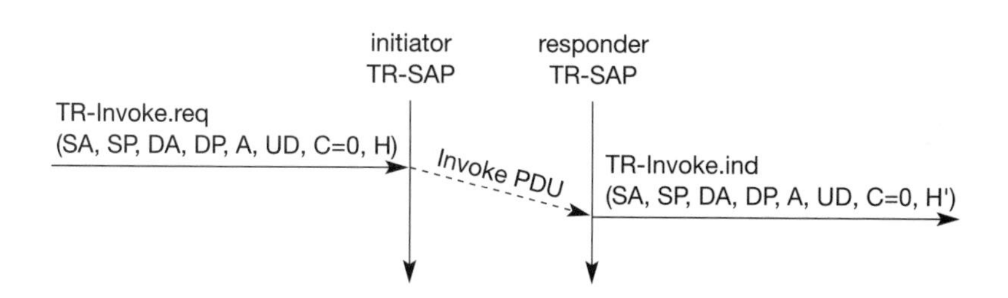

Figure 10.14 Basic transaction, WTP class 0

The WTP entity at the initiator sends an invoke PDU which the responder receives. The WTP entity at the responder then generates a **TR-Invoke.ind** primitive with the same parameters as on the initiator's side, except for H' which is now the local handle for the transaction on the responder's side.

In this class, the responder does not acknowledge the message and the initiator does not perform any retransmission. Although this resembles a simple datagram service, it is recommended to use WDP if only a datagram service is required. WTP class 0 augments the transaction service with a simple datagram-like service for occasional use by higher layers.

10.3.4.2 WTP class 1

Class 1 offers a reliable transaction service but without a result message. Again, the initiator sends an invoke PDU after a **TR-Invoke.req** from a higher layer. This time, class equals '1', and no user acknowledgement has been selected as shown in Figure 10.15. The responder signals the incoming invoke PDU via the **TR-Invoke.ind** primitive to the higher layer and acknowledges automatically without user intervention. The specification also allows the user on the responder's side to acknowledge, but this acknowledgement is not required. For the initiator the transaction ends with the reception of the acknowledgement. The responder keeps the transaction state for some time to be able to retransmit the acknowledgement if it receives the same invoke PDU again indicating a loss of the acknowledgement.

If a user of the WTP class 1 service on the initiator's side requests a user acknowledgement on the responder's side, the sequence diagram looks like Figure 10.16. Now the WTP entity on the responder's side does not send an acknowledgement automatically, but waits for the **TR-Invoke.res** service primitive from

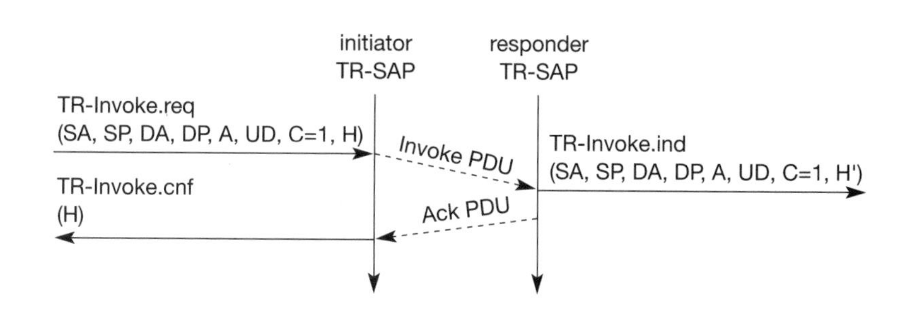

Figure 10.15 Basic transaction, WTP class 1, no user acknowledgement

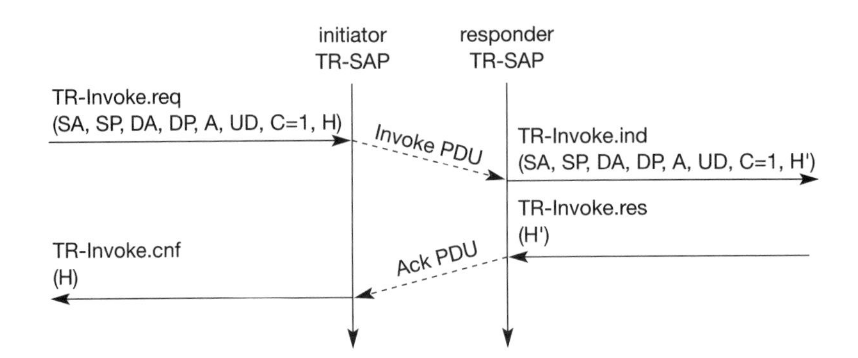

Figure 10.16 Basic transaction, WTP class 1, with user acknowledgement

the user. This service primitive must have the appropriate local handle H′ for identification of the right transaction. The WTP entity can now send the ack PDU. Typical uses for this transaction class are reliable push services.

10.3.4.3 WTP class 2

Finally, class 2 transaction service provides the classic reliable request/response transaction known from many client/server scenarios. Depending on user requirements, many different scenarios are possible for initiator/responder interaction. Three examples are presented below.

Figure 10.17 shows the basic transaction of class 2 without-user acknowledgement. Here, a user on the initiator's side requests the service and the WTP entity sends the invoke PDU to the responder. The WTP entity on the responder's side indicates the request with the **TR-Invoke.ind** primitive to a user. The responder now waits for the processing of the request, the user on the responder's side can finally give the result UD* to the WTP entity on the responder

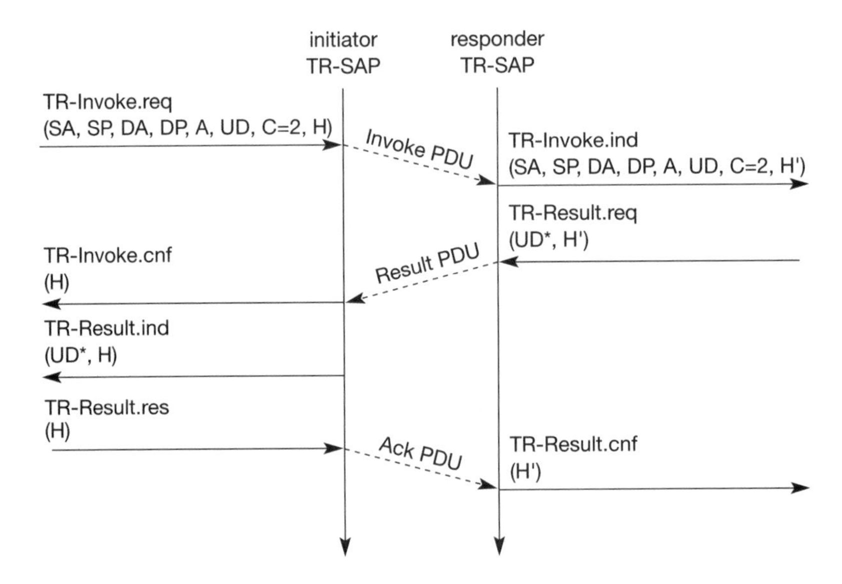

Figure 10.17 Basic transaction, WTP class 2, no user acknowledgement

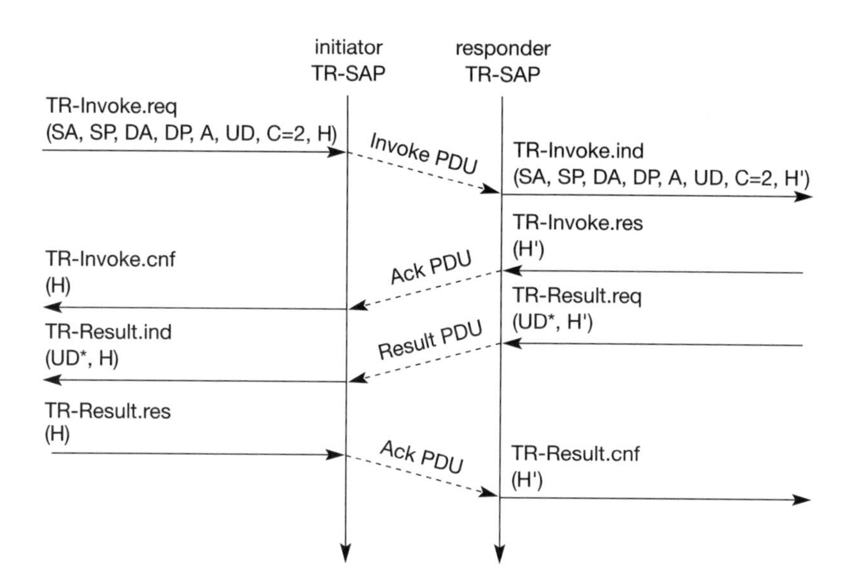

Figure 10.18 Basic transaction, WTP class 2, with user acknowledgement

side using **TR-Result.req.** The **result PDU** can now be sent back to the initiator, which implicitly acknowledges the invoke PDU. The initiator can indicate the successful transmission of the invoke message and the result with the two service primitives **TR-Invoke.cnf** and **TR-Result.ind**. A user may respond to this result with **TR-Result.res**. An acknowledgement PDU is then generated which finally triggers the **TR-Result.cnf** primitive on the responder's side. This example clearly shows the combination of two reliable services (TR-Invoke and TR-Result) with an efficient data transmission/acknowledgement.

An even more reliable service can be provided by user acknowledgement as explained above. The time-sequence diagram looks different (see Figure 10.18). The user on the responder's side now explicitly responds to the Invoke PDU using the **TR-Invoke.res** primitive, which triggers the **TR-Invoke.cnf** on the initiator's side via an **ack PDU**. The transmission of the result is also a confirmed service, as indicated by the next four service primitives. This service will likely be the most common in standard request/response scenarios as, e.g., distributed computing.

If the calculation of the result takes some time, the responder can put the initiator on "hold on" to prevent a retransmission of the invoke PDU as the initiator might assume packet loss if no result is sent back within a certain timeframe. This is shown in Figure 10.19. After a time-out, the responder automatically generates an acknowledgement for the Invoke PDU. This shows the initiator that the responder is still alive and currently busy processing the request. After more time, the result PDU can be sent to the initiator as already explained.

WTP provides many more features not explained here, such as concatenation and separation of messages, asynchronous transactions with up to 2^{15} transactions outstanding, i.e., requested but without result up to now, and segmentation/reassembly of messages (WAP Forum, 2000d).

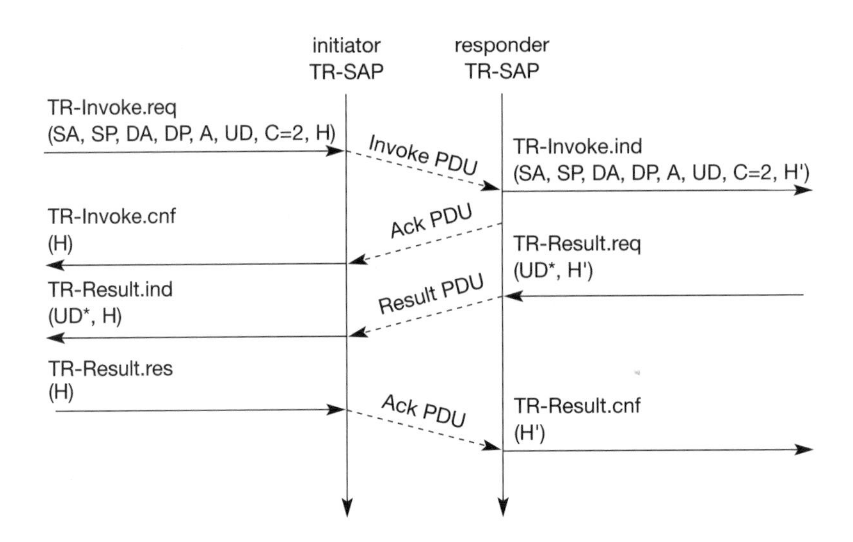

Figure 10.19
WTP class 2 transaction with "hold on", no user acknowledgement

10.3.5 Wireless session protocol

The **wireless session protocol (WSP)** has been designed to operate on top of the datagram service WDP or the transaction service WTP (WAP Forum, 2000e). For both types, security can be inserted using the WTLS security layer if required. WSP provides a shared state between a client and a server to optimize content transfer. HTTP, a protocol WSP tries to replace within the wireless domain, is stateless, which already causes many problems in fixed networks (see section 10.2.1). Many web content providers therefore use cookies to store some state on a client machine, which is not an elegant solution (see section 10.2.3). State is needed in web browsing, for example, to resume browsing in exactly the same context in which browsing has been suspended. This is an important feature for clients and servers. Client users can continue to work where they left the browser or when the network was interrupted, or users can get their customized environment every time they start the browser. Content providers can customize their pages to clients' needs and do not have to retransmit the same pages over and over again. WSP offers the following general features needed for content exchange between cooperating clients and servers:

- **Session management:** WSP introduces sessions that can be **established** from a client to a server and may be long lived. Sessions can also be **released** in an orderly manner. The capabilities of **suspending** and **resuming** a session are important to mobile applications. Assume a mobile device is being switched off – it would be useful for a user to be able to continue operation at exactly the point where the device was switched off. Session lifetime is independent of transport connection lifetime or continuous operation of a bearer network.

- **Capability negotiation:** Clients and servers can agree upon a common level of protocol functionality during session establishment. Example parameters to negotiate are maximum client SDU size, maximum outstanding requests, protocol options, and server SDU size.
- **Content encoding:** WSP also defines the efficient binary encoding for the content it transfers. WSP offers content typing and composite objects, as explained for web browsing.

While WSP is a general-purpose session protocol, WAP has specified the **wireless session protocol/browsing (WSP/B)** which comprises protocols and services most suited for browsing-type applications. In addition to the general features of WSP, WSP/B offers the following features adapted to web browsing:

- **HTTP/1.1 functionality:** WSP/B supports the functions HTTP/1.1 (Fielding, 1999) offers, such as extensible request/reply methods, composite objects, and content type negotiation. WSP/B is a binary form of HTTP/1.1. HTTP/1.1 content headers are used to define content type, character set encoding, languages etc., but binary encodings are defined for well-known headers to reduce protocol overheads.
- **Exchange of session headers:** Client and server can exchange request/reply headers that remain constant over the lifetime of the session. These headers may include content types, character sets, languages, device capabilities, and other static parameters. WSP/B will not interpret header information but passes all headers directly to service users.
- **Push and pull data transfer:** Pulling data from a server is the traditional mechanism of the web. This is also supported by WSP/B using the request/response mechanism from HTTP/1.1. Additionally, WSP/B supports three push mechanisms for data transfer: a confirmed data push within an existing session context, a non-confirmed data push within an existing session context, and a non-confirmed data push without an existing session context.
- **Asynchronous requests:** Optionally, WSP/B supports a client that can send multiple requests to a server simultaneously. This improves efficiency for the requests and replies can now be coalesced into fewer messages. Latency is also improved, as each result can be sent to the client as soon as it is available.

As already mentioned, WSP/B can run over the transaction service WTP or the datagram service WDP. The following shows several protocol sequences typical for session management, method invocation, and push services.

10.3.5.1 WSP/B over WTP

WSP/B uses the three service classes of WTP presented in section 10.3.4.1 to 10.3.4.3 as follows. Class 0 is used for unconfirmed push, session resume, and session management. Confirmed push uses class 1, method invocation, session resume, and session management class 2. The following time sequence charts will give some examples.

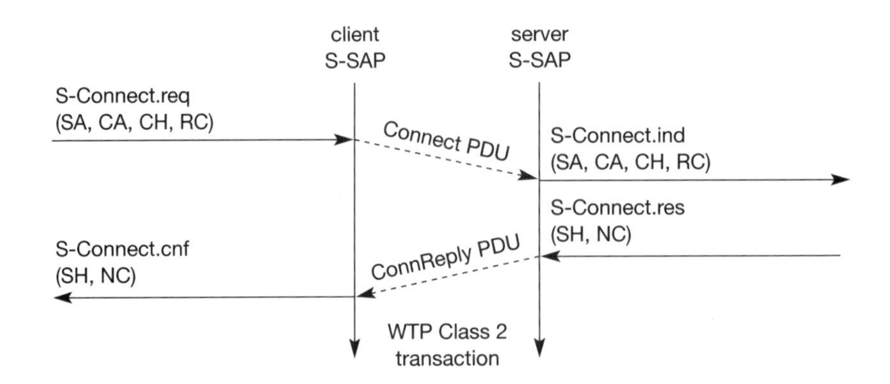

Figure 10.20 WSP/B session establishment

The first example (seeFigure 10.20) shows the session establishment of WSP/B using WTP class 2 transactions. With the **S-Connect.req** primitive, a client can request a new session. Parameters are the **server address (SA)**, the **client address (CA)**, and the optional **client header (CH)** and **requested capabilities (RC)**. The session layer directly uses the addressing scheme of the layer below. TR-SAP and S-SAP can be directly mapped. A client header can comprise user-to-user information compatible with HTTP message headers according to Fielding (1999). These headers can be used, e.g., for caching if they are constant throughout the session. Interpretation is up to the user of this service. The capabilities are needed for the capability negotiation between server and client as listed in the features of WSP above.

WTP transfers the **connect PDU** to the server S-SAP where an **S-Connect.ind** primitive indicates a new session. Parameters are the same, but now the capabilities are mandatory. If the server accepts the new session it answers with an **S-Connect.res**, parameters are an optional **server header (SH)** with the same function as the client header and the **negotiated capabilities (NC)** needed for capability negotiation.

WTP now transfers the **connreply PDU** back to the client; **S-Connect.cnf** confirms the session establishment and includes the **server header** (if present) and the **negotiated capabilities** from the server. WSP/B includes several procedures to refuse a session or to abort session establishment.

A very useful feature of WSP/B **session suspension** and **session resume** is shown in Figure 10.21. If, for example, a client notices that it will soon be unavailable, e.g., the bearer network will be unavailable due to roaming to another network or the user switches off the device, the client can suspend the session. Session suspension will automatically abort all data transmission and freeze the current state of the session on the client and server side. A client suspends a session with **S-Suspend.req**, WTP transfers the **suspend PDU** to the server with a class 0 transaction, i.e., unconfirmed and unreliable. WSP/B will signal the suspension with **S-Suspend.ind** on the client and server side. The only parameter is the **reason R** for suspension. Reasons can be a user request or a suspension initiated by the service provider.

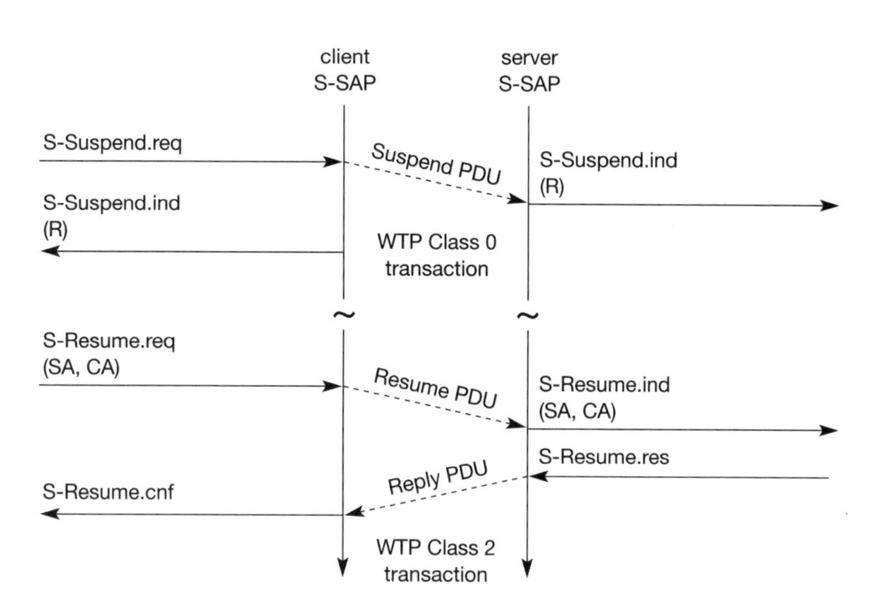

Figure 10.21 WSP/B session suspension and resume

As also shown in Figure 10.21, a client can later resume a suspended session with **S-Resume.req**. Parameters are **server address (SA)** and **client address (CA)**. If SA and CA are not the same as before suspending this session, it is the responsibility of the service user to map the addresses accordingly so that the same server instance will be contacted. Resuming a session is a confirmed operation. It is up to the server's operator how long this state is conserved.

Terminating a session is done by using the **S-Disconnect.req** service primitive (Figure 10.22). This primitive aborts all current method or push transactions used to transfer data. Disconnection is indicated on both sides using **S-Disconnect.ind**. The **reason R** for disconnection can be, e.g., network error, protocol error, peer request, congestion, and maximum SDU size exceeded. S-Disconnect.ind can also include parameters that redirect the session to another server where the session may continue.

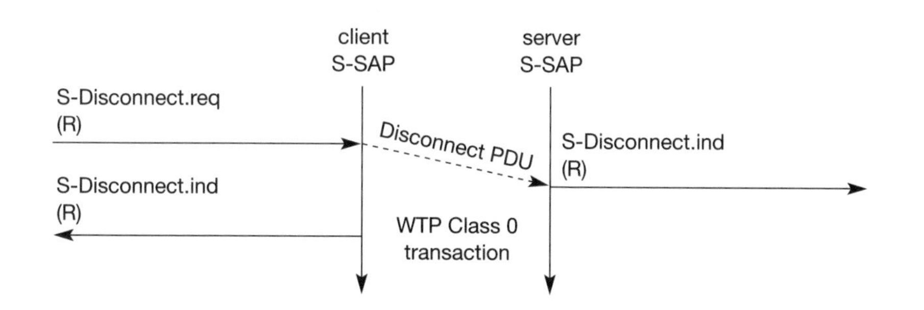

Figure 10.22 WSP/B session termination

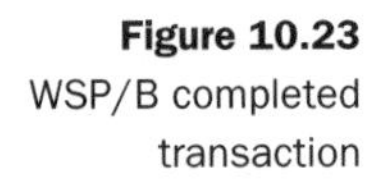

Figure 10.23
WSP/B completed transaction

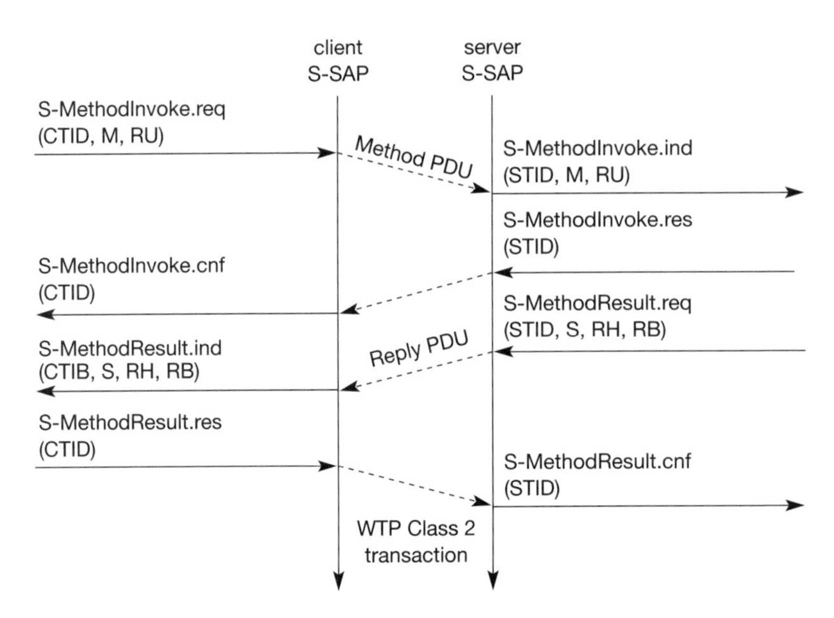

The **S-MethodInvoke** primitive is used to request that an operation is executed by the server. The result, if any, is sent back using the **S-MethodResult** primitive (Figure 10.23). A client requests an operation with **S-MethodInvoke.req**. Parameters are the **client transaction identifier CTID** to distinguish between pending transactions, the **method M** identifying the requested operation at the server, and the **request URI** (Uniform Resource Identifier (Berners-Lee, 1994a) **RU**. URLs, such as http://www.xyz.int/ are examples of URIs (Berners-Lee, 1994b). Additional headers and bodies can be sent with this primitive.

The WTP class 2 transaction service now transports the **method PDU** to the server. A method PDU can be either a get PDU or a post PDU as defined in Fielding (1999). **Get PDUs** are used for HTTP/1.1 GET, OPTIONS, HEAD, DELETE and TRACE methods, and other methods that do not send content to the server. A **post PDU** is used for HTTP/1.1 POST and PUT and other methods that send content to the server.

On the server's side, **S-MethodInvoke.ind** indicates the request. In this case a **server transaction identifier STID** distinguishes between pending transactions. The server confirms the request, so WSP/B does not generate a new PDU but relies on the lower WTP layer (see Figure 10.24).

Similarly, the result of the request is sent back to the client using the **S-MethodResult** primitive. Additional parameters are now the **status (S)**, the **response header (RH)**, and the **response body (RB)**. Again, WSP/B stays close to HTTP/1.1 and so the Status S corresponds to the HTTP status codes in Fielding (1999). One famous example for a **status code** is **404**, indicating that the server could not find the web page specified in the request, typically a sign of an outdated bookmark or a lazy managed web server. But most of the time a server returns **200** indicating that everything is okay. Header and body, too, are equivalent to the HTTP header and body, therefore the response body typically carries the code of the web page if the status is 200.

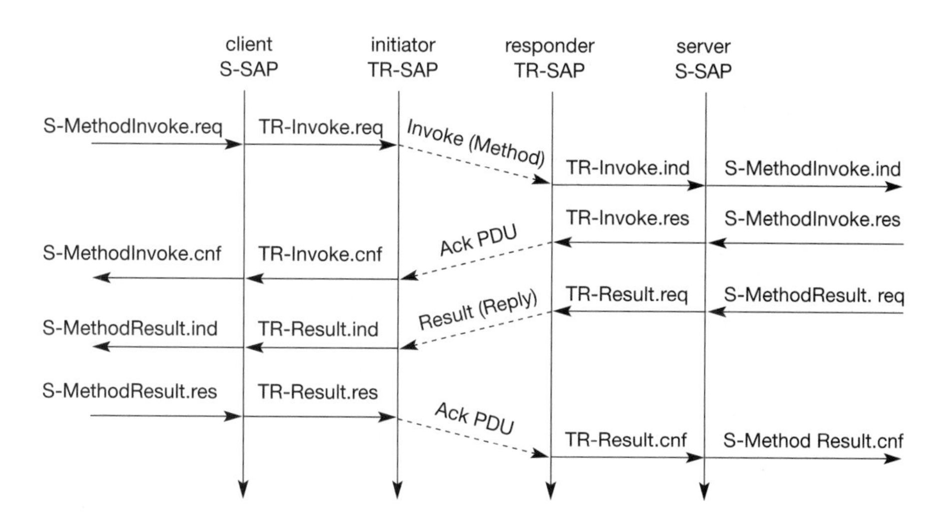

Figure 10.24
WSP utilization of WTP as lower layer

WSP does not introduce PDUs or service primitives just for the sake of symmetric and aesthetic protocol architecture. Figure 10.24 shows how WSP (thus also WSP/B) uses the underlying WTP services for its purposes. The **S-MethodInvoke.req** primitive triggers the **TR-Invoke.req** primitive, the parameters of the WSP layer are the user data of the WTP layer. The **invoke PDU** of the WTP layer carries the **method PDU** of the WSP layer inside. In contrast to a pure-layered communication model, the lower WTP layer is involved in the semantics of the higher layer primitives and does not consider them as pure data only.

For the confirmation of its service primitives the WSP layer has none of its own PDUs but uses the **acknowledgement PDUs** of the WTP layer as shown. **S-MethodInvoke.res** triggers **TR-Invoke.res**, the **ack PDU** is transferred to the initiator, here **TR-Invoke.cnf** confirms the invoke service and triggers the **S-MethodInvoke.cnf** primitive which confirms the method invocation service. This mingling of layers saves a lot of redundant data flow but still allows a separation of the tasks between the two layers.

WSP neither provides any sequencing between different requests nor does it restore any sequence between responses. Figure 10.25 shows four requests on the client's side (**S-MethodInvoke_i.req**). WSP may deliver them in any order on the server's side as indicated by **S-MethodInvoke_i.ind** (the confirmation primitives **S-MethodInvoke.res** and **S-MethodInvoke.cnf** have been omitted for clarity). The user on the server's side may need different amounts of time to respond to the requests, e.g., if some requested data has to be fetched from disk while other data is already available in memory. Therefore, the responses **S-MethodResult_i.req** may be in arbitrary order as the WSP service only delivers them to the client S-SAP where they finally appear as **S-MethodResult_i.ind**. This may be completely independent from the original order of the requests.

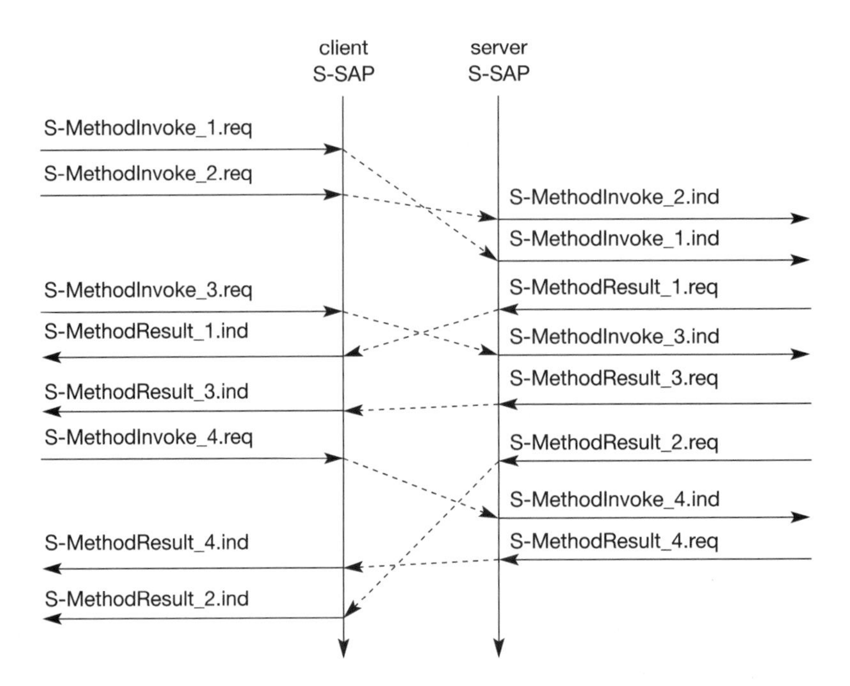

Figure 10.25 WSP/B asynchronous, unordered requests

Up to now all service primitives allowed the client to pull data from a server. With the help of push primitives, a server can push data towards a client if allowed. The simplest push mechanism is the non-confirmed push as shown in Figure 10.26. The server sends unsolicited data with the **S-Push.req** primitive to the client. Parameters are the **push header (PH)** and the **push body (PB)** again, these are the header and the body known from HTTP. The unreliable, unconfirmed WTP class 0 transaction service transfers the **push PDU** to the client where **S-Push.ind** indicates the push event.

A more reliable push service offers the **S-ConfirmedPush** primitive as shown in Figure 10.27. Here the server has to determine the push using a **server push identifier (SPID)**. This helps to distinguish between different pending pushes. The reliable WTP class 1 transaction service is now used to transfer the **confpush PDU** to the client. On the client's side a **client push identifier (CPID)** is used to distinguish between different pending pushes.

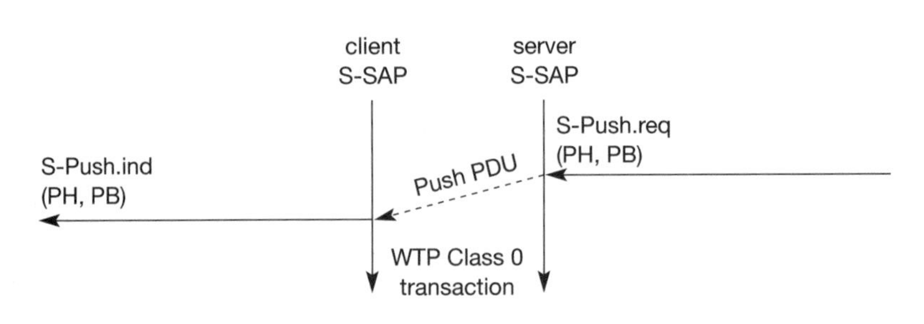

Figure 10.26 WSP/B non-confirmed push

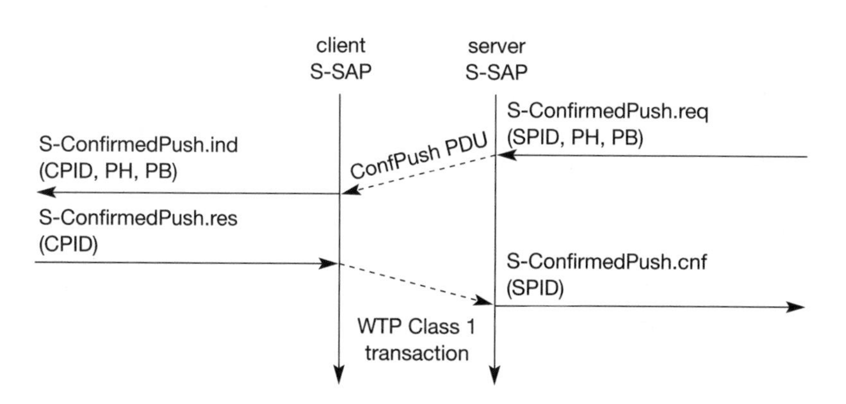

Figure 10.27
WSP/B confirmed push

Additionally, WSP/B provides many ways to abort any operation such as session establishment, method invocation or data push.

10.3.5.2 WSP/B as connectionless session service

There are cases where the overhead of session establishment and release, confirmed method invocation and all associated states is simply too much and a high degree of reliability is not required. In these cases, It is possible to runWSP/B on top of the connectionless, unreliable WDP service. As an alternative to WDP, WTLS can always be used if security is required. The service primitives are directly mapped onto each other. Example applications could be periodic pushes of weather data from a remote sensor device to a client.

Figure 10.28 shows the three service primitives available for connectionless session service: **S-Unit-MethodInvoke.req** to request an operation on a server, **S-Unit-MethodResult.req** to return results to a client, and **S-Unit-Push.req** to push data onto a client. Transfer of the PDUs (**method**, **reply** and **push**) is done with the help of the standard unreliable datagram transfer service of WDP.

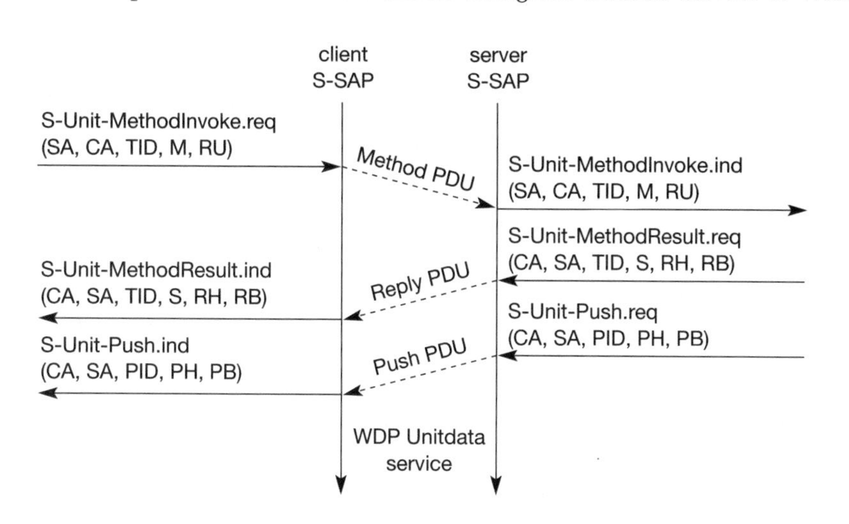

Figure 10.28
WSP/B as connectionless session service

Besides the **server address (SA)**, the **client address (CA)**, the **method (M)**, and the **request URI (RU)**, which have already been explained above, the user of the **S-Unit-MethodInvoke.req** primitive can determine a **transaction identifier (TID)** to distinguish between different transactions on the user level. TID is communicated transparently from service user to service user.

The function of the **S-Unit-MethodResult** primitive remains the same as explained above: the **status (S)**, **response header (RH)**, and **response body (RB)** represent the result of the operation. The **S-Unit-Push** primitive has the parameters **client address (CA)**, **server address (SA)**, **push identifier (PID)**, **push header (PH)**, and **push body (PB)**.

Although WSP already offers many services, e.g., specifies binary encodings for headers and content etc., there are many unsolved problems. Examples are: the provisioning of QoS, (i.e., how can certain quality parameters be applied to transactions and sessions?) multi-cast support (i.e., how can a multicast session be created?) or isochronous multimedia objects (i.e., the support of very strict time bounds for session services). Moreover, management of all services is still an open field.

10.3.6 Wireless application environment

The main idea behind the **wireless application environment (WAE)** is to create a general-purpose application environment based mainly on existing technologies and philosophies of the world wide web (WAP Forum, 2000g). This environment should allow service providers, software manufacturers, or hardware vendors to integrate their applications so they can reach a wide variety of different wireless platforms in an efficient way. However, WAE does not dictate or assume any specific man-machine-interface model, but allows for a variety of devices, each with its own capabilities and probably vendor-specific extras (i.e., each vendor can have its own look and feel). WAE has already integrated the following technologies and adapted them for use in a wireless environment with low power handheld devices. HTML (Raggett, 1998), JavaScript (Flanagan, 1997), and the handheld device markup language HDML (King, 1997) form the basis of the **wireless markup language (WML)** and the scripting language **WMLscript**. The exchange formats for business cards and phone books **vCard** (IMC, 1996a) and for calendars **vCalendar** (IMC, 1996b) have been included. **URLs** from the web can be used. A wide range of mobile telecommunication technologies have been adopted and integrated into the **wireless telephony application (WTA)** (WAP Forum, 2000f).

Besides relying on mature and established technology, WAE focuses on devices with very limited capabilities, narrow-band environments, and special security and access control features. The first phase of the WAE specification developed a whole application suite, especially for wireless clients as presented in the following sections. Future developments for the WAE will include extensions for more content formats, integration of further existing or emerging technologies, more server-side aspects, and the integration of intelligent telephone networks.

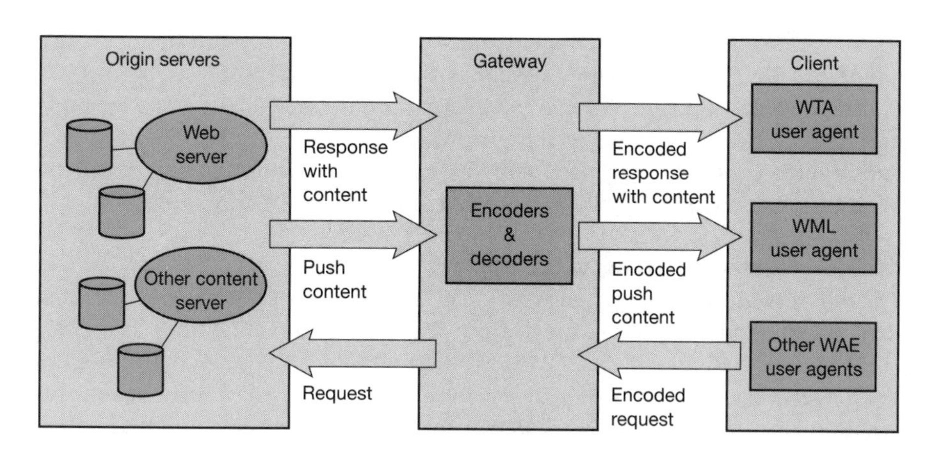

Figure 10.29
WAE logical model

One global goal of the WAE is to minimize over-the-air traffic and resource consumption on the handheld device. This goal is also reflected in the logical model underlying WAE (Figure 10.29) showing some more detail than the general overview in Figure 10.10. WAE adopts a model that closely follows the www model, but assumes additional gateways that can enhance transmission efficiency. A **client** issues an encoded request for an operation on a remote server. Encoding is necessary to minimize data sent over the air and to save resources on the handheld device as explained together with the languages WML and WMLscript.

Decoders in a **gateway** now translate this encoded request into a standard request as understood by the **origin servers**. This could be a request to get a web page to set up a call. The gateway transfers this request to the appropriate origin server as if it came from a standard client. Origin servers could be standard web servers running HTTP and generating content using scripts, providing pages using a database, or applying any other (proprietary) technology. WAE does not specify any standard content generator or server, but assumes that the majority will follow the standard technology used in today's www.

The origin servers will respond to the request. The gateway now encodes the response and its content (if there is any) and transfers the encoded response with the content to the client. The WAE logical model not only includes this standard request/response scheme, but it also includes push services. Then an origin server pushes content to the gateway. The gateway encodes the pushed content and transmits the encoded push content to the client.

Several user agents can reside within a client. User agents include such items as: browsers, phonebooks, message editors etc. WAE does not specify the number of user agents or their functionality, but assumes a basic **WML user agent** that supports WML, WMLscript, or both (i.e., a 'WML browser'). Further domain-specific user agents with varying architectures can be implemented. Again, this is left to vendors. However, one more user agent has been specified with its fundamental services, the **WTA user agent**. This user agent handles access to, and interaction with, mobile telephone features (such as call control). As over time many vendor dependent user agents may develop, the standard defines **a user agent profile (UAProf)**, which describes the capabilities of a user agent (WAP Forum, 2000aa). Capabilities may be related to hardware or software. Examples are: display size, operating system, browser version, processor, memory size, audio/video codec, or supported network types. The basic languages WML and WMLScript, and the WTA will be described in the following three sections.

10.3.7 Wireless markup language

The **wireless markup language (WML)** (WAP Forum, 2000j) is based on the standard HTML (Raggett, 1998) known from the www and on HDML (King, 1997). WML is specified as an XML (W3C, 1998a) document type. When designing WML, several constraints of wireless handheld devices had to be taken into account. First of all, the wireless link will always have only a very limited capacity compared to a wire. Current handheld devices have small displays, limited user input facilities, limited memory, and only low performance computational resources. While the bandwidth argument will remain for many years, it currently seems that the gap between mobile and fixed devices regarding processing power is getting narrower. Today's CPUs in PDAs have performance figures close to desktop CPUs just a few years ago.

WML follows a deck and card metaphor. A WML document is made up of multiple **cards**. Cards can be grouped together into a **deck**. A WML deck is similar to an HTML page, in that it is identified by a URL and is the unit of content transmission. A user navigates with the WML browser through a series of WML cards, reviews the contents, enters requested data, makes choices etc. The WML browser fetches decks as required from origin servers. Either these decks can be static files on the server or they can be dynamically generated.

It is important to note that WML does not specify how the implementation of a WML browser has to interact with a user. Instead, WML describes the intent of interaction in an abstract manner. The user agent on a handheld device has to decide how to best present all elements of a card. This presentation depends much on the capabilities of the device.

WML includes several basic features:

- **Text and images:** WML gives, as do other mark-up languages, hints how text and images can be presented to a user. However, the exact presentation of data to a user is up to the user agent running on the handheld device. WML only provides a set of mark-up elements, such as emphasis elements (bold, italic, etc.) for text, or tab columns for tabbing alignment.
- **User interaction:** WML supports different elements for user input. Examples are: text entry controls for text or password entry, option selections or controls for task invocation. Again, the user agent is free to choose how these inputs are implemented. They could be bound to, e.g., physical keys, soft keys, or voice input.

- **Navigation:** As with HTML browsers, WML offers a history mechanism with navigation through the browsing history, hyperlinks and other inter-card navigation elements.
- **Context management:** WML allows for saving the state between different decks without server interaction, i.e., variable state can last longer than a single deck, and so state can be shared across different decks. Cards can have parameters defined by using this state without access to the server over the narrow-band wireless channel.

The following paragraph gives a simple example of WML; the reader is referred to the standard or Singhal (2001) for a full reference and in-depth discussion of the language.

First, a reference to XML is given where WML was derived from. Then, after the keyword `wml` the first `card` is defined. This first card of the deck "displays" a text after loading ("displaying" could also mean voice output etc.). As soon as a user activates the `do` element (a button or voice command), the user agent displays the second card. On this second card, the user can select one out of three pizza options. Depending on the choice of the user, `PIZZA` can have one of the values `Mar`, `Fun`, or `Vul`. If the user proceeds to the third card without choosing a pizza, the value Mar is used as default. Again, describing these options with WML does not automatically mean that these options are displayed as text. It could also be possible that the user agent reads the options through a voice output and the user answers through a voice input. WML only describes the intention of a choice. The third card finally outputs the value of `PIZZA`.

```
<?xml version="1.0"?>
<!DOCTYPE wml PUBLIC "-//WAPFORUM//DTD WML 1.1//EN"
          "http://www.wapforum.org/DTD/wml_1.1.xml">
<wml>
  <card id="card_one" title="Simple example">
    <do type="accept">
      <go href="#card_two"/>
    </do>
    <p>
      This is a simple first card!
      <br/>
      On the next one you can choose ...
    </p>
  </card>
  <card id="card_two" title="Pizza selection">
    <do type="accept" label="cont">
      <go href="#card_three"/>
    </do>
    <p>
```

```
          ... your favourite pizza!
          <select value="Mar" name="PIZZA">
            <option value="Mar">Margherita</option>
            <option value="Fun">Funghi</option>
            <option value="Vul">Vulcano</option>
          </select>
        </p>
      </card>
      <card id="card_three" title="Your Pizza!">
        <p>
          Your personal pizza parameter is <b>$(PIZZA)</b>!
        </p>
      </card>
    </wml>
```

WML may be encoded using a compact binary representation to save bandwidth on the wireless link. This compact representation is based on the binary XML content format as specified in WAP Forum (2000k). The binary coding of WML is only one special version of this format; the compact representation is valid in general for XML content. The compact format allows for transmission without loss of functionality or of semantic information. For example, the URL prefix `href=_http://`, which is very common in URLs, will be coded as 4B. The code for the `select` keyword is 37 and `option` is 35. These single byte codes are much more efficient than the plain ASCII text used in HTML and today's www.

10.3.8 WMLScript

WMLScript complements to WML and provides a general scripting capability in the WAP architecture (WAP Forum, 2000h). While all WML content is static (after loading on the client), WMLScript offers several capabilities not supported by WML:

- **Validity check of user input:** before user input is sent to a server, WMLScript can check the validity and save bandwidth and latency in case of an error. Otherwise, the server has to perform all the checks, which always includes at least one round-trip if problems occur.
- **Access to device facilities:** WMLScript offers functions to access hardware components and software functions of the device. On a phone a user could, e.g., make a phone call, access the address book, or send a message via the message service of the mobile phone.
- **Local user interaction:** Without introducing round-trip delays, WMLScript can directly and locally interact with a user, show messages or prompt for input. Only, for example, the result of several interactions could be transmitted to a server.

- **Extensions to the device software:** With the help of WMLScript a device can be configured and new functionality can be added even after deployment. Users can download new software from vendors and, thus, upgrade their device easily.

WMLScript is based on JavaScript (Flanagan, 1997), but adapted to the wireless environment. This includes a small memory footprint of the simple WMLScript bytecode interpreter and an efficient over-the-air transport via a space efficient bytecode. A WMLScript compiler is used to generate this bytecode. This compiler may be located in a gateway (see Figure 10.29) or the origin servers store pre-compiled WMLScript bytecode.

WMLScript is event-based, i.e., a script may be invoked in response to certain user or environment events. WMLScript also has full access to the state model of WML, i.e., WMLScript can set and read WML variables.

WMLScript provides many features known from standard programming languages such as functions, expressions, or `while`, `if`, `for`, `return` etc. statements. The language is weakly-typed, i.e., any variable can contain any type (such as integer, float, string, boolean) – no explicit typing is necessary. WMLScript provides an automatic conversion between different types if possible. Parameters are always passed by value to functions.

Here is a simple example for some lines of WMLScript: the function `pizza_test` accepts one value as input. The local variable `taste` is initialized to the string `"unknown"`. Then the script checks if the input parameter `pizza_type` has the value `"Mar"`. If this is the case, taste is set to `"well... "`, otherwise the script checks if the `pizza_type` is `"Vul"`. If this is the case, taste is set to `"quite hot"`. Finally, the current value of taste is returned as the value of the function `pizza_test`.

```
function pizza_test(pizza_type) {
    var taste = "unknown";
    if (pizza_type = "Mar") {
     taste = "well... ";
    }
    else {
     if (pizza_type = "Vul") {
           taste = "quite hot";
     };
    };
    return taste;
};
```

The WMLScript compiler can compile one or more such scripts into a **WMLScript compilation unit**. A handheld wireless device can now fetch such a compilation unit using standard protocols with HTTP semantics, such as WSP (see section 10.3.5). Within a compilation unit, a user can call a particular function using standard **URLs** with a **fragment anchor**. A fragment anchor is specified by the URL, a hash mark (#), and a fragment identifier. If the URL of the compilation unit of the example script was: `http://www.xyz.int/myscr`, a user could call the script and pass the parameter `"Vul"` via `http://www.xyz.int/myscr#pizza_test("Vul")`.

The WAP Forum has specified several **standard libraries** for WMLScript (WAP Forum, 2000i). These libraries provide access to the core functionality of a WAP client so they, must be available in the client's scripting environment. One exception is the float library, which is optional and is only useful if a client can support floating-point operations. The following six libraries have been defined so far:

- **Lang:** This library provides functions closely related to WMLScipt itself. Examples are `isInt` to check if a value could be converted into an integer or `float` to check if floating-point operations are supported.
- **Float:** Many typical arithmetic floating-point operations are in this library (which is optional as mentioned before). Example functions are `round` for rounding a number and `sqrt` for calculating the square root of a given value.
- **String**: Many string manipulation functions are in this library. Examples are well-known functions such as `length` to return the length of a string or `subString` to return a substring of a given string. Nevertheless, this library also provides more advanced functions such as `find` to find a substring within a string or `squeeze` to replace several consecutive whitespaces with only one.
- **URL:** This library provides many functions for handling URLs with the syntax defined in Fielding (1995):

  ```
  <scheme>://<host>:<port>/<path>;<parameters>?<query>#<fragment>
  ```

  for example: `http://www.xyz.int:8080/mypages;5;2?j=2&p=1#crd`. The function `getPath` could now extract the path of this URL, i.e., `"mypages"`, `getQuery` has the query part `"j=2&p=1"` as return value, and `getFragment` delivers the fragment used in the URL, i.e., `"crd"`.
- **WMLBrowser:** This library provides several functions typical for a browser, such as `prev` to go back one card or `refresh` to update the context of the user interface. The function `go` loads the content provided as parameters:

  ```
  var my_card =
  "http://www.xyz.int/pizzamatic/apps.dck#start";
  var my_vars = "j=4&k=7";
  WMLBrowser.go(my_card, my_vars);
  ```

- **Dialogs:** For interaction with a user, this library has been defined. An example function is `prompt` which displays a given message and prompts for user input.

An additional library is the **WMLScript Crypto Library** (WAP Forum, 2000s). This library contains, for example, functions for signing text in addition to the security functions provided by WTLS. The required keys can be stored on the **wireless identity module (WIM)** which could be part of the mobile phone's SIM (WAP Forum, 2000t).

10.3.9 Wireless telephony application

Browsing the web using the WML browser is only one application for a handheld device user. Say a user still wants to make phone calls and access all the features of the mobile phone network as with a traditional mobile phone. This is where the **wireless telephony application (WTA)**, the **WTA user agent** (as shown in Figure 10.29), and the **wireless telephony application interface WTAI** come in. WTA is a collection of telephony specific extensions for call and feature control mechanisms, merging data networks and voice networks (WAP Forum, 2000l).

The WTA framework integrates advanced telephony services using a consistent user interface (e.g., the WML browser) and allows network operators to increase accessibility for various special services in their network. A network operator can reach more end-devices using WTA because this is integrated in the wireless application environment (WAE) which handles device-specific characteristics and environments. WTA should enable third-party developers as well as network operators to create network-independent content that accesses the basic features of the bearer network. However, most of the WTA functionality is reserved for the network operators for security and stability reasons.

WTA extends the basic WAE application model in several ways:

- **Content push:** A WTA origin server can push content, i.e., WML decks or WMLScript, to the client. A push can take place without prior client request (see sections 10.3.10 and 10.3.11). The content can enable, e.g., the client to handle new network events that were unknown before. An example is given in Figure 10.31.
- **Access to telephony functions:** The **wireless telephony application interface (WTAI**, WAP Forum, 2000m) provides many functions to handle telephony events (call accept, call setup, change of phone book entries etc.).
- **Repository for event handlers:** The repository represents a persistent storage on the client for content required to offer WTA services. Content are either channels or resources. Examples for resources are WML decks, WMLScript objects, or WBMP pictures. Resources are loaded using WSP or are pre-installed. A channel comprises references to resources and is associated with a lifetime. Within this lifetime, it is guaranteed that all resources the channel points to are locally available in the repository. The motivation behind the repository is the necessity to react very quickly for time-critical events (e.g., call accept). It would take too long to load content from a server for this purpose.
- **Security model**: Mandatory for WTA is a security model as many frauds happen with wrong phone numbers or faked services. WTA allows the client to only connect to trustworthy gateways, which then have to check if the servers providing content are authorized to send this content to the client. Obviously, it is not easy to define trustworthy in this context. In the beginning, the network operator's gateway may be the only trusted gateway and the network operator may decide which servers are allowed to provide content.

These libraries have been defined in the WTAI specification (WAP Forum, 2000m) and allow for the creation of telephony applications using the WTA user agent. Library functions can be used from WML decks or WMLScript.

Three classes of libraries have been defined:

- **Common network services:** This class contains libraries for services common to all mobile networks. The **call control** library contains, e.g., functions to set up, accept, and release calls. **Network text** contains functions to send, read, and delete text messages. **Phonebook** allows for the manipulation of the local phonebook entries (e.g. read, write, delete). Finally, the library **miscellaneous** contains, e.g., a function to indicate incoming data, e-mail, fax, or voice messages.
- **Network specific services:** Libraries in this class depend on the capabilities of the mobile network. Additionally, this class might contain operator-specific libraries.
- **Public services:** This class contains libraries with publicly available functions, i.e., functions third-party providers may use, not just network operators. One example is "make call" to set up a phone call.

Functions in these libraries all follow the same simple syntax. For the use in WML, a URI is used (Berners-Lee, 1994a).

```
wtai://<library>/<function>;<parameters>;!<results>
```

The first parameter `<library>` indicates the name of the library, e.g., `cc` for WTACallControl, `wp` for WTAPublic. This is followed by the function name `<function>`, e.g., `sc` for "setup call", `mc` for "make call". If required, parameters may follow. These could be phone numbers etc. Finally, one or more results could be returned. These results set variables in the user agent context.

Within a WML card, the URI for calling a certain number could now be as follows:

```
wtai://wp/mc;07216086415
```

The same functions can also be used in WMLScript. Here calling a function follows the same scheme as calling any other function within WMLScript:

```
<returnvalue> = functionname(parameters);
```

The `returnvalue` is needed if an application requires this value for further operation. The `functionname` is again derived from the name of the library and the name of the function within the library. Finally, one or more values can be passed to the function.

The same example for making a call would now be:

```
WTAPublic.makeCall("07216086415");
```

The execution of a new service in the user agent can happen in several ways:

- Based upon a user request or another service a URI plus content stored in the repository can be used. This is (roughly) comparable with the access to cached items in a web browser.
- The user agent can also access a URI via the WTA server. In this case, the URI plus content was not stored in the repository.
- A server can push a message to the client as explained in the context of service indication (see section 10.3.11).

Finally, any network event can trigger the execution of a new service. The client has to translate the event into a URI which then can be processed by the WTA user agent. For example, an incoming call plus required parameters may be accepted via "Accept Call" (`wtai://vc/ac`).

Figure 10.30 gives an overview of the WTA logical architecture. The components shown are not all mandatory in this architecture; however, firewalls or other origin servers may be useful. A minimal configuration could be a single server from the network operator serving all clients. The **client** is connected via a mobile network with a **WTA server**, other telephone networks (e.g., fixed PSTN), and a **WAP gateway**. A WML user agent running on the client or on other user agents is not shown here. The client may have voice and data

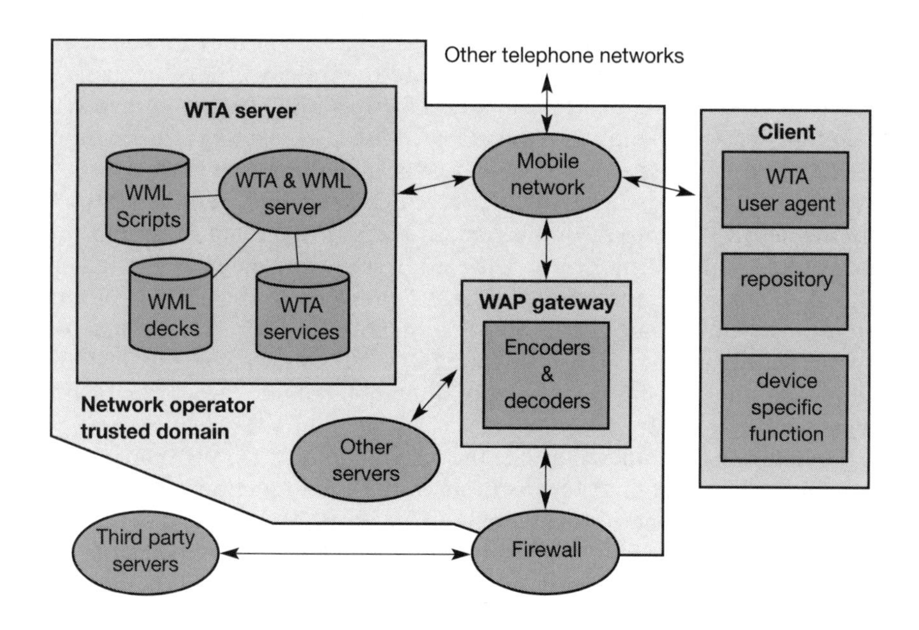

Figure 10.30
WTA logical architecture

connections over the mobile network. Other origin servers within the trusted domain may be connected via the WAP gateway. A firewall is useful to connect third-party origin servers outside the trusted domain.

One difference between WTA servers and other servers besides security is the tighter control of QoS. A network operator knows (more or less precisely) the latency, reliability, and capacity of its mobile network and can have more control over the behavior of the services. Other servers, probably located in the internet, may not be able to give as good QoS guarantees as the network operator. Similarly, the WTA user agent has a very rigid and real-time context management for browsing the web compared to the standard WML user agent.

Figure 10.31 shows an exemplary interaction between a WTA client, a WTA gateway, a WTA server, the mobile network (with probably many more servers) and a voice box server. Someone might leave a message on a voice box server as indicated. Without WAP, the network operator then typically generates an SMS indicating the new message on the voice box via a little symbol on the mobile phone. However, it is typically not indicated who left a message, what messages are stored etc. Users have to call the voice box to check and cannot choose a particular message.

In a WAP scenario, the voice box can induce the WTA server to generate new content for pushing to the client. An example could be a WML deck containing a list of callers plus length and priority of the calls. The server does not push this deck immediately to the client, but sends a push message containing a single URL to the client. A short note, e.g., "5 new calls are stored", could accompany the push message. The WTA gateway translates the push URL into a service indication (see section 10.3.11) and codes it into a more compact binary format. The WTA user agent then indicates that new messages are stored.

If the user wants to listen to the stored messages, he or she can request a list of the messages. This is done with the help of the URL. A WSP get requests the content the URL points to. The gateway translates this WSP get into an HTTP get and the server responds with the prepared list of callers. After displaying the content, the user can select a voice message from the list. Each voice message in this example has an associated URL, which can request a certain WML card from the server. The purpose of this card is to prepare the client for an incoming call. As soon as the client receives the card, it waits for the incoming call. The call is then automatically accepted. The WTA server also signals the voice box system to set up a (traditional) voice connection to play the selected voice message. Setting up the call and accepting the call is shown using dashed lines, as these are standard interactions from the mobile phone network, which are not controlled by WAP.

The following examples illustrate the integration of WTA, WTAI library calls, WML and WMLScript, and how authors can use the functions of WTA (WAP Forum, 2000m). Imagine you are watching a show on TV. At the end of the show, you may vote for your personal champion. The traditional method is that each candidate gets an associated phone number and you have to dial the number –

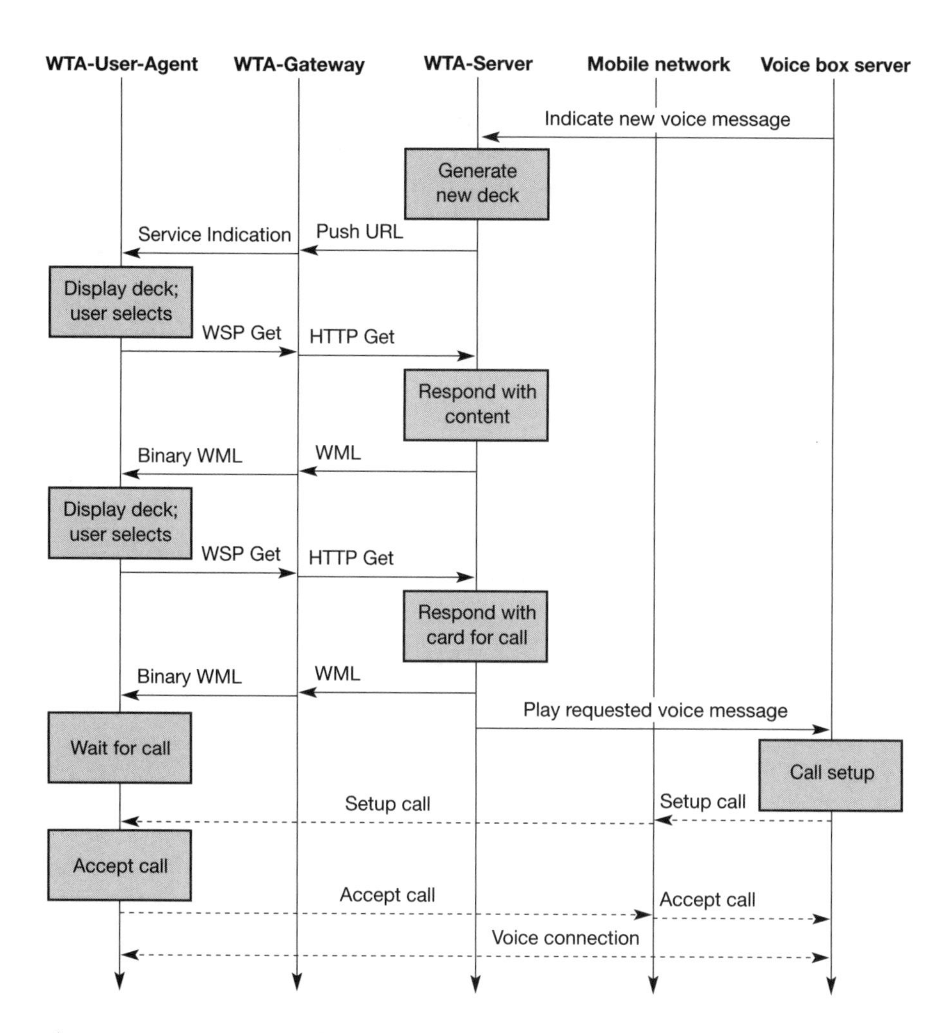

Figure 10.31 WTA example: voice message

this is quite error-prone if long numbers are used. Using WAP with WTA, the network operator could push a deck with several cards onto your handheld device and present a simple choice for voting as explained in the following paragraphs.

The first example consists of WML only and comprises two cards. The first card tells you to vote. If you accept, then the second card will be displayed. You can choose between Mickey, Donald, and Pluto. You do not have to dial a number but directly select the name of your champion. The variable `dialno` will be set to the value (i.e., the phone number) associated with your champion. Now note the URI within the `do` tags. This URI tells the system to use the function `mc` (make call) from the library `wp` (WTAPublic) in the WTAI. This function places a call to the specified number (here `dialno`). A traditional system can be used for counting calls.

Altogether, this WML deck lets the user select a champion and then automatically calls the correct number. The user does not have to know what the correct numbers are. A general problem of WML is the lack of dynamic behavior. What happens in our example, if anything goes wrong with the call, e.g., the number is busy? To check for errors and report them to a user, WMLScript has to be used in addition to WML.

```
<?xml version="1.0"?>
<!DOCTYPE wml PUBLIC "-//WAPFORUM//DTD WML 1.1//EN"
          "http://www.wapforum.org/DTD/wml_1.1.xml">
<wml>
  <card id="card_one" title="Tele voting">
     <do type="accept">
       <go href="#card_two"/>
     </do>
     <p> Please choose your candidate! </p>
  </card>
  <card id="card_two" title="Your selection">
    <do type="accept">
      <go href="wtai://wp/mc;$dialno"/>
    </do>
    <p> Your selection:
      <select name="dialno">
        <option value="01376685">Mickey</option>
        <option value="01376686">Donald</option>
        <option value="01376687">Pluto</option>
      </select>
    </p>
  </card>
</wml>
```

The following shows the same example, voting for a champion, but with WMLScript and WML for better error handling and reporting. Again, the network operator could push this code onto a handheld device. First, the function `voteCall` is defined. This function takes a number as input and then, in the second line, sets up a call to this number. Now the library function for setting up a call, `WTAVoiceCall.setup`, is used, not the URI as in the example before with just WML. Two values are passed to this function, the first is the phone number to dial and the second is a value indicating that the call should be kept after the current context is removed. The advantage of using this function and not the URI is the simple handling of the return value, here stored in `j`.

Depending on the value of `j`, the next lines can prepare a message the WML browser can display to the user. The function to set up a call is specified in a way that a negative return value indicates an error. The value itself represents the **WTAI error code**. Predefined error codes are, for example, –5 for "called part is busy", –6 for "network is busy", or –7 for "no answer" (i.e., the call setup timed out). If the return value is not negative, the variable `Message` of the browser is set to the string `"Called"`, the variable `No` to the value of `Nr`, i.e., the called number. Otherwise an error has occurred and `Message` is set to the string `"Error"` and `No` is set to the error code stored in `j`.

Now a WML deck follows, similar to the first example, but with some important differences. Again, text is displayed by the first card. After accepting, the WML browser displays the choice of the three candidates as before. Again, the user can make a choice and the phone number associated with the candidate is stored in the variable `dialno`. In this case, no URI for the WTAI is loaded as in the example before, but the browser loads a URI pointing to a function. The WMLScript is located in the compilation unit `myscripts`; the name of the script is `voteCall`. The value of the variable `dialno` is passed to this function.

It is important to note that the script now controls the execution. The second card does not forward control to the third card. In the example, this third card, which has to display a message, is called by the WMLScript function with the line `WMLBrowser.go ("showResult")`. This loads the card `showResult` in the WML browser. This third card displays some text and the values of the variables `Message` and `No`. These values have been set before in the WMLScript.

```
function voteCall(Nr) {
  var j = WTACallControl.setup(Nr,1);
  if (j>=0) {
    WMLBrowser.setVar("Message", "Called");
    WMLBrowser.setVar("No", Nr);
  }
  else {
    WMLBrowser.setVar("Message", "Error!");
    WMLBrowser.setVar("No", j);
  }
  WMLBrowser.go("showResult");
  }
```

```
<?xml version="1.0"?>
<!DOCTYPE wml PUBLIC "-//WAPFORUM//DTD WML 1.1//EN"
                    "http://www.wapforum.org/DTD/wml_1.1.xml">
<wml>
   <card id="card_one" title="Tele voting">
     <do type="accept"> <go href="#card_two"/> </do>
     <p> Please choose your candidate! </p>
    </card>
    <card id="card_two" title="Your selection">
      <do type="accept">
```

```
        <go href="/myscripts#voteCall($dialno)"/>
      </do>
      <p> Your selection:
        <select name="dialno">
          <option value="01376685">Mickey</option>
          <option value="01376686">Donald</option>
          <option value="01376687">Pluto</option>
        </select>
      </p>
    </card>
    <card id="showResult" title="Result">
      <p> Status: $Message $No </p>
    </card>
  </wml>
```

This very simple example showed the interaction of WML and WMLScript together with the WTAI. Sure, error codes or phone numbers should not be displayed to a customer, but these codes and numbers should be translated into plain text with the help of WMLScript.

While WTAI is valid for many different mobile networks, the WAP Forum has specified several WTAI extensions valid only for specific networks. For example, extensions have been defined for the Pacific Digital Cellular system (PDC, (WAP Forum, 2000n)), the Global System for Mobile Communications (GSM, (WAP Forum, 2000o)), and the IS-136 TDMA cellular network (WAP Forum, 2000p). A typical GSM function, for example, is joining a multiparty call. The WTAI extension for GSM has the function `WTAGSM.multiparty` for WMLScript in its library and adds the URI `wtai://gsm/jm`.

10.3.10 Push architecture

Compared to the early versions of WAP, version 1.2 introduced a new push architecture (WAP Forum, 2000v) together with several protocols (WAP Forum, 2000y and 2000z), message formats (WAP Forum, 2000x), and a special gateway (WAP Forum, 2000w). The very general architecture was already introduced in Figure 10.29; Figure 10.32 shows the push architecture alone together with the names introduced in the standard. Clients pulling content from servers are typical for today's www. In a push context the server initiates the message transfer, not the client. The server is called **push initiator (PI)** and transfers content via a **push proxy gateway (PPG)** to a client. The **push access protocol (PAP)** controls communication between PI and PPG. The **push over the air (OTA) protocol** is used between PPG and the client. If the PI is able to use the push OTA protocol, it can directly communicate with the client, too. Example usage scenarios for push messages are news, road conditions, e-mail indication etc.

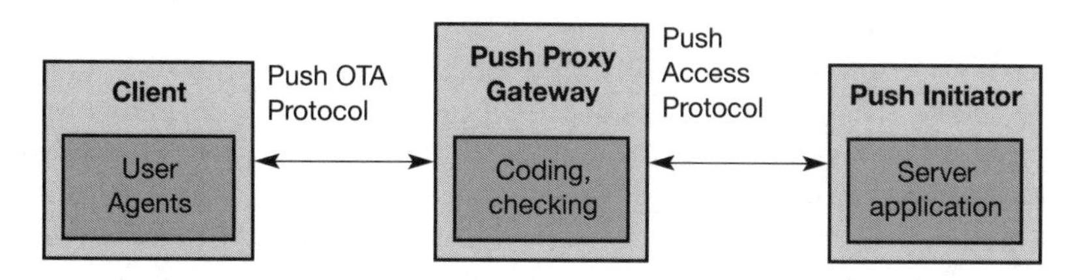

Figure 10.32
WAP push architecture with proxy gateway

10.3.10.1 Push proxy gateway

Similar to the WTA gateway for pull services, the PPG provides many functions to transform protocol messages and content exchanged between server and client (WAP Forum, 2000w). The PPG accepts push messages from a PI and checks if this message can be forwarded to the client. Checking comprises mapping of a client address onto a format valid in the mobile network. Additionally, it may be necessary to locate the client first. The content may be transcoded to gain greater efficiency during transmission with the push OTA protocol. If required, the PPG can send success or failure messages to the PI. A PPG can also multicast messages to a group of receivers.

10.3.10.2 Push access protocol

The push access protocol (PAP) transfers content from a PI to the PPG (WAP Forum, 2000z). Although PAP was developed independently from the underlying transport service, PAP over HTTP is the first implementation. HTTP POST messages are used for transmission. PAP offers the following operations:

- **Push submission:** Delivery of push messages from a PI to the PPG for forwarding to a client. The response indicates the initial acceptance or rejection of the forwarding request.
- **Result notification:** The PPG notifies the PI about the result of the push request. With the help of this notification, the server knows if the message reached the client, if the client deleted the message, if an error occurred at the client or if the message arrived too late.
- **Push cancellation:** A PI can try to cancel a push request before delivery to a client. The result of the PPG indicates if the cancellation was successful or not. Cancellation may be used, e.g., for periodic news messages buffered at the PPG because the client is currently unavailable. In this case, only the most recent message may be of interest.
- **Status query:** With the help of PAP, the PI can query the status of a push message. The result could indicate, if the PPG already delivered the message or is still trying to reach the client.
- **Client capabilities query:** The PI might be interested in the client's capabilities (WAP Forum, 2000aa). Additionally, a PPG can inform the PI about its capabilities regarding content transformation (e.g., JPEG coded pictures into WBMP if the client supports WBMP only).

10.3.10.3 Push OTA protocol

The push OTA protocol is a very simple protocol used on top of WSP (WAP Forum, 2000y). The protocol offers delivery of push messages, selection of transport services for push messages and authentication of PIs. The push service can be connection-oriented or connectionless based on a bidirectional or unidirectional bearer. The standard bearer for push messages in GSM is SMS as this is typically the only way to reach a mobile phone from within a network.

The following service primitives have been defined:

- **po-push:** unacknowledged push of content within a push session via a connection-oriented service (maps to S-Push in WSP);
- **po-confirmedpush:** acknowledged push of content within a push session via a connection-oriented service (maps to S-ConfirmedPush in WSP);
- **po-pushabort:** rejection of a push by the client;
- **po-unit-push:** unacknowledged push of content via a connectionless session service (maps to S-Unit-Push in WSP);

The following service primitives are also available for push session management:

- **pom-connect:** Creation of a push session by the client (maps to S-Connect in WSP);
- **pom-suspend:** suspension of a push session (maps to S-Suspend in WSP);
- **pom-resume:** Resume of a push session (maps to S-Resume in WSP);
- **pom-disconnect:** Termination of a push session (maps to S-Disconnect in WSP);
- **pom-sessionrequest:** Request for a push session by a server.

Each client has a predefined unsecured (mandatory) and WTLS secured (optional) WDP port for connectionless push services. For unidirectional bearers even the ports for sessions may be predefined so, sessions may already be set-up by default.

10.3.11 Push/pull services

As mobile devices have relatively low computing power (compared to desktop machines), the device may be too busy to accept and execute a push message from a PI. In this case, the PI can only indicate a service to the client but not transfer the whole push content. WAP specified a **service indication (SI)** for this purpose (WAP Forum, 2000ab). The simplest version of a SI comprises a URI pointing to the service and a short message. The PPG converts message formats between the PI and the client.

It is up to the user whether to use the service immediately or later. As soon as the service is used (no matter when), a classical client pull follows. The client requests data from the server with the help of WSP and the indicated URI. The server then offers the requested service. The PPG transforms content and protocols (e.g., WSP to HTTP and WML to binary WML).

A push message may comprise additional attributes indicating, e.g., the importance or the lifetime of the message. The following example shows a simple SI message containing a special pizza offer, which is only valid for five minutes. If a customer wants to use this offer, he or she has to request the indicated service via the URL. After five minutes, this push message will be automatically deleted.

```
<?xml version="1.0"?>
  <!DOCTYPE si PUBLIC "-//WAPFORUM//DTD SI 1.0//EN"
    "http://www.wapforum.org/DTD/si.dtd">
<si>
  <indication
   href="http://www.piiiizza4u.de/offer/salad.wml"
   created="2002-10-30T17:45:32Z"
   si-expires="2002-10-30T17:50:31Z">
   Salad special: The 5 minute offer
 </indication>
</si>
```

Another variant of delayed service usage can be described using **service loading (SL)** as specified in WAP Forum (2000ac). In this case, the PI also sends a short push message to the client. The client may be busy. In the example shown, the message contains a URI only, which points to a service. In contrast to SI, the client's user agent decides when to submit the URI, i.e., using the service. Although the service is accessed using a traditional pull scheme it looks for the user like a push service. The user does not notice the delay between the arrival of the push message and the pull. A disadvantage of SL is the higher number of messages sent over the air. The following example shows how a PI can indicate a certain URI to the client. No interaction with a user is intended.

```
<?xml version="1.0"?>
  <!DOCTYPE sl PUBLIC "-//WAPFORUM//DTD SL 1.0//EN"
    "http://www.wapforum.org/DTD/sl.dtd">
<sl
  href="http://www.piiiizza4u.de/offer/salad.wml">
</sl>
```

10.3.12 Example stacks with WAP 1.x

After presenting different aspects of WAP 1.x, this last subsection deals with the scope of standardization efforts using sample configurations as shown in Figure 10.33. WAP tries to use existing technologies and philosophies as much as possible, mainly from the internet. The simplest protocol stack, stack number 3, does not require new protocols or implementations. If an application needs

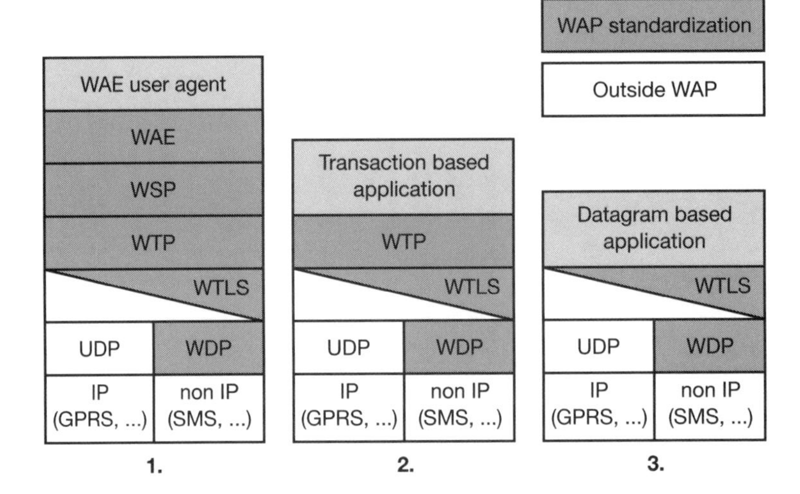

Figure 10.33
Sample protocol stack according to WAP 1.x

only unreliable datagram service without security, WAP offers a way to use UDP if the bearer network provides IP service (as this is the case for, e.g. GPRS). Based on this very simple stack, more and more complex stacks can be configured by adding security with WTLS or a reliable transaction service with WTP. Applications for distributed computing such as CORBA could use this reliable data transfer service. Currently, these applications mostly use TCP. However, TCP might not always be a good choice in a wireless mobile environment as demonstrated in chapter 9. WAP could provide an alternative solution.

The typical WAP application, i.e., a WAP user agent such as a WML or a WTA user agent, may require the full stack of protocols as shown in stack 1. These user agents run in the WAE and rely on, e.g., the WSP push service for pushing WTA events from a WTA server to the client.

10.4 i-mode

The i-mode service was introduced in Japan by the mobile network operator NTT DoCoMo in 1999. While other network operators in Japan (e.g., KDDI) use WAP, NTT DoCoMo decided to use its own system which is roughly based on the web protocols and content formats known from the www. Example services offered by i-mode are e-mail, web access (with certain restrictions), and picture exchange. The system soon became a big success with more than 30 million users only three years after its introduction. In comparison to i-mode, WAP was often cited as a failure, and operators outside Japan took over i-mode to participate in the success.

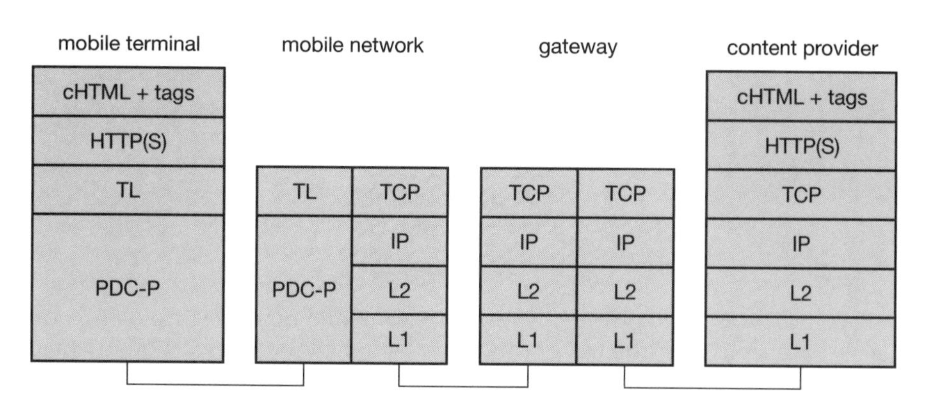

Figure 10.34
i-mode protocol stacks for Japan

Figure 10.34 shows the i-mode protocol stack as used in Japan. The packet-oriented PDC-P provides the bearer service between the mobile terminal and the operator's network. Typical data rates are 9.6 kbit/s, while enhanced versions offer 28.8 kbit/s. On top of the bearer service, i-mode uses a special connection-oriented transport layer protocol with stop-and-go flow control, ARQ, push services (ARIB standard RCR STD27X). Within the operator's network and between the operator's gateway and a content provider, i-mode uses standard Internet protocols (TCP/IP over different layer 1 and layer 2 protocols). On top of the transport services, i-mode uses HTTP (with or without security if supported by the handset) as known from the www. i-mode applications can use an e-mail service or display pages described in **compact HTML (cHTML)**, which is a subset of HTML (W3C, 1998b), plus some proprietary extra **tags** for emoticons, telephony and e-mail (the HTML is then sometimes called i-HTML).

Compact HTML supports animated color pictures (GIF format), but no frames, image maps or style sheets. However, the supported subset of HTML is large enough to display many web pages from the internet. This is a big advantage compared to WAP 1.x, which requires pages written in WML. To utilize i-mode fully many pages have to be redesigned. Many portals today offer HTML, cHTML, and WML content for different devices.

Why was i-mode a big success in Japan while WAP was not, e.g., in Europe? Many important factors in Japan are independent of the technology. For many people, i-mode was their first contact with the internet. i-mode is very simple to use, so, many people use it as PC replacement for e-mailing, chatting etc. i-mode was never announced as "Internet on the mobile phone," as WAP was in the beginning. Users simply experienced a new service that came with color displays and easy to use mobile phones.

One big difference from a technology point of view is the use of a packet-oriented bearer from the beginning in i-mode. WAP started with connection-oriented bearers (CSD in GSM), which ended in very poor user experience. A connection had either to be permanently open to support real interactive web browsing (which is expensive), or a new connection had to be established each

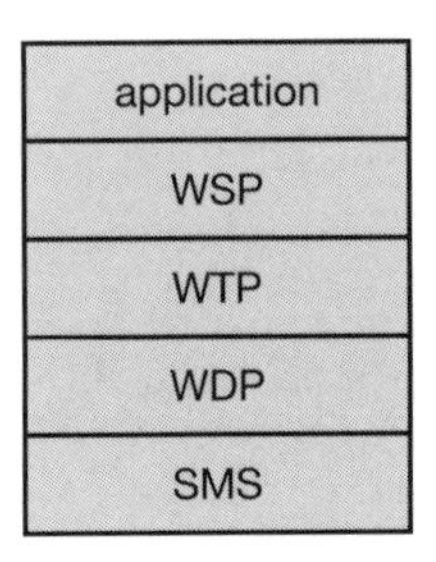

Figure 10.35 i-mode push architecture based on SMS

time content was loaded (which takes a very long time). Although WAP is completely independent of the bearer this fatal combination of connection-oriented bearer, and interactive browsing behavior led to the popular misconception that the whole WAP concept was a failure. Today, using WAP over GPRS, which is packet-oriented, shows what a difference the bearer can make.

Figure 10.35 shows a simplified protocol stack for an i-mode push based on SMS as used in Europe. As the figure shows, i-mode push in Europe uses WAP protocols! This stack is used, for example, for sending an SMS indicating that a new e-mail has arrived. If the user wants to read the message, the client sends an HTTP GET. The server responds with the email.

Figure 10.36 shows the possible mix of protocol components from different architectures. i-mode can be based on any bearer, e.g., GPRS, PDC-P, or a UMTS data service. Typically, most new packet-oriented bearers offer IP services. On top of that, i-mode uses TCP with a wireless profile (see chapter 9) between the user equipment and the gateway. Gateways fulfill many purposes in the i-mode service: translation of WTCP/TCP, address translation, protection of the user equipment etc. However, in contrast to WAP 1.x, the gateway does not break the security association between user equipment and server (SSL is simply tunneled through). cHTML/HTTP is added for browsing. Protocols like IP, TCP, HTTP etc. stem from the fixed internet and cHTML plus proprietary tags was introduced by NTT DoCoMo. The adaptation of TCP is performed by the IETF, and all protocols together will be part of the WAP 2.0 architecture described in section 10.6.

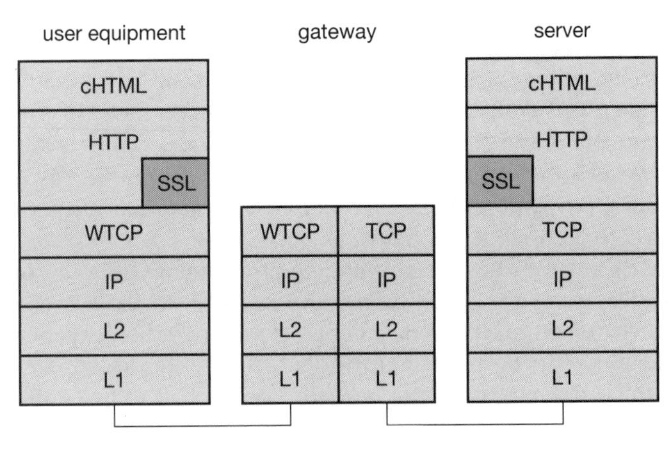

Figure 10.36 i-mode protocol stack using internet/WAP2.0 protocols

Table 10.1 Example technical requirements for i-mode systems

Functions	Status	Requirement
Web access to portal or Internet	M	cHTML plus tags (i-mode compatible HTML or i-HTML)
Internet e-mail	M	HTTP/1.1 (between UE and gateway)
End-to-end security	O	SSL (versions 2 and 3), TLS v1
Java applications	O	i-mode Java
Ringing tone download	M	SMF, MFi
Image download	M	GIF (O: JPEG)
Voice call notification	M	3GPP standard
Content charge billing	M	Operator dependent
Third party payment collection	M	Operator dependent
Reverse billing	O	Operator dependent
Subscriber ID transmission to the content provider on each content access	M	Operator dependent
Numbers of bytes per e-mail	M	Operator and handset dependent (e.g., 1 kbyte, 10 kbyte)

Table 10.1 lists the typical technical requirements i-mode systems have to fulfill. The status indicates mandatory (M) or optional (O) mechanisms.

However, i-mode is more than a pure technology – it is primarily a business model, which is independent of technology. Content providers share the revenue with the network provider. This makes i-mode quite attractive compared to the early WAP approaches which lacked any model for payment. The network provider bills customers; content providers get more than 80 per cent of the revenue. This model is independent of bearers, so, works as well with PDC-P as with, e.g., GSM/GPRS (using GPRS, i-mode over GSM is a lot faster than over PDC-P).

10.5 SyncML

A set of protocols and a markup language for synchronization of data in mobile scenarios is provided by the SyncML framework (SyncML, 2002), (Hansmann, 2003). The SyncML initiative is supported by companies like Ericsson, IBM, Motorola, Nokia, Openwave, Panasonic, Starfish, and Symbian. SyncML provides vendor independent mechanisms not only for synchronization of data, but also for the administration of devices and applications. The WAP 2.0 framework, which is described in the next section, chose SyncML as a synchronization mechanism.

Synchronization, as already explained in the context of file systems (see section 10.1), provides a major service for mobile users. Not only e-mails and calendar data, business spreadsheets, text documents, programs etc. all have to be synchronized. A common standard for synchronization simplifies application design and usage of synchronization mechanisms. SyncML enhances servers and clients with **sync server agents** and **sync client agents** respectively. The agents execute the synchronization protocol. The server also has a **sync engine** that is responsible for data analysis and conflict detection (the same conflicts already described in the context of file systems in section 10.1 may appear here, too).

The synchronization protocol may run over HTTP, WSP, or the object exchange protocol OBEX. However, many more protocols such as SMTP or TCP/IP could be used. SyncML does not make many assumptions about the data structures. Each set of data must have a unique identifier. Clients and servers can use their individual identifiers for data sets. However, servers have to know the mapping between the identifiers. Clients and servers have to log changes and must be able to exchange these logs.

Several modes are specified for synchronization. Two-way synchronization exchanges change logs between server and client. If, for example, a client crashed and has lost all change information, a special slow synchronization can be used. This synchronization mode first transfers all data from the client to the server. The server then compares all data and sends the necessary changes back to the client. Several variants of one-way synchronization are available. In this case, only one party (client or server) is interested in change logs.

The messages exchanged for synchronization are based on XML. Tags have been specified to `<add>`, `<copy>`, `<delete>`, and `<replace>` data sets. Operations can be made `<atomic>` (i.e., either all or no change operations may be applied) or applied in a certain `<sequence>`. If a conflict occurs (e.g., the same data set has been changed on the client and the server) SyncML does not specify a conflict resolution strategy. Instead, several recommendations for conflict resolution are given. Data sets can be mixed, the client may override server changes (or vice versa), a duplicate of the data set can be generated, or a failure of synchronization is signaled. These examples show that SyncML has no general solution for the synchronization problem.

10.6 WAP 2.0

In July 2001, version two of the wireless application protocol (WAP 2.0) was published by the WAP Forum (the first WAP 2.0 devices have been available since the end of 2002). All standards can be downloaded from the web (OMA, 2002). It can be stated that WAP 2.0 is roughly the sum of WAP 1.x, i-mode, Internet protocols, and many mobility specific enhancements. WAP 2.0 continues to support **WAP 1.x** protocols, but additionally integrates **IP**, **TCP** (with a wireless profile), **TLS**, and **HTTP** (wireless profiled). WAP 2.0 browsers support

WML as well as **XHTML with a mobile profile (XHTMLMP)**. XHTML is the extensible hypertext mark-up language developed by the W3C (2002) to replace and enhance the currently used HTML. WAP 2.0 uses the **composite capabilities/preference profiles (CC/PP)** framework for describing user preferences and device capabilities. CC/PP provides the technical basis for the UAProf device profile function described in section 10.3. These examples from the WAP 2.0 architecture already demonstrate the fusion of many different concepts that led to the new architecture.

Figure 10.37 gives an overview of the WAP 2.0 architecture as specified in WAP Forum (2001a). The **protocol framework** consists of the following four components:

- **Bearer networks:** Similar to WAP 1.x, many bearers are supported. Typical bearers today are **GPRS** in GSM networks, **SMS** for push services. Third generation networks will directly offer **IP** services.
- **Transport services:** These services offer an end-to-end abstraction on top of different bearers. Transport services can be either connection-oriented or connectionless. For reliable, connection-oriented services **TCP** with a wireless profile can be used as described in chapter 9, (WAP Forum, 2001b). **WDP** or **UDP** (in case of an IP bearer) can be used for unreliable, connectionless (datagram) services.

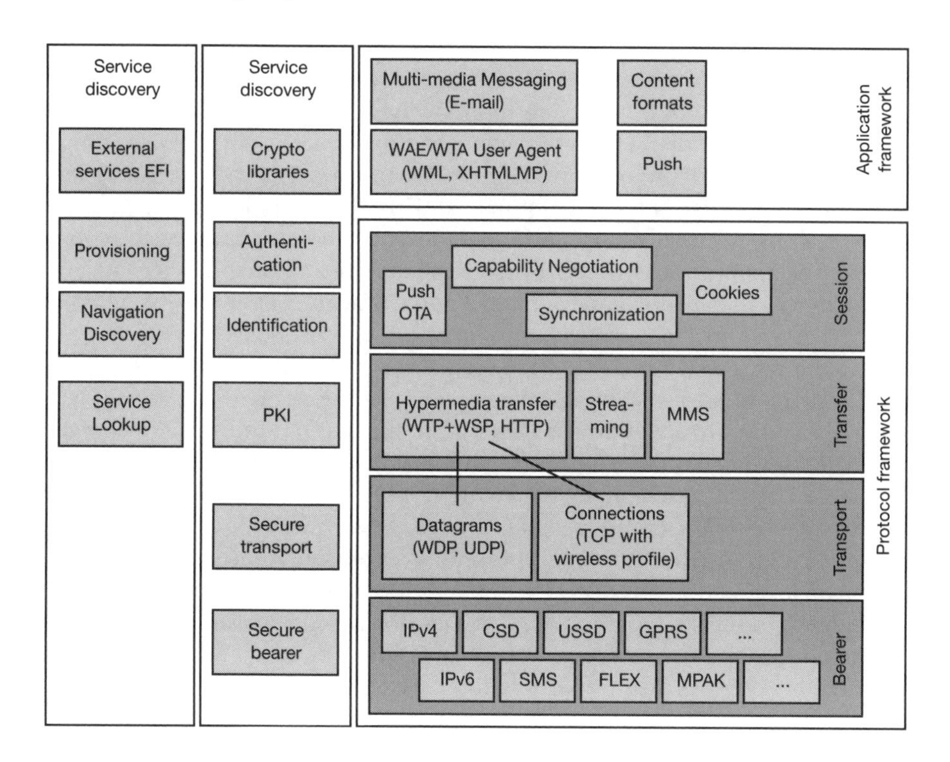

Figure 10.37
WAP 2.0 architecture

- **Transfer services:** Examples for transfer protocols are a wireless profiled **HTTP** (WAP Forum, 2001c), the combination of WTP/WSP, streaming protocols, and message transfer protocols. While the **hypermedia transfer** protocols can be used for web browsing, **streaming** protocols provide tight time-bounds to support isochronous data (audio and video). The **multi-media messaging service (MMS)** transfers asynchronous multi-media content (WAP Forum, 2001d). MMS supports different media types, such as JPEG, GIF, text, and AMR coded audio. There is no fixed upper bound for the message size. Depending on the network operator and device capabilities, typical sizes are 30–100 kbyte. MMS is already offered by many providers.
- **Session services:** As discussed in the beginning of this chapter, mobile devices need a shared state between network elements to operate efficiently. CC/PP can be used for **capability negotiation**. This includes information about client, server, and proxy capabilities, and allows for customization of content (WAP Forum, 2001e). The **push OTA** service offers reliable and unreliable push services as described in section 10.3.10 (WAP Forum, 2001f). The **cookie** service has been adopted from the Internet. This service can establish state on a client that survives multiple hypermedia transfers. Finally, a **synchronization** service has been defined for synchronizing replicated data. The SyncML framework as presented in section 10.5 is used for this purpose (SyncML, 2002), (Hansmann, 2003), (WAP Forum, 2001g).

The **application framework** comprises the basic applications needed for browsing, support of different content formats, e-mail service etc. This framework was developed to establish an interoperable application environment for many different vendors, service providers, and network operators (WAP Forum, 2002). The **WAE/WTA user agent** supports WML known from WAP 1.x as well as XHTML mobile profile (WAP Forum, 2001h). This includes scripting, stylesheets, telephony applications, and programming interfaces adapted to mobile devices. The applications for **multi-media messaging** and **push** services are also located in this framework. Finally, many **content formats** have to be supported: color images, audio, video, calendar information, phone book entries etc.

Mobile devices require extensive **security services**. Just imagine someone pushing many messages on a mobile device that let the device access certain web pages repeatedly. This might cause heavy air traffic that has to be paid for by the user! The user might not even notice anything from this attack as the mobile device is always-on, exchanging data without user interaction. The security services have to cover the traditional aspects of security, e.g., privacy, authentication, integrity, and non-repudiation. **Cryptographic libraries** are needed for signing data at the application level (WAP Forum, 2001i). **Authentication** services offer several mechanisms to authenticate servers, proxies, and clients at different levels. For example, at the transport layer WTLS or TLS may be used. The wireless identity module (WIM) provides the functions needed for user **identification** and authentication (WAP Forum, 2001j). Many business models rely on customer identification. In general, **public key infrastructures (PKI)** can be used for the management of public-key cryptography and certificate exchange (WAP Forum, 2001k). WTLS provides **secure transport** over datagram protocols, while TLS is used for connection-oriented transport protocols (TCP). Finally, some **bearer** networks already provide **security** functions, e.g., IPSec for IP networks.

Service discovery is particularly important for mobile devices. External functions or services on a device (i.e., outside the WAP specification, typically vendor dependent) can be discovered via the **external functionality interface (EFI)** as specified in WAP Forum (2001l). For many network services, a device needs additional parameters to get access (e.g., bootstrap information, user agent behavior, smart card specification, content type information). The device can get these parameters via the **provisioning** service (WAP Forum, 2001m). While surfing through the network using hypermedia documents the device may need to discover new network services. The **navigation discovery** provides a secure way to do this. The **service lookup** provides for the discovery of parameters needed for a certain service with the help of a directory. The domain name system (DNS) is an example mapping a name onto an IP address. It should be noted that the components listed for security services and service discovery cannot be assigned to only one layer. Instead, the services may span several layers in the protocol architecture. Up to now, not all of the services have been fully specified and, thus, the reader should check for changes in the specifications.

Figure 10.38 shows four examples of **protocol stacks** using WAP 2.0 components. The stacks in the upper left corner show the classical WAP 1.x configuration with a WAP gateway translating between internet and WAP protocols. The upper right corner illustrates protocol stacks similar to those used in i-mode. The WAP proxy translates the unchanged HTTP/TCP from the internet into profiled versions of these protocols. The lower left corner shows a WAP proxy tunneling HTTP over TLS. This architecture offers true end-to-end security which contrasts with the example in the upper left corner that breaks the security association in the gateway. The example in the lower right corner finally shows that WAP 2.0 does not necessarily need a proxy. The WAP device can directly access internet content. Although proxies are not required in WAP 2.0, they may enhance the performance (profiled TCP, profiled HTTP, compression etc.).

10.7 Summary

The application of the internet that attracts an ever-growing number of people is the www. As chapters 1 and 4 show, wireless and mobile communication, especially in the wide area, has an increasing number of subscribers worldwide. It is only logical to combine www and mobile communications. This chapter dealt with several problems, which combining the two areas causes, e.g., low available bandwidth in combination with inefficient protocols or HTML as a description language, which is not adapted to the requirements of portable devices.

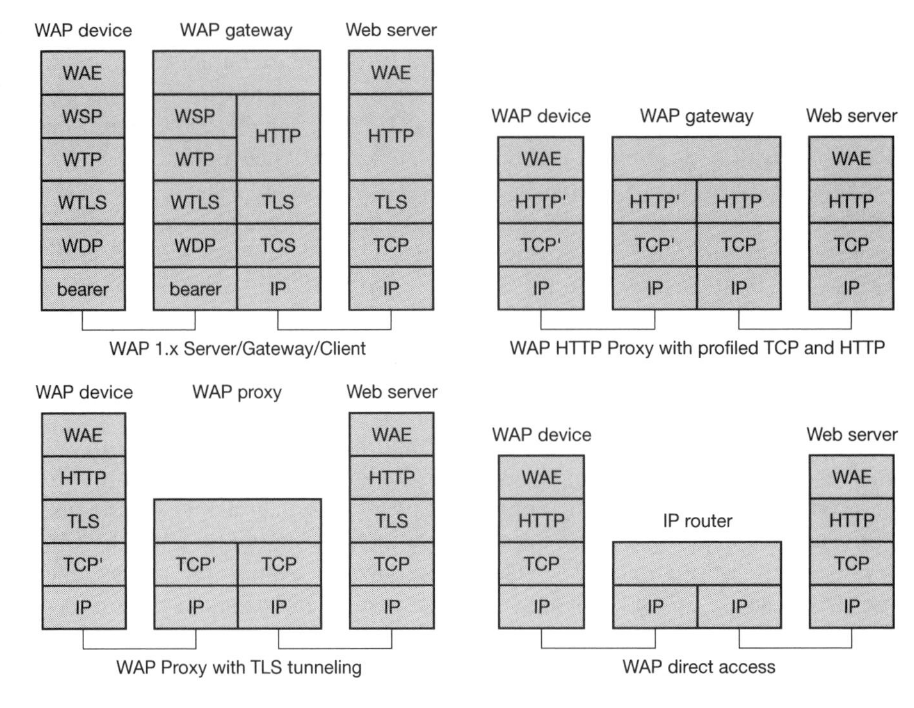

Figure 10.38 Example protocol stacks according to WAP 2.0

The major part of the chapter presented the WAP 1.x as a framework comprising several communication layers, a markup and scripting language, and a connection to the telephony network. Due to the influences of the internet and i-mode, this framework evolved to WAP version 2.0. Devices following this standard have just hit the market. This new WAP version integrates a lot of Internet technology while still offering special telephony services that are not available in today's standard Internet devices.

Although there have been a lot of developments in the context of mobile devices, the big problems of weakly connected devices, as described in the late nineties, remain. Important issues for all architectures and research projects supporting mobile web browsing are pre-fetching and caching of content (Jiang, 1998), content transformation and adaptation to the capabilities of portable devices (Brewer, 1998), (Fox, 1998), (Han, 1998), and using a browser while disconnected (Mazer, 1998). Many problems of efficient and convenient synchronization are unsolved.

As this chapter is about support for mobility, many more topics could be discussed. The following is an example collection of additional topics:

- **Message queuing:** Instead of using connections for data transfer, the whole communication system could be based on message queuing. This approach supports disconnected devices via buffering of messages until devices can be reached again. Typical applications are asynchronous updates, message distribution, and news distribution. Example products are Microsoft Message Queuing (MSMQ, Microsoft, 2002) and IBM WebSphere (MQSeries, IBM, 2002). Message queuing is not that useful for real-time or interactive applications.

- **Database support:** Many mobile devices carry information. Altogether, this can be seen as a large, distributed database. However, standard database products are rather heavyweight and run on big machines. The problems of distributed databases are similar to the problems of file systems. Example products supporting mobile devices are Oracle 9i Lite (Oracle, 2002) or IBM Everyplace (IBM, 2002).
- **Operating systems and execution environments:** Mobile devices need operating systems and execution environments for applications. While in the early days of mobile phones only very specialized, proprietary operating systems were used, the trend is towards more desktop-like systems. Examples for operating systems are Symbian OS (Symbian, 2002) or Windows Smartphone (Microsoft, 2002). To support platform independent application development, mobile devices, too, come with virtual machines. Examples are Java 2 Platform Micro Edition (J2ME) from Sun (2002) and the Common Language Runtime (CLR) from Microsoft (2002). Applications written for a virtual machine should run everywhere the virtual machine is available. In the context of mobile phones, 3GPP (2002) calls these application environments **mobile execution environments (MExE)**. Several versions of MExE (so-called classmarks) have been defined: classmark 1 supports WAP, classmark 2 supports personal Java, classmark 3 supports J2ME with certain libraries and profiles (Mobile Information Device Profile/Connected Limited Device Configuration; MIDP/CLDC), and classmark 4 supports the Common Language Infrastructure (CLI) as standardized by ECMA (2002). MExE frameworks sit on top of the vendor operating system and should form a secure sandbox for application execution.
- **Positioning/location based services:** Applications that take the current location of a user into account, is a big and promising field for research. This makes a real difference to applications for fixed systems. Positioning comprises many technologies from satellites (GPS) and cell ID based systems to triangulation of devices.
- **Service discovery:** A lot more could be written about service discovery. While this book presented several approaches suitable for certain environments (e.g., SDP for Bluetooth, 2002, see section 7.5), a more scalable and general mechanism for service discovery within a site is the IETF's **service location protocol (SLP)** as described in RFC 2608 (Guttman, 1999). The next step towards the discovery of whole business services takes the **universal description, discovery and integration (UDDI)** project (UDDI, 2002).

10.8 Review exercises

1. Why is strong consistency of file systems problematic in a wireless and mobile environment? What are the alternatives?
2. How do conventional file systems react to disconnected systems? Try unplugging a computer that has mounted a file system via a network.
3. What advantages has the statelessness of HTTP? In what situations is state useful and how is it provided today? Where is long-term state stored, where is short-term?
4. Which properties of HTTP waste bandwidth? What is the additional problem using HTTP/1.0 together with TCP? How does HTTP/1.1 improve the situation?
5. How does caching improve access time and reduce bandwidth requirements? What are locations for a cache and their specific advantages?
6. What are problems of caches in real life? What type of content can be cached, which content causes problems? What are the additional problems with client mobility?
7. What discrepancies exist between the possibilities of HTML and the realities of wireless handheld devices? What are the proposed solutions? What is the role of plug-ins today and how do they influence the usability of web pages?
8. Name mechanisms to improve web access for handheld devices. What is their common problem and what led finally to the development of WAP?
9. What are typical enhancements to the basic client/server architecture of the web? Reconsider these enhancements for a mobile wireless user with web access over a mobile phone network. What are efficient locations for the enhancements?
10. What are the primary goals of the WAP Forum efforts and how are they reflected in the initial WAP protocol architecture?
11. What migration paths does WAP 1.x offer for Internet and telephony applications and their protocols? Compare with WAP 2.0.
12. Is WDP a fixed protocol and why does WAP not define a SAP which WDP can use?
13. Why does WAP define its own security layer and does not rely on the security provided by the mobile phone network? What problems does the WAP security layer cause? Think of end-to-end security.
14. Name the advantages and disadvantages of user acknowledgements in WTP. What are typical applications for both cases?
15. Which WTP class reflects the typical web access best? How is unnecessary overhead avoided when using WSP on top of this class for web browsing?
16. What problems of HTTP can WSP solve? Why are these solutions especially needed in wireless mobile environments?

17 Why does WSP/B not put responses into the same order as the requests? Think, for example, of requests for different items on a web page.

18 What advantages does a connectionless session service offer compared to a simple datagram service?

19 What are the enhancements of WAE to the classic client/server model of the web? What are functions of this enhancement?

20 What is the fundamental difference of WML compared to HTML? Why can this difference be important for handheld devices? What is specified in addition to save bandwidth?

21 Why has a scripting language been added to WML? How can this language help saving bandwidth and reducing delay?

22 What are typical telephony events and how are they integrated into WAP? How can a user access features of mobile phones via the web browser?

23 What is the role of a WTA server? What are the different ways of integrating WTA servers into the WAP architecture?

24 What is the difference between WAP service indication and service loading? What applications could use these services? What is a push good for anyway?

25 Name key differences between WAP 1.x and i-mode. What were problems in the early WAP days and why was i-mode that successful in Japan?

26 Why is a common synchronization framework useful? What problems remain?

27 What are major differences between WAP 2.0 and WAP 1.x? What influenced the WAP 2.0 development?

28 Compare the presented protocol stacks for WAP 2.0 and give application examples.

10.9 References

3GPP (2002) Third Generation Partnership Project, http://www.3gpp.org/.

Adobe (2002) Adobe Systems Inc., http://www.adobe.com/.

Angin, O., Campbell, A.T., Kounavis, M.E., Liao, R.R.F. (1998) 'The MobiWare toolkit: Programmable support for adaptive mobile networking,' *IEEE Personal Communications*, 5(4).

Berners-Lee, T. (1994a) *Universal Resource Identifiers in WWW, a unifying syntax for the expression of names and addresses of objects on the network as used in the world wide web*, RFC 1630, for an update see RFC 2396.

Berners-Lee, T. (1994b) *Uniform Resource Locators* (URL), RFC 1738, updated by RFC 1808, RFC 2368, RFC 2396.

Berners-Lee, T., Fielding, R., Frystyk, H. (1996) *Hypertext Transfer Protocol – HTTP/1.0*, RFC1945.

Bickmore, T., Schilit, B. (1997) 'Digestor: device independent access to the world wide web,' proc. sixth International World Wide Web Conference, Santa Clara, CA, USA.

Blue Squirrel (2002) *WebWhacker v5.0*, Blue Squirrel Corporation, http://www.bluesquirrel.com/.

Bluetooth (2002) Bluetooth Special Interest Group, http://www.bluetooth.com/.

Brewer, E.A., Katz, R.H., Chawathe, Y.,; Gribble, S.D.; Hodes, T., Nguyen, G., Stemm, M., Henderson, T., Amit, E., Balakrishnen, H., Fox, A., Padmanabhan, V., Seshan, S. (1998) 'A network architecture for heterogeneous mobile computing,' *IEEE Personal Communications*, 5(5).

Certicom (2002) Certicom Corporation, http://www.certicom.com/.

Conta, A., Deering, S. (1998) *Internet Control Message Protocol (ICMPv6) for the Internet Protocol Version 6 (IPv6) Specification*, RFC 2463.

Dierks, T., Allen, C. (1999) *The TLS protocol version 1.0*, RFC 2246.

Diffie, W., Hellman, M. (1976) 'New directions in cryptography,' *IEEE Transactions on Information Theory*, 22(6).

ECMA (2002) ECMA International – European association for standardizing information and communication systems, http://www.ecma.ch/.

ETSI (2002) European Telecommunications Standards Institute http://www.etsi.org/.

Fielding, R. (1995) *Relative Uniform Resource Locators*, RFC 1808, updated by RFC2368, RFC2396.

Fielding, R., Gettys, J., Mogul, J., Frystyk, H., Masinter, L., Leach, P., Berners-Lee, T. (1999) *Hypertext Transfer Protocol – HTTP/1.1*, RFC 2616, updated by RFC 2817.

Flanagan, D. (1997) *JavaScript: the definitive guide*. O'Reilly.

Floyd, R., Housel, B., Tait, C. (1998) 'Mobile web access using eNetwork Web Express,' *IEEE Personal Communications*, 5(5).

Fox, A., Brewer, E. (1996a) 'Reducing WWW latency and bandwidth requirements by real-time distillation,' proc. Fifth International World Wide Web Conference, Paris, France.

Fox, A., Gribble, S., Brewer, E.A., Amir, E. (1996b) 'Adapting to network and client variability via on-demand dynamic distillation,' proc. ASPLOS'96, Cambridge, MA, USA.

Fox, A., Gribble, S.D., Chawathe, Y., Brewer, E.A. (1996b) 'Adapting to network and client variation using infrastructure proxies: Lessons and perspectives,' *IEEE Personal Communications*, 5(4).

Guedes, V., Moura, F. (1995) 'Replica Control in Mio-NFS,' proc. ECOOP'95 Workshop on Mobility and Replication, Aarhus, Denmark.

Guttman, E., Perkins, C., Veizades, J., Day, M. (1999) *Service Location Protocol, Version 2*, RFC 2608, updated by RFC 3224.

Han, R., Bhagwat, P., LaMaire, R., Mummert, T., Perret, V., Rubas, J. (1998) 'Dynamic adaptation in an image transcoding proxy for mobile web browsing,' *IEEE Personal Communications*, 5(6).

Hansmann, U., Mettala, R., Purakayastha, A., Thompson, P. (2003) *SyncML: Synchronizing and Managing Your Mobile Data*. Prentice Hall.

Heidemann, J.S., Page, T.W., Guy, R.G., Popek, G.J. (1992) 'Primarily disconnected operation: experiences with Ficus,' proc. Second Workshop on the Management of Replicated Data, Monterey, IEEE Computer Society Press, CA, USA.

Honeyman, P., Huston, L.B. (1995) 'Communications and consistency in mobile file systems', *IEEE Personal Communications*, 2(6).

Housel, B., Lindquist, D. (1996) 'WebExpress: A system for optimizing web browsing in a wireless environment,' proc. ACM/IEEE MobiCom'96 conference, Rye, NY, USA.

Howard, J.H., Kazar, M.L., Menees, S.G., Nichols, D.A., Satyanarayanan, M., Sidebotham, R.N., West, M.J. (1988) 'Scale and performance in a distributed file system,' *ACM Transactions on Computer Systems*, 6(1).

Huston, L. B., Honeyman, P. (1993) 'Disconnected operation for AFS', proc. USENIX Symposium on Mobile and Location-Independent Computing, Cambridge, MA, USA.

IBM (2002) IBM Corporation, http://www.ibm.com/.

IETF (2002) Internet Engineering Task Force, http://www.ietf.org/.

IMC (1996a) *vCard – the electronic business card*, Internet Mail Consortium.

IMC (1996b) *vCalendar – the electronic calendaring and scheduling format*, Internet Mail Consortium.

Jiang, Z., Kleinrock, L. (1998) 'Web prefetching in a mobile environment,' *IEEE Personal Communications*, 5(5).

Joseph, A.D., Tauber, J., Kaashoek, M.F. (1997a) 'Mobile Computing with the Rover Toolkit,' *IEEE Transactions on Computers: Special issue on Mobile Computing*, 64(3).

Joseph, A.D., Tauber, J., Kaashoek, M.F. (1997b) 'Building reliable mobile-aware applications using the Rover toolkit,' *Wireless Networks*, J.C. Baltzer, 3(5).

Kaufman, C., Perlman, R., Speciner, M. (1995) *Network security – private communication in a public world*. Prentice Hall.

Khare, R. (1999) 'W* Effect Considered Harmful,' *IEEE Internet Computing*, 3(4).

King, P., Hyland, T. (1997) *Handheld device markup language specification*. Unwired Planet.

Kistler, J.J., Satyanarayanan, M. (1992) 'Disconnected operation in the Coda file system,' *ACM Transactions on Computer Systems*, 10(1).

Krishnamurthy, B., Mogul, J., Kristol, D. (1999) 'Key Differences between HTTP/1.0 and HTTP/1.1,' proc. Eighth Intl. WWW Conference, Toronto, Canada.

Kristol, D., Montulli, L. (2000) *HTTP state management mechanism*, RFC 2965.

Kumar, P., Satyanarayanan, M. (1993) "Supporting application-specific resolution in an optimistically replicated file system," proc. Fourth Workshop on Workstation Operating Systems, Napa, CA, USA.

Liljeberg, M., Alanko, T., Kojo, M., Laamanen, H., Raatikainen, K. (1995) "Optimizing world wide web for weakly connected mobile workstations: An indirect approach," proc. Second International Workshop on Services in Distributed and Networked Environments, SDNE'95, Whistler, B.C., Canada.

Liljeberg, M., Helin, H., Kojo, M., Raatikainen, K., Mowgli WWW (1996) "Improved usability of WWW in mobile WAN environments," proc. IEEE Global Internet 1996, London, England.

Lotus (2002) *Domino Offline Services (was: Weblicator)*, IBM corporation, Lotus software, http://www.lotus.com/.

LoVerso, J., Mazer, M. (1997) 'Caubweb: detaching the web with Tcl,' proc. Fifth Annual USENIX Tcl/Tk Workshop, Boston, MA, USA.

Lu, Q., Satyanarayanan, M. (1994) 'Isolation-only transactions for mobile computing,' *Operation Systems Review*, 28(2).

Mazer, M.S., Brooks, C.L. (1998) 'Writing the web while disconnected,' *IEEE Personal Communications*, 5(5).

Microsoft (2002) Microsoft Corporation, http://www.microsoft.com/.

Mummert, L.B., Ebling, M. R., Satyanarayanan, M. (1995) 'Exploiting weak connectivity for mobile file access,' proc. Fifteenth Symposium on Operating System Principles, Copper Mountain Resort, CO, USA.

Netscape (2002) Netscape Corporation, http://www.netscape.com/.

OMA (2002) Open Mobile Alliance, http://www.openmobilealliance.org/.

Openwave (2002) Openwave Systems Inc., http://www.openwave.com/.

Oracle (2002) Oracle Corporation, http://www.oracle.com/.

Popek, G.J., Guy, R.G., Page, T.W. (1990) 'Replication in the Ficus distributed file system,' proc. Workshop on the Management of Replicated Data, Los Alamitos, IEEE Computer Society Press, CA, USA.

Postel, J. (1980) *User Datagram Protocol*, RFC 768.

Postel, J. (1981a) *Internet Protocol*, RFC 791, updated by RFC 1349.

Postel, J. (1981b) *Internet Control Message Protocol*, RFC 792, updated by RFC 950.

Raggett, D., LeHors, A., Jacobs, I. (1998) *HTML 4.0 specification*, W3C recommendation, REC-html40-19980424.

Rivest, R., Shamir, A., Adleman, L. (1978) 'A method for obtaining digital signatures and public-key cryptosystems,' *Communications of the ACM*, 21(2).

Satyanarayanan, M., Kistler, J.J., Mummert, L.B., Ebling, M.R., Kumar, P., Lu, Q. (1993) 'Experiences with disconnected operation in a mobile computing environment,' proc. USENIX Symposium on Mobile and Location-Independent Computing, Cambridge, MA, USA.

Schilit, B., Douglis, F., Kristol, D.M., Krzyzanowski, P., Sienicki, J., Trotter, J.A. (1996) 'TeleWeb: loosely connected access to the world wide web,' proc. Fifth International World Wide Web Conference, Paris, France.

Schneier, B. (1996) *Applied cryptography Protocols, Algorithms and source code in C* 2nd edn. John Wiley & Sons.

Singhal, S., Bridgman, T., Suryanarayana, L., Mauney, D., Alvinen, J., Bevis, D., Chan, J., Hild, S. (2001) *The Wireless Application Protocol*. Addison-Wesley.

Sun (2002) Sun Microsystems, http://www.sun.com/.

Symbian (2002) Symbian Ltd., http://www.symbian.com/.

SyncML (2002) SyncML Initiative, http://www.syncml.org/.

UDDI (2002) Universal Description, Discovery and Integration project, http://www.uddi.org/.

W3C (1998a) *Extensible Markup Language (XML) 1.0 Specification*, World Wide Web Consortium, W3C recommendation, REC-xml-19980210.

W3C (1998b) *Compact HTML for small information appliances*, World Wide Web Consortium, NOTE-compactHTML-19980209.

W3C (2002) World Wide Web Consortium, http://www.w3c.org/.

WAP Forum (2000a) *Wireless application protocol architecture specification*, WAP Forum, http://www.wapforum.org/.

WAP Forum (2000aa) *User Agent Profile Specification*, WAP Forum, http://www.wapforum.org/.

WAP Forum (2000ab) *Service Indication Specification*, WAP Forum, http://www.wapforum.org/.

WAP Forum (2000ac) *Service Loading Specification*, WAP Forum, http://www.wapforum.org/.

WAP Forum (2000b) *Wireless datagram protocol specification*, WAP Forum, http://www.wapforum.org/.

WAP Forum (2000c) *Wireless transport layer security protocol*, WAP Forum, http://www.wapforum.org/.

WAP Forum (2000d) *Wireless transaction protocol specification*, WAP Forum, http://www.wapforum.org/.

WAP Forum (2000e) *Wireless session protocol specification*, WAP Forum, http://www.wapforum.org/.

WAP Forum (2000f) *Wireless application environment specification*, WAP Forum, http://www.wapforum.org/.

WAP Forum (2000g) *Wireless application environment overview*, WAP Forum, http://www.wapforum.org/.

WAP Forum (2000h) *WMLscript language specification*, WAP Forum, http://www.wapforum.org/.

WAP Forum (2000i) *WMLscript standard libraries specification*, WAP Forum, http://www.wapforum.org/.

WAP Forum (2000j) *Wireless markup language specification*, WAP Forum, http://www.wapforum.org/.

WAP Forum (2000k) *Binary XML content format specification*, WAP Forum, http://www.wapforum.org/.

WAP Forum (2000l) *Wireless telephony application specification*, WAP Forum, http://www. wapforum.org/.

WAP Forum (2000m) *Wireless telephony application interface specification*, WAP Forum, http://www.wapforum.org/.

WAP Forum (2000n) *Wireless telephony application interface specification*, PDC specific addendum, WAP Forum, http://www.wapforum.org/.

WAP Forum (2000o) *Wireless telephony application interface specification*, GSM specific addendum, WAP Forum, http://www.wapforum.org/.

WAP Forum (2000p) *Wireless telephony application interface specification*, IS-136 specific addendum, WAP Forum, http://www.wapforum.org/.

WAP Forum (2000q) *WAP over GSM USSD specification*, WAP Forum, http://www.wapforum.org/.

WAP Forum (2000r) *Wireless control message protocol specification*, WAP Forum, http://www.wapforum.org/.

WAP Forum (2000s) *WMLScript Crypto Library Specification*, WAP Forum, http://www.wapforum.org/.

WAP Forum (2000t) *Wireless Identity Module*, WAP Forum, http://www.wapforum.org/.

WAP Forum (2000u) *WDP and WCMP Adaptation for access of a WAP Proxy Server to a Wireless Data Gateway*, WAP Forum, http://www.wapforum.org/.

WAP Forum (2000v) *Push Architectural Overview*, WAP Forum, http://www.wapforum.org/.

WAP Forum (2000w) *Push Proxy Gateway Service Specification*, WAP Forum, http://www.wapforum.org/.

WAP Forum (2000x) *Push Message Specification*, WAP Forum, http://www.wapforum.org/.

WAP Forum (2000y) *Push OTA Protocol Specification*, WAP Forum, http://www.wapforum.org/.

WAP Forum (2000z) *Push Access Protocol Specification*, WAP Forum, http://www.wapforum.org/.

WAP Forum (2001a) *WAP Architecture*, WAP Forum, WAP-210-WAPArch-20010712, http://www.wapforum.org/.

WAP Forum (2001b) *Wireless profiled TCP Specification*, WAP Forum, WAP-225-TCP-20010331, http://www.wapforum.org/.

WAP Forum (2001c) *Wireless profiled HTTP Specification*, WAP Forum, WAP-229-HTTP-20010329, http://www.wapforum.org/.

WAP Forum (2001d) *Multimedia Messaging Service Architecture Overview*, WAP Forum, WAP-205-MMSArchOverview-20010425, http://www.wapforum.org/.

WAP Forum (2001e) *User Agent Profiling Specification*, WAP Forum, WAP-248-UAProf-20011020, http://www.wapforum.org/.

WAP Forum (2001f) *Push Architectural Overview*, WAP Forum, WAP-250-PushArchOverview-20010703, http://www.wapforum.org/.

WAP Forum (2001g) *WAP Synchronization Specification*, WAP Forum, WAP-234-SYNC-20010530, http://www.wapforum.org/.

WAP Forum (2001h) *XHTML Mobile Profile Specification*, WAP Forum, WAP-277-XHTMLMP-20011029, http://www.wapforum.org/.

WAP Forum (2001i) *WMLScript Crypto API Library Specification*, WAP Forum, WAP-161-WMLScriptCrypto-20010620, http://www.wapforum.org/.

WAP Forum (2001j) *Wireless Identity Module Specification*, WAP Forum, WAP-260-WIM-20010712, http://www.wapforum.org/.

WAP Forum (2001k) *WAP Public Key Infrastructure Specification*, WAP Forum, WAP-217-WPKI-20010424 and WAP-217_103-WPKI-20011102, http://www.wapforum.org/.

WAP Forum (2001l) *External Functional Interfaces Specification*, WAP Forum, WAP-231-EFI-20011217, http://www.wapforum.org/.

WAP Forum (2001m) *Provisioning architecture overview*, WAP Forum, WAP-182-ProvArch-20010314, http://www.wapforum.org/.

WAP Forum (2002) *Wireless Application Environment Specification*, WAP Forum, WAP-236-WAESpec-20020207, http://www.wapforum.org/.

Outlook 11

11.1 The architecture of future networks

What will future mobile communication networks look like? Which wireless access will we use for what applications? A lot has already been written about fourth generation or next generation mobile communication system, e.g., (Jamalipour, 2001), (Mähönen, 2001), (Arroyo-Fernandez, 2001), Lu (2002). However, up to now, companies have not recouped their investment in 3G systems. Several architectures compete for the future of mobile communications.

While we cannot outline the precise architecture for future networks, we can think of useful scenarios for mobile users. Remember Figures 1.1 and 1.2. Many different devices accessed different networks while on the move or inside vehicles. This very demanding scenario leads to the general concept of **overlaying networks**. Figure 11.1 shows four example areas that different networks may cover: in-car/in-house/personal, campus, metropolitan, and regional. Users on the move, like in Figure 1.2, always want to use the 'best' (fastest, cheapest, most secure, company owned etc.) network. However, coverage is limited so the wireless devices have to perform handover. If the device stays in the same wireless system (e.g., GSM) the **handover** is called **horizontal**. If the device switches between networks on different layers, the handover is called **vertical**.

Figure 11.2 gives an overview of some wireless access technologies presented in this book and compares them related to bandwidth and device mobility. Wide area systems, such as GSM, cdma2000 and UMTS, allow for relatively high mobility, but suffer from low data rates. Local area systems, such as IEEE 802.11 and HiperLAN2 have not been designed for high device mobility. Going to higher bandwidth at higher speed is difficult and expensive. Research may push towards more sophisticated systems. However, it is uncertain if someone wants to pay for it (WLAN access points every 30 m are technically feasible, but too expensive for wide area coverage).

The next (or fourth) generation of mobile communication systems is not simply pushing the physical/economic border towards higher data rates at higher relative speed. The following **key features** based on user requirements, technical limitations, and the current development in Internet technology may mark future networks (WWRF, 2002), (DRIVE, 2001), (BRAIN, 2001):

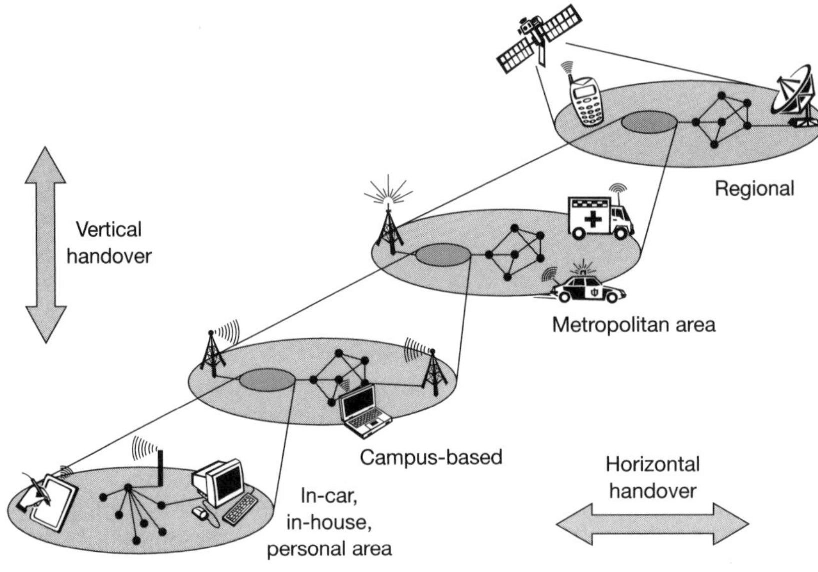

Figure 11.1 Wireless overlay networks

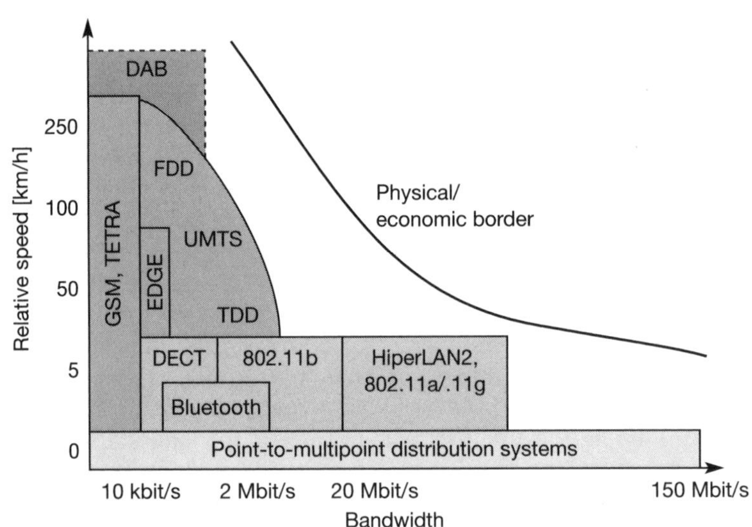

Figure 11.2 Wireless access technologies

- **Improved radio technology and antennas:** Wireless devices will use several **smart antennas** to benefit from multi-path propagation. Base stations can use **beam forming** to follow single wireless devices, using space division multiplex in a very fine-grained way. These technologies will strongly increase the capacity of wireless cells. Instead of just using a single antenna on each side of the wireless channel, future systems will be **multiple-input multiple-output (MIMO)** transmission systems using multiple antennas on both sides. **Software defined radios (SDR)** allow for multiple-system devices, reconfigurable senders/receivers, and the adaptation of radio transmission to application requirements. SDR emulate the RF hardware by software. However, today the computing power of standard processors is not high enough for this task. The RF component of UMTS alone is estimated to require the processing power of 10,000 GIPS (giga instructions per second). **Dynamic spectrum allocation**, i.e., allocating spectrum on demand, will further increase the available capacity per radio cell. MAC schemes, and coding can be improved. The device can then choose the 'best' coding and modulation scheme, spectrum, access technology etc. among a large set of combinations to achieve, e.g., low cost, high bandwidth, low jitter, high security etc.

- **Core network convergence:** While there will be many different access technologies in the future, the core becomes more and more **IP-based**. Many different networks already use Internet protocols so a common network layer is already available. The hardware systems tend to be cheaper due to the mass market. The future core network uses IPv6 and offers **quality of service** based on, e.g., the differentiated services approach combined with MPLS for simpler traffic engineering. Macro mobility technologies like **mobile IP** support vertical handover, while fast and seamless horizontal handover remains in the access networks for a long time.
- **Ad-hoc technologies:** Driven by the demand for spontaneous communication, different ad-hoc communication scenarios will show up. Many different communication layers will support ad-hoc communication: layer two for the spontaneous creation of links, layer three for efficient routing, and the application layer for service discovery, automatic configuration, authentication etc. Multihop ad-hoc technologies can furthermore extend the range of devices or lower interference/extend battery lifetime by transmitting via neighbor nodes instead of transmitting directly to a base station with high transmission power.
- **Simple and open service platform:** Current second generation mobile systems rely heavily on intelligent networks (IN) for service provisioning (e.g., call forwarding, multi-party call, toll-free numbers etc.). Future networks will use Internet technology for this purpose. This pushes the "intelligence" to the edges of the network and leaves the core network simple. This also enables many more companies to offer new services compared to IN solutions where the network operator controls everything.

Figure 11.3 gives a simplified view of an IP-based mobile communication system with gateways to legacy networks. A user's device may connect to different technologies (UMTS, GSM, public and private WLANs) or may receive broadcasted data via a satellite. The IP-based core of a network provider (many may exist) connects to traditional circuit switched networks and an SS7 signaling system. Second generation networks will be around for many more years due to the sheer amount of equipment installed and its millions of users. The IP

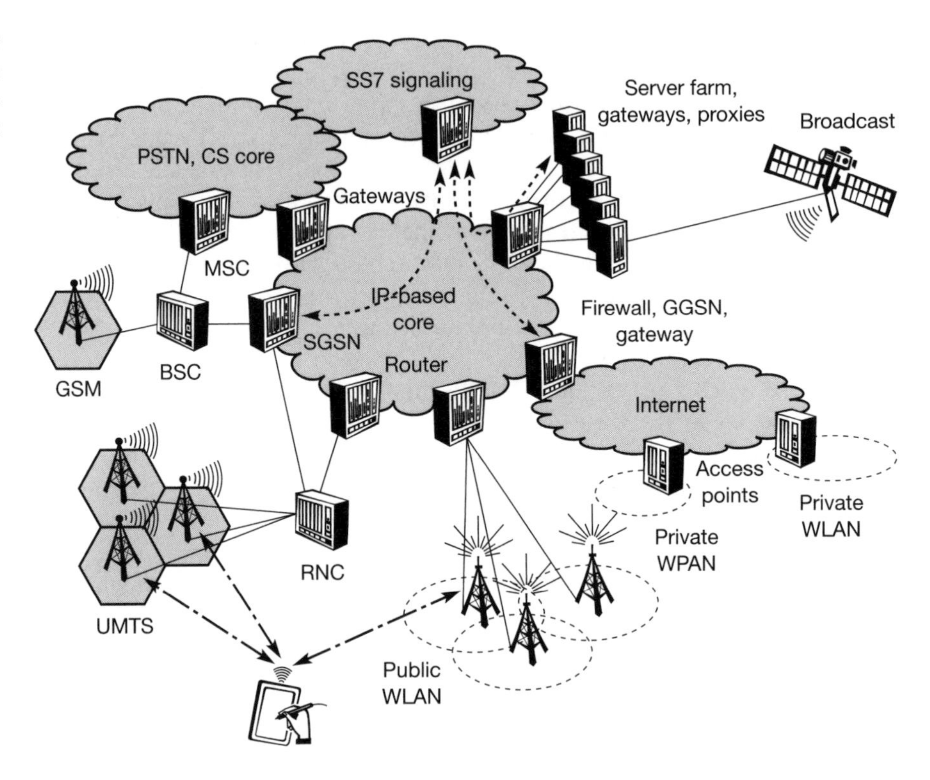

Figure 11.3
Example IP-based fourth generation mobile communication network

core needs many servers for name lookup, device configuration, authentication etc. Proxies and gateways have to translate addresses, content formats, and protocols (Wisely, 2002).

Although the figure may look perfect, there are many problems on the way towards an all-IP system (if it will ever come true). Today, most of the internet still relies on the best-effort protocols from the eighties. Integrated services and IPv6 have been developed in the mid-nineties, but only few systems use it. Due to scalability problems of integrated services the differentiated services approach has been developed – again, it will take a long time before it is really deployed. Today, no one can really say what will happen as soon as the internet offers quality of service at a large scale. The simplicity of the Internet may be gone then. Researchers have to take care that featurism does not migrate into the IP core but stays at the edge.

As soon as Internet technology is used for mission critical applications (e.g., emergency calls), reliability and security are of utmost importance. While denial of service attacks already today are an annoying threat, the reader can imagine what happens, if hackers attack an IP-based worldwide telephone network. Compared to today's telephone technology, Internet technology has a huge base of potential and skilled attackers. Up to now it has not been proved that

Internet technology is really simpler compared to classical technology as soon as all the QoS, accounting, and security features are integrated and the reliability has to be 99.9999 per cent.

Finally, we can discuss many interesting technologies and dream of a perfect mobile and wireless world, but someone has to pay for it. Current charging models, in particular if volume or time based, do not fit the users' requirements. Users understand pay-per-application, but not per byte or per minute. Companies often look for the 'killer application' for fast return on investment. This book showed that there is no single network, no single technology, and no single application that can guarantee success. The user's choice of many different networks, seamless access, and 'always best connected' will determine the future success of mobile and wireless communications.

11.2 References

Arroyo-Fernandez, B., DaSilva, J., Fernandes, J., Prasad, R. (2001) Life After Third-Generation Mobile Communications, collection of articles, *IEEE Communications Magazine*, 39(8).

BRAIN (2001) *Broadband Radio Access for IP based Networks*, IST-1999–10050, http://www.cordis.lu/, http://www.ist-brain.org/.

DRIVE (2001) Dynamic Radio for IP-Services in Vehicular Environments, IST-1999–12515, http://www.cordis.lu/, http://www.ist-drive.org/.

Jamalipour, A., Tekinay, S. (2001) Fourth Generation Wireless Networks and Interconnection Standards, collection of articles, *IEEE Personal Communications*, 8(5).

Lu, W. (2002) Fourth-Generation Mobile Initiatives and Technologies, *IEEE Communications Magazine*, 40(3).

Mähönen, P., Polyzos, G. (2001) European R&D on Fourth-Generation Mobile and Wireless IP Networks, collection of articles, *IEEE Personal Communications*, 8(6).

Wisely, D., Eardley, P., Burness, L. (2002) *IP for 3G, Networking Technologies for Mobile Communications*. Wiley & Sons.

WWRF (2002) Wireless World Research Forum, http://www.wireless-world-research.org/, http://www.ww-rf.org/.

Appendix 1 – Acronyms

3GPP	Third Generation Partnership Project
AAL	ATM Adaptation Layer
ABR	Available BitRate
ACF	Association Control Function
ACH	Access feedback CHannel
ACID	Atomicity, Consistency, Isolation, Durability
ACK	ACKnowledgement
ACL	Asynchronous Connectionless Link
ACT	Ad-hoc Controller Terminal
ADA	Alias DA
ADSL	Asymmetric Digital Subscriber Line
AES	Advanced Encryption Standard
AESA	ATM End System Address
AFS	Andrew File System
AGCH	Access Grant CHannel
AIB	Alias Information Base
AID	Acknowledgement IDentifier
AIDCS	AID CheckSum
AK-HCPDU	AcKnowledgement HCPDU
AM	Amplitude Modulation
AMA	Active Member Address
AMES	ATM Mobility Extension Service
AMPS	Advanced Mobile Phone System
ANSI	American National Standards Institute
AODV	Ad-hoc On-demand Distance Vector
AP	Access Point
APC	AP Controller
APCF	AP Control Function
APCM	AP Connection Management
APCP	AP Control Protocol
APT	AP Transceiver
ARIB	Association of Radio Industries and Broadcasting
ARQ	Automatic Repeat reQuest

ARQN	ARQ sequence Number
ASA	Alias SA
ASCH	Association Control CHannel
ASCI	Advanced Speech Call Items
ASCII	American Standard Code for Information Interchange
ASK	Amplitude Shift Keying
ASP	Active Server Page
Assoc	Association
AT	ATtention sequence
ATIM	Ad-hoc TIM
ATM	Asynchronous Transfer Mode
ATMC	ATM Connection function
AuC	Authentication Centre
AUS	AUthentication Server
Auth	Authentication
BCA	Borrowing Channel Allocation
BCCH	Broadcast CCH
BCH	Broadcast CHannel
BCH	Bose-Chaudhuri-Hocquenghem
BER	Bit Error Rate
BFSK	Binary FSK
B-ISDN	Broadband ISDN
BLI	Block Length Indicator
BLIR	Block Length Indicator Replica
BLIRCS	BLIR CheckSum
BMP	BitMaP
BNEP	Bluetooth Network Encapsulation Protocol
BPSK	Binary PSK
BRAN	Broadband Radio Access Networks
BSC	Base Station Controller
BSS	Base Station (Sub)system
BSS	Basic Service Set
BSSAP	BSS Application Part
BSSGP	BSS GPRS Protocol
BTS	Base Transceiver Station
BTSM	BTS Management
BW	BackWard
CAC	Channel Access Control
CAC	Connection Admission Control
CAC	Channel Access Code
CAMEL	Customized Application for Mobile Enhanced Logic
CATV	Community Antenna Television
CBR	Constant BitRate
CC	Call Control

CC	Country Code
CC/PP	Composite Capabilities/Preference Profiles
CCA	Clear Channel Assessment
CCCH	Common CCH
CCF	Call control and Connection Function
CCH	Control CHannel
CCIR	Consultative Committee for International Radiocommunication
CCK	Complementary Code Keying
CD	Compact Disk
CDM	Code Division Multiplexing
CDMA	Code Division Multiple Access
CDPD	Cellular Digital Packet Data
CDV	Cell Delay Variation
CEPT	European Conference for Posts and Telecommunications
CGI	Common Gateway Interface
CGSR	Clusterhead-Gateway Switch Routing
cHTML	compact HTML
CID	Channel ID
CIDR	Classless InterDomain Routing
CIF	Common Interleaved Frame
CKSN	Ciphering Key Sequence Number
CL	Convergence Layer
CLDC	Connected Limited Device Configuration
CLI	Common Language Infrastructure
CLMS	Connectionless Message Service
CM	Call Management
CM	Centralized Mode
CN	Core Network
CN	Correspondent Node
Cnf	Confirmation
COA	Care-Of Address
Codec	Coder/decoder
COFDM	Coded OFDM
COMS	Connection-Oriented Message Service
CORBA	Common Object Request Broker Architecture
COS	Cross Over Switch
CPM	Continuous Phase Modulation
CPU	Central Processing Unit
CRC	Cyclic Redundancy Check
CS	CheckSum
CS	Coding Scheme
CSCW	Computer Supported Cooperative Work
CSD	Circuit Switched Domain

CSMA	Carrier Sense Multiple Access
CSMA/CA	CSMA with Collision Avoidance
CSMA/CD	CSMA with Collision Detection
CT	Cordless Telephone
CTS	Clear To Send
CU	Capacity Unit
CVSD	Continuous Variable Slope Delta
CW	Contention Window
DA	Destination Address
DAB	Digital Audio Broadcasting
DAC	Device Access Code
DAMA	Demand Assigned Multiple Access
D-AMPS	Digital-AMPS
DBPSK	Differential Binary PSK
DC	Direct Current
DCA	Dynamic Channel Allocation
DCC	DLC user Connection Control
DCCH	Dedicated CCH
DCF	Distributed Coordination Function
DCS	Digital Cellular System
DCS	Dynamic Channel Selection
DDIB	Duplicate Detection Information Base
DECT	Digital Enhanced Cordless Telecommunications
DES	Data Encryption Standard
DFS	Dynamic Frequency Selection
DFWMAC	Distributed Foundation Wireless MAC
DH	Diffie, Hellman
DHCP	Dynamic Host Configuration Protocol
DIFS	DCF IFS
Disassoc	Disassociation
DLC	Data Link Control
DM	Direct Mode
DNS	Domain Name System
DPCCH	Dedicated Physical Control CHannel
DPCH	Dedicated Physical CHannel
DPDCH	Dedicated Physical Data CHannel
DQPSK	Differential QPSK
DRNC	Drift RNC
DS	Distribution System
DSDV	Destination Sequence Distance Vector
DSL	Digital Subscriber Loop
DSMA	Digital Sense Multiple Access
DSR	Dynamic Source Routing
DSSS	Direct Sequence Spread Spectrum

DT-HCPDU	DaTa-HCPDU
DTIM	Delivery TIM
DTMF	Dual Tone Multiple Frequency
DUCC	DLC User Connection Control
DV	Distance Vector
DVB	Digital Video Broadcasting
DVB-C	DVB-Cable
DVB-S	DVB-Satellite
DVB-T	DVB-Terrestrial
DVD	Digital Versatile Disk
DVTR	Digital Video Tape Recorder
ECDH	Elliptic Curve Diffie Hellman
ECMA	ECMA International – European association for standardizing information and communication systems (was European Computer Manufacturers Association until 1994)
ECN	Explicit Congestion Notification
EDGE	Enhanced Data rates for [Global|GSM] Evolution
EDTV	Enhanced Definition TV
EFI	External Functionality Interface
EFR	Enhanced Full Rate
EHF	Extremely High Frequency
EIR	Equipment Identity Register
EIRP	Equivalent Isotropic Radiated Power
EIT	Event Information Table
EMAS	End-user Mobility-supporting ATM Switch
EMAS-E	EMAS-Edge
EMAS-N	EMAS-Network
EMF	ElectroMagnetic Fields
EMS	Enhanced Message Service
ESS	Extended Service Set
ETSI	European Telecommunications Standards Institute
EY-NPMA	Elimination-Yield Non-preemptive Priority Multiple Access
FA	Foreign Agent
FACCH	Fast Associated Dedicated CCH
FCA	Fixed Channel Allocation
FCC	Federal Communications Commission (USA)
FCCH	Frame CCH
FCCH	Frequency Correction CHannel
FCH	Frame CHannel
FDD	Frequency Division Duplex
FDM	Frequency Division Multiplexing
FDMA	Frequency Division Multiple Access
FEC	Forward Error Correction
FFT	Fast Fourier Transform

FHSS	Frequency Hopping Spread Spectrum
FIB	Fast Information Block
FIC	Fast Information Channel
FM	Frequency Modulation
FOMA	Freedom Of Mobile multi-media Access
FPLMTS	Future Public Land Mobile Telecommunication System
FR	Frame Relay
FR	Full Rate
FSK	Frequency Shift Keying
FSLS	Fuzzy Sighted Link State
FSR	Fisheye State Routing
FT	Fixed Radio Termination
FW	ForWard
GEO	Geostationary (or Geosynchronous) Earth Orbit
GFSK	Gaussian FSK
GGSN	Gateway GSN
GIF	Graphics Interchange Format
GIPS	Giga Instructions Per Second
GMM	Global Multimedia Mobility
GMSC	Gateway MSC
GMSK	Gaussian MSK
GP	Guard Period
GPRS	General Packet Radio Service
GPS	Global Positioning System
GPSR	Greedy Perimeter Stateless Routing
GR	GPRS Register
GRE	Generic Routing Encapsulation
GSM	Groupe Spéciale Mobile, Global System for Mobile communications
GSN	GPRS Support Node
GTP	GPRS Tunneling Protocol
GWL	GateWay Link
HA	Home Agent
HAP	High-Altitude Platform
HBR	High Bit-Rate
HC	HIPERLAN CAC
HCPDU	HIPERLAN CAC PDU
HCQoS	HIPERLAN CAC QoS
HCSAP	HIPERLAN CAC SAP
HCSDU	HIPERLAN CAC SDU
HDA	Hashed Destination HCSAP Address
HDACS	HDA CheckSum
HDB	Home Data Base
HDLC	High level Data Link Control

HDML	Handheld Device Markup Language
HDTP	Handheld Device Transport Protocol
HDTV	High Definition TV
HEC	Header Error [Control | Check]
HEO	Highly Elliptical Orbit
HF	High Frequency
HI	HBR-part Indicator
HIB	Hello Information Base
HID	HIPERLAN IDentifier
HIPERLAN	High-PERformance LAN
HLR	Home Location Register
HM	HIPERLAN MAC
HMPDU	HIPERLAN MAC PDU
HMQoS	HIPERLAN MAC QoS
HO	HandOver
HO-HMPDU	HellO-HMPDU
HP	HIPERLAN PHY
HR	Half Rate
HSCSD	High Speed Circuit Switched Data
HSDPA	High Speed Downlink Packet Access
HSR	Hierarchical State Routing
HTML	HyperText Markup Language
HTTP	HyperText Transfer Protocol
IAC	Inquiry Access Code
IAPP	Inter Access Point Protocol
IBSS	Independent BSS
ICMP	Internet Control Message Protocol
ICO	Intermediate Circular Orbit
ID	IDentifier
IEEE	Institute of Electrical and Electronics Engineers
IETF	Internet Engineering Task Force
IFS	Inter Frame Spacing
ILR	Interworking Location Register
IMEI	International Mobile Equipment Identity
IMF	Identity Management Function
IMS	IP-based Multi-media Services
IMSI	International Mobile Subscriber Identity
IMT	International Mobile Telecommunications
IN	Intelligent Network
IOT	Isolation Only Transactions
IP	Internet Protocol
IR	Infra Red
IrDA	Infra red Data Association
IS	Interim Standard

ISDN	Integrated Services Digital Network
ISI	InterSymbol Interference
ISL	Inter Satellite Link
ISM	Industrial, Scientific, Medical
ISMA	Inhibit Sense Multiple Access
ISO	International Organization for Standardization[1]
I-TCP	Indirect TCP
ITU	International Telecommunication Union
ITU-R	ITU Radiocommunication sector
ITU-T	ITU Telecommunication sector
IV	Initialization Vector
IWF	InterWorking Function
JCT	Japanese Cordless Telephone
JDC	Japanese Digital Cellular
JPEG	Joint Photographic Experts Group
KID	Key IDentifier
L2CAP	Logical Link Control and Adaptation Protocol
LA	Location Area
LAI	Location Area Identification
LAN	Local Area Network
LAP	Lower Address Part
LAPC	Link Access Procedure for the C-Plane
LAPD	Link Access Procedure for the D-channel
$LAPD_m$	LAPD for mobile
LBR	Low Bit-Rate
LC	Link Controller
LCCH	Link Control CHannel
LCH	Long transport CHannel
LED	Light Emitting Diode
LEO	Low Earth Orbit
LF	Low Frequency
LI	Length Indicator
LIR	Least Interference Routing
LLC	Logical Link Control
LM	Link Manager
LMP	Link Manager Protocol
Loc	Location
LOS	Line-Of-Sight
LR-WPAN	Low-Rate WPAN
LRU	Last Recently Used

1 This is *not* the 'International Standards Organization' or the 'International Standardization Organization' or whatever some authors write. ISO is not an acronym, but is derived from the Greek word *isos* which means equal as used (as prefix) in isometric, isomorphic etc. (http://www.iso.ch/).

LS	Location Server
LS	Link State
M-QoS	Mobile QoS
MAC	Medium Access Control
MACA	Multiple Access with Collision Avoidance
MANET	Mobile Ad-hoc NETwork
MAP	Mobile Application Part
MATM	Mobile ATM
MBS	Mobile Broadband System
MBWA	Mobile Broadband Wireless Access
MCC	Mobile Country Code
MCI	Multiplex Configuration Information
MCM	MultiCarrier Modulation
MEO	Medium Earth Orbit
MExE	Mobile Execution Environment
MF	Medium Frequency
MFi	Melody Format for i-mode
MH	Mobile Host
MHEG	Multi-media and Hypermedia information coding Experts Group
MIB	Management Information Base
MIDI	Musical Instrument Digital Interface
MIDP	Mobile Information Device Profile
MIMO	Multiple Input Multiple Output
MKK	Radio Equipment Inspection and Certification Institute (Japan)
ML	MSDU Lifetime
MM	Mobility Management
MMF	Mobility Management Function
MMS	Multi-media Messaging Service
MN	Mobile Node
MNC	Mobile Network Code
MOC	Mobile Originated Call
MOT	Multi-media Object Transfer
MPEG	Moving Pictures Expert Group
MPLS	Multi Protocol Label Switching
MQ	Message Queuing
MS	Mobile Station
MS	Mobile Switch
MSAP	MAC SAP
MSC	Mobile (Services) Switching Centre
MSC	Mobile Switch Controller
MSC	Main Service Channel
MSDU	MAC SDU
MSIN	Mobile Subscriber Identification Number

MSISDN	Mobile [Station (International)|Subscriber] ISDN Number
MSK	Minimum Shift Keying
MSRN	Mobile [Station|Subscriber] Roaming Number
MSS	Mobile Satellite [Service|System]
MT	Mobile Terminal
MT	Mobile Termination
MTC	Mobile Terminated Call
M-TCP	Mobile TCP
MTSA	Mobile Terminal Security Agent
MUL	Mobile User Link
NA-TDMA	North American-TDMA
NAV	Net Allocation Vector
NAT	Network Address Translator
NDC	National Destination Code
NEMO	NEtwork MObility
NFS	Network File System
NIB	Neighbor Information Base
NIT	Network Information Table
NMAS	Network Mobility-supporting ATM Switch
NMT	Nordic Mobile Telephone
NNI	Network-to-Network Interface
NNI+M	NNI+Mobility
NRL	Normalized Residual HMPDU Lifetime
NSA	Network Security Agent
NSS	Network and Switching Subsystem
NTSC	National Television Standards Committee
OBEX	OBject EXchange
OFDM	Orthogonal FDM
OHG	Operators Harmonization Group
OLSR	Optimized Link State Routing
OMC	Operation and Maintenance Centre
OSI	Open Systems Interconnection
OSS	Operation Subsystem
OTA	Over The Air
OVSF	Orthogonal Variable Spreading Factor
PACS	Personal Access Communications System
PACS-UB	PACS-Unlicensed Band
PAD	PADding
PAD	Program Associated Data
PAL	Phase Alternating Line
PAP	Push Access Protocol
PBCC	Packet Binary Convolutional Coding
PC	Personal Computer
PCF	Point Coordination Function

PCH	Paging Channel
PCM	Pulse Code Modulation
PCS	Personal Cellular System
PCS	Personal Communications Service
PDA	Personal Digital Assistant
PDC	Pacific Digital Cellular
PDCP	Packet Data Convergence Protocol
PDF	Portable Document Format
PDN	Public Data Network
PDO	Packet Data Optimized
PDTCH	Packet Data TCH
PDU	Protocol Data Unit
PEP	Performance Enhancing Proxy
PHS	Personal Handyphone System
PHY	PHYsical layer
PI	Push Initiator
PIFS	PCF IFS
PIN	Personal Identity Number
PKI	Public Key Infrastructure
PLCP	Physical Layer Convergence Protocol
PLI	Padding Length Indicator
PLL	Phase Lock Loop
PLMN	Public Land Mobile Network
PLW	PLCP-PDU Length Word
PM	Phase Modulation
PMA	Parked Member Address
PMD	Physical Medium Dependent
POS	Personal Operating Space
POTS	Plain Old Telephone Service
PPG	Push Proxy Gateway
PPM	Pulse Position Modulation
PPP	Point-to-Point Protocol
PRACH	Physical Random Access Channel
PRMA	Packet Reservation Multiple Access
PS	Power Saving
PSD	Packet Switched Domain
PSF	PLCP Signaling Field
PSK	Phase Shift Keying
PSM	Protocol/Service Multiplexor
PSN	PDU Sequence Number
PSPDN	Public Switched Packet Data Network
PSTN	Public Switched Telephone Network
PT	Portable radio Termination
PTM	Point-to-Multipoint

PTP	Point-to-Point
PTP-CLNS	PTP-ConnectionLess Network Service
PTP-CONS	PTP-Connection Oriented Network Service
PUK	PIN Unblocking Key
QAM	Quadrature AM
QoS	Quality of Service
QPSK	[Quadrature | Quaternary] PSK
RA	Receiver Address
RACH	Random Access Channel
RAL	Radio Access Layer
RAND	RANDom number
RBCH	RLC Broadcast CHannel
RCH	Random CHannel
Req	Request
Res	Response
RF	Radio Frequency
RFCOMM	RF COMMunications
RFID	RF IDentification
RFC	Request For Comments
RFCH	Random access Feedback CHannel
RIB	Route Information Base
RIP	Routing Information Protocol
RL	Residual Lifetime
RLC	Radio Link Control
RLP	Radio Link Protocol
RM	Resource Management
RNC	Radio Network Controller
RNS	Radio Network Subsystem
ROM	Read Only Memory
RPC	Remote Procedure Call
RR	Radio Resource
RRC	Radio Resource Control
RRM	Radio Resource Management
RSA	Rivest, Shamir, Adleman
RSS	Radio SubSystem
RT	Radio Transceiver
RTR	Radio Transmission and Reception
RTS	Request To Send
RTT	Radio Transmission Technologies
RTT	Round Trip Time
S-DMB	Satellite-Digital Multimedia Broadcasting
SA	Source Address
SAAL	Signalling AAL
SACCH	Slow Associated Dedicated CCH

SAMA	Spread Aloha Multiple Access
SAP	Service Access Point
SAT	SIM Application Toolkit
SATM	Satellite ATM Services
SC	Sanity Check
SC	Synchronization Channel
SCF	Service Control Function
SCH	Short transport CHannel
SCH	Synchronization Channel
SCO	Synchronous Connection-Oriented link
SCPS	Space Communications Protocol Standards
SCPS-TP	SCPS-Transport Protocol
SDCCH	Stand-alone Dedicated CCH
SDM	Space Division Multiplexing
SDMA	Space Division Multiple Access
SDP	Service Discovery Protocol
SDR	Software Defined Radio
SDT	Service Description Table
SDTV	Standard Definition TV
SDU	Service Data Unit
SEC-SAP	Security SAP
SEQN	SEQuence Number
SFD	Start Frame Delimiter
SFN	Single Frequency Network
SGSN	Serving GSN
SH	Supervisory Host
SHF	Super High Frequency
SI	Service Indication
SIFS	Short IFS
SIG	SIGnaling
SIM	Subscriber Identity Module
SIP	Session Initiation Protocol
SL	Service Loading
SMF	Standard MIDI File
SMRIB	Source Multipoint Relay Information Base
SMS	Short Message Service
SN	Subscriber Number
SNACK	Selective Negative ACKnowledgement
SNAP	Sub-Network Access Protocol
SNDCP	Subnetwork Dependent Convergence Protocol
SRES	Signed Response
SRNC	Serving RNC
SS	Supplementary Service
SS7	Signalling System No. 7

S-SAP	Session-SAP
SSL	Secure Sockets Layer
STA	STAtion
SUMR	Satellite User Mapping Register
SW	Short Wave
SwMI	Switching and Management Infrastructure
T	Terminal
TA	Transmitter Address
TBRPF	Topology Broadcast based on Reverse Path Forwarding
TCH	Traffic CHannel
TCH/F	TCH Full rate
TCH/FS	TCH/F Speech
TCH/H	TCH Half rate
TCH/HS	TCH/H Speech
TC-HMPDU	Topology Control-HMPDU
TCP	Transmission Control Protocol
TCS BIN	Telephony Control protocol Specification – BINary
TD-CDMA	Time Division-CDMA
TDD	Time Division Duplex
TDM	Time Division Multiplexing
TDMA	Time Division Multiple Access
TDT	Time and Data Table
TE	TErminal
TEDDI	TErms and Definitions Database Interactive
TETRA	Terrestrial Trunked Radio
TFI	Transport Format Identifier
TFO	Tandem Free Operation
TFTS	Terrestrial Flight Telephone System
TI	Type Indicator
TIB	Topology Information Base
TIM	Traffic Indication Map
TINA	Telecommunication Information Networking Architecture
TLLI	Temporary Logical Link Identity
TLS	Transport Layer Security
TM	Traffic Management
TMN	Telecommunication Management Network
TMSI	Temporary Mobile Subscriber Identity
TOS	Type Of Service
TPC	Transmit Power Control
TR-SAP	Transaction SAP
T-SAP	Transport SAP
TSF	Timing Synchronization Function
T-TCP	Transaction TCP
TTC	Telecommunications Technology Council

TTL	Time To Live
TV	TeleVision
U-NII	Unlicensed National Information Infrastructure
UBCH	User Broadcast CHannel
UBR	Unspecified BitRate
UD	User Data
UDCH	User Data CHannel
UDP	User Datagram Protocol
UE	User Equipment
UHF	Ultra High Frequency
UIM	User Identification Module
UMCH	User Multicast CHannel
UMTS	Universal Mobile Telecommunications System
UN	United Nations
UNI	User-to-Network Interface
UNI+M	UNI+Mobility
UP	User Priority
UPT	Universal Personal Telecommunications
URI	Uniform Resource Identifier
URL	Uniform Resource Locator
USAT	UMTS SAT
UTRA	Universal Terrestrial Radio Access
UTRAN	UTRA Network
UUID	Universally Unique ID
UWB	Ultra WideBand
UWC	Universal Wireless Communications
V+D	Voice and Data
VAD	Voice Activity Detection
VBR	Variable BitRate
VBR-nrt	VBR non real-time
VBR-rt	VBR real-time
VC	Virtual Circuit
VCC	Visitor Country Code
VDB	Visitor Data Base
VHE	Virtual Home Environment
VHF	Very High Frequency
VLF	Very Low Frequency
VLR	Visitor Location Register
VNDC	Visitor National Destination Code
W3C	World Wide Web Consortium
WAE	Wireless Application Environment
WAN	Wide Area Network
WAP	Wireless Application Protocol
WATM	Wireless ATM

WCAC	Wireless Connection Admission Control
W-CDMA	Wideband-CDMA
WCMP	Wireless Control Message Protocol
W-CTRL	Wireless ConTRoL
WDP	Wireless Datagram Protocol
WEP	Wired Equivalent Privacy
WHO	World Health Organization
WIM	Wireless Identity Module
WLAN	Wireless LAN
WLL	Wireless Local Loop
WML	Wireless Markup Language
WMLScript	Wireless Markup Language Script
WMT	Wireless Mobile Terminal
WPAN	Wireless Personal Area Network
WP-CDMA	Wideband Packet-CDMA
WRC	World Radio Conference
WSP	Wireless Session Protocol
WSP/B	Wireless Session Protocol/Browsing
WTA	Wireless Telephony Application
WTAI	Wireless Telephony Application Interface
WTLS	Wireless Transport Layer Security
WTP	Wireless Transaction Protocol
WWAN	Wireless WAN
WWRF	Wireless World Research Forum
WWW	World Wide Web
XHTML	eXtensible HTML
XHTMLMP	XHTML Mobile Profile
XML	eXtensible Markup Language
XOR	eXclusive OR
ZRP	Zone Routing Protocol

Appendix 2 – Glossary

- **Ad-hoc network:** Ad-hoc networks do not need any infrastructure to operate. In particular, they do not need a base station controlling medium access. This type of network allows for spontaneous communication without previous planning between mobile devices. Some devices may even have forwarding capabilities to extend coverage.

  Example: Bluetooth is the most prominent ad-hoc network for spontaneous communication between different peripherals, such as mobile phones, PDAs, notebooks, etc. Each device can communicate with any other device. Within one piconet a maximum number of eight devices can be active at the same time. Communication between piconets takes place with the help of devices jumping back and forth between the networks.

- **Base station (access point):** Typical infrastructure-based wireless networks provide the access to the fixed network via a base station. The base station may act as a repeater, bridge, router, or even as gateway to filter, translate, retransmit etc. messages. At least the physical layer is present in a base station, which modulates and demodulates the signals. Most base stations also include medium access control.

  Example: WLANs following IEEE 802.11 specify base stations with bridging functionality. These bridges connect mobile and wireless devices to the fixed network and separate traffic within the fixed or wireless network from the traffic flowing from the fixed into the wireless part or vice versa. All communication between mobile devices has to take place via the base station. Additionally, the base station may control time critical services via a polling scheme. Within mobile phone networks the functions of modulation/demodulation and medium access control are quite often split into two or more entities (BTS and BSC/MSC in GSM).

- **Downlink:** The term 'downlink' denoting a certain direction for communication between two devices has its origin in satellite systems. In this case the link from the satellite down to the Earth is called downlink. The downlink is typically the direction of communication where a single station controls medium access, i.e., no competition or collision can takes place in that direction. The communication system has to separate the downlink from uplinks with the help of a multiplexing scheme (time, frequency, code, and space) in order to avoid collisions.

Example: Satellites determine in classical satellite communication systems at what time and frequency they send which data on the downlink. The separation of the up- and downlink happens via different frequencies (FDD, Frequency Division Duplex). The situation is similar in GSM and UTRA-FDD-systems. The base station can always send data on the downlink to the terminals without collisions. Further examples are TDD schemes used in Bluetooth, DECT or USB, where only a master, a base station, or a root may send on the downlink.

- **End-system (terminal, host, node):** End-systems contain applications and protocols for the application and transport layers. Users can access services from communication networks or distributed applications via an end-system. End-systems terminate layer 4 connections and should work independently of the underlying network technology.

Example: Mobile phones represent end-systems in mobile phone networks, for wireless LANs PDAs, laptops or notebooks can act as end-systems. Radios are end-systems for DAB and TV-sets for DVB. Sure, mainframes and PCs can also act as end-systems for communication systems.

- **Handover (handoff):** The seamless handover of a connection between two or more base stations using the same wireless technology is the key feature of today's cell-based mobile phone systems. As a single sender can not cover a whole country mobile phone systems use many base stations with each base station creating a radio cell. As soon as a user changes a cell due to changes in signal strength, interference, or load balancing, the system has to redirect all connections or forward all data. The shorter the interruption of the service is – up to the ideal case of no service interruption and soft handover – the better is the quality of service for higher layers or a user.

Example: All mobile phone systems, such as GSM, cdmaOne, UMTS, cdma2000 etc. support the seamless handover between base stations. While GSM supports hard handovers only (with service interruption of up to 100 ms), CDMA systems, such as cdmaOne and W-CDMA, support soft handover without any service interruption. Mobile IP, too, has to support redirection of data to a new foreign agent after the change of network access. In this case the additional delay caused by the redirection is of particular interest. To avoid this additional delay, Mobile IP may use optimizations such as rerouting of the whole packet flow.

- **Infrastructure:** Infrastructure based networks need ahead planning in contrast to ad-hoc networks. This planning is required, e.g., for setting up base stations together with a connection to the fixed network and for installing services for naming, data forwarding, authentication etc.

Example: All big wireless networks of today are based on an infrastructure. GSM requires a radio subsystem, a switching system, several databases and connections to other networks. WLANs often require bridges for roaming support, medium access control, and packet filtering.

- **Interworking unit:** In general, these components forward data and can operate on different layers in the reference model. Repeaters simply convert and regenerate signals on the physical layer to connect different media or to extend coverage. Storage of data or protocol processing is not possible in repeaters. Bridges connect different links on layer two. These components may adapt different protocols, filter data packets, or store data for a short amount of time. Simple routing functionality is available, too. At the networking layer routers connect different subnets. The main purpose of routers is data forwarding according to internal tables. The main difficulty is finding an optimal route between communication partners. Different parameters and metrics may help finding this optimum (distance in hops, interference level, cost, load etc.). Gateways can connect different networks at transport and application level. Interconnection of different networks often implies the conversion of data formats, content extraction, or content conversion.

  Example: Repeaters typically extend the coverage of a sender. For example, repeaters in trains increase the signal strength of a mobile phone system inside the train to enable phone usage during traveling. Access points of wireless LANs are typically bridges, which connect the wired with the wireless network and filter traffic. At the IP layer the routers in the Internet take care of packet forwarding. This is also the layer where Mobile IP extends IP for mobility support. Several gateways are needed for the support of Internet applications on mobile phones. These gateways may convert content formats, downscale pictures, adapt protocols, or connect traditional phone networks to push services.

- **Mobile node:** A mobile node can have the role of an end-system (e.g., mobile terminal, mobile host, mobile station) or intermediate system (e.g., mobile router). A key feature is the mobility of the node, which can be defined in many different ways according to the application scenario. Using vehicles much larger computers can act as mobile node, while humans accept only small, lightweight devices.

  Example: Mobile nodes can act as a router on board of an aircraft to coordinate communication between the passengers and a base station on the ground. Mobile phones are mobile end-systems for Internet access with the help of WAP or standard Internet protocols.

- **Multiplex:** Different multiplex technologies can be used to control medium access of devices. This helps separating communication in space, time, frequency, or code. Communication systems can avoid interference between two transmissions by applying appropriate multiplexing schemes.

  Example: Traditional fixed networks typically apply space division multiplexing by using different wires for different connections (one phone line per household). Mobile phone systems such as GSM use space (different cells), time (different time-slots), and frequency (different carrier frequencies) division multiplexing for the separation of users. UMTS additionally applies code division multiplex. Future wireless systems will also use beam forming technologies (i.e., space division multiplexing) to increase user density.

- **Protocol:** Protocols determine the rules of communication between two or more communication partners. Protocols control communication horizontally between communication partners on the same layer (communication protocol) or vertically between two layers at the service access point within a system (interface protocol). All partners have to obey the protocol to guarantee successful communication.

  Example: Many different end-systems use TCP at layer 4 in the internet for reliable end-to-end data transmission. Mobile IP can support data forwarding to mobile nodes at layer 3. Applications or higher layers can request session services at the service access point of the session layer WSP in the WAP architecture. However, this requires the knowledge of the right primitives and their application in the right order. A user of the service must react to events according to the protocol.

- **Quality of Service (QoS):** According to the ITU-T standard E.800 QoS is defined as "The collective effort of service performances, which determine the degree of satisfaction of a user of this service." Services can have qualitative (security mechanisms, manageability etc.) and quantitative (bandwidth, jitter, delay etc.) QoS parameters. QoS parameters are measured at a service access point.

  Example: The packet switched service GPRS of GSM offers a data transfer service with specified QoS parameters. Quantitative parameters are the average delay of a packet of certain size (measured in seconds) or the bit error rate. Qualitative service statements are, for example, that GPRS does not support isochronous data delivery and does not give guarantees for data transmission.

- **Roaming:** In contrast to handover, roaming typically denotes a more complex, more time-consuming change of the network access. Roaming typically takes time as it can comprise a change in network technology, several database requests, authentication procedures, forwarding of data, relaying of connections, or a change of the provider (which then involves accounting mechanisms).

  Example: National roaming in GSM enables the change of the mobile phone provider within a country. International roaming enables the use of foreign networks. Quite often roaming requires setting up the connection again after changing the network. However, some providers already offer seamless cross border roaming. Roaming between different systems, such as satellite, UMTS, WLAN, will gain increased importance in the future. This helps increasing coverage and bandwidth at hot-spots, while possibly lowering communication cost due to the choice of the most adequate system in a certain situation.

- **Service:** A service describes the sum of all functions with certain properties that are offered at a service access point located at the border between two layers. In general, a service is defined between two arbitrary objects that are in a certain relation to each other. Services may be described using service level agreements (SLA).

  Example: The transport layer of the internet offers a reliable service for data transfer between two end-systems at the socket interface if the protocol TCP is used.

- **Uplink:** The term 'uplink' denoting a certain direction for communication between two devices has its origin in satellite systems. In this case the link from the stations on the Earth up to a satellite is called uplink. The uplink is also the direction of communication where collisions may take place, i.e., competition of different medium accesses may destroy transmissions if no entity coordinates the accesses. The communication system has to separate the uplinks from the downlink with the help of a multiplexing scheme (time, frequency, code, and space) in order to avoid collisions.

  Example: Wireless local area networks following IEEE 802.11 permit medium access for devices that want to send a message to a base station almost anytime. This mechanism may cause collisions on the medium. Devices following DECT or UTRA-TDD may send data on the uplink only at predefined points in time. The base station assigns for this purpose a certain time-slot to each device. This avoids all collisions. DECT, Bluetooth, and UTRA-TDD use Time Division Duplex (TDD) for the separation of uplink and downlink.

- **Wired network:** The traditional network technology is based on wires or fibers and can offer much higher data rates compared to wireless technology. Wired networks can guide all signals along the wire and predetermine the propagation of signals precisely. It is much simpler to use space division multiplexing with wires compared to antennas and radio cells or beam forming technologies.

  Example: Fiber optics reach already today data rates of several Tbit/s, while delay and bit error rates are very low. Wireless networks still operate in the range of several Mbit/s with much higher error rates.

- **Wireless network:** Wireless networks differ mainly in their physical layer from their fixed counterpart. Data transmission takes place using electromagnetic waves which propagate through space, are reflected, scattered, attenuated etc. This type of transmission requires modulation of data onto carrier frequencies as it is not possible to transmit baseband signals. Additionally, the data link layer differs as wireless networks typically require more complex medium access control mechanisms. Compared to fixed networks, wireless networks offer low data rates and exhibit high error rates or higher latencies due to complex error correction mechanisms.

  Example: Wireless networks following the IEEE 802.11 family of standards can replace fixed networks in many situations. This allows for much higher user flexibility while still similar services are available. However, bandwidth is much lower (e.g., approx. 34 Mbit/s half duplex with 802.11a compared to 100 Mbit/s full duplex with Ethernet) and delay is higher compared to fixed networks following IEEE 802.3.

Index

universal mobile telecommunications system (UMTS) 3, 13, 30, 136–49
  core network 151–3
  handover 154–6
  radio interface 143–9
  releases and standardization 141–2
  system architecture 142–3
  UMTS (universal) terrestrial radio access *see* UTRA 138

UTRA network (UTRAN) 142, 149–51